AF772134

VOLUME ONE HUNDRED

ADVANCES IN BOTANICAL RESEARCH

Past, Current and Future Topics

ADVANCES IN BOTANICAL RESEARCH

Series Editor

Jean-Pierre Jacquot
Membre de l'Institut Universitaire de France, Unité Mixte de Recherche INRA, UHP 1136 "Interaction Arbres Microorganismes", Université de Lorraine, Faculté des Sciences, Vandoeuvre, France

VOLUME ONE HUNDRED

Advances in BOTANICAL RESEARCH

Past, Current and Future Topics

Volume Editor

JEAN-PIERRE JACQUOT
Unité Mixte de Recherche INRA,
UHP 1136 "Interaction Arbres Microorganismes",
Université de Lorraine, Faculté des Sciences,
54500 Vandoeuvre, France

ELSEVIER

ACADEMIC PRESS
An imprint of Elsevier

Academic Press is an imprint of Elsevier
125 London Wall, London, EC2Y 5AS, United Kingdom
The Boulevard, Langford Lane, Kidlington, Oxford OX5 1GB, United Kingdom
50 Hampshire Street, 5th Floor, Cambridge, MA 02139, United States
525 B Street, Suite 1650, San Diego, CA 92101, United States

First edition 2021

Copyright © 2021 Elsevier Ltd. All rights reserved.

No part of this publication may be reproduced or transmitted in any form or by any means, electronic or mechanical, including photocopying, recording, or any information storage and retrieval system, without permission in writing from the publisher. Details on how to seek permission, further information about the Publisher's permissions policies and our arrangements with organizations such as the Copyright Clearance Center and the Copyright Licensing Agency, can be found at our website: www.elsevier.com/permissions.

This book and the individual contributions contained in it are protected under copyright by the Publisher (other than as may be noted herein).

Notices
Knowledge and best practice in this field are constantly changing. As new research and experience broaden our understanding, changes in research methods, professional practices, or medical treatment may become necessary.

Practitioners and researchers must always rely on their own experience and knowledge in evaluating and using any information, methods, compounds, or experiments described herein. In using such information or methods they should be mindful of their own safety and the safety of others, including parties for whom they have a professional responsibility.

To the fullest extent of the law, neither the Publisher nor the authors, contributors, or editors, assume any liability for any injury and/or damage to persons or property as a matter of products liability, negligence or otherwise, or from any use or operation of any methods, products, instructions, or ideas contained in the material herein.

ISBN: 978-0-12-821692-7
ISSN: 0065-2296

For information on all Academic Press publications
visit our website at https://www.elsevier.com/books-and-journals

Publisher: Zoe Kruze
Acquisitions Editor: Sam Mahfoudh
Developmental Editor: Jhon Michael Peñano
Production Project Manager: James Selvam
Cover Designer: Victoria Pearson

Typeset by SPi Global, India

Working together to grow libraries in developing countries

www.elsevier.com • www.bookaid.org

Contents

Contributors

Sophia Bałdysz
Molecular Virology Research Unit, Institute of Experimental Biology, Faculty of Biology, Adam Mickiewicz University, Poznań, Poland

Catherine Bennetau-Pelissero
UMR1212 ARNA, University of Bordeaux, Inserm, CNRS, Equipe TAMS, Bordeaux, France

Nathalie Bourgougnon
Laboratoire de Biotechnologie et Chimie Marines, Université Bretagne Sud, EA3884, UBS, IUEM, Vannes, France

Anne-Sophie Burlot
Olmix SA • ZA du Haut du Bois, Bréhan, France

James Callow
Emeritus Professor of Plant Science, University of Birmingham, Birmingham, United Kingdom

Corinne Cassier-Chauvat
Laboratory of Biology and Biotechnology of Cyanobacteria, Université Paris-Saclay, CEA, CNRS, Institute for Integrative Biology of the Cell (I2BC), Gif-sur-Yvette, France

Franck Chauvat
Laboratory of Biology and Biotechnology of Cyanobacteria, Université Paris-Saclay, CEA, CNRS, Institute for Integrative Biology of the Cell (I2BC), Gif-sur-Yvette, France

Ridong Chen
APT Therapeutics Inc, Naperville, IL, United States

Frederic De Lamotte
UMR AGAP Institut, Univ Montpellier, CIRAD, INRAE, Institut Agro, Montpellier, France

Michel Delseny
Laboratoire Génome et Développement des Plantes, UMR CNRS 5096, Université de Perpignan Via Domitia, Perpignan, France

Laurence Drouard
Institut de biologie moléculaire des plantes-CNRS, Université de Strasbourg, Strasbourg, France

Michel Droux
Microbiologie, Adaptation, Pathogénie, UMR5240 CNRS-UCBL-INSA-Bayer CropScience, Domaine Scientifique de la Doua, Université Lyon 1, Villerbanne, France

Anne-Marie Duchêne
Institut de biologie moléculaire des plantes-CNRS, Université de Strasbourg, Strasbourg, France

Carolina Escobar
Facultad de Ciencias Ambientales y Bioquímica, Área de Fisiología Vegetal, Universidad de Castilla-La Mancha, Toledo, Spain; International Research Organization for Advanced Science and Technology (IROAST), Kumamoto University, Kumamoto, Japan

Carmen Fenoll
Facultad de Ciencias Ambientales y Bioquímica, Área de Fisiología Vegetal, Universidad de Castilla-La Mancha, Toledo, Spain

Aymeric Goyer
Department of Botany and Plant Pathology, Oregon State University, Hermiston Agricultural Research and Extension Center, Hermiston, OR, United States

Anna Goździcka-Józefiak
Molecular Virology Research Unit, Institute of Experimental Biology, Faculty of Biology, Adam Mickiewicz University, Poznań, Poland

Bertrand Hirel
INSTITUT Jean-Pierre BOURGIN. Departement Adaptation des Plantes à l'Environnement INRAE, Versailles Cedex, France

Michael Hodges
Institute of Plant Sciences Paris-Saclay (IPS2), CNRS, INRAE, University Paris-Saclay, University of Paris, Gif Sur Yvette, France

Emmanuelle Issakidis-Bourguet
Institute of Plant Sciences Paris-Saclay (IPS2), UMR Université Paris-Sud—CNRS 9213—INRAE 1403, Orsay Cedex, France

Anne-Gaëlle Jacquin
The Algonauts Project, www.algonauts.org, San Francisco, CA, United States

Jean-Pierre Jacquot
Université de Lorraine, INRAE, IAM (Interaction Arbres Microorganismes), Nancy; Laboratoire Génome et Développement des Plantes, UMR CNRS 5096, Université de Perpignan Via Domitia, Perpignan, France

Stanislav Kopriva
Institute for Plant Sciences, Cluster of Excellence on Plant Sciences (CEPLAS), University of Cologne, Cologne, Germany

Stephane D. Lemaire
Sorbonne Université, CNRS, UMR 7238, Laboratoire de Biologie Computationnelle et Quantitative, Institut de Biologie Paris Seine, Paris, France

Martine Lemaire-Chamley
INRAE, Université de Bordeaux, UMR Biologie du Fruit et Pathologie, Villenave d'Ornon, France

Myroslawa Miginiac-Maslow
Retired from Institute of Plant Sciences Paris-Saclay (IPS2), UMR Université Paris-Sud—CNRS 9213—INRA 1403, Orsay Cedex, France

Oskar Musidlak
Molecular Virology Research Unit, Institute of Experimental Biology, Faculty of Biology, Adam Mickiewicz University, Poznań, Poland

Robert Nawrot
Molecular Virology Research Unit, Institute of Experimental Biology, Faculty of Biology, Adam Mickiewicz University, Poznań, Poland

Pierre Pétriacq
UMR1332 BFP, University of Bordeaux, INRAE; Bordeaux Metabolome, MetaboHUB, PHENOME-EMPHASIS, Villenave d'Ornon, France

Dominique Rolin
UMR1332 BFP, University of Bordeaux, INRAE; Bordeaux Metabolome, MetaboHUB, PHENOME-EMPHASIS, Villenave d'Ornon, France

Nicolas Rouhier
Université de Lorraine, INRAE, IAM (Interaction Arbres Microorganismes), Nancy, France

Tatsuo Sugiyama
Emeritus Researcher of RIKEN, Nagoya, Japan

Akira Suzuki
INRAE, Institut Jean-Pierre Bourgin, UMR1318, AgroParisTech, Versailles Cedex, France

Sarada D. Tetali
Department of Plant Sciences, School of Life Sciences, University of Hyderabad, Hyderabad, Telangana, India

Jean Vidal
Retired from Institute of Plant Sciences Paris-Saclay (IPS2), UMR Université Paris-Sud—CNRS 9213—INRA 1403, Orsay Cedex, France

Alicja Warowicka
Department of Animal Physiology and Developmental Biology, Institute of Experimental Biology, Faculty of Biology, Adam Mickiewicz University, Poznań, Poland

Martyna Węglewska
Molecular Virology Research Unit, Institute of Experimental Biology, Faculty of Biology, Adam Mickiewicz University, Poznań, Poland

David Wendehenne
Agroécologie, AgroSup Dijon, CNRS, INRAE, University Bourgogne Franche-Comté, Dijon, France

Foreword

Advances in Botanical Research (ABR) is turning 100! In terms of human life expectancy that would obviously be a very high number ... Needless to say, we are extremely proud to have been able to generate 100 volumes of this book series over nearly half a century. For ABR the editorial tasks are divided as follows: series editors select volume editors who in turn contact colleagues who will write the individual chapters. It was a bit hard for me to find out all the colleagues who have served as series editors in the very early times of ABR as I have never met them physically, but following an exchange with Michel Delseny I found that RD (Reginald) Preston and Harold Woolhouse more or less coordinated the first 10 volumes. Then, probably the most influential series editor was Jim Callow who coordinated volumes 11–44 and was also involved in volume 10. After this Michel Delseny and Jean-Claude Kader took over in 2006 and coordinated 18 volumes during their 6-year tenure. Starting at volume 63 Pierre Gadal and myself were the series editors, i.e., from 2012 to 2020. The ABR staff contacted me not long ago to probe whether I could plan "something exceptional" to celebrate this 100th birthday. Well I thought it was a very good idea and I started to work on that project. It happened that unfortunately shortly before and briefly after I had started to take over this project two former coeditors Jean-Claude Kader and Pierre Gadal have deceased. Michel Delseny and myself thought that volume 100 should by all means pay an extensive tribute to the scientific and editorial work of these two outstanding colleagues and part of volume 100 is devoted to that. After I succeeded in contacting him, Jim Callow very gently agreed to write comments about the early volumes of ABR and his own interactions with RD Preston and Harold Woolhouse and I want to thank him here heartily for that. Chapter 1 also describes briefly the contents of Chapters 44–100. Then Michel Delseny recalls in Chapter 2 some aspects of Jean-Claude scientific contributions and their experience as coeditors. Likewise, in Chapter 3 I describe our editorial contributions together with Pierre and a larger section details with the help of other former colleagues the now "historical" achievements of the Gadal lab which was very influential in France in photosynthesis, more specifically concerning CO_2 fixation and its redox regulation, but also very much in understanding nitrogen assimilation. In line with this description of the Pierre Gadal's lab contribution,

Akira Suzuki provides in Chapter 4, a review article concerning glutamate synthase and amino acid synthesis in plants. The other 10 chapters are contributed by the former editors of volumes (52, 63, 65, 67, 71, 73, 77, 80, 93, 95, 98) who have accepted to recollect their editorial experience, summarize their former volumes, and discuss the developments that have occurred since in their respective fields.

The result is an unusual "hybrid" volume which I hope will attract the attention of the readers thanks to its historical perspective but also to the very up to date new scientific data relating to plant biology, physiology, development, metabolomics, and genomics. Before concluding this foreword, I would like to thank all members of the Elsevier editorial staff, without them none of that would have been possible. For my own tenure they include Kristi Gomez, Zoe Kruze, Narmada Thangavelu, Tara Hoey, Preethy Simonraj, Helene Kabes, Mary Zimmermann, Kirsten Shankland, Hannah Colford, Sam Mahfoudh, James Selvam, Peter Llewellyn, Hilal Johnson, and Jhon Michael Peñano. Sincere thanks to all of them and apologies to those that I have omitted. And of course, a lot of thanks to the ca. 4000 chapter-contributors originating from more than 30 different countries. I believe that this number is a testimony of the attractivity of the series over this half-century.

Happy birthday to ABR, despite its apparent old age it's still alive and kicking, and happy reading to all!

JEAN-PIERRE JACQUOT

CHAPTER ONE

Editorial activities for Advances in Botanical Research

James Callow[a,*], Michel Delseny[b,*], and Jean-Pierre Jacquot[b,*]
[a]Emeritus Professor of Plant Science, University of Birmingham, Birmingham, United Kingdom
[b]Laboratoire Génome et Développement des Plantes, UMR CNRS 5096, Université de Perpignan Via Domitia, Perpignan, France
[*]Corresponding authors: e-mail address: j.a.callow@bham.ac.uk; delseny@univ-perp.fr; j2p@univ-lorraine.fr

Contents

Abstract

The history of the series is described by three successive Editors. Each one reports how he has been approached, his conception of the different volumes and gives an overview of the topics which have been treated. The first period, from creation of the series in 1963 by Professor Reginald Preston to 2006, was covered by Professor James (Jim) Callow who became Editor after Professor Harold Woolhouse, in 1985. During this period, 44 volumes were published, essentially eclectic, covering most of the fields in plant biology. Under James Callow's editorship, the frequency of issues increased up to two/year and a few thematic volumes were published. When he decided to stop in 2006, Dr. Jean Claude Kader and Dr. Michel Delseny were appointed. They maintained a rate of two to three issues/year, alternating eclectic and thematic volumes. They together published 16 volumes and resigned in 2012. Finally, they were replaced by Professors Jean-Pierre Jacquot and Pierre Gadal. They published only thematic volumes and the rate increased to four–five issues/year.

1. The early years (volumes 1–43)

By James Callow

Professor James Callow (Emeritus Professor of Plant Science, University of Birmingham) writes about the early history of Advances in Botanical Research.

The "Advances" was first edited by Professor R.D. (Reginald Dawson) Preston. Initially trained as a physicist, Preston joined the Botany Department

Advances in Botanical Research, Volume 100
ISSN 0065-2296
https://doi.org/10.1016/bs.abr.2021.01.001
Copyright © 2021 Elsevier Ltd
All rights reserved.

at the University of Leeds in 1929, where he developed interests in applying a range of physical techniques to investigate the molecular organization of different types of plant cell walls, ranging from xylem tracheids, to the cellulosic walls of the green alga *Cladophora* and *Valonia*. A Lectureship followed in 1936, a Readership in 1949, and in 1953 he was appointed to the Chair of Plant Biophysics in the Department of Botany. He was elected to the Royal Society the year after, and in 1962 became Head of the newly formed Astbury Department of Biophysics. The first volume of Advances in Botanical Research was published under his Editorship in 1963 and contained seven articles ranging from palaeobotany and palynology, to the use of electron paramagnetic resonance in studies of photosynthesis, thus establishing the essentially eclectic character of the series.

Subsequent volumes followed rather sporadically, volume 3 not appearing until 1970. However, the series was energized by the involvement of Professor H.W. Woolhouse (known to everyone as "Harold"). Woolhouse had been appointed to the Chair of Botany at the University of Leeds in 1969. A man with considerable energies and wide interests, ranging from the physiology of leaf senescence to the ecophysiology of heavy metal tolerance, he teamed up with Preston to jointly edit volume 4, which appeared in 1978. Preston then took a back seat so that Woolhouse was the sole editor of volume 5 (1977) (no this is not a mistake, the record shows that volume 5 was published before volume 4)!

Woolhouse maintained the eclectic character of the Advances as sole Editor over a further 5 volumes until 1983, and during his tenure he was very clear about the type of article he was looking for. Specifically, in his invitation letter to potential authors, he emphasized that he was not interested in review articles of the "Annual Reviews in...." type. Rather he wanted authors to pursue topics in depth, to speculate, to incorporate novel data and proofs where appropriate, and without regard to page length. Notably, volume 10 contained only two articles, but one of these ("Light-Harvesting Processes in Algae" by A.W. Larkum and J. Barrett) occupied no less than 219 pages!

In 1980 Woolhouse left Leeds to take up the position of Director at the John Innes Institute, Norwich where his vision, energy and commitment established the Institute at the international forefront of basic research in plant science, with a particular emphasis on the burgeoning field of plant molecular genetics. However, such commitments led him to seek to divest other responsibilities, and in 1985 he asked Professor James (Jim) Callow to join him as co-Editor of volume 11 (1985). Callow, previously a Lecturer in the Botany Dept. at Leeds University, had by then been appointed to the Chair of Botany in the Department of Plant Sciences in the University of Birmingham in 1983. He had also been a contributor to volume 10, in which he authored an article on recognition processes in plant-pathogen interactions. Then in 1986 Callow became sole Editor, or more correctly, "Editor-in-Chief" since one of the innovations introduced by Woolhouse was a supporting Editorial Board. For many years the Board would meet, most enjoyably, at the London HQ of Academic Press, in Oval Road, the main business on the Agenda being to "brainstorm" ideas for future contributors.

A significant development in 1995 was the incorporation of the sister review series, "Advances in Plant Pathology," into Advances in Botanical Research, a move welcomed by Callow since it coincided with his own interest in plant-pathogen interactions. Volume 21 was the first of the expanded series (Advances in Botanical Research incorporating Advances in Plant Pathology), edited by John Andrews and Inez Tommerup. There followed a number of other volumes specifically on topics in plant pathology, and volume 23 (1996) is noteworthy since it was the first volume in the history of Advances in Botanical Research to focus on a specific theme, rather than being an eclectic collection of unrelated articles. This proved to be a very worthwhile innovation, since over the ensuing years, in addition to the regular volumes, a number of excellent thematic volumes were published with "Guest Editors," among which we may mention "The Plant Vacuole" (volume 25, 1997, edited by R.A.Leigh and D. Sanders), "Trichomes" (volume 31, 1997, edited by D.L. Hallahan and J.C. Gray), "Cereal Genomics" (volume 34, 2001, edited by P.R. Shewry, K.J. Edwards and P.A. Lazzeri).

Volume 28 (1998) is particularly poignant since the volume was dedicated to Harold Woolhouse who died in 1996 aged 63. By that time he had left the John Innes Institute (1990) to tackle new challenges at the Waite Institute in Adelaide, as its new Director. A tribute to him in volume 28 records a remarkable career which inspired a whole generation of plant scientists and laid important foundations for the future of plant research.

The final volume with James Callow as Editor-in-Chief, was another thematic volume, "Developmental Genetics of the Flower" (volume 44, 2006, edited by D.E. Soltis, J.H. Leebensmack and P.S. Soltis). At that point, Callow was acquiring new roles and responsibilities, and he felt that after 21 years it was time that the series acquired new leadership, hence the appointment of Jean-Claude Kader and Michel Delseny as new

Editors-in-Chief. It is a tribute to them that the next volume (45) quickly followed, in 2007. Their involvement proved to be an inspired choice since 13 years later we are now up to volume 100—well done!

Fig. 1 Picture of James Callow.

2. The middle ages (volumes 44–62)

By Michel Delseny, Emeritus Director of Research at CNRS, University of Perpignan Via Domitia, 66,860 Perpignan, France.

Our adventure in Advances in Botanical Research began in the autumn 2006, when Jean-Claude called me and asked whether I would accept to co-edit with him the ABR series. He had just been approached by Elsevier to take over from James Callow. We did not know much about this series and it seemed to us an exciting challenge at a time when both of us were close to retirement, end of 2006 for him and early 2008 for me. Jean-Claude and I were old friends and we had known each other since many years. This friendship and complicity were certainly crucial in my decision to accept his proposition and in the way we managed the series for approximately 6 years.

We first met as early as 1967, when I was a student in Paris University and Ecole Normale Supérieure de Saint Cloud. He has just arrived from Tunisia, his native country, with his wife, to work as an assistant in the plant lipid laboratory, under the leadership of Paul Mazliak, who was also one of my plant physiology professors. I used to visit his lab from time to time because some of my friends were preparing their master degree in his group, while I was preparing mine in a near-by plant cytology laboratory. Jean-Claude was rapidly recruited as a CNRS (Centre National de la Recherche Scientifique) scientist to work with Paul Mazliak. After completion of my degrees, I was also recruited by CNRS, to work in a newly established laboratory at the University of Perpignan, in the south of France. Both of us were working on completely different topics: he was interested in identifying the various lipids in plant cells and in understanding their metabolism and function, whereas I was studying nucleic acids metabolism in germinating seeds. Nevertheless, we kept in touch, meeting each other during the annual meetings of the French Plant Physiology Society, in which Jean-Claude was already very active. This period of the 1970s was marked by the emergence of plant molecular biology and we had frequent discussions on the respective aspects of molecular biology, biochemistry and physiology in plant research. I completed my training by a post-doc at the John Innes Institute in Norwich on cauliflower mosaic virus. I was there when Harold Woolhouse was appointed as a Director of the Institute, and I had the privilege of discussing with him quite often at that time. Back in France I maintained contacts with Jean-Claude and, in 1985 we started a collaboration on a protein, purified in Jean-Claude"s group, the lipid transfer protein (LTP), and we isolated the corresponding maize cDNA in collaboration with Pere Puigdomenech in Barcelona. This has been the beginning of a warm friendship and a long-term collaboration on genes involved in lipid metabolism, including ACP-thioesterases, fatty acid elongase and lysophosphatidic acyl transferases. At this time, both of us were Directors of Research at CNRS and responsible for our labs. Meanwhile our research work was progressing with the discovery of additional members of the LTP family and with other advances in our respective fields. We both became members of the national committee of CNRS, which was in charge of recruiting and promoting CNRS scientists, as well as evaluating CNRS and CNRS-associated laboratories. In addition to reviewing many plant physiology or biochemistry manuscripts for various journals, Jean-Claude became the chief editor of Plant Physiology and Biochemistry, the journal of the French Plant Physiology Society, while I became editor of Plant Molecular Biology for 5 years and, later on, co-editor of Plant Science. In 1998, Jean-Claude was appointed at the CNRS headquarters where he was in charge of plant science. He shared his time between CNRS and his lab. Thus, we gained a rather large knowledge not only of our respective specialized fields, but also, more broadly, of many aspects of plant biology and plant scientific community.

From the beginning of our careers, we used to travel and attend many scientific meetings in which we met and exchanged with numerous colleagues (Fig. 1). Both, Plant Physiology and Biochemistry and Plant Science, were journals from the Elsevier group. Therefore, journal managers in Elsevier had not too much difficulty in identifying us as potential new editors of ABR and, when they approached us, we were prepared to accept the challenge of editing a review series in the broad field of botany, although our expertise was more in physiology, biochemistry and genomics than in systematics.

ABR, at that time was publishing two to three volumes a year, some being eclectic volumes, with a variety of topics, while others were thematic volumes, focusing on a specific field. We decided that we should continue this custom and in 6 years we published 10 eclectic and 8 thematic volumes. All published manuscripts have been invited and most of the time, they were selected after we listened or met one of the authors during a conference or a seminar to which we attended. Before inviting an author, we always discussed the opportunity and decided who will approach the authors and then follow the progress through to publication. Although the manuscripts were invited, we sent all of them to one or two external reviewers and they were reviewed in addition by at least one of us. We tried to focus on rapidly progressing fields such as development, stress biology or genomics. We also tried to promote new models, at a time when *Arabidopsis thaliana* was the obvious one, or new research fields, as well as young scientists.

We also used to meet regularly the managing editors, at least once a year: they often attended important scientific meetings, such as those of ASPB (American Society of Plant Biologists), FESPB (Federation of European Societies of Plant Biology) and PAG (Plant and Animal Genome) in which they have a booth where they present Elsevier publications. We had several meetings with them, essentially in San Diego, during PAG, because one of the managing-editor, Kristi Gomez, was based in this city and both Jean-Claude and I regularly attended this yearly conference. Several times, Tara Hoey and Narmada Thangavelu joined us for meeting. This was a unique opportunity to discuss our ideas and projects, in a good restaurant or in front of a margarita and to improve the series.

In a distinct chapter I shall review some of the topics we selected for the eclectic volumes. Reviewing progress after thematic volumes is much more difficult and this is more the responsibility of our co-editors. Indeed, some of our authors have become co-editors of the series of new thematic volumes. The period covered by our editorship is marked by the explosion of genomics. The *Arabidopsis thaliana* genome was the first one to be determined in 2000 and that of rice had just been completed when we took over after James Callow in 2005. A major consequence was a tremendous increase in our understanding of developmental processes and cell biology in these two species, and consequently in other plants. Control of pathogens was a major issue for agriculture, but it was also realized that the field should be enlarged to plant-microbe interactions. Another key issue, which was already clearly identified, was the response of plants to various environmental changes. Finally, we anticipated that after *Arabidopsis* and rice, many more genomes will be sequenced and we decided to introduce reviews on specific plants, describing their traits, their systematics and phylogeny, and the most recent approaches to sequence their genomes.

These were subjects in which we were interested, although most of the time we were not specialists. We very much enjoyed editing this articles and volumes, exchanging with authors, many of whom became our friends. Most of the time, we satisfied our curiosity and we learnt a lot in many domains we were not familiar with. I had great pleasure revisiting this period and having a quick look at most recent developments. I feel sad being alone to commemorate this period because Jean-Claude unfortunately passed away on July 2018. I am sure he would have enjoyed, as well, this exercise and would have been impressed, as I have been, by the outstanding advances in such a short time. My curiosity and enthusiasm for plant science are still intact and I wish that many colleagues can profit from the work carried out by the successive editors and authors of Advances in Botanical Research. I wish a long life to Advances in Botanical Research and another bunch of exciting and stimulating articles in the coming years Fig. 2.

Fig. 2 Picture of Jean-Claude Kader (left) and Michel Delseny (right) visiting the National Cheng Kung University in Tainan (Taiwan) in 1994.

3. Recent evolution of Advances in Botanical Research (volumes 63–100, 2012–2021)

By Jean-Pierre Jacquot

Professor Emeritus, Université de Lorraine, INRAE, IAM (Interaction Arbres Microorganismes), F-54000 Nancy, France.

Following the Delseny-Kader tenure, I was asked in 2011 if I was willing to serve as the next series editor for Advances in Botanical Research. After having been invited by Jean-Claude and Michel, I had already served as a volume editor published in 2009 (volume 52, Oxidative stress and redox regulation in plants). Incidentally Nicolas Rouhier and myself have produced for ABR 100 (a follow-up chapter "News on the redox front" by Rouhier and Jacquot of volume 52). Anyway, after editing volume 52, I knew a bit more about the job and the difficulties to chase down chapter contributors and coerce them in delivering in due time. Looking back at the correspondence at that time I believe that we had this volume wrapped up nearly in time and that may have impressed my colleagues and led them to suggest my name to succeed them. Of course, it might have just been a punishment for delivering a tiny bit late, I have actually never asked the reason, sometimes it's probably better not to know... Anyway, I gave it some thought and realized that it would be more comfortable to have a partner to help me in this task. It happened at that time that my former PhD supervisor and lab director Pierre Gadal had retired a few years ago and basically he was getting bored and didn't really know what to do with his professional life after his retirement. Of course, I knew about the tenacity of Pierre about scientific matters and of his multiple connections especially in France and I thought he would be a great help for this duty. I was very happy as he readily accepted the offer. In addition, that gave me the opportunity of coming back from Nancy to visit him in Orsay and we had nice meetings and chats over the years that followed. So shortly after Pierre came on board we had a short meeting, both of us together with Zoe Kruze who was at the time the ABR series manager in Elsevier. A nice meeting it was where in a couple hours in a Paris restaurant she explained to us what our duties would be and off it went. Before we took our turn as series editors Jean-Claude and Michel had put together a variety of thematic but also of more generalist issues (*i.e.*, a mix of different reviews on various scientific topics). Together with Pierre we decided to follow another strategy and we initiated only thematic issues. Our names are featured as series editors for volumes 61 and 62 on lignin and medicinal plants but these very fine contributions were actually initiated by the Kader-Delseny binome. As I write those lines volumes are in preparation until number 101, so we will have together with Pierre before his passing in November 2019 initiated 39 volumes altogether.

Much of this series (11 volumes altogether) was devoted to genomics which was doing giant strides at that time (still doing actually). It started with the description of mitochondrial genomes in volume 63 coordinated by Laurence Maréchal-Drouart. This volume gave valuable information about the mitochondrial genome evolution in plants using a variety of models along the evolution tree. It also addressed gene transfer to the nucleus but also horizontally from other organisms without neglecting macromolecule trafficking to mitochondria. Laurence also kindly accepted to write a follow-up of volume 63 for the anniversary volume 100 (chapter "Last news on plant mitochondria" by Duchêne and Drouard). The second genomic volume (64) coordinated by Gwenael Piganeau discussed genomics of algae, primary, secondary and tertiary endosymbiosis, more generally symbiosis in microalgae with quite a bit of information on diatoms. Volume 65 coordinated by Corinne and Franck Chauvat deals with genomics of cyanobacteria. Franck and Corinne have actually also written a follow-up chapter of volume 65 in the anniversary volume (chapter "Genomics of cyanobacteria: New insights and lessons for shaping our future" by Chauvat and Cassier-Chauvat) and I will thus not discuss it any further. The 4th volume dealing with genomes (volume 66) by Thomas Beatty described genome evolution in photosynthetic bacteria. It presented data concerning Heliobacteria, green bacteria, purple bacteria, Rhodobacter, and more generally anoxygenic phototrophs. Particular attention was paid to specific pigments as light harvesting antennae and bacteriophytochromes. A very interesting chapter described the regressive evolution of photosynthesis occurring in Roseobacter species. The fifth volume (69) on the genomes of cereals and various herbaceous plants by Andrew Paterson first described the sequencing techniques available at the time. It then provided a survey of the literature on Arabidopsis and various monocots and dicots including cereals. Historical and prehistorical evolution were also addressed and some data were included on population genomics and the plant microbiome. Francis Martin has coordinated a very fine volume on the genomics of fungi (volume 70). It describes data on mycorrhizal fungi, and pathogenic fungi as mildew, rust or smut. A very interesting chapter addresses the genome of the truffle which has been published shortly after. Overall there are 13 different chapters covering many different fungi with different lifestyles in this volume that is a mine of information for those interested in fungi and their evolution. Volume 74 by Christophe Plomion and Anne-Françoise Adam-Blondon addresses the genomics of trees. Fruit and forest trees are discussed and the domestication and conservation of fruit trees is also addressed in terms of genomics. Environmental factors and adaptation of trees is also discussed in two chapters and the genetic control of bud break in another one. Noteworthy, after this volume was published many tree sequences were made available including the one of oak recently. Another exciting volume (78 coordinated by Stefan Rensing) on genomics discusses genome evolution from Charophytes to ferns via Bryophytes and Leucophytes. Eight different chapters cover some unusual plants including *Physcomitrella patens*, one of the few plants amenable to gene replacement before the advent of the CrispR Cas9 technology. Volume 84 coordinated by Yoshihisa Hirakawa devoted to secondary endosymbiosis nicely completes data presented in volume 63 on mitochondria. The secondary endosymbiosis volume centers on plastids and their complexity in secondary endosymbiosis such as Cryptomonads and several others. Interestingly, apicoplasts, vestiges of chloroplasts found in non-photosynthetic species as Plasmodium or Toxoplasma are also included in this very interesting volume. Plastid genome evolution is discussed in more depth in volume 85 edited by Shu-Miaw Chaw and Robert Jansen. It covers the genome evolution of plastids from early organisms via algae and diatoms to finally reach land plants including gymnosperms and flowering plants. The combination of volumes 84 and 85 actually brings a very comprehensive picture on the evolution of plastid genomes. Philippe Gallusci, Etienne Bucher and Marie Mirouze have produced volume 88 (a nice number as it represents in the French system the department of the Vosges where I was born) which discusses epigenetics and breedings with connections to genomics. Various aspects of epigenetics are considered in this volume including a historical perspective, a description of the mechanisms involved in epigenetic regulation, and several chapters dealing with potential agronomical applications. The last volume which touches briefly on genomics has been engineered by Francisco Canovas entitled "Molecular Physiology and Biotechnology of trees." The two first chapters of this volume discuss genomics of forest trees and especially of conifers. Nitrogen usage and cycling is discussed in the context of mycorrhizae. Three other chapters deal with wood development, embryology of conifers and meristems. The next three chapters discuss pectins, transgenesis in poplar and phytoremediation with trees, all very timely topics. In conclusion of this first part of our editorial work with Pierre, I believe that the 11 volumes that we have finally produced give a very comprehensive view of the genomes of photosynthetic organisms from photosynthetic bacteria to all kinds of plants including at the top of the evolution trees Angiosperms, but also fungi. It has also covered in depth the much smaller, but nevertheless also essential mitochondrial and plastid genomes. I believe that this part is certainly a big success of our editorial tenure and I hope the readers will indeed feel like it too.

The next broad center of interest that we have explored could be defined as plant molecules and 11 different chapters fall in that category (61, 62, 67, 68, 71, 76, 80, 90, 91, 92, 93, 99). As explained before volume 62 edited by Lie-Fen Shyur (Recent Trends in Medicinal Research) was commanded by Jean-Claude Kader and Michel Delseny as series editors. It presents several examples of metabolites extracted from a variety of medicinal plants including several traditional Chinese herbs and their health effects. Examples include taxol and beta glucans and their effect on cancer cells, but also other cases involving arthritis, stroke, immunomodulation, and chronic diseases. A couple chapters also detail the potential negative health effects of plant metabolites and the necessary legal regulations for their use. Volume 61 by Lise Jouanin and colleagues (Lignins—Biosynthesis, Biodegradation and Bioengineering) was actually initiated by Michel and Jean-Claude as series editors. In 9 different chapters it describes the biosynthesis, biodegradation and bioengineering of lignin, an important component of plant cell wall. Lignification is described in conifers and grasses and the involvement of P450 and laccases evaluated. Lignin degradation and the evolutionary history of lignins is also addressed in volume 61. Since then lignin has been used in biotechnological applications thanks to its remarkable chemical stability. Dominique Rolin and colleagues in volume 67 (Metabolomics coming of age with its technological diversity) discuss the advantages and setbacks of various metabolomics techniques (NMR, LC-MS, 2D NMR, mass spec, and GCMS essentially). The application of these methods to various compounds including lipids is discussed as well as the possibility to follow plant metabolism. Finally modeling and data storage is discussed in this volume. Volume 67 is to be completed soon by a companion volume 98 organized by Pierre Petriacq and Alain Bouchereau. Volume 98 (Plant Metabolomics 2) will deal more with applications of metabolomics while volume 67 dealt more with technical aspects. In the present anniversary volume Pierre Petriacq has produced a chapter "When dietary supplements meet metabolomics: A fast-evolving field" summarizing volumes 67 and 91 and discussing recent developments and achievements of plant metabolomics.

Nathalie Giglioli-Guivarch has coordinated volume 68 "New light on alkaloid biosynthesis and future prospects" which discusses the biosynthetic pathways leading to the synthesis of these molecules essential for plant defense. In there the synthesis of various alkaloids is described including indolic compounds as monoterpenes. Metabolic engineering of these compounds is also addressed as well as the development of new production systems including bioreactors. Nathalie Bourgougnon has been very active for our series coordinating two volumes that deal to some extent with various molecules, volume 71 "Sea Plants" and volume 95 "Seaweeds around the world: state of art and perspectives." Overall these two volumes contain nearly 30 chapters and they discuss many aspects of seaweeds with examples in a wide variety of species and a wide variety of compounds including parietal compounds proteins and glycoderivatives. Biotechnological and cosmetic applications are also discussed at large. Nicely enough Nathalie has also produced a chapter for the anniversary volume (chapter "Algae for global sustainability" by Bourgougnon et al.) and I will thus not detail those any further although they have plenty of merit for those interested in marine algae and their use and also for the general readers. Volume 76 coordinated by David Craik is devoted to plant cyclotides, fascinating cyclic peptides found only in some plant species as Violaceae and Rubiaceae for example. These unusual molecules are generated from genes as precursor proteins which are made cyclic enzymatically post-synthesis and truncation. The structural properties of cyclotides which often contain disulfide bonds (cysteine knot) are also discussed in this issue. Their use in biotechnology and medicine is of course extremely interesting and promising due to the very unusual structure of these peptides. A different type of molecules is glucosinolates found essentially in Brassicaceae. Stan Kopriva and his colleagues have in volume 80 soberly entitled "Glucosinolates" discussed the nature of those products (they contain a glucose-like moiety linked to derivatives of amino acids with a high sulfur content), their biosynthesis and application as insect repellants and others. Stan Kopriva has also produced a separate follow-up chapter "Glucosinolates revisited" in ABR 100. The two following volumes concerning molecules in a broad sense have been coordinated by Bernhard Grimm (volumes 90, 91, Metabolism, Structure and Function of Plant Tetrapyrroles: Introduction, Microbial and Eukaryotic chlorophyll synthesis and catabolism, Metabolism, Structure and Function of Plant Tetrapyrroles: Control Mechanisms and Analysis of Chlorophyll-binding proteins). Altogether in 17 chapters overall Grimm has succeeded to produce a very comprehensive picture of the biosynthesis and metabolism of chlorophyll and its association in plant protein complexes. Of course, tetrapyrroles (hemes and chlorophylls) play an essential role in electron transfer chains in mitochondria and chloroplasts but they are also essential for photosynthesis in allowing the capture of the light energy and its conversion into chemical energy, a process essential to maintain biological life on earth. Hormones have been traditionally a large field of research in plants. Annie Marion-Poll has put out with her colleagues a very fine volume on abscisic acid (92, Abscisic acid in plants). In 10 chapters they decribe the metabolism of this molecule (synthesis, degradation) its plant receptors, its role in guard cell closure, its interaction with other hormones, its role in germination and in interactions and signaling. A must read for hormone buffs! Another volume that deals with molecules has been engineered by Robert Nawrot and his collaborators (93, Latex, laticifers and their molecular components—from functions to possible applications). Interestingly, laticifer containing plants as *Chelidonium majus* produce latex with various properties and colors (yellow sap in Chelidonium) and various medical properties (antiviral, wart killing for the Chelidonium sap). The diversity and distribution of laticifers is discussed in this volume and the variety of proteins and low molecular weight compounds in these saps discussed. Obviously, the very well-known latex producing species Hevea and Euphorbia are also discussed at length. Robert Nawrot has kindly accepted to write a follow-up chapter of volume 93 for the anniversary volume (chapter "Traditional use and perspectives for the application of plant latex and its constituents in agriculture, medicine, and industry" by Nawrot et al.). Finally volume 101 edited by Fabrice Rebeillé and Eric Maréchal to be published after the anniversary volume will discuss another category of molecules essential in plants, lipids.

We have not limited our interest to genomes/genomics and molecules and another group of volumes has investigated interactions in biological system involving plants (73, 75, 81, 82). Volume 73 edited by Carolina Escobar and Carmen Fenoll describes Plant-Nematode interactions (Plant-Nematode interactions—A View on Compatible Interactions). This volume contains 14 chapters which describe in detail morphological alterations induced by nematodes on plant cells, the signaling involved in these interactions, the plant defense system and the use of biotechnology to fight nematode invasions. Escobar and Fernoll have written a follow-up of this volume for ABR 100 (chapter "Compatible interactions between plants and endoparasitic nematodes" by Escobar and Fenoll). Volume 75 coordinated by Harsh Bais and Janine Sherrier describes Plant Microbe interactions, centering essentially on Rhizobia and legumes. Of course, nitrogen fixation in legumes occurs in Rhizobia present in the root nodules and this is essential for their growth and productivity. Several features of the Rhizobium association are discussed (exchange of information, signaling, development). Other models are also discussed as the very concerning human pathogen/plant interactions, soil microbial community, nematodes again etc.... Nicolas Sauvion, Denis Thiéry and Paul-André Calatayud have coordinated volume 81 "Insect-Plant Interactions in a Crop Protection Perspective." Interestingly this volume starts with a very original chapter providing a paleontological and evolutionary perspective of the field. Herbivore insects are evaluated but also the role of plants for providing shelter for entomophagous insects is discussed. The effect of the environment on these associations is mentioned and the chemical aspect of these relations is detailed. The last volume of this smaller series on interactions was produced by Guillaume Bécard (82, How Plants Communicate with their Biotic Environment). This very interesting volume describes many facets of plant communication including the one with pollinators, parasitic plants, mycorrhizae for forest species, Rizobacteria, pathogens and so on and so forth. One chapter addresses the very timely topic of tree communication via volatile compounds. Communication with herbivores and nematodes is also included and these chapters are a complement to volumes 73 and 81.

Two volumes pay tribute to development and signaling. Fabio Fornara (volume 72, The Molecular Genetics of Floral Transition and Flower Development) and his teammates have produced 10 chapters that describe photoperiodic regulation, the effect of vernalization, the signal involved in flowering, the genetic control of flower development in Arabidopsis and grasses as well as some structural aspects of the inflorescence. Of course signaling is heavily involved in flowering, and NO signaling has been discussed by Wendehenne and co-authors in volume 77 (Nitric Oxide and signaling in Plants). Nitrosylation by NO is detailed in this volume as well as the formation of nitrosothiols. The involvement of NO in nitrogen fixation and metabolism, response to stress and hypersensitive response is also largely debated. In addition, the interactions of NO with ROS and hydrogen sulfide as well as with calcium linked signaling are also detailed. A follow-up chapter of volume 77 is present in ABR 100 (chapter "Nitric oxide synthase in plants" by Wendehenne).

The remaining six volumes tackle subjects of more diverse nature. Volume 79 edited by Bruno Robert is devoted to Artificial Photosynthesis. The topic is of course of extremely high interest in view of the energy limitations that humankind is bound to experience in the future. This volume discusses in seven chapters how the biological knowledge accumulated on photosynthesis can be used to build artificial models based largely on chemistry but not only. Bioinspired photocatalysis and direct conversion from cyanobacteria are evaluated. Of course, one of the hopes to transform light energy into usable fuels would be to improve hydrogen production from algae and the progress in this domain is also discussed. The last chapter summarizes the international and European initiatives in this fast moving section of plant science/chemistry. Another timely topic is addressed in volume 83 Phytoremediation edited by Jaco Vangronsveld and Ann Cuypers. Indeed, in the post-industrial revolution many polluted sites containing mostly heavy metals or explosives have been abandoned and there is a pressant need to decontaminate those. Several tries have been done using microbes (bacteria essentially) or plants associated or not with mycorrhizae. Metal hyperaccumulators have also been selected and air phytoremediation, *i.e.,* generating plants exhaling volatiles containing heavy metals is also a possibility. All those aspects are discussed in volume 83 as well as some molecular and physiological details of these reactions.

During our tenure ABR has not shied away from tackling controversial subjects including the one concerning GMO's. We have provided as a tribune volume 86 (Transgenic Plants and Beyond) edited by Marcel Kuntz to discuss the controversies surrounding GMO's and their acceptability. In this volume the techniques used to generate GMO's are detailed and many applications discussed. The safety of food derived from GMO's is also discussed. New editing techniques are also addressed as well as the potential problems that could arise about insect pollination. We believe that the exceptional quality of this volume should help defuse the fear for GMO's that will probably be essential for human survival in the future.

The three last volumes tackle problems of a more specialized nature. Christophe Maurel and colleagues in volume 87 (Membrane Transport in Plants) discuss several transporters embedded in the cell membranes (ABC transporters, aquaporins, heavy metal pumps, metal and hormone transporters). The involvement of transporters in nitrate uptake, guard cell functioning, pollen fertilization, xylem loading and mycorrhizal exchanges is also described in separate chapters. Volume 87 is especially interesting in view of the giant strides made about the structure of membrane embedded proteins in the last decade involving in particular electron microscopy. Volume 96 coordinated by Toru Hisabori also addresses the fascinating membrane bound ATP synthases (ATP synthase in Photosynthetic Organisms). As I write this synthesis the Hisabori et al. contribution is not yet published, but there is no doubt that it will be an outsatnding contribution too. Finally, volume 97 of ABR will be coordinated by Robert Verpoorte and entitled "Entectic Solvents and Stress in Plants." I was fortunate enough a few years ago to listen to a talk by Robert and I was stunned by the quality and originality of the work, very classical in nature (basically evaluating the quality of chemical mixtures), far from molecular biology and so on and so forth but explaining so many things in a such a simple way that I immediately asked him if he wanted to contribute and he accepted readily. I really look forward to see this one it should look like a UFO in today's world. To conclude this list which looks a bit like "une liste à la Prévert" (this is for French readers with a bit of knowledge in literature), I have to mention that volumes 98 and 99 are already scheduled and they will deal with Metabolomics (Petriacq and Bouchereau Eds) and ligninolytic fungi (Morel-Rouhier and Sormani Eds). Likewise, volumes 102 and 103 are also underway.

So overall it has been nearly 8 years as series editors with Pierre, 10 years overall till volume 100 and 39 volumes produced we have gone a long way... I hope that the readers will realize that there was a logic in our choices with two long series on genomes and molecules and shorter series devoted to stress signaling, development and more specific topics that were all very timely. I hope that we have been able to cover large span of plant biology/biochemistry/molecular biology and I am actually quite pleased with the results (I am actually very often quite pleased with myself so this probably doesn't mean much anyway). To conclude this chapter I would like to stress that I have very much enjoyed working with Pierre as series editors. He has been very active in detecting potential volume editors but also as I have mentioned earlier he was definitely a very tenacious person. I have mentioned also that sometimes you have to chase chapter contributors to get them to deliver. Well that happened to be the case also for two volume editors that I will not name. As ABR does not have a budget to hire hitmen that would come and break your legs if you don't keep your promises, in these two cases I had to rely mostly on Pierre to "coerce" them. Well I don't know how he did it, was that persuasion, blackmail, whatever it worked. In the end I suspect that it was linked to his scientific reputation and connections in the plant research community. Since Pierre has disappeared I thoroughly miss him and I have convinced several colleagues of the former Gadal lab to join me to write a chapter paying tribute to the research that was done in the Gadal lab and this is one of the chapters of volume 100 as well (Figs. 3 and 4).

Fig. 3 Jean-Pierre Jacquot (left) with Jean-Claude Kader (right). *Photo taken ca 2011 when Jean-Claude asked me if I would agree to be the next series editor. I accepted because he had a very nice necktie.*

Fig. 4 From left to right Myroslawa Miginiac-Maslow, Stéphane Lemaire, Jean-Pierre Jacquot and Pierre Gadal. *Photo taken at IBP Orsay in 1997 before I moved back to Nancy. I also have a very nice necktie and Tshirt there as this was the celebration of my last day in Orsay.*

CHAPTER TWO

Revisiting ABR editing in the period 2006–2012 and recent developments

Michel Delseny*
Laboratoire Génome et Développement des Plantes, UMR CNRS 5096, Université de Perpignan Via Domitia, Perpignan, France
*Corresponding author: e-mail address: delseny@univ-perp.fr

Contents

Abstract

With my colleague Jean-Claude Kader, who unfortunately prematurely passed away in July 2018, we edited 18 issues of *Advances in Botanical Research*, including 10 eclectic and 8 thematic volumes. The topics discussed in these articles are briefly described and recent advances in these various fields are reviewed. The main topics concerned developmental biology, cell biology and metabolism, plant-microorganism interactions, environmental and stress biology and, finally, monographs and reviews in genomics.

1. Introduction

Writing a commemorative review is always challenging, particularly in a scientific journal. However, when Jean-Pierre Jacquot, the present editor, proposed me to review my experience of editing *Advances in Botanical Research*, in order to celebrate the 100th volume of the series, I immediately accepted. This was a unique opportunity to explain how, with Jean-Claude Kader, we managed the series and to evaluate the impact of the topics we

Advances in Botanical Research, Volume 100
ISSN 0065-2296
https://doi.org/10.1016/bs.abr.2021.01.002
Copyright © 2021 Elsevier Ltd
All rights reserved.

selected. After discussions, with Jean-Pierre Jacquot, we decided to split my retrospect contribution in two parts. This section reports the views of the different editors and their experience with ABR. For my part it explains how Jean-Claude Kader and myself met in the late 60s, became friends and collaborators, and finally became editors of this series. It also explains how we worked in selecting articles and inviting co-editors for thematic volumes. Section 2 summarizes some of the topics on which we focused and examines how these fields have evolved and progressed during the last few years, after we retired from the editorship in 2012.

The period covered by our editorship was marked by the explosion of plant genomics. The *Arabidopsis thaliana* genome was the first plant genome to be determined (Arabidopsis Genome Initiative, 2000) and that of rice had just been completed (International Rice Genome Sequencing Project, 2005) when we took over after Jim Callow. A major consequence was a tremendous increase in our understanding of developmental processes and cell biology in these two species, and consequently in other plants. Control of pathogens was a major issue for agriculture, but it was also realized that the field should be enlarged to plant-microbe interactions. Another key issue, which was already clearly identified, was the response of plants to various environmental changes. Finally, we anticipated that after Arabidopsis and rice, many more genomes were bound to be sequenced and we decided to introduce reviews on specific plants, describing their traits, their systematics and phylogeny, and the most recent approaches to sequence their genomes. It is certainly not possible to present all the chapters we invited in detail and to update the state of the art for all of them. I have therefore focused on a few aspects we were interested in and simply made reference to the others.

2. Reviews in plant developmental biology

Most of the articles on developmental biology addressed two major topics: root development and flower biology.

Root architecture and development gained much knowledge from studies on Arabidopsis (Jovanovic et al., 2008). This article described the combined effects of plant hormones auxin and cytokinins on root development as well as the roles of ethylene, ABA and brassinosteroids in response to environment and in controlling transcription factors. It also examined the effects of water stress, and of soil concentrations in nitrates and phosphates. A further level of control on root development is at the level of microRNA. Although most of the reported information was derived from Arabidopsis, the authors also extended the information to legume roots. An update of root development was also published 2 years later (Ingram & Malamy, 2010). Another article focused on the role of nitric oxide in controlling lateral root formation (Molina-Favero et al., 2008) and its crosstalk with plant hormones. The importance of nitric oxide was further recognized by our successors since they invited a thematic volume on this topic (Wendehenne, 2016) in which many updates can be found. Several soil bacteria (PGPR—Plant Growth-Promoting Rhizobacteria) promote root growth and they act by producing nitric oxide. Advances in plant interactions with rhizobacteria, including PGPR, were described in another thematic volume (Bais & Sherrier, 2015). Other aspects of root development were also discussed in articles in the series dealing more specifically with interaction with microbes and response to environmental stress. Since the publication of these chapters, numerous reviews have been published (Choe & Lee, 2017; Hochholdinger, Yu, & Marcon, 2018; Petricka, Winter, & Benfey, 2012). The hormone interactions have been investigated in much more detail (Du & Scheres, 2018; Fonouni-Farde, Diet, & Frugier, 2016; Hu, Vandenbussche, & Van Der Straeten, 2017; Wei & Li, 2016) and a new class of hormones, strigolactones, has been discovered (Sun, Tao, Gu, & Zhang, 2016). Gravitropism and its relationship with auxin transport and signaling has been elucidated (Su, Gibbs, Jancewicz, & Masson, 2017). Reactive oxygen species have been identified as key regulators in the control of root development (Tsukagoshi, 2016). The mode of action of PGPR on root development has been reviewed (Bais & Sherrier, 2015; Verbon & Liberman, 2016).

Flower biology was another key aspect of plant development. The genes controlling flower patterning were already known from pioneer studies in Arabidopsis (Drews, Weigel, & Meyerowitz, 1991). They were also discussed in some of the monographs on specific species, such as orchids (Tsay, Hsiao, Pan, et al., 2008) or bamboos (Das, Bhattacharya, Singh, et al., 2008). One of the invited chapters reviewed the progress in identifying the gene regulatory network involved specifically in carpel development (Ferrandiz, Fourquin, Prunet, et al., 2010). Flower morphology and symmetry was an important issue. The floral symmetry regulatory gene network, initially identified in *Antirrhinum majus*, was reviewed (Citerne, Jabbour, Nadot, & Damerval, 2010). This domain changed rapidly, particularly concerning the evolutionary background, with the development of genomics. An update of the field was recently published (Spencer & Kim, 2018), as well as an update of the development of inflorescence, flower and meristem determinacy in grasses (Bommert & Whipple, 2018). Flowering time, in most plants depends on photoperiod and circadian clock. The main regulatory factors in this pathway were identified and described (Fekih, Nefissi, Miyata, Ezura, & Mizoguchi, 2009). Among them were genes involved in histone methylation and demethylation. Since this review, many studies in Arabidopsis, rice and other plants have focused on the control of flowering time by environmental factors, including clock, photoperiod, temperature, drought, salinity, nutrients, hormones and chemicals or pathogens (Cho, Yoon, & An, 2017; Fornara, 2014). A chapter in the thematic volume on rapeseed breeding (Gupta, 2007) dealt with self-incompatibility in Brassica (Fujimoto & Nishio, 2007). We also invited a review on pollen germination (Wang, Huang, & Jauh, 2010) and a further one on sex determination (Chuck, 2010). Updates on role of calcium in pollen germination (Zheng, Su, Xiao, & Tian, 2019), on sex determination (Harkess & Leebens-Mack, 2017; Pannell, 2017) and on the molecular dialog, involving protein kinases and small peptides, controlling pollen tube guidance (Stegmann & Zipfel, 2017) have been recently reported. Similarly, a review of the pollen tube plasma membrane ion transporters was published (Pertl-Obermeyer, Lackner, Dunlop, & Obermeyer, 2018).

Other aspects of development were the molecular mechanisms underlying vascular development (Jung, Kim, Seo, & Park, 2008), the clock control over plant development (Baudry & Kay, 2008) and the impact of phototropism and gravitropism on development (Molas & Kiss, 2009).

3. Reviews in plant cell biology and metabolism

Several reviews were invited on various aspects of plant cell biology and metabolism. Important classes of proteins were described. They include the expansins (Choi, Kim, & Lee, 2008) and the Late Embryogenesis Abundant (LEA) proteins (Shih, Hoekstra, & Hsing, 2008), as well as plant lectins (Van Damme, Lannoo, & Peumans, 2008), aquaporins (Postaire, Verdoucq, & Maurel, 2008), cytochrome P450 (Schuller & Rupasinghe, 2011) and Arabidopsis histone methyl transferases (Pontvianne, Bevins, & Pikaard, 2010). Recent progress in the field of lectins concern the mechanisms of signal recognition at the cell surface, the detection of symbionts and pathogens, the monitoring of cell wall structure and cell growth, and their role in plant immunity (Bellande, Bono, Savelli, Jamet, & Canut, 2017; Van Holle & Van Damme, 2018). Roles, regulation, and evolution of aquaporins have been further explored (Verdoucq & Maurel, 2018). It has been shown that oscillations in phosphorylation of aquaporins regulate the circadian rhythm of root hydraulics (Prado, Cotelle, Li, et al., 2019). Advances in the field of cytochrome P450 have been made concerning the understanding of their biological function, as recently illustrated in the case of CYP706A3 from Arabidopsis, which reduces volatile terpene emission in flowers, thus protecting them from insects and affecting flower-associated bacterial communities. There is also progress in understanding the evolution of these widespread enzymes and their functional role in plant adaptation to their environment (Alber, Renault, Basilio-Lopez, et al., 2019; Boachon, Burdloff, Ruan, et al., 2019). Plant histone methyl transferases are now involved in most epigenetic regulations, such as flowering, control of transposable elements, stress response and plant immunity (Lee et al., 2016). Structural analysis of certain histone methyl transferases provided insights into the mechanics of DNA methylation (Li, Harris, Zhong, et al., 2018). Plant epigenetics has also been recently reviewed in a thematic volume of this series (Mirouze, Bucher, & Gallusci, 2018).

Several articles concerned aroma volatiles and metabolism. A first one (Defilippi, Manriquez, Luengwilai, & Gonzalez-Agüero, 2009) addressed the question of fruits aromas and reviewed their metabolism. A second one (Vanavichit & Yoshihashi, 2010) explored the rice fragrance, and finally the more recent (Tamogami, Agrawal, & Rakwal, 2011) concerned a class of molecules, jasmonates, which were initially discovered as a jasmine perfume, but which are indeed also key signaling molecules. Each of these reviews examined biosynthetic pathways and the main genes involved. All these studies require sophisticated analytical chemistry and advanced metabolomics, which have been described later on in these series (Rolin, 2013). A major progress in the field of jasmonate has been the identification of the perception mechanism by the co-receptor COI1-JAZ (Sheard, Tan, Mao, et al., 2010). The role of jasmonate in signaling and manipulation by pathogens and insects has been reviewed (Zhang, Liu, Li, et al., 2017; Zhang, Zhang, Melotto, Yao, & He, 2017). The elucidation of the biosynthesis pathway of rose scent was another landmark in the field of volatiles (Magnard, Roccia, Caissard, et al., 2015).

Several cellular structures or processes were discussed: the plant nucleolus, protein turnover and programmed cell death. The plant nucleolus structure and the process of ribosome synthesis (Saez-Vasquez & Medina, 2008) were described. Since then, nucleolus proteome has been analyzed and comparison of plant genomes with yeast and animal genomes allowed the identification of hundreds of proteins involved in the biosynthesis and assembly of plant ribosome and to analyze their regulation. Several pre-rRNA processing pathways were resolved by analyzing gene expression in mutants (Saez-Vasquez & Delseny, 2019). The role of rDNA and nucleolus in shaping chromosomes organization and 3D structure of the nucleus was recognized (Picart-Picolo, Picault, & Pontvianne, 2019). Programmed cell death (PCD) is an important phenomenon in animal cells but it was still in infancy in plants. The available data were presented (Kacprzyk, Daly, & McCabe, 2011). Two recent overviews of this topic analyzed the role of PCD in plant development and control of immunity and the mechanisms which are involved in plant cells, compared to animal cells (Kabbage, Kessens, Bartholomay, & Williams, 2017; Locato & De Cara, 2018). The chapter on protein turnover in grass leaves (Irving, Suzuki, Ishida, & Makino, 2010) addressed the question of the recovery of nitrogen from the most abundant protein in leaves, Rubisco, during normal development and senescence in leaves and focused on protein degradation in chloroplasts. Recently, the protein degradation rate was determined for >1200 individual Arabidopsis leaf proteins, showing half-lives ranging from several hours to several months (Li et al., 2017). Factors required for the outer envelope membrane protein degradation were identified in Arabidopsis: they involve the ubiquitin E3 ligase SP1 and allow retro-translocation of ubiquitinated substrates to the cytosol where they are degraded by the 26S proteasome (Ling et al., 2019).

4. Reviews on plant-microorganism interactions

Interactions of plants with microorganisms constitute an essential domain in botany and plant sciences. It is dominated by interactions with pathogens, which cause considerable losses in agriculture, and by the symbiotic relationship with bacteria or fungi. Some aspects of these interactions have already been mentioned, because they interfere with growth and development (Molina-Favero et al., 2008).

Physiology of legume nitrogen fixating nodules was analyzed (Qureshi, Muneer, Bashir, Ahmad, & Iqbal, 2010), with emphasis on new developments, such as the role of flotillins in formation of the infection threads, gene regulation and proteomics of the nodule. The response of the nodule to various stresses was also discussed. Two reviews recently addressed the question of immunity of the legumes to rhizobium infection (Berrabah,

Ratet, & Gourion, 2019; Cao, Halane, Gassmann, & Stacey, 2017). Sequencing of at least 10 legume genomes, the last one being pea (Kreplak, Madoui, Capal, et al., 2019), as well as several Rhizobia strains considerably added to our knowledge of genes controlling this agronomically important symbiosis and on the evolution of nitrogen fixating symbiosis. An update of the proteomic approach, also including stress conditions, has been recently made available (Larrainzar & Wienkoop, 2017). Bacteria are not the only organisms involved in symbiosis and fungi are equally important. The field of arbuscular mycorrhizal fungi was reviewed by Gianinazzi-Pearson, Séjalon-Delmas, Genre, Jeandroz, and Bonfante (2008). Recent progress in deciphering the mechanisms of establishment of mycorrhiza was described, including the roles of flavonoid, strigolactones and Myc factors. Updates on commonalities in communication in symbioses and plant growth-promoting bacteria were presented in a recent thematic volume in the ABR series, specifically addressing the question on how plants communicate with their biotic environment (Bécard, 2017). Another chapter in this volume also examined plant to plant communication through common mycorrhizal networks. Contribution of ectomycorrhiza to tree nutrition was reviewed in a chapter of another ABR issue on molecular physiology and biotechnology of forest trees (Canovas, 2019).

The mechanisms of plant innate immunity were the subject of a thematic volume (Van Loon, 2009) with 15 chapters, covering many aspects of the responses of plant to pathogens. In particular, the concept of PAMP (Pathogen-Associated Molecular Patterns) recognized by receptors, leading to PAMP-triggered immunity, the concept of hypersensitive reaction (HR) and the ways in which pathogens suppress the plant defenses were illustrated. Several examples of plant-pathogen interactions were described in more detail. The interplay between the signaling pathways involving salicylic acid, jasmonates and ethylene was discussed as well as the role of iron and volatile organic compounds. A chapter also examined the interaction with plant growth-promoting bacteria. In addition, we invited a review on the molecular mechanisms of rice resistance to bacterial blight, caused by the bacteria *Xanthomonas oryzae*, pathovar *oryzae* (Lee et al., 2011). This chapter described the disease, the diversity of the bacterial strains, and the disease resistance genes, either dominant or recessive, which have been genetically identified. When this article was published, out of 34 known disease resistance genes, only 6 had been cloned! Yet, their mode of action was not always understood. A recent paper reports that 41 genes are identified and that 11 have been cloned (Cao et al., 2018). It discussed the mode of action of dominant versus recessive genes and pointed to an autophagy-like cell death mechanism for dominant genes and vacuolar-mediated cell death for recessive genes. Numerous papers dealing with plant disease resistance have been published since these early papers and tremendous progress has been achieved in our understanding of the interactions between plants and fungi or bacterial pathogens. A volume in this series documented advances in plant-microbe interactions, particularly plant bacteria (Bais & Sherrier, 2015). The signaling cascades initiated by receptor recognition have been reviewed recently (Kourelis & Van der Hoorn, 2018; Tang, Wang, & Zhou, 2017) and the advances and perspectives in engineering disease resistance have been discussed (Dong & Ronald, 2019).

Interactions between plants and nematodes are also an important issue in agriculture (Abad & Williamson, 2010). Life cycles of root-knot and cyst nematodes were described as well as the various effectors involved in penetration and suppression of plant defenses. The information derived from sequencing the genomes of two root-knot nematodes was discussed and served as a basis for numerous studies which were further analyzed in a specific volume of this series (Escobar & Fenoll, 2015) and in a recent review (Gillet, Bournaud, Antonino de Souza Junior, & Grossi de Sa, 2017).

The last type of pathogenic interaction we focused on is the plant parasitism (Irving & Cameron, 2009). The anatomical interactions of the host with its parasitic plant were described and different classes of parasites were recognized, some feeding on xylem and others on phloem. The development of the parasite and the host defense mechanisms were described. Finally, the ecological implications of parasite–host physiology were discussed. A major breakthrough occurred with the sequencing of the genome of Striga, one of the most devastating parasitic plants (Yoshida, Kim, Wafula, et al., 2019). It reveals expansion of the strigolactone receptor gene family as well as evidence for horizontal gene transfers from the host plant. The volume on how plants communicate with their environment (Bécard, 2017) also contains a chapter on communication with parasitic nematodes, an update on the interactions between parasitic and host plants and a review on the roles of allelopathy and allochemicals in plant defense. Finally, a chapter described the defense of marine macroalgae against pathogens (Cosse, Leblanc, & Potin, 2008).

Several topics were missing in this list of organisms interacting with plants, such as pathogenic fungi, insects, viruses. This gap has been at least partially corrected in the most recent issues of ABR (Bécard, 2017; Martin, 2014; Sauvion, Thiéry, & Calatayud, 2017).

5. Reviews in plant environmental and stress biology

Adaptation of plants to their environment is a key factor of their evolutionary success and an important issue in the context of climate change, erosion of soils and dramatic changes in agriculture practice in the 20th and now 21st century. Many sectors are concerned and only a few of them could be addressed during the time we were responsible for editing ABR. Two thematic volumes were entirely devoted to stress responses. The first one dealt with oxidative stress and redox regulation (Jacquot, 2009). Oxidative stress begins with the production of reactive oxygen species (ROS). This volume contained 15 chapters including various aspects of oxidative stress, such as the role of ROS in growth and development, the mechanisms of oxidation of proteins, role of the thioredoxin and glutaredoxin pathways, metabolism of glutathione. Several examples of situations in which oxidative stress are important were analyzed such as the exposition of UV-B light and role of antioxidants, the production of ROS in the chloroplast, during establishment of Rhizobium-legume symbiosis, or during lignin degradation by the fungus *Phanerochaete chrysosporium*. Since then, it has become obvious that redox regulation is central in plant development and response to environmental cues (Mittler, 2017). ROS were considered as toxic for living cells but they also appear to be necessary for the progression of several basic biological processes including cell proliferation, cell differentiation, and response to environmental cues. Their production induces many signaling cascades determining the final responses. A recent development in the field is the interaction between redox metabolism and plant epigenetics which turns out to be an essential level of control for plant adaptation to environmental changes (Shen, Issakidis-Bourguet, & Zhou, 2016). Other aspects of oxidative stress discussed in different ABR issues are the role of nitric oxide (Molina-Favero et al., 2008), the response and tolerance to heavy metals (Chaffai & Koyama, 2011), the impact and dynamics of iron (Briat, 2008) and low oxygen tolerance (Licausi & Perata, 2009).

A second major theme in stress physiology is the response to drought and salinity. We have invited several articles in this field, before deciding that a thematic volume should address the question. These early articles reported a description of aquaporins and their role in water transport

(Postaire et al., 2008), an analysis of the response of halophytes to salinity (Riadh, Wided, Hans-Werner, & Chedly, 2010) and a review on the ecological water-use strategies of succulent plants (Ogburn & Edwards, 2010). Another article examined the situation of culture systems and their optimization to limit stress caused by plant nutrient solutions (Gorbe & Calatayud, 2010). The thematic volume (Turkan, 2011) extended the domain and contained 13 chapters. They described the various strategies of plant adaptation to drought and salinity stresses, the molecular mechanisms of adaptation, including ion transport in halophytes, the role of abscisic acid in triggering plant response to drought or salt, the protective metabolites and the impact of drought and salinity on photosynthetic activity. The molecular genetics of drought stress and approaches to engineer salinity and water stress tolerance were discussed in other chapters. This field also evolved rapidly and numerous articles have been published during the last 10 years. Several reviews can be found in the recent ABR thematic volume on abscisic acid (Seo & Marion-Poll, 2019).

Finally, two other aspects of environmental physiology were discussed: the first one concerns the perception of light and the syndrome of shade avoidance. Several transcription factors which modify the hormone sensitivity and growth response were described (Martinez-Garcia et al., 2010). The second one reviewed cold signaling and cold tolerance (Ruelland, Vaultier, Zachowski, & Hurry, 2009). Shade avoidance mechanisms were updated in two recent reviews (Fraser, Hayes, & Franklin, 2016; Yang & Li, 2017). More recently, miRNA (Xie, Liu, Wang, et al., 2017) and histone modification (Peng, Li, Zhou, et al., 2018) were also recognized as part of the regulatory network. Advances and future challenges in signaling and mechanisms underlying cold tolerance have been recently summarized (Ding, Shi, & Yang, 2019).

6. Monographs and reviews in genomics

Because the series was entitled *Advances in Botanical Research*, we tried to include some papers on systematics of some emerging species, asking specialists to outline their peculiarities and their interest for various aspects of plant biology. We initiated our role as co-editor of ABR with a thematic volume on rapeseed breeding (Gupta, 2007). This volume had 18 chapters reviewing the evolutionary history of this crop, the breeding methods, oil and meal quality, use of molecular markers for breeding and cultivar identification, self-incompatibility, use of haploid and double haploid technology, breeding for apetalous rapeseed, resistance to herbicides or to black leg, a disease due to the fungus *Leptosphaeria maculans*. Other chapters dealt with development of alloplasmic rape, with the interactions between honeybees and rapeseed, with glucosinolate metabolism, gene flow with wild species, rapeseed biotechnology and oil technology. Although a genetic map of rapeseed already existed at that time, only a few of the important genes for agronomy and breeding of this crop were identified and cloned. Since this volume was published, considerable progress has been made in most of the subjects. A reference genome of rapeseed was published in 2014 (Chaloub, Denoeud, Liu, et al., 2014) and, more recently, genomes from a world-wide collection of a 1000 germplasms have been re-sequenced, shedding light on domestication of this species and its evolution (Wu, Liang, Yan, et al., 2019). Many genes have been identified and introduced in improved varieties including disease resistance genes to black leg disease (Raman, Raman, Coombes, et al., 2016), using a variety of techniques. Several genomes of the fungal pathogen *Leptosphaeria* have been sequenced (Dutreux, Da Silva, d'Agata, et al., 2018). Glucosinolate pathways have been elucidated (Pietro, Lopez, & Simal-Gandara, 2019) as well as fatty acid metabolism (Qu, Jia, Fu, et al., 2017). A volume in this series updates our knowledge on glucosinolates (Kopriva, 2016).

We invited Taiwanese colleagues to write a review on the molecular biology of orchid flowers (Tsay et al., 2008), because this is fascinating model. They had made valuable contributions to the characterization of genes involved in flower morphology and scent metabolism and they had initiated genomic studies by sequencing ESTs, preparing BAC libraries and molecular tools for functional genomics such as VIGS (Virus-Induced Gene Silencing). They have developed their work and reported the first genome sequence of an orchid species, *Phalaenopsis equestris* (Cai, Liu, Vanneste, et al., 2015), and more recently, the genome of *Apostasia shenzhenica* and transcriptome data for one of each of the five orchid subfamilies (Zhang, Liu, et al., 2017; Zhang, Zhang, et al., 2017), thus boosting orchid research. Now several genomes of orchid species, including Dendrobium and Vanilla are available. The role of transposable element in the dark purple spots in some Phalaenopsis flowers was elucidated (Hsu, Su, Jeng, et al., 2019).

In the same issue as the one on Phalaenopsis we invited a review on citrus fruit quality (Tadeo, Cercos, Colmenero-Flores, et al., 2008). It focused essentially on the physiology of fruiting and ripening, composition of the fruit, smell, abscission and conservation. Citrus genomics was just beginning with an EST program. Since then, important advances have been made in the understanding of fruit ripening and of fruit abscission. This progress is due to the development of molecular biology and biochemical approaches to characterize ethylene receptors, although citrus is a non-climacteric fruit (Chen, Grimplet, David, et al., 2018). Important factors in fruit abscission have also been characterized (Merelo, Agusti, Arbona, et al., 2017). The genomes of lemon, mandarin and orange were determined (Wu, Prochnik, Jenkins, et al., 2014) and further sequencing of related species and varieties allowed one to reconstruct the evolutionary history of Citrus domestication (Wu, Terol, Ibanez, et al., 2018).

In the same volume, a monography of bamboo was published (Das et al., 2008). Bamboo is a fascinating genus too, with ~1300 species, either herbaceous or woody, distributed all over the world but mostly in tropical and subtropical era. This is the only grass that can exist as a forest. Some species are characterized by extremely fast growth. Based on flowering time, they can be divided into three categories: some are flowering annually, others flower sporadically and irregularly and the majority flower synchronously at long time intervals ranging from 3 to 120 years. Analysis of DNA content and chromosome numbers indicated different ploidy levels. This article was basically a systematic paper, discussing the phylogeny and classification of bamboo, using anatomical and biological characters as well as molecular markers such as microsatellites and other PCR derived markers. Here again a tremendous progress has been made by the sequencing of one of the woody species, *Phyllostachys heterocycla* (Peng, Lu, Li, et al., 2013). It revealed that bamboo had many more cellulose synthase (*CesA*) genes and cellulose synthase-like (*Csl*) genes than any other grass (19 *CesA* and 38 *Csl* compared to 11 and 34 in rice and 10 and 29 in Arabidopsis). The gene number in this superfamily is similar to that in poplar, a woody species. More recently four genomes were determined (Guo, Ma, Yang, et al., 2019) from different ploidy levels bamboo species and it clarified the phylogeny of the genus: there are one diploid herbaceous lineage and three distinct polyploid woody ones which result from alloploidy between three extinct diploid parents. Of course, the availability of genomes and transcriptome sequences generated many specific gene studies, including flowering genes, disease resistance genes and many others such as aquaporins.

Along the same lines, we invited review papers on two potential bioenergy crops, *Jatropha curcas* (Carels, 2009) and Miscanthus species (Heaton, Dohleman, Miguez, et al., 2010). Both articles reviewed the biology of the two species, their advantages and drawbacks, as well as the perspectives of developing them as a crop. Genomics and sequencing have allowed one to progress. A Jatropha genome draft was reported as early as 2011 (Sato, Hirakawa, Isobe, et al., 2011) and an improved sequence a few years later (Wu, Zhou, Cheng, et al., 2015). Similarly, *Miscanthus sinensis* genome sequence was published (Ma, Jensen, Alexandrov, et al., 2012). As a result, numerous traits concerning cell wall composition and bioconversion, were identified and now serve as a background in breeding programs (Van der Weijde, Kamei, Severing, et al., 2017). A recent article reviews major progress and perspectives in using various crops as a source of bioenergy (Clifton-Brown, Harfouche, Casler, et al., 2019).

Another aspect of plant genomics was the development of research on three tropical crops: cocoa, coffee, and sugar cane. The review on cocoa (Micheli, Guiltinan, Peres-Gramacho, et al., 2010) was already genomics orientated. One year later, the same group of authors reported a draft genome sequence of *Theobroma cacao*, and, more recently an improved version (Argout, Martin, Droc, et al., 2017). This work has provided information on genes involved in disease resistance, seed quality, aroma and self-incompatibility, and is now the basis for further breeding. This reference genome has also been very useful to investigate the evolutionary history and domestication of this tree by Amazonian people. Finally, a recent review provides an overview of cocoa production and recent advances in biotechnology for cocoa improvement (Wickramasuriya & Dunwell, 2018). The coffee story (De Kochko, Akaffou, Andrade, et al., 2010) is fairly similar to the cocoa one. The complete genome was reported (Denoeud, Carretero-Paulet, Dereeper, et al., 2014) and opened the way to further improvement strategies (Tran, Lee, Furtado, Smyth, & Henry, 2016). Sugarcane is both a food and energy crop (Manners, 2011). Modern sugarcane results from an initial cross between two octoploid species with different haploid chromosome numbers, a male *Saccharum spontaneum* ($n=10$) and a female *Saccharum officinarum* ($n=8$). After several generations of breeding, the varieties have chromosome numbers ranging from 100 to 120, 80% of which have a *S. officinarum* origin, 10–15% a *S. spontaneum* one and 5–10% are recombinant chromosomes. Recently the genome of the *S. spontaneum* parent has been reported (Zhang, Zhang, Tang, et al., 2018) and the challenges in exploiting sugar cane genomic and transcriptomic data have been discussed (Thirugnanasambandam, Hoang, & Henry, 2018).

The monography on *Artemisia* addressed another topic (Valles et al., 2011). It reported the description of a genus with more than 150 species, its phylogeny and the ethnobotanical use of some of the species, including medicinal, food or beverage. The genetic diversity and genome information were fairly limited at the time this paper was published, but the species had great potential in treatment of malaria. A draft genome has recently been published and it shed new insight on the biosynthesis of artemisin, the anti-malarial drug, opening the way to the production of artemisin in transgenic plants (Shen, Zhang, Liao, et al., 2018).

Finally, a chapter dealing with genome evolution of plant pathogenic or symbiotic fungi was invited (Aguileta, Hood, Refrégier, & Giraud, 2009). This chapter examined the evolution of genome organization, chromosomal rearrangements, evolution of gene families, rapid evolving genes, horizontal transfers, transposable elements, and suppression of recombination around mating type loci. This field also evolved very rapidly, with a lot more fungal genomes being sequenced. Such studies provided new insights on population biology of fungus invasion (Gladieux, Feurtey, Hood, et al., 2015) and on the mechanisms which allow a fungus species to adapt specifically to its plant host, as illustrated in a recent example (Badoin, Gladieux, Gouzy, et al., 2017). It is worth mentioning that this aspect of plant genomics has been the subject of several thematic volumes in the following years: herbaceous land plants (Paterson, 2014), trees (Plomion & Adam-Blondon, 2015) and fungi (Martin, 2014).

7. Conclusion

Before concluding, I have to cite four thematic volumes which I have not commented in this review: two volumes reviewed our knowledge on vitamins (Rebeillé & Douce, 2011a, 2011b), one volume updated our knowledge on lignin biosynthesis, biodegradation and bioengineering (Jouanin & Lapierre, 2012) and a last volume presented advances and challenges in medicinal plants (Shyur & Lau, 2012). This volume is reviewed in a chapter of this issue, as well as recent advances and trends in the field of medicinal plant research.

I have tried to illustrate a few topics we selected with Jean-Claude Kader a few years ago. These were subjects in which we were interested, although most of the time we were not specialists. We very much enjoyed editing these articles and volumes, exchanging with authors, many of whom became our friends and it has been an exciting experience. I had great pleasure revisiting this period and having a quick look at most recent developments. I feel sad being alone to commemorate this period and I am sure Jean-Claude would have enjoyed, as well, this exercise and would have been impressed, as I have been, by the outstanding advances in such a short time. The most spectacular development is certainly in the field of plant genomics. Many of the current achievements were not even imaginable at the time the articles were written and it is difficult to foresee the future developments. I expect they will occur in the field of the interactions between plants and their microbiote, in the field of ecology and also in the applications of all these findings to a more efficient and sustainable agriculture. I wish a long life to *Advances in Botanical Research* and another bunch of exciting and stimulating articles in the coming years.

Acknowledgments

The author would like to thank Thomas Roscoe for revising the English language of this manuscript. This article is dedicated to the memory of my regretted friend and co-editor Jean-Claude Kader who passed away on July 2018.

References

Abad, P., & Williamson, V. (2010). Plant nematode interaction: A sophisticated dialogue. *Advances in Botanical Research*, *53*, 147–192.

Aguileta, G., Hood, M. E., Refrégier, G., & Giraud, T. (2009). Genome evolution in plant pathogenic and symbiotic fungi. *Advances in Botanical Research*, *49*, 151–193.

Alber, A. V., Renault, H., Basilio-Lopez, A., et al. (2019). Evolution of coumaroyl conjugate 3-hydroxylases in land plants: Lignin biosynthesis and defense. *The Plant Journal*, *99*, 924–936.

Arabidopsis Genome Initiative. (2000). Analysis of the genome sequence of the flowering plant *Arabidopsis thaliana*. *Nature*, *408*, 796–815.

Argout, X., Martin, G., Droc, G., et al. (2017). The cacao Criollo genome v2.0: An improved version of the genome for genetic and functional genomic studies. *BMC Genomics*, *18*, 730. https://doi.org/10.1186/s12864-017-4120-9.

Badoin, H., Gladieux, P., Gouzy, J., et al. (2017). Widespread selective sweeps throughout the genome of model plant pathogenic fungi and identification of effector candidates. *Molecular Ecology*, *26*, 2041–2062.

Bais, H., & Sherrier, J. (Eds.). (2015). Plant microbe interactions. *Advances in Botanical Research*, *75*, 1–310.

Baudry, A., & Kay, S. (2008). Clock control over plant gene expression. *Advances in Botanical Research*, *48*, 69–105.

Bécard, G. (Ed.). (2017). How plants communicate with their biotic environment. *Advances in Botanical Research*, *82*, 3–390.

Bellande, K., Bono, J. J., Savelli, B., Jamet, E., & Canut, H. (2017). Plant lectins and receptor-like kinases: How do they sense the outside. *International Journal of Molecular Sciences*, *18*. https://doi.org/10.3390/ijms18061164. pii: E1164.

Berrabah, F., Ratet, P., & Gourion, B. (2019). Legume nodules: Massive infection in the absence of defense induction. *Molecular Plant-Microbe Interactions*, *32*, 35–44.

Boachon, B., Burdloff, Y., Ruan, J. X., et al. (2019). A promiscuous CYP706A3 reduces terpene volatile emission from Arabidopsis flowers, affecting florivores and the floral microbiome. *Plant Cell*, *31*, 2947–2972.

Bommert, P., & Whipple, C. (2018). Grass inflorescence architecture and meristem determinacy. *Seminars in Cell & Developmental Biology*, *79*, 37–47.

Briat, J. F. (2008). Iron dynamics in plants. *Advances in Botanical Research*, *46*, 137–180.
Cai, J., Liu, X., Vanneste, K., et al. (2015). The genome sequence of the orchid *Phalaenopsis equestris*. *Nature Genetics*, *47*, 65–72.
Canovas, F. M. (Ed.). (2019). Molecular physiology and biotechnology of forest trees. *Advances in Botanical Research*, *89*, 2–321.
Cao, Y., Halane, M. K., Gassmann, W., & Stacey, G. (2017). The role of plant innate immunity in the legume rhizobium symbiosis. *Annual Review of Plant Biology*, *68*, 535–561.
Cao, J., Zhang, M., Xiao, J., Li, X., Yuan, M., & Wang, S. (2018). Dominant and recessive major R genes lead to different types of host cell death during resistance to *Xanthomonas oryzae* in rice. *Frontiers in Plant Science*, *9*, 1711. https://doi.org/10.3389/fpls.2018.01711.
Carels, N. (2009). *Jatropha curcas*: A review. *Advances in Botanical Research*, *50*, 40–86.
Chaffai, R., & Koyama, H. (2011). Heavy metal tolerance in *Arabidopsis thaliana*. *Advances in Botanical Research*, *60*, 1–49.
Chaloub, B., Denoeud, F., Liu, S., et al. (2014). Plant genetics. Early allopolyploid evolution in the post-neolithic *Brassica napus* oilseed genome. *Science*, *345*, 950–953.
Chen, Y., Grimplet, J., David, K., et al. (2018). Ethylene receptors and related proteins in climacteric and non-climacteric fruits. *Plant Science*, *276*, 63–72.
Cho, L. H., Yoon, J., & An, G. (2017). The control of flowering time by environmental factors. *The Plant Journal*, *90*, 708–719.
Choe, G., & Lee, J. Y. (2017). Push-pull strategy in the regulation of postembryonic root development. *Current Opinion in Plant Biology*, *35*, 158–164.
Choi, D., Kim, J. H., & Lee, Y. (2008). Expansins in plant development. *Advances in Botanical Research*, *47*, 47–97.
Chuck, G. (2010). Molecular mechanisms of sex determination in monoecious and dioecious plants. *Advances in Botanical Research*, *54*, 53–83.
Citerne, H., Jabbour, F., Nadot, S., & Damerval, C. (2010). The evolution of floral symmetry. *Advances in Botanical Research*, *54*, 85–137.
Clifton-Brown, J., Harfouche, A., Casler, M. D., et al. (2019). Breeding progress and preparedness for mass-scale development of perennial lignocellulose biomass crops switchgrass, miscanthus, willow and poplar. *Global Change Biology. Bioenergy*, *11*, 118–151.
Cosse, A., Leblanc, C., & Potin, P. (2008). Dynamic defense of marine macroalgae against pathogens: From early activated to gene regulated responses. *Advances in Botanical Research*, *46*, 221–266.
Das, M., Bhattacharya, S., Singh, P., et al. (2008). Bamboo taxonomy and diversity in the era of molecular markers. *Advances in Botanical Research*, *47*, 226–268.
De Kochko, A., Akaffou, S., Andrade, A. C., et al. (2010). Advances in Coffea genomics. *Advances in Botanical Research*, *53*, 23–63.
Defilippi, B. G., Manriquez, D., Luengwilai, K., & Gonzalez-Agüero, M. (2009). Aroma volatiles: Biosynthesis and mechanisms of modulation during fruit ripening. *Advances in Botanical Research*, *50*, 1–37.
Denoeud, F., Carretero-Paulet, L., Dereeper, A., et al. (2014). The coffee genome provides insight into the convergent evolution of caffeine biosynthesis. *Science*, *345*, 1181–1184.
Ding, Y., Shi, Y., & Yang, S. (2019). Advances and challenges in uncovering cold tolerance regulatory mechanisms in plants. *The New Phytologist*, *222*, 1690–1704.
Dong, O. X., & Ronald, P. C. (2019). Genetic engineering for disease resistance in plants: Recent progress and future perspectives. *Plant Physiology*, *180*, 26–38.
Drews, G. N., Weigel, D., & Meyerowitz, E. M. (1991). Floral patterning. *Current Opinion in Genetics & Development*, *2*, 174–178.
Du, Y., & Scheres, B. (2018). Lateral root formation and the multiple roles of auxin. *Journal of Experimental Botany*, *69*, 155–167.

Dutreux, F., Da Silva, C., d'Agata, L., et al. (2018). *De novo* assembly and annotation of three Leptosphaeria genomes using Oxford Nanopore MinION sequencing. *Science Data, 5,* 180235. https://doi.org/10.1038/sdata 2018.235.

Escobar, C., & Fenoll, C. (Eds.). (2015). Plant nematode interactions. A view of compatible interrelationships. *Advances in Botanical Research, 73,* 1–376.

Fekih, R., Nefissi, R., Miyata, K., Ezura, H., & Mizoguchi, T. (2009). Roles of circadian clock and histone methylation in the control of floral repressors. *Advances in Botanical Research, 50,* 199–225.

Ferrandiz, C., Fourquin, C., Prunet, N., et al. (2010). Carpel development. *Advances in Botanical Research, 55, 1–73.*

Fonouni-Farde, C., Diet, A., & Frugier, F. (2016). Root development and endosymbioses: DELLAs lead the orchestra. *Trends in Plant Science, 21,* 898–900.

Fornara, F. (Ed.). (2014). The molecular genetics of floral transition and flower development. *Advances in Botanical Research, 72,* 1–363.

Fraser, D. P., Hayes, S., & Franklin, K. A. (2016). Photoreceptor crosstalk in shade avoidance. *Current Opinion in Plant Biology, 33,* 1–7.

Fujimoto, R., & Nishio, T. (2007). Self-incompatibility. *Advances in Botanical Research, 45,* 139–154.

Gianinazzi-Pearson, V., Séjalon-Delmas, N., Genre, A., Jeandroz, S., & Bonfante, P. (2008). Plants and arbuscular mycorrhizal fungi: Cues and communication in the early step of symbiotic interactions. *Advances in Botanical Research, 46,* 181–219.

Gillet, F. X., Bournaud, C., Antonino de Souza Junior, J. D., & Grossi de Sa, M. F. (2017). Plant-parasitic nematodes: Towards understanding molecular players in stress responses. *Annals of Botany, 119,* 775–789.

Gladieux, P., Feurtey, A., Hood, M. E., et al. (2015). The population biology of fungal invasions. *Molecular Ecology, 24,* 1969–1986.

Gorbe, E., & Calatayud, A. (2010). Optimization of nutrition in soilless systems: A review. *Advances in Botanical Research, 53,* 193–245.

Guo, Z. H., Ma, P. F., Yang, G. Q., et al. (2019). Genome sequences provide insights into the reticulate origin and unique traits of woody bamboos. *Molecular Plant, 12,* 1353–1365.

Gupta, S. K. (Ed.). (2007). Rapeseed breeding. *Advances in Botanical Research, 45,* 1–543.

Harkess, A., & Leebens-Mack, J. (2017). A century of sex determination in flowering plants. *The Journal of Heredity, 108,* 69–77.

Heaton, E. A., Dohleman, F. G., Miguez, A. F., et al. (2010). Miscanthus: A promising biomass crop. *Advances in Botanical Research, 56,* 75–137.

Hochholdinger, F., Yu, P., & Marcon, C. (2018). Genetic control of root system development in maize. *Trends in Plant Science, 23,* 79–88.

Hsu, C. C., Su, C. J., Jeng, M. F., et al. (2019). A *HORT1* retrotransposon insertion in the *PeMYB11* promoter causes Harlequin/black flowers in Phalaenopsis orchids. *Plant Physiology, 180, 1535–1548.*

Hu, Y., Vandenbussche, F., & Van Der Straeten, D. (2017). Regulation of seedling growth by ethylene and the ethylene-auxin crosstalk. *Planta, 245,* 467–489.

Ingram, P. A., & Malamy, J. E. (2010). Root system architecture. *Advances in Botanical Research, 55,* 75–117.

International Rice Genome Sequencing Project. (2005). The map-based sequence of the rice genome. *Nature, 436,* 793–800.

Irving, L. J., & Cameron, D. D. (2009). You are what you eat: Interactions between root parasitic plants and their host. *Advances in Botanical Research, 50,* 87–138.

Irving, L. J., Suzuki, Y., Ishida, H., & Makino, A. (2010). Protein turnover in grass leaves. *Advances in Botanical Research, 54,* 139–182.

Jacquot, J. P. (Ed.). (2009). Oxidative stress and redox regulation in plants. *Advances in Botanical Research*, *52*, 1–495.

Jouanin, L., & Lapierre, C. (Eds.). (2012). Lignins: Biosynthesis, biodegradation and bioengineering. *Advances in Botanical Research*, *61*, 2–391.

Jovanovic, M., Lefebvre, V., Laporte, P., Gonzalez-Rizzo, S., Lelandais-Brière, C., Frugier, F., et al. (2008). How the environment regulates root architecture in dicots. *Advances in Botanical Research*, *46*, 35–74.

Jung, J. H., Kim, S. G., Seo, P. J., & Park, C. M. (2008). Molecular mechanisms underlying vascular development. *Advances in Botanical Research*, *48*, 1–68.

Kabbage, M., Kessens, R., Bartholomay, L. C., & Williams, B. (2017). The life and death of a plant cell. *Annual Review of Plant Biology*, *28*, 375–404.

Kacprzyk, J., Daly, C. T., & McCabe, P. F. (2011). The botanical dance of death: Programmed cell death in plants. *Advances in Botanical Research*, *60*, 168–261.

Kopriva, S. (Ed.). (2016). Glucosinolates. *Advances in Botanical Research*, *80*, 1–350.

Kourelis, J., & Van der Hoorn, R. A. L. (2018). Defended to the nines: 25 years of resistance gene cloning identifies nine mechanisms for R protein function. *Plant Cell*, *30*, 285–299.

Kreplak, J., Madoui, M. A., Capal, P., et al. (2019). A reference genome for pea provides insight into legume genome evolution. *Nature Genetics*, *51*, 1411–1422.

Larrainzar, E., & Wienkoop, S. (2017). A proteomic view on the role of legume symbiotic interactions. *Frontiers in Plant Science*, *8*, 1267. https://doi.org/10.3389/fpls.2017.01267.

Lee, S., Fu, F., Xu, S., Lee, S. Y., Yun, D. J., & Mengiste, T. (2016). Global regulation of plant immunity by histone lysine methyl transferases. *Plant Cell*, *28*, 1640–1661.

Lee, S. W., Han, M., Park, C. J., Seo, Y. S., Bartley, L. E., & Jeon, J. S. (2011). The molecular mechanisms of rice resistance to the bacterial blight pathogen *Xanthomonas oryzae* pathovar *oryzae*. *Advances in Botanical Research*, *60*, 51–87.

Li, X., Harris, C. J., Zhong, Z., et al. (2018). Mechanistic insights into plant SUVH family H3K9 methyltransferases and their binding to context-biased non-CG DNA methylation. *Proceedings of the National Academy of Sciences of the United States of America*, *115*, E8793–E8802.

Li, L., Nelson, C. I., Trösch, J., Castleden, I., Huang, S., & Millar, A. H. (2017). Protein degradation rate in *Arabidopsis thaliana* leaf growth and development. *Plant Cell*, *29*, 207–228.

Licausi, F., & Perata, P. (2009). Low oxygen signaling and tolerance in plants. *Advances in Botanical Research*, *50*, 139–198.

Ling, Q., Broad, W., Trosch, R., Töpel, M., Demiral Sert, T., Lymperopoulos, P., et al. (2019). Ubiquitin-dependent chloroplast-associated protein degradation in plants. *Science*, *363*. https://doi.org/10.1126/science.aav4467. pii: eaav4467.

Locato, V., & De Cara, L. (2018). Programmed cell death in plants: An overview. *Methods in Molecular Biology*, *1743*, 1–8.

Ma, X. F., Jensen, E., Alexandrov, N., et al. (2012). High resolution genetic mapping by genome sequencing reveals genome duplication and tetraploid genetic structure of the diploid *Miscanthus sinensis*. *PLoS One*, 7, e33821.

Magnard, J. L., Roccia, A., Caissard, J. C., et al. (2015). Plant volatiles. Biosynthesis of monoterpene scent compounds in roses. *Science*, *349*, 81–83.

Manners, J. M. (2011). Functional genomics of sugarcane. *Advances in Botanical Research*, *60*, 89–168.

Martin, F. (Ed.). (2014). Fungi. *Advances in Botanical Research*, *70*, 2–472.

Martinez-Garcia, J. F., Galstyan, A., Sala-Martret, M., Cifuentes-Esquivel, N., Gallemi, M., & Bou-Torrent, J. (2010). Regulatory components of shade avoidance syndrome. *Advances in Botanical Research*, *53*, 65–116.

Merelo, P., Agusti, J., Arbona, V., et al. (2017). Cell wall remodeling in abscission zone cells during ethylene promoted fruit abscission in citrus. *Frontiers in Plant Science*, *8*, 126. https://doi.org/10.3389/fpls2017.00126.

Micheli, F., Guiltinan, M., Peres-Gramacho, K., et al. (2010). Functional genomics of cacao. *Advances in Botanical Research*, *55*, 119–177.

Mirouze, M., Bucher, E., & Gallusci, P. (Eds.). (2018). Plant epigenetics coming of age for breeding applications. *Advances in Botanical Research*, *88*, 1–454.

Mittler, R. (2017). ROS are good. *Trends in Plant Science*, *22*, 11–19.

Molas, M. L., & Kiss, J. Z. (2009). Phototropism and gravitropism in plants. *Advances in Botanical Research*, *49*, 1–34.

Molina-Favero, C., Creus, C. M., Lanteri, M. L., Correa-Aragunde, N., Lombardo, M. C., Barassi, C. A., et al. (2008). Nitric oxide and plant growth promoting rhizobacteria: Common features influencing root growth and development. *Advances in Botanical Research*, *46*, 1–33.

Ogburn, R. M., & Edwards, E. J. (2010). The ecological water-use strategies of succulent plants. *Advances in Botanical Research*, *55*, 179–225.

Pannell, J. R. (2017). Plant sex determination. *Current Biology*, *27*, R191–R197.

Paterson, A. H. (Ed.). (2014). Genomes of herbaceous land plants. *Advances in Botanical Research*, *69*, 2–413.

Peng, M., Li, Z., Zhou, N., et al. (2018). Linking phytochrome-interacting factor to histone modification in plant shade avoidance. *Plant Physiology*, *176*, 1341–1351.

Peng, Z., Lu, Y., Li, L., et al. (2013). The draft genome of the fast-growing non-timber forest species moso bamboo (*Phyllostachys heterocycle*). *Nature Genetics*, *45*, 456–461.

Pertl-Obermeyer, H., Lackner, P., Dunlop, J. W. C., & Obermeyer, G. (2018). The pollen plasma membrane permeome converts transmembrane ion transport into speed. *Advances in Botanical Research*, *87*, 215–265.

Petricka, J. J., Winter, C. M., & Benfey, P. N. (2012). Control of Arabidopsis root development. *Annual Review of Plant Biology*, *63*, 563–590.

Picart-Picolo, A., Picault, N., & Pontvianne, F. (2019). Ribosomal RNA genes shape chromatin domains associating with the nucleolus. *Nucleus*, *10*, 67–72.

Pietro, M. A., Lopez, C. J., & Simal-Gandara, J. (2019). Glucosinolates: Molecular structure, breakdown, genetic, bioavailability, properties and healthy and adverse effects. *Advances in Food and Nutrition Research*, *90*, 305–350.

Plomion, C., & Adam-Blondon, A. F. (Eds.). (2015). Land plants-trees. *Advances in Botanical Research*, *74*, 1–306.

Pontvianne, F., Bevins, T., & Pikaard, C. S. (2010). Arabidopsis histone lysine methyltransferases. *Advances in Botanical Research*, *53*, 1–22.

Postaire, O., Verdoucq, L., & Maurel, C. (2008). Aquaporins in plants: From molecular structure to integrated functions. *Advances in Botanical Research*, *46*, 75–136.

Prado, K., Cotelle, V., Li, G., et al. (2019). Oscillating aquaporin phosphorylation and 14-3-3 proteins mediate the circadian regulation of leaf hydraulics. *Plant Cell*, *31*, 417–429.

Qu, C., Jia, L., Fu, F., et al. (2017). Genome-wide association mapping and identification of candidate genes for fatty acid composition in *Brassica napus*, using SNP markers. *BMC Genomics*, *18*, 232. https://doi.org/10.1186/s12864-017-3607-8.

Qureshi, M. I., Muneer, S., Bashir, H., Ahmad, J., & Iqbal, M. (2010). Nodule physiology and proteomics of stressed legumes. *Advances in Botanical Research*, *56*, 1–48.

Raman, H., Raman, R., Coombes, N., et al. (2016). Genome wide-association study identifies new loci for resistance to *Leptosphaeria maculans* in canola. *Frontiers in Plant Science*, 7, 1513. https://doi.org/10.3389/fpls.2016.01513.

Rebeillé, F., & Douce, R. (Eds.). (2011a). Biosynthesis of vitamins in plants. Part A: Vitamins A,B1, B2, B3, B5. *Advances in Botanical Research, 58, 1–287.*

Rebeillé, F., & Douce, R. (Eds.). (2011b). Biosynthesis of vitamins in plants. Part B: Vitamins B6, B8, B9, C, E, K. *Advances in Botanical Research, 59*, 1–293.
Riadh, K., Wided, M., Hans-Werner, K., & Chedly, A. (2010). Responses of halophytes to environmental stresses with special emphasis to salinity. *Advances in Botanical Research, 53*, 117–145.
Rolin, D. (Ed.). (2013). Metabolomics coming of age with its technological diversity. *Advances in Botanical Research, 67*, 2–693.
Ruelland, E., Vaultier, M. N., Zachowski, A., & Hurry, V. (2009). Cold signalling and cold acclimation in plants. *Advances in Botanical Research, 49*, 35–150.
Saez-Vasquez, J., & Delseny, M. (2019). Ribosome biogenesis in plants: From functional 45S ribosomal DNA organization to ribosome assembly factors. *Plant Cell, 31*, 1945–1967.
Saez-Vasquez, J., & Medina, F. J. (2008). The plant nucleolus. *Advances in Botanical Research, 47*, 1–46.
Sato, S., Hirakawa, H., Isobe, S., et al. (2011). Sequence analysis of the genome of an oil-bearing tree, *Jatropha curcas L. DNA Research, 18*, 65–76.
Sauvion, N., Thiéry, D., & Calatayud, P. A. (Eds.). (2017). Insect plant interactions in a crop protection perspective. *Advances in Botanical Research, 81*, 1–406.
Schuller, M. A., & Rupasinghe, S. G. (2011). Molecular and structural perspectives on Cytochome P450s in plants. *Advances in Botanical Research, 60*, 263–307.
Seo, M., & Marion-Poll, A. M. (Eds.). (2019). Abscisic acid in plants. *Advances in Botanical Research, 92*, 1–370.
Sheard, L. B., Tan, X., Mao, H., et al. (2010). Jasmonate perception by inositol-phosphate--potentiated COI1-JAZ co-receptor. *Nature, 468*, 400–405.
Shen, Y., Issakidis-Bourguet, E., & Zhou, D. X. (2016). Perspectives on the interactions between metabolism, redox, and epigenetics in plants. *Journal of Experimental Botany, 67*, 5291–5300.
Shen, Q., Zhang, L., Liao, Z., et al. (2018). The genome of *Artemisia annua* provides insight into the evolution of Asteraceae family and artemisin biosynthesis. *Molecular Plant, 11*, 776–788.
Shih, M. D., Hoekstra, F. A., & Hsing, Y. I. (2008). Late embryogenesis abundant proteins. *Advances in Botanical Research, 48*, 211–255.
Shyur, L. F., & Lau, A. S. Y. (Eds.). (2012). Recent trends in medicinal plant research. *Advances in Botanical Research, 62*, 2–430.
Spencer, V., & Kim, M. (2018). Re"CYC"ling molecular regulators in the evolution and development of flower symmetry. *Seminars in Cell & Developmental Biology, 79*, 16–26.
Stegmann, M., & Zipfel, C. (2017). Complex regulation of plant sex by peptides. *Science, 358*, 1544–1545.
Su, S. H., Gibbs, N. M., Jancewicz, A. L., & Masson, P. H. (2017). Molecular mechanisms of root gravitropism. *Current Biology, 27*, R964–R972.
Sun, H., Tao, J., Gu, P., & Zhang, Y. (2016). The role of strigolactones in root development. *Plant Signaling & Behavior, 11*, e1110662. https://doi.org/10.1080/15592324. 2015. 1110662.
Tadeo, F. R., Cercos, M., Colmenero-Flores, J. M., et al. (2008). Molecular physiology of development and quality of citrus. *Advances in Botanical Research, 47*, 148–223.
Tamogami, S., Agrawal, G. K., & Rakwal, R. (2011). Jasmonates to jasmolites in plants: Past, present and future. *Advances in Botanical Research, 60*, 309–348.
Tang, D., Wang, G., & Zhou, J. M. (2017). Receptor kinases in plant-pathogen interactions: More than pattern recognition. *Plant Cell, 29*, 618–637.
Thirugnanasambandam, P. P., Hoang, N. V., & Henry, R. J. (2018). The challenge of analyzing the sugarcane genome. *Frontiers in Plant Science, 9*, 616. https://doi.org/10.3389/fpls.2018.00616.

Tran, H. T., Lee, L. S., Furtado, A., Smyth, H., & Henry, R. J. (2016). Advances in genomics for the improvement of quality in coffee. *Journal of the Science of Food and Agriculture*, *96*, 3300–3312.
Tsay, W. C., Hsiao, Y. Y., Pan, Z. J., et al. (2008). Molecular biology of orchid flowers: With emphasis on Phalaenopsis. *Advances in Botanical Research*, *47*, 100–145.
Tsukagoshi, H. (2016). Control of root growth and development by reactive oxygen species. *Current Opinion in Plant Biology*, *29*, 57–63.
Turkan, I. (Ed.). (2011). Plant responses to drought and salinity stress: Development in a post-genomic era. *Advances in Botanical Research*, *57*, 1–493.
Valles, J., Garcia, S., Hidalgo, O., Martin, J., Pellicer, J., Sanz, M., et al. (2011). Biology, genome evolution, biotechnological issues and research including applied perspectives in Artemisia (Asteraceae). *Advances in Botanical Research*, *60*, 350–419.
Van Damme, E. J. M., Lannoo, N., & Peumans, W. J. (2008). Plant lectins. *Advances in Botanical Research*, *48*, 107–209.
Van der Weijde, T., Kamei, C. L. A., Severing, E. I., et al. (2017). Genetic complexity of miscanthus cell wall composition and biomass quality for biofuels. *BMC Genomics*, *18*, 406. https://doi.org/10.1186/s12864-017-3802-7.
Van Holle, S., & Van Damme, E. J. M. (2018). Signaling through plant lectins: Modulation of plant immunity and beyond. *Biochemical Society Transactions*, *46*, 217–233.
Van Loon, L. C. (Ed.). (2009). Plant innate immunity. *Advances in Botanical Research*, *51*, *1–754*.
Vanavichit, A., & Yoshihashi, T. (2010). Molecular aspects of fragrance and aroma in rice. *Advances in Botanical Research*, *56*, 49–73.
Verbon, E. H., & Liberman, L. M. (2016). Beneficial microbes affect endogenous mechanisms controlling root development. *Trends in Plant Science*, *21*, 218–229.
Verdoucq, L., & Maurel, C. (2018). Plant aquaporins. *Advances in Botanical Research*, *87*, 25–56.
Wang, H. J., Huang, J. C., & Jauh, G. Y. (2010). Pollen germination and tube growth. *Advances in Botanical Research*, *54*, 1–52.
Wei, Z., & Li, J. (2016). Brassinosteroids regulate root growth, development and symbiosis. *Molecular Plant*, *9*, 81–100.
Wendehenne, D. (Ed.). (2016). Nitric oxide and signaling in plants. *Advances in Botanical Research*, *77*, 1–374.
Wickramasuriya, A. M., & Dunwell, J. M. (2018). Cacao biotechnology: Current status and futur prospects. *Plant Biotechnology Journal*, *16*, 4–17.
Wu, D., Liang, Z., Yan, T., et al. (2019). Whole-genome resequencing of a worldwide collection of rapeseed accessions reveals the genetic basis of ecotype divergence. *Molecular Plant*, *12*, 30–43.
Wu, G. A., Prochnik, S., Jenkins, J., et al. (2014). Sequencing of diverse mandarin, pummelo and orange genomes reveals complex history of admixture during domestication. *Nature Biotechnology*, *32*, 656–662.
Wu, G. A., Terol, J., Ibanez, V., et al. (2018). Genomics of the origin and evolution of citrus. *Nature*, *554*, 311–316.
Wu, P., Zhou, C., Cheng, S., et al. (2015). Integrated genome sequence and linkage map of physic nut (*Jatropha curcas L*), a biofuel plant. *Plant Journal*, *81*, 810–821.
Xie, Y., Liu, Y., Wang, H., et al. (2017). Phytochrome interacting factors directly suppress MIR156 expression to enhance shade avoidance syndrome in Arabidopsis. *Nature Communications*, *8*, 348. https://doi.org/10.1038/s41467-017-00404-y.
Yang, C., & Li, L. (2017). Hormonal regulation in shade avoidance. *Frontiers in Plant Science*, *8*, 1527. https://doi.org/10.3389/fpls.2017.01527.
Yoshida, S., Kim, S., Wafula, E. K., et al. (2019). Genome sequence of *Striga asiatica* provides insight into the evolution of plant parasitism. *Current Biology*, *29*, 3041–3052.

Zhang, G. Q., Liu, K. W., Li, Z., et al. (2017). The Apostasia genome and the evolution of orchids. *Nature*, *549*, 379–383.

Zhang, L., Zhang, F., Melotto, M., Yao, J., & He, S. Y. (2017). Jasmonate signaling and manipulation by pathogens and insects. *Journal of Experimental Botany*, *68*, 1371–1385.

Zhang, J., Zhang, X., Tang, H., et al. (2018). Allele-defined genome of the autopolyploid sugarcane *Saccharum spontaneum L. Nature Genetics*, *50*, 1565–1573.

Zheng, R. H., Su, S., Xiao, H., & Tian, H. Q. (2019). Calcium: A critical factor in pollen germination and tube elongation. *International Journal of Molecular Sciences*, *20*, E420. https://doi.org/10.3390/ijms20020420.

CHAPTER THREE

Scientific contributions of Pierre Gadal and his lab—A tribute to Pierre Gadal (1938–2019)

Ridong Chen[a], Michel Droux[b], Aymeric Goyer[c], Bertrand Hirel[d], Michael Hodges[e], Emmanuelle Issakidis-Bourguet[f], Jean-Pierre Jacquot[g], Frederic De Lamotte[h,*], Stephane D. Lemaire[j], Martine Lemaire-Chamley[i], Myroslawa Miginiac-Maslow[k], Tatsuo Sugiyama[l], Akira Suzuki[m], and Jean Vidal[k]

[a]APT Therapeutics Inc, Naperville, IL, United States
[b]Microbiologie, Adaptation, Pathogénie, UMR5240 CNRS-UCBL-INSA-Bayer CropScience, Domaine Scientifique de la Doua, Université Lyon 1, Villerbanne, France
[c]Department of Botany and Plant Pathology, Oregon State University, Hermiston Agricultural Research and Extension Center, Hermiston, OR, United States
[d]INSTITUT Jean-Pierre BOURGIN. Departement Adaptation des Plantes à l'Environnement INRAE, Versailles Cedex, France
[e]Institute of Plant Sciences Paris-Saclay (IPS2), CNRS, INRAE, University Paris-Saclay, University of Paris, Gif Sur Yvette, France
[f]Institute of Plant Sciences Paris-Saclay (IPS2), UMR Université Paris-Sud—CNRS 9213—INRAE 1403, Orsay Cedex, France
[g]Université de Lorraine, INRAE, IAM (Interaction Arbres Microorganismes), Nancy, France
[h]UMR AGAP Institut, Univ Montpellier, CIRAD, INRAE, Institut Agro, Montpellier, France
[i]INRAE, Université de Bordeaux, UMR Biologie du Fruit et Pathologie, Villenave d'Ornon, France
[j]Sorbonne Université, CNRS, UMR 7238, Laboratoire de Biologie Computationnelle et Quantitative, Institut de Biologie Paris Seine, Paris, France
[k]Retired from Institute of Plant Sciences Paris-Saclay (IPS2), UMR Université Paris-Sud—CNRS 9213—INRA 1403, Orsay Cedex, France
[l]Emeritus Researcher of RIKEN, Nagoya, Japan
[m]INRAE, Institut Jean-Pierre Bourgin, UMR1318, AgroParisTech, Versailles Cedex, France
*Corresponding author: e-mail address: frederic.de-lamotte@inrae.fr

Contents

Advances in Botanical Research, Volume 100
ISSN 0065-2296
https://doi.org/10.1016/bs.abr.2021.01.012
Copyright © 2021 Elsevier Ltd
All rights reserved.

Abstract

This chapter is dedicated to the memory of Pierre Gadal a former Series Editor for *Advances in Botanical Research*. Its aim is to describe the research that has been done in his laboratory covering essentially the period 1970–2000 and provides first-hand testimony about how it has been done. Besides the scientific output concerning CO_2 and nitrogen assimilation and their regulation described in this chapter Pierre Gadal has been very influential in plant science in France through his interactions with the University system but also with CNRS and INRA communities. Through the organization of the doctorate school in University of Paris-Sud he has helped train hundreds of students, and more specifically concerning his own research group, more than 10 of his former students have become CNRS researchers, many with high positions and responsibilities, a similar number INRA researchers and Professors and Assistant Professors in the University system. Fourteen different colleagues have contributed to this chapter, each with his own perception, heart and guts, a further testimony to the exceptional human and scientific qualities of Pierre.

1. Pierre Gadal's career

Pierre Gadal was born in a modest family of farmers in Brittany in France, and he initially spoke only the Gaelic dialect. It was only in primary school that he got trained in French (amazingly enough his French grammar and spelling was perfect despite the slow take-off). After high school he went to University probably in Rennes and then started a PhD on galls and the metabolism of tannins in plants in the laboratory of Pr A. Brunel in Toulouse. Pierre Gadal has then become assistant professor in Université Paul Sabatier in Toulouse. He later became a full professor at Université de Nancy I in Lorraine (1973–1981) and finally at Université de Paris-Sud Orsay until the end of his career in 2003. He has headed a CNRS-associated lab (Physiologie Végétale Moléculaire) in Nancy and later in Orsay to finally become the initiator and the first director of Institut de Biotechnologie des Plantes (CNRS-Université Paris-Sud), an institute which played a crucial role for the structuration of plant biology around Paris. Pierre Gadal has had numerous pedagogic and administrative responsibilities, in particular he has supervised the training of PhD students at Paris-Sud during many years. Pierre Gadal has been very influential in the domain of plant physiology in France and was awarded a state prize from the French Academy of Sciences in 1998 as a crowning of his scientific career. In 2000 he was elected as a Corresponding Member of the American Society of Plant Biologists. When he officially retired in 2003, a special issue of the *Journal Plant Physiology and Biochemistry* has been dedicated to him. His scientific contributions have gained international recognition with more than 200 articles published in peer-reviewed journals. More specifically his major scientific contributions were linked to the elucidation of the regulation of CO_2 fixation in plants through the study of PEP carboxylase and NADP-malate dehydrogenase, but also to the study of the pathways allowing nitrogen assimilation in plants (mostly glutamine synthetase and GOGAT). His latest interests concerned the carbon-nitrogen relations in plants involving isocitrate dehydrogenase. Pierre Gadal and his group have been leaders in plant biochemistry, introducing in the late 80s the use of genetic engineering and structural biology. After his retirement, Pierre Gadal has been a series editor for *Advances in Botanical Research*, a book series by Elsevier, supervising volumes 63–86. As an emeritus he was also a very highly appreciated advisor for students belonging to the doctoral school of Orsay Plant Sciences, helping them find challenging post-doctoral stays. He died in October 2019 after fighting a difficult disease. Reactions to his departure have been unanimous: this was a major actor of plant biology who has disappeared. All the students and colleagues who had the chance to work with him have been impressed by his charisma and enthusiasm. He was always stimulating us to go forward and in the same time he paid extreme attention to his colleagues helping them as much as he could, irrespective of their social positions. He will be thoroughly missed by all of us. In the following section of this chapter several of the members of the Gadal lab pay tribute to Pierre by recalling how science was done there and the major achievements that were produced.

2. Mechanisms and regulation of CO_2 fixation in C_3 and C_4 plants

2.1 PEP carboxylase

2.1.1 Jean Vidal's recollections

This short text summarizes the scientific contributions from P. Gadal's laboratory at the University of Nancy and then at the University of Paris-Sud in Orsay, by the people cited in the below acknowledgements. It is intended to illustrate, via some selected memories, a long period of interaction with this outstanding plant scientist and teacher. To avoid enumerating a long list of scientific details (already published elsewhere in International scientific journals) I have decided to briefly revisit, in an informal, elliptical style, some milestones of the corresponding periods. In this way, the people cited below and I express our deep gratitude to Professor Pierre Gadal (PG).

Toulouse, October 1973: Centre de Physiologie Végétale, Université Paul Sabatier (Director, G. Brunel); on the evening of the same day I had presented my thesis defense: "My application for a Professor position has been positively considered at Nancy University, would you be interested to move there as a CNRS researcher and to participate in the installation and development of a new Plant Physiology laboratory?," PG asked the newly graduated PhD student that I was. I could not believe what I heard but I managed to answer "Yes of course!!!" with a stammering voice. This short dialog with PG was the starting point of my long (actually up to my retirement in 2010) professional collaboration with him. Until then, I had been living in a dilettante way, with no concrete objective for my career, although I remember when regarding the functioning of biological systems, I was constantly haunted by the fundamental, mechanistic, quasi existentialist question: How does it work? In what follows, and essentially thanks to PG, it will appear that my wishes were ultimately granted.

Before I develop my personal story further, I think it is indispensable to underline the seminal impulse of Professor G. Cavalié, director of my thesis work centered on the study of two generally coupled enzymes of primary plant metabolism, namely, phospho*enol*pyruvate carboxylase (PEPC) and NADP-dependent malate dehydrogenase (NADP-MDH). Later, he allowed this scientific project to be transferred to Nancy.

Nancy, October 1974: The fact that most of the laboratory had to be entirely set up raised the paradoxical feeling of concern (energy and time-consuming for, perhaps, little scientific yield and rewards) and unalterable enthusiasm when considering our contribution to this noble goal. Again, PG was leading us with his infectious, optimistic personality, conferring enormous doses of confidence for the future to the people around him and, thus, provoking a strong impetus; nothing and no one could have stopped the machine.

As expected, the formerly classical botany laboratory had neither the necessary equipment to do modern biochemistry nor space for controlled plant cultures. It was indeed a desert, but we never felt discouraged; on the contrary, this adverse situation turned out to be a very stimulating period since the first year was devoted to laying down the foundations of the new laboratory, setting up minimal working conditions, together with the two very first students of the nascent group (J. Maréchal and M-C. Rio) and, I must say, the precious help of colleagues in the Biology Department. Cultivating, centrifuging, measuring and more in neighboring laboratories was the rule; in fact it was just necessary to have a good ice machine and buckets to transport fragile materials!!! What an intense pleasure it was—the inauguration of a cold room for plant extraction and experimentations—the arrival of a newly purchased spectrophotometer. Our efforts converged toward the biochemical characterization of the two above-mentioned enzymes and the regulatory mechanisms impacting on their activity. I remember that this period was exceptionally stimulating and enriched by permanent, informal discussions with Pierre as our scientific mentor. At the same time, he was organizing laboratory activities, including boring administrative tasks, internal meetings and useful collaborations nearby and around the world, while also teaching plant physiology. His charisma made his course so attractive that it was quickly attended by an evergrowing number of young students (see J-P. Jacquot, this issue). This had the advantage of getting more students interested in plant sciences and, more importantly to us, in research options developed by the group. Already during the second year of the laboratory's existence, this option took shape with the addition of three more students: C. Guiz, J-P. Jacquot, and N. Galvaing, all surmised as promising elements by Pierre (H. Membre, F. Martin A. Suzuki, and B. Hirel would follow shortly later). In the ensuing years, together with J-P. Jacquot, exciting research on NADP-MDH was developed, with the identification of a redox-activating factor, later to be named thioredoxin. This was largely amplified by J-P. by interactions and stays abroad, most notably in the group of Professor Bob Buchanan (a world leader in plant enzyme redox regulation) at UC-Berkeley in the USA. During the next period at Orsay (see below), J-P. developed a research group that rapidly expanded to produce a wealth of new and exciting results so that they quickly reached the pinnacle of the thioredoxin field (cf J-P. contribution, this issue).

As for myself, I concentrated on PEPC in the context of C_4 photosynthesis and light, and the fascinating application of immunological techniques to plant proteins that led to a technical breakthrough that brought considerable progress. Up until then, to clarify functional parameters of proteins and to allow us to make hypotheses about their metabolic impact, we purified enzymes by a series of very time-consuming chromatography techniques. High purity (but never perfect purity) was achieved in this way, but this was not the guarantee that alterations did not occur and generate super artifacts and thus lead to invalid conclusions. PG had been harassing me with the question of the production of PEPC antibodies. Consider it done, I just had to contact our next-door neighbor, Professor G. Godbillon at the Animal Physiology laboratory, and thanks to him I learned how to prepare the precious matter and how to use it (a savoir-faire that was perfected later by a stay in J. Daussant's laboratory at POVAR, CNRS Meudon). It has to be said that this was a particularly enjoyable and productive period during which, together with our recently recruited, excellent technician J. Gerard, we set up a fast and efficient one-step purification procedure (immunochromatography with water elution) of the photosynthesis related C_4 form of PEPC that was collected quasi free of contamination. This also opened up the rewarding period of in situ immunolocalization studies. The actors brought together by Pierre were newly recruited C. Perrot (later Catherine Perrot-Rechenmann), and G. and A. Burlet at "Hôpital Universitaire de Brabois," who had a very good knowledge of such approaches on human or animal samples, both on their theoretical and technical aspects. The question at the time was: could immunolocalization be applied to plant systems? The answer was a resounding "Yes!," and by using this immunological tool, Catherine clarified another highly debated question of the time, namely the tissue, cell and subcellular localization of C_4-PEPC. She produced the first, very impressive, truly artistic, but nevertheless precise images showing the cytosolic localization of the enzyme in Sorghum leaf mesophyll cells. This result was of paramount importance since it allowed to place this metabolic component at its correct position in the cell, and consequently inferred the organization and functioning of the C_4 photosynthesis-linked metabolic network in the entire leaf.

We also established that the light-induced accumulation of the very abundant C_4-PEPC in the mesophyll cells was due to neosynthesis and, as a corollary (what was another important issue at the time was whether the chromatographic-based enzyme diversity was true or due to experimental artifacts) the occurrence of PEPC isoforms thereby likely implicating a corresponding multigene family. Meanwhile, a long-term interest of PG had been achieved via the installation of a new research activity, devoted to the study of plant nitrogen metabolism, alongside that of carbon. He had been eagerly waiting for such a possibility, as he thought that a comprehensive, integrated approach of primary plant metabolism necessarily involved both aspects (refer to nitrogen contributors, this issue). The CNRS soon recognized the merits of PG's vision by attributing his team a CNRS-associated group label, which was a deserved and important distinction for him and his lab.

Orsay, October 1981: But everything has an end. Pierre was offered a position at the faculty of Sciences (Université Paris-Sud, Orsay) as the replacement of Professor Moyse (Laboratory of "Physiologie Végétale"). We all followed him there, even though it was quite difficult to leave behind what had been painstakingly built-up at Nancy II; a nice scientific environment, lovely people (both inside—M. Jacquemin, B. Maudinas, G. Chemardin, J.F. Pierre, H. Courbet, R. Beauval and the administrative staff—and outside—groups within the Biology department) and the gorgeous Lorraine region. The foreigner that I was, strangely and unexpectedly felt uprooted.

History repeats itself! Curiously enough for such a well-respected University (dating from the origins of the faculty in 1956), we first had to renovate the old laboratory in Building 430. Even the electricity did not meet the safety norms. Again, PG was particularly active and efficient in striving to insert his group into this new, particularly rich, biochemical and physicochemical environment of a much broader dimension. When functionality was finally attained, the time had come to enter a new scientific era of molecular biology and genetics. For lack of space, I must make a very long story short. This era was initiated by M. Thomas and C. Crétin, pursued and nicely developed later by Claude's student, L. Lepiniec. Martine was concentrating her interests on the molecular mechanisms controlling light-activated synthesis of the C_4-PEPC protein. Briefly, using sophisticated techniques including monoclonal antibodies (in collaboration with J.M. Bidart, IGR, Villejuif) and in vitro transcription analyses (collaboration with Dr. E. Mosinger, University of BERN), she elegantly showed that leaf greening was accompanied by an increase in the production of C_4-PEPC encoding mRNA, and that this was phytochrome-controlled.

On his side, Claude had been learning abroad (Max Planck Institute in Munich with Professor Hahlbrock) how to develop and use a variety of new techniques that any research group had to possess to face the challenges of the moment and notably, DNA cloning, genomic libraries, production of recombinant proteins and plant transformation that were successfully applied to the PEPC theme. Claude conducted his research with a technician, Eliane Keryer, and a PhD student, Loïc Lepiniec. Three genes of the sorghum PEPC family were isolated, sequenced and their expression clarified. Their nucleotide sequence allowed revealing their deduced amino acid sequences and specific domains, like the N-terminus containing a conserved serine residue which was receiving special attention at the time in the other part of the PEPC group. Biolistic experiments and reporter gene expression allowed the tissue expression and the specificity of PEPC gene promoters to be analyzed. Several years later, in relation to this, a Polish PhD student, S. Rydz (Warzaw University, Professor A. Richter) detected protein factors with an affinity for the C_4 gene promoter that were possibly regulated by protein phosphorylation.

However, in August 1993 a very dramatic and sad event occurred when Claude passed away. It was a great loss, both scientific and human, that struck everybody who knew him.

Then in 1994 came the moment to move from the banks of the Yvette river to the heights of the Saclay plateau where stood a magnificent new building called the "Institut de Biotechnologie des Plantes" or IBP for short. This Institute was in all aspects, from the idea of its creation to the raising of its walls, a PG project with the later collaboration of Professor F. Quétier. At the time, the vision of PG was met with some indigenous resistance and some strong opposition from the CNRS department of Life sciences, but PG was not deterred in his demonstration of a vision for the future of Plant Sciences at Orsay and his strong character. This big building with an elegant glass greenhouse housed around 150 people including 100 researchers/teachers working in three domains of plant sciences; development, plant-microbe interactions and metabolism. In fact, it was so attractive, especially when compared to the rest of the campus, that the faculty used to send, for years, foreign delegations to visit the "fleuron" of the campus buildings. PG became the director of the IBP, which was awarded an UMR label by CNRS.

After the closing of the "Institut de Physiologie Végétale" (CNRS Gif/Yvette) at around the end of the 1990s, several O. Queiroz group members joined the PEPC group at the IBP. J. Brulfert (with PhD students Tahar Taybi and Nadia Bakrim) brought her excellent knowledge and problematics of CAM PEPC, Kim Tran (and Van Le Bui) developed useful and original techniques for plant transformation, while J-N. Pierre mastered a very precious technique offering the potential for in-cell integrated approaches. We had shown that sorghum C_4-PEPC became phosphorylated in illuminated sorghum leaves by an elusive PEPC kinase (PEPCk). The identification of the kinase was an important challenge. Making use of recombinant protein technologies, V. Pacquit, N. and N. Bakrim, M. Nhiri and C. Echevarria (University of Sevilla) were studying C_4-PEPC and its phosphorylation in vitro, showing that calcium-dependent kinases could be candidates for the role, at variance with Professor R. Chollet's group (Lincoln, Nebraska, USA) which was proposing a calcium-independent kinase. The long debate came to an end when Professor H. Nimmo (Glasgow, UK), and quasi simultaneously R. Chollet, and J. Cushman (Reno, Nevada, USA) achieved the cloning of the authentic PEPC kinase in 1999 and 2000, respectively. Indeed, this kinase did not require calcium to phosphorylate the target serine of the PEPC N-terminus. In addition, it was amazingly small for a Ser/thr-kinase (only 32kDa, untiringly neo synthesized every day at the beginning of the light phase, nevertheless of incredibly low abundance, and dedicated to PEPC. However, signaling systems are very intricate and particularly complex and biological knowledge is never complete so it can be hypothesized that our calcium-kinase was perhaps useful in some, as yet unknown, specific physiological condition. Valerie also produced the first C_4-PEPC crystals (collaboration with C. Cambillau of the CRMC2, Marseille) for X-ray diffraction analyses, but K. Izui (Kyoto University, Japan) was the first to publish the first, eagerly awaited, maize C_4-PEPC protein structure. It was with an extreme deep emotion (the dream had come true) that I, along with others, discovered the details (folding, N and C-termini, phosphorylation and active site domains, homotetramer) of the wonderful PEPC structure. It was certainly a nice protein structure, actually a dimer of dimers, but above all, it was an extremely useful tool to inspire the design of future experimental approaches. It was also the time to begin what was to become a long lasting and particularly rewarding collaboration with R. Chollet (Lincoln, East Campus then G. W. Beadle Center), an eminent US Plant biochemist who was (and still is) a reference in C_3/C_4 photosynthetic enzymes and mechanisms, and C. Echevarria, who had just obtained a professor position and taken over as head of the laboratory of "Fisiologia vegetal," Sevilla). With them, a very active research network was built-up that allowed various short sabbatical visits and a dense exchange of PhD students and postdocs, all of them carrying out their thesis work on PEPC and/or PEPCk and thereby leading to an important number of publications.

The technique developed by J-N. Pierre opened up an exciting period dominated by cell biology to notably study how the light signal controls the synthesis of PEPC kinase in the cell. Again, it is difficult to try to describe the intense pleasure we felt at deciphering some key components of the transduction machinery. By using mesophyll cell protoplasts and both pharmacological and confocal microscopy approaches (S. Brown CNRS, Gif/Yvette), J-N. Pierre, N. Giglioli-Guivarc'h, and later S. Coursol and

J. Grisvard, identified one by one a good part of the cascade components controlling PEPC kinase synthesis in the physiological context of the entire C_4 leaf. It was amazing to see how they were easily deducible from our knowledge of the spatial organization of C_4-photosynthesis in terms of both tissue concentric organization and metabolite exchange. Obviously much remained to be done but a paradigmatic, integrated scheme was now drawn to stimulate further investigations.

PG retired in 2003 and a tribute was given to him by way of a Special Issue of *Plant Physiology and Biochemistry* in June–July 2003 (see reference list). He then spent several years giving a hand to people (both teachers and PhD students) of the doctoral school and performing editorial work for *Advances in Botanical Research*. It was still possible to meet him to discuss our most recent research. The rapid progress of plant genetics, genomics and molecular genetic techniques had made it necessary to constantly update the level and variety of investigations. Now, manipulating plants for biotechnological applications had become routine. It also provided fundamental research with extremely efficient and sophisticated tools to examine biological mechanisms with the holistic philosophy of Pierre, at the most integrated level possible. For instance, switching to the model C_3 plant *Arabidopsis thaliana* allowed the isolation of PEPCk KO mutants from seed libraries in order to clarify the physiological role of leaf C_3-PEPC phosphorylation in situ. A. Gousset-Dupont made a good contribution to this topic, showing that this post-translational modification (PTM) was particularly needed to resist salt stress conditions. This approach on a C_3 plant was completed by B. Lebouteiller on PEPC mutants and overexpressors of the four genes encoding PEPC in Arabidopsis leaf and seeds. In collaboration with a private company, Biogemma (P. Perez, D. Gerentes), M. Jeanneau produced transgenic maize (over)expressing sorghum C_4-PEPC in all organs. Interestingly, these plants were significantly more resistant to drought when grown in a greenhouse. Unfortunately, when grown in the field, they were continuously cut-down by people opposed to transgenic plants or GMOs, and so the company (and us too) finally abandoned these experiments! At IBP, P. Meimoun had the honor of being the PhD student involved in bringing to an end the long story of PEPC. Before closing finally the doors of the PEPC laboratory at IBP, this "last of the Mohicans," with the excellent assistance of F. Ambard-Bretteville, added an interesting contribution to C_3 PEPC phosphorylation via a generic study of the Arabidopsis phosphoproteome.

Today, clearly biological questions regarding plant metabolism and its regulation have shifted to other targets. Forgotten is the time when

PEPC alone was attracting dozens of researchers worldwide to meetings (such as the one we had organized at the IBP as a satellite of the Xth International Photosynthesis congress, Montpellier 1995). To my knowledge, PEPC is still actively investigated by the groups of B. Plaxton (Kingston, Canada), C. Echevarria (Sevilla) and J. Cushman (Reno). It is both satisfying and reassuring to see that many unexpected findings were being revealed, among which I will underline only the PTM of the enzyme by regulatory mono-ubiquitination and the phosphorylation of C_3-PEPC in seeds. Although a considerable amount of data has been gathered from reductionism-based studies, the precise role of these PTMs remains obscure and the potential of more global, integrated approaches may help improve our understanding. So much is yet to be discovered and original properties might be expected to emerge. The great lesson learnt from practicing research along these many years, is that the permanent quest for/and acquisition of relevant techniques, concepts and bioinformatics-based models that invade the field, requires weaving inevitable and pertinent collaborations. Lastly, on a more biotechnological point of view, I also think that certain enzymes, having a particularly high capacity for fixing CO_2 have the potential to contribute in some way or another to fixing the rapidly growing challenge of climate change.

Above, I used the term of "mentor" to describe the profound influence of Pierre Gadal upon a group of young researchers hungry for plant science. Indeed, Pierre had all of the characteristics defining this word: an experienced advisor, attentive and wise and who deserves one's entire confidence. Although involved in teaching duties, laboratory organization and national and international responsibilities, PG constantly paid attention to the work performed in his laboratory. He was a continual driving force, always available for discussions, always introducing pertinent new ideas, concepts and technological approaches, always demanding rigor (the need to multiply controls and experimental rehearsals), interpretation and dissemination of results. The manner was a dialectic ping-pong until a consensus was reached. It was particularly efficient in coming to the "substantifique moelle" of the question to be examined. In this way, he undoubtedly left his indelible mark on each one of us. To draw a bit more of the shape of his "interior statue," I shall now underline what I consider prominent traits a great leader must possess. In terms of human relationships, he had a friendly attitude and gave his colleagues freedom to undertake what they believe was important, which in his mind was the guarantee that they would give their best. When working in Orsay, he decided to cultivate some land located in front of the greenhouses and for many seasons I joined him working the soil and sweating a lot to finally collect nice tomatoes and some other vegetables. This story is to highlight a dimension of humility through the osmosis with the earth that PG had kept, while this was often forgotten by many colleagues of his rank. Finally, what best defines the personality of PG is the famous, oxymoron-like slogan "la force tranquille." Overall, I recall the excellent atmosphere that prevailed in the laboratory and the many researchers that were fed on his philosophy can testify to the benefits of their encounter with PG. Among the impressive list of people cited above, most students that came to your laboratory are now working in teaching and research, worldwide, and keeping his lessons alive. In October 2019, Pierre sadly passed away "*Acta est fabula.*" Most of the citations concerning the work described in the text can be found in the following review articles: Chollet, Vidal, and O'Leary (1996), Echevarria and Vidal (2003), Lepiniec, Thomas, and Vidal (2003), and Vidal and Chollet (1997).

Acknowledgements: Hereafter are listed the researchers who have contributed to the writing, to some extent or extensively, of the PEPC story as described above. J. Maréchal, M-C. Rio, J. Gérard, C. Perrot-Rechenmann, M. Thomas, C. Crétin, L. Lepiniec, E. Keryer, R. Chen, J. Brulfert, T. Taybe, N. Bakrim, K. Tran, V-L. Bui, V. Pacquit, S. Rydz, N. Bakrim, M. Nhiri, C. Echevarria, J-N. Pierre, N. Giglioli-Guivarc'h, S. Coursol, J-L. Prieto, J. Grisvard, A. Gousset-Dupont, B. Lebouteiller, P. Meimoun, and F. Ambard-Bretteville.

Thanks are due to researchers, teachers, technicians, and administrators from PG's laboratories and from the many (inter)national collaborators that we had over the years.

I am indebted to my long-time colleague, Michael Hodges, who corrected a certain number of our PEPC articles, including this one, a very good touchstone against which to refine interpretations, and who became a specialist of this plant enzyme and others.

2.1.2 *Tatsuo Sugiyama's recollections*

By *Tatsuo Sugiyama* (*Emeritus Professor of Nagoya University / Emeritus researcher at RIKEN*).

As a researcher, Pierre has made great strides in studying the mechanisms and regulation of CO_2 fixation in higher plants and in the mechanism and regulation of nitrogen metabolism. His research trend was similar to mine. That's why the exchange with him has been going on for 40 years.

One of the most impressive moments in the exchange was a relationship of trust with his disciple, Dr. Akira Suzuki. That involved Bob Buchanan, Professor of University of California at Berkeley. That occurred in 1986. My lab was looking for an associate professor. Bob was aware of this and participated as a judge on Akira's PhD defense, and recommended to me that Akira fit the requirements. In response, I asked Akira directly to accept my invitation. He replied politely "Professor Pierre Gadal has nominated me as a public servant for my job. I am sorry but, unfortunately, I must refuse your kind invitation."

After that I had the opportunity to stay in Gadal's lab as a guest professor for a month. In the meantime, he struggled with the government to expand the research institute. So, I in fact conducted a collaboration work with Jean Vidal by doing the isolation of corn nuclei to be used for a run-on experiment. Anyhow, I was able to enjoy my stay enough.

It is unforgettable that this visit was triggered by Dr. Claude Cretin's participating in a collaborative work in my lab in Nagoya. Unfortunately, the young, talented researcher died after returning from Japan with fruitful results.

Pierre made a great contribution to the scientific collaboration between France and Japan in his later years. In 2004, a joint seminar was held in Nara, Japan's ancient capital, where he played the role of organizer on the French side. After the stay of Pierre and his wife in Japan, my wife and I began a private exchange with them and enjoyed the summer at Pierre's and his wife's villa in Brittany several times. Pierre then told us an interesting story about a wooden barrel in the corner of the living room of the villa. He told us, "The barrel was for my mother to kill and butcher the pig once a year, to salt the meat, and when I was a child, I stared in prayer so that mother would not fail to make valuable food." At other times, we drove to Bordeaux to help us deepen our understanding of French differentiation.

I will never forget the scientific achievements Pierre Gadal has left behind and our private warm-hearted exchange.

2.2 Redox regulation and the thioredoxin systems

As stated above P. Gadal oriented part of its research toward the study of the regulation of enzymes involved in carbon and nitrogen metabolisms. In line with the admonition of Ephraim Racker: "Do not waste clean thinking on dirty enzymes," the first efforts were initially concentrated on enzyme purification. The following testimonies explain the various steps that were used for this achievement.

2.2.1 Jean-Pierre Jacquot recollections

The early Nancy years [1975–1981] NADP-MDH activation, thioredoxin heterogeneity discovery.

In the Gadal lab, the earliest experiments concerning the redox regulation of chloroplast enzymes have dealt primarily with the elucidation of the NADP-malate dehydrogenase light activation process. It all started in 1975 at the University Henri Poincaré of Nancy where I performed the experimental part of my DEA (Master) study. Chloroplastic NADP-malate dehydrogenase had been partially characterized in C_4 plants by Hal Hatch and colleagues and it was already known to be light activated, i.e., basically inactive in the dark and active in the light and responsive to thiol reagents as dithiothreitol (DTT) (Johnson & Hatch, 1970). At the time, the study of the light activation of chloroplast enzymes was already well underway in Berkeley where Bob Buchanan, Peter Schürmann and Ricardo Wolosiuk had obtained evidence for the existence of two regulatory proteins involved in the activation process of FBPase that they had baptized ARPa (later to become ferredoxin-thioredoxin reductase) and ARPb (later thioredoxin) required together with photoreduced ferredoxin (Buchanan, Schürmann, & Kalberer, 1971; Schürmann, Wolosiuk, Breazeale, & Buchanan, 1976). As was often the case at the time, I had two supervisors a senior professor Pierre Gadal and a more junior CNRS researcher Jean Vidal. The Gadal lab was just starting at the time, this was the third year after Pierre Gadal's arrival as a professor in Nancy after transferring from Toulouse. That year two other students started their Master project at the same time and we were distributed three different tasks, as I recall the other two dealt with glutamate synthetase and glutamate dehydrogenase. The project I was given was described as follows by Jean Vidal: I can't purify the NADP-MDH, the enzyme requires dithiothreitol to be active but the activity vanishes as soon as I try to purify it so there must be a protein factor required to activate it that I am losing along the purification. This was definitely visionary and it made sense in the light of the experiments done in Berkeley on the FBPase, but still I thought that the project was very risky because I could think of a zillion reasons explaining why an enzyme could be unstable in solution and thus I was not too happy to have been landed this project. In short, I thought it was too hypothetical and prone to failure. I couldn't have been more wrong as these initial experiments started in early 1976 would ultimately shape my career for the next 40 years or so. The first idea was to try to separate a protein factor (later Trx) required for the DTT -linked activation of NADP-MDH by gel filtration and to this effect we tried first to use Ultrogel ACA 44 or Sephadex G50 columns. At the time we had very little experience in protein purification and we started by using French bean leaves (a very poor biochemical material that turned immediately black when the polyphenols it contained oxidized, irrespective of the presence of antioxidants). As a result, the columns that we were using were quickly damaged and needed to be regenerated with cycles of base/acid treatments, but this also degraded the matrix polymers. On top of that we were using very small quantities of plant material in the first experiments (ca. 1/4th of a g of lyophilized material equivalent to ca 2 g fresh material). We would realize only much later that in order to purify reasonable quantities of highly purified proteins, it was actually necessary to use much higher quantities in the kg range for leaves. Nevertheless, we have very quickly determined that the NADP-MDH was likely present in the fractions eluting at the void volume and that "species" present in a fraction of lower molecular weight assisted the reactivation of the enzyme by DTT. Meanwhile a paper had described the purification of a LAF (light activating factor) of RubisCO from ethanol extracts (Wildner & Criddle, 1969) and we tried to adapt that with very limited success. At that time, it happened that Jean Vidal left for a congress and I was left on my own for a week or so. Somehow having read in the literature that some proteins are thermostable I tried my luck with the fraction putatively containing the regulatory protein and it was very exciting to discover that this regulatory activity was indeed thermostable at 80 °C for several minutes. Obviously, it offered the great perspective of separating easily the activating factor from the NADP-MDH by simply heating the sample. When Jean came back he was less than impressed and his first analysis was that the component in the activating fraction was probably not of protein nature as most proteins are not thermostable. After discussion, we set out at treating the activating factor with a protease called pronase and subsequently destroyed the pronase component by heating. In turn, we tested the activity of the protein fraction, which turned out to be totally destroyed. This clever experiment designed by Jean confirmed that the fraction that we had isolated indeed contained a protein. We could publish these results in December of the same year in FEBS Letters (Jacquot, Vidal, & Gadal, 1976). This was fortunate as Kagawa and Hatch published very similar results using a C_4 plant model instead. Interestingly their contribution indicated that the activating protein could also serve to deactivate, something that basically would not be explored further for another 40 years (Kagawa & Hatch, 1977). On the other hand, their contribution had not recognized the thermostability of the protein which would later help identify it as thioredoxin in the Buchanan and Holmgren labs. Although not extremely cited, our *FEBS Letters* paper in retrospect was very important as it opened the way to the study of redox regulation in plants in France.

Having obtained evidence for the existence of this regulatory protein it became of importance to verify that it was indeed also located in the chloroplast as the target enzyme and to generalize its existence to other photosynthetic organisms. By isolating chloroplasts, we could indeed demonstrate the chloroplastic localization of the regulatory proteins in spinach, a species renowned for being adequate for chloroplast isolation (it was also a much better biochemical material to work with compared to bean and much of the subsequent biochemical work in Berkeley, Neuchâtel and Nancy was done later with spinach) (Jacquot, Vidal, & Gadal, 1977). We then set out at studying the presence of these regulatory proteins in Sorghum, a C_4 plant. In agreement with the results of Kagawa and Hatch we found that sorghum like corn indeed contained protein activating factors and we could show also that there were apparently multiple forms that could be chromatographically separated (Vidal, Jacquot, Membre, & Gadal, 1978). We confirmed a bit later the ubiquitous presence of the regulatory protein (thioredoxin) by studying its presence in the photosynthetic bacterium *Rhodopseudomonas capsulata* (Jacquot, Maudinas, & Gadal, 1979). Once again in this purple sulfur bacterium there was evidence for multiple forms of thioredoxin. In between Bob Buchanan, Ricardo Wolosiuk, Arne Holmgren and their colleagues had provided firm evidence that the protein factors were in fact thioredoxins (Holmgren, Buchanan, & Wolosiuk, 1977; Wolosiuk & Buchanan, 1977). A physiological significance for the existence of multiple thioredoxins in plants became clear after we started cooperating with Peter Schürmann in Neuchâtel. The idea was to compare the properties of the thioredoxins that were activating the NADP-MDH and those which activated fructose 1,6 bisphosphatase, the model enzyme in Berkeley and Neuchâtel. To this end I visited Peter in Neuchâtel during a week to learn about FBPase activation and measurement. This was basically the first scientific trip that I made in my career and I was enchanted both by Neuchâtel and its lake and Peter and his Swiss-German coolness efficiency and kindness. Most of all I learnt a lot about protein purification there, how to use larger quantities of leaves, scaling up the columns, etc. After this week of training, Peter handed me an FBPase sample which I accidentally dropped in the sink in Nancy after discarding the ice and recovered after the week-end at room temperature there. I would then do all the experiments with this sample which tells that FBPase is really a very robust enzyme... Anyway, when I probed the fractions obtained after gel filtration in low ionic strength I could show that FBPase and NADP-MDH were activated by different thioredoxin fractions, the one that activated FBPase eluting at molecular weights roughly twice the size of that activating the NADP-MDH. We could further purify these two thioredoxins by ion exchange chromatography. Jean Vidal and myself were more than a tiny bit happy about those results, these experiments were actually done at feverish pace as we were seeing real discovery unfolding there and we celebrated that very honorably. So we got in touch with Peter Schürmann either by snail mail (email and the internet did nor exist) or by phone, but calling to another country was very expensive at the time so we probably relied on the post. Peter was actually less than excited as he had done replicate experiments in Neuchâtel with a partially purified NADP-MDH sample that I brought there. He had indeed obtained drastically different results in his parallel experiments, with NADP-MDH and FBPase being activated by the same fractions. Damned, all our hopes were shattered... So, we reviewed carefully the experimental conditions that we used, respectively, and Peter noticed that he was using a buffer with a much higher ionic strength for gel filtration compared to me and we hypothesized that this might be a reason for the difference. In any case we decided that I should visit again in Neuchâtel with the purified fractions exhibiting specificity versus the two target enzymes so that we could repeat the experiments together. The week that we spent there together working on the DTT activation of the enzymes was certainly one of the most exciting in my career. It started under nice auspices as in between Peter had repeated the chromatographies in low ionic strength buffer and had replicated our earlier results obtained in Nancy. We realized much later that thioredoxin f dimerizes in low ionic strength explaining the discrepancy. In short, we obtained evidence that leaves contained two different types of thioredoxin m and one type of thioredoxin f. In roots there was a single thioredoxin able to activate only the NADP-MDH (Jacquot, Vidal, Gadal, & Schürmann, 1978). This paper was definitely a second breakthrough for the Nancy group in the redox field and it was well received and granted us some more international recognition. The results were confirmed independently in Berkeley nearly at the same time (Wolosiuk, Crawford, & Buchanan, 1979). It was already mentioned that thioredoxin was known in other organisms as *E. coli* and human as a protein required for the regeneration of ribonucleotide reductase and hence deoxyribonucleotide and DNA synthesis. At the time colleagues working with those models had evidence for a single thioredoxin in there. I believe that they suspected for some time that our results were flawed as we had not used protease inhibitors, so there could be a protein degradation and generation of artifacts. While this was absolutely plausible, mounting evidence originating from plants indicated that the heterogenous nature of the thioredoxins was a reality and then genome sequencing changed everything. Nevertheless, it has taken nearly another 20 years to obtain firm evidence for the existence of multiple thioredoxins and glutaredoxins in *E. coli* and other non-plant species (Lillig et al., 1999; Miranda-Vizuete, Damdimopoulos, Gustafsson, & Spyrou, 1997) but of course overall their contributions to the field were essential and they produced superb science.

So, going back to 1978 we then got interested by kinetic details of the activation and published a paper that featured a lot of mathematics describing the activation that did not excite too much the biologist community (Vidal, Jacquot, Gadal, & Vidal, 1978). At the same time, we then turned our attention to the more physiological light-dependent regulation. As Peter had purified the ferredoxin-thioredoxin reductase we now had all the components of the ferredoxin-thioredoxin system in hand and could reconstitute the light system so I took my suitcase again and went back to Neuchâtel for yet another week. We could simplify the assay by working in sealed Eppendorf tubes under nitrogen a considerable improvement over the Warburg vessel used previously in Berkeley and Neuchâtel (Schürmann & Jacquot, 1979). This improved and lighter system confirmed the specificities of thioredoxins m and f. Together with Jean Vidal, a first light-dependent reconstituted system was then set up in Nancy including a crude version of ferredoxin-thioredoxin reductase and partially purified ferredoxin. We obtained also evidence for the in vivo light-dependent activation of the NADP-MDH by directly lysing chloroplasts into the reaction medium where they burst open because of osmotic shock (Vidal, Jacquot, & Gadal, 1980).

Meanwhile Pierre Gadal had attended the international photosynthesis congress in Reading in 1977 where he met Bob Buchanan for the first time. He came back with information that Bob was apparently a nice person to work with and an invitation for someone of our group for a sabbatical/post doc in Berkeley. At the time Jean Vidal and myself were both tenured CNRS researchers in the Gadal lab and both potential candidates for a 1-year visit. As Jean Vidal was senior he was naturally asked first but for a variety of reasons, mostly family-linked he declined to go. I couldn't believe my luck! At the time it was not a big problem for me and my family to go to

Berkeley and I accepted enthusiastically as California was really a magnet culturally and intellectually. In Berkeley Bob asked me to purify the NADP-MDH from corn and I accepted although this would later create "border" problems in the Nancy lab that took a long time to heal. Anyway, the stay was very successful as I was able to purify the enzyme to homogeneity and study some of its kinetic properties (Jacquot, Buchanan, Martin, & Vidal, 1981). This was the same preparation that we used later in Nancy to generate specific antibodies against the enzyme that would in turn be used to isolate cDNAs from a λgt11 library in Orsay by Claude Crétin. This sequence would then be used to generate an expression system in *E. coli* that allowed the structure-function study of that important enzyme.

And how did Pierre Gadal fit into that? Well to state it simply he was the one who boosted everything. First by selecting the subjects that he would like his lab to work on and that dealt essentially with carbon and nitrogen assimilation first. Although he was not a fantastic orator Pierre Gadal also had a knack for captivating and recruiting students of interest through his lectures that carried an unmistakable air of sincerity. Pierre Gadal was also a superb organizer and well connected in the world of French plant science. This helped the group created in Nancy expand notably through an association with CNRS which allowed the recruiting of five tenured researchers over a 10-year span. Pierre Gadal paid extreme attention to his younger colleagues and for example he accompanied me to Paris when I needed to interview for my CNRS job. This was very helpful as I had very little traveled before. Another key intervention is that he was the one who got initially in touch with Bob Buchanan and helped us connect with the big players in the field. A third one was the joint organization of the first thioredoxin congress by Pierre and Bob Buchanan in Berkeley in 1980, sponsored by NASA and CNRS. This was the first time I got in contact with Arne Holmgren the other king of that domain on the bacterial/mammal side. Many other congresses of that type would follow including one that I co-organized later in Roscoff in the Jacques Monod series. On a daily basis Pierre was at that time following our research projects very closely, the magic sentence being every Friday "So how has it been, got new results?" Also, he has been throughout his career constantly pushing us to publish our results in peer-reviewed journals and in English, which, in the middle of the 70s was not yet so common in France. The latter section dealing with the work in Orsay will detail further his organization skills and scientific visions that proved essential for turning later to molecular biology techniques. Still, on a more personal basis it was fun to work in the Gadal's group, we did many things together, much sports including jogging some biking, an occasional odd football tournament in Nayemont-les-Fosses and quite a bit of ping-pong as we had a table next adjacent to the lab. Among the contests that we were having were vertical jump tests where we had to jump as high as possible and we recorded the results with blue chalk on a beam hanging down the ceiling. From that time, I can indeed testify that white men can't jump... Anyway, our performances were pretty much related to our heights, Jean being the most altitudinally challenged, experiencing the most difficulties. Pierre was substantially taller than most of us and when he saw us doing that he came and touched the ceiling on his first try and left with a shrug meaning "this type of contest is too easy for me." Humiliated we were, thank you for this one Pierre. Another of the contests that we had routinely was chest band pull with Jean much more at ease there with his shorter limbs (besides he was very strong anyway)... We were also occasionally testing ourselves at the arm wrestling contest, Jean Vidal was very good at it but nobody wanted to play that game with Pierre Gadal worried that we were that he would destroy our shoulders...

Pierre Gadal was in general a rather reserved person never communicating on the informal "tu" basis but always maintaining the more formal "vous." Nevertheless, Pierre himself played ping-pong quite enthusiastically and I remember a Saturday afternoon where he was teaming in a double with Jean Vidal against myself and Bertrand Hirel a game that became so serious that one of us (BH) ended up with a broken tooth (from a racket incident and not from a fight...). Except for incidents of that type working there was good old fun and Pierre was clever enough to mix professionalism with sort of a family-like atmosphere. I believe that was a big success for this group, it provided enthusiasm (it was kind of standard to work 10 h a day including week-ends when the research adrenaline kicked in) and results followed.

The later Orsay years [1981–1997] Protein sequence determinations, gene cloning, site-directed mutagenesis and structural output.

In 1981 Pierre Gadal applied successfully for a professor position in Orsay near Paris and asked his staff whether they would agree to move there as the CNRS-associated lab was to be transferred at the University of Paris-Sud. Essentially all of the colleagues active in research agreed to the move and I was one of them. Moving to Orsay was accompanied by a reorganization of the team and it was there that the nucleus of the redox team took form with the beginning of a long-standing cooperation with Myroslawa Miginiac-Maslow and later Paulette Decottignies. Together with Myroslawa and

Michel Droux we started by isolating ferredoxin-thioredoxin reductase and refining and simplifying the light-dependent ferredoxin-thioredoxin system. This sequence of events will be described in more detail in the recollection of Myroslawa and Michel. One decisive technical advance was the arrival of Paulette Decottignies and Pierre Le Maréchal and their know-how in protein sequencing with the help of Jean-Marie Schmitter initially. This more or less coincided with us turning to a new biochemical model *Chlamydomonas reinhardtii* referred to later in this recollection as Chlamy (Huppe, de Lamotte, Jacquot, & Buchanan, 1990). Initially we thought that Chlamy would be a fantastic tool for generating mutants of the FTR system but this was not to be the case as the genes that we were interested in were nuclear ones and gene targeting in Chlamy worked well but only for chloroplastic genes at the time. Anyway, Chlamy turned out to be a beautiful biochemical material to work with, generating enzyme extracts much cleaner than what we got from spinach, corn or bean. This property combined with the fact that the genome of Chlamy was more simple than the one of land plants (a property that was not known in the early 90s) allowed us (after many efforts and a lot of work) to finally obtain substantial amounts of pure ferredoxin, ferredoxin-NADP reductase, chloroplastic thioredoxin m and cytosolic thioredoxin h. Paulette and Pierre were subsequently able to cut the proteins in pieces with the help of various proteases including trypsin, chymotrypsin and SV3 and to isolate and purify the peptides by HPLC. Edman sequencing of the intact proteins straight from the N-terminus and also of the various peptides was rather tedious and needed an iron organization but this was fine for those chemist-oriented researchers and they were able to fully sequence those four proteins (they range in size from 94 to 320 amino acids per polypeptide) (Decottignies et al., 1990; Decottignies, Schmitter, Dutka, Jacquot, & Miginiac-Maslow, 1991, Decottignies, Le Maréchal, Jacquot, Schmitter, & Gadal, 1995; Schmitter et al., 1988). In parallel to those efforts, Claude Crétin had isolated a cDNA coding for sorghum NADP-MDH by screening a λgt11 expression library with antibodies and I had been able together with Eliane Keryer to introduce it into an expression plasmid and to generate some recombinant protein in *E. coli*. Actually seeing that additional tiny protein band in the *E. coli* extracts was one of the most memorable achievements in my scientific life. This observation was also accompanied by the appearance of a new oxalocetate reducing activity, NADP-dependent, and requiring reduction to be active in the *E. coli* extracts and that was definitely the cherry on the cake (Jacquot et al., 1991). Champagne on that day! Meanwhile we were also active on other fronts learning molecular biology techniques and site-directed mutagenesis. We started with Frederic de Lamotte and Philippe Minard by engineering a mutation on the *E. coli* thioredoxin gene (D61N) that allowed the new protein to acquire a thioredoxin f-like property (de Lamotte-Guéry et al., 1991). In parallel, together with Mariana Stein and John Rogers we started isolating by PCR the cDNA fragments corresponding to the coding sequences of ferredoxin and thioredoxin m (called Ch2 then), based on the previously determined protein sequences (Jacquot, Stein, Hodges, & Miginiac-Maslow, 1992; Rogers et al., 1992). This was facilitated by using preferential codons for the Chlamy nuclear genes that heavily favor those ending with G or C, A and T being in general not used in this organism. We could thus limit the number of degeneracies in the forward and reverse oligonucleotides and this proved to be crucial for the success of this PCR cloning. The initial fragments that we had generated concerned only the mature part of the polypeptides generated after cleavage occurs following import into chloroplasts. Nevertheless, in a later step we were able using either antibodies or the 32P labeled PCR fragments to isolate full-length cDNA clones from λgt11 libraries that featured the 5′ and 3′ UTR and also the sequence coding for the transit peptide needed for import. In a subsequent step, those fragments were labeled to isolate the corresponding clones from a genomic library together with the one coding for thioredoxin h (Stein et al., 1995; Stein, Chedozeau, & Jacquot, 1996; Stein, Jacquot, & Miginiac-Maslow, 1993).

So in a few years we had gone a long way to determine the primary structure of several of the enzymes of the ferredoxin-thioredoxin system. This was very quickly followed by developments that led to the elucidation of 3D structures of those proteins. The initial work with the NADP-MDH had gotten us introduced to genetic engineering and the production of recombinant proteins, and rather quickly we adopted a new type of expression plasmid from the pET series that was a lot more efficient. This has allowed us to produce now massive amounts of multiple thioredoxins including pea thioredoxin f with Michael Hodges, NADPH-thioredoxin reductase together with Yves Meyer in Perpignan, FBPase with Lopez-Gorge and Anna Chueca in Spain that would then lead to solving several 3D structures (Hodges et al., 1994; Jacquot et al., 1994, 1995; Lopez-Jaramillo et al., 1997; Rivera-Madrid et al., 1995). A whole new area of research opened after we shifted to that new expression tool and we were able to perform site-directed mutagenesis to determine which cysteines were essential for the NADP-MDH regulation (see the description of

Myroslawa for that) but also for FBPase (Jacquot et al., 1997). Likewise, we could determine which acidic residues govern the tight association between ferredoxin and ferredoxin-thioredoxin reductase (Jacquot, Lopez-Jaramillo, et al., 1997).

3D structures of full-length enzymes have been then rapidly produced by X-ray crystallography for NADPH-thioredoxin reductase with Hans Eklund and Shaodong Dai in Uppsala (Dai et al., 1996) and FBPase with Jacqueline Cherfils in Gif sur Yette (Chiadmi, Navaza, Miginiac-Maslow, Jacquot, & Cherfils, 1999). Using NMR and through a very fruitful cooperation with Jean-Marc Lancelin we have been able to solve structures for thioredoxin m, thioredoxin h as well as for transit peptides required for protein translocation (Lancelin, Stein, & Jacquot, 1993; Lancelin et al., 1994; Mittard et al., 1995).

This describes essentially the main output of my stint in Orsay, but there were also quite a number of side projects to which we participated. Among others they include the discovery of thioredoxin in human (Jacquot et al., 1990; Wollmann et al., 1988), the description of thioredoxins in *Dictyostelium discoideum* (Wetterauer, Jacquot, & Véron, 1992) and the interactions of ferredoxin with cellular components other than FTR, such as PSI (Fischer, Hippler, Sétif, Jacquot, & Rochaix, 1998) and glutamate synthase (García-Sánchez, Gotor, Jacquot, & Vega, 2000).

What about the social life in the Gadal Orsay lab? Well it was equally fun compared to Nancy, our interests moved to playing bilboquet, lots of biking and jogging in the woods surrounding the beautiful Orsay campus. We also initiated a new one, rope jump, at which we became pretty good but on one occasion Keishiro Wada visiting us stunned us by showing how a guy close to 60 could do it squatting on his haunches, a very difficult thing to do. Another of those humiliations there... Intellectually we kept stimulating ourselves playing a French version of Trivial Pursuit, Frédéric was very good in the pink category there (the vainest of them all) ... Of course, we still played quite a bit of ping-pong as we had three tennis tables in the basement of the building underneath the greenhouse. Pierre Gadal then in his 50s was still very keen on playing and we could communicate a lot that way as otherwise he became very busy with administrative duties (he was actually a superb organizer of science). Nevertheless, the best tennis table player was by far Stéphane Lemaire as he handily won the first tournament of our Institute. Also, with Frederic de Lamotte it was fun to try to pull a Nelson (double Nelson in French) on one another especially when the colleague was pipetting and could hardly defend himself. This slowed Frédéric a bit for the cloning of his mutants but he was so bright and dedicated that he succeeded anyway. As supervisors with Philippe Minard we had to kick his ass a couple of times to further motivate him, not that we enjoyed that too much but it worked in the end.

In 1997 after more than 20 years as a CNRS researcher I moved to another position as a Professor at University of Nancy I later to become Université de Lorraine. The work that was subsequently done in Nancy focused on other redox aspects including the study of thiol peroxidases and their relations to thioredoxin and glutaredoxin, and more recently the assembly of iron sulfur clusters, a process in which glutaredoxin is heavily involved. All of the work that was done in this later step of my career would not have been possible without the Orsay experience where we learnt how to combine biochemistry with molecular biology, became acquainted to PCR and had the first access to genomics data on *Arabidopsis thaliana*. Fortunately, the enthusiasm for research that was the trademark of the Gadal's lab has been transferred to the Nancy redox group later headed by Nicolas Rouhier and the results have followed.

Biology is a science that develops incredibly rapidly especially at the technical level. Of course with the tools generated today what we had done in the 90s could be done in a matter of weeks or months as genomes are now sequenced and we do not have to chase for sequences any more with those "primitive" tools. Most of this was initiated before genomics and PCR and therefore the rhythm of scientific production was quite slow compared to today's output. At first restriction enzymes were scarce, genomics and proteomics inexistent, DNA sequencing had to be done manually with the Sanger method and 35S using long acrylamide gels that were a pain to prepare, etc. In retrospect, was that all in vain? Well there are some facts that remain of high interest even after all these new developments. For instance, the route that we had to follow included the isolation of proteins before we could isolate the genes. Taking Chlamy ferredoxin as an example it is one of the few instances where we know exactly where the cleavage or the preprotein takes place as we know the real N-terminus of the chloroplastic processed protein. Nowadays people "guess" where the cleavage is supposed to take place by bioinformatics but this is basically never tested experimentally (although proteomics can now provide data about that). Incidentally, when you produce a recombinant protein, cleaving the presequence in the right position can be very important for the production of a soluble recombinant protein. Another example of the interest of this approach is ferredoxin-NADP reductase. When we confronted the results of protein sequencing to those of DNA sequencing it became clear that this protein is in vivo methylated at three positions on Lysine residues, a property that we could never have deduced from the sole DNA sequence (Decottignies et al., 1995). Besides, and more importantly our work has given insight on how disulfide bonds put a physical strain on regulatory enzymes and how reduction relaxes this strain and allows them to shift into catalytically competent conformations. A couple of Figs. 1 and 2 extracted from a rather recent review paint the molecular details for many of the redox-regulatory enzymes connected to photosynthesis (Gütle et al., 2017). A common feature to all these enzymes is that their redox-regulatory site is in general distant from the active site and the regulatory cysteines are not involved in catalysis. Yet they influence at a distance the opening of active sites and their accessibility to substrates. Of course, this is crucial for the regulation of photosynthesis and probably agronomical yields. A final comment is that the raw data of genomics have little meaning as long as you don't know what the function of a given gene/protein is and that is best obtained from biochemical experience. So, in this sense all the work generated in the Holmgren lab in Sweden, the Buchanan lab in California and, hopefully, to some extent our lab in Nancy/Orsay/Nancy was not futile after all.

2.2.2 Frédéric de Lamotte's recollections of the Orsay time [1985–1990]

Pierre Gadal was my professor at University of Paris-Sud, Orsay. I attended his Plant Physiology courses during my bachelor and master. I also performed the experimental part of my master in his lab, then my PhD research under the brilliant supervision of Jean-Pierre Jacquot. Pierre Gadal was an unusual professor. During his lectures, I had the impression of being in front of Science, of glimpsing the "big picture." I felt like I was intelligent (even though I was disappointed about it more than once afterwards...).

When I joined the group, the main elements of the Redox cascade had already been identified and purified in higher plants by "the first generation" as Jean-Pierre described earlier. The next step was to focus on one organism, *Chlamydomonas reinhardtii* referred to later as Chlamydomonas.

The choice was guided by several reasons. At that time, it was not yet established that the FTR system was present in all photosynthetic organisms. In addition, Chlamydomonas, a unicellular eukaryotic organism, was easily amenable to microbiological techniques and its genetics was known. Moreover, we were hoping to finally have an organism that could supply all the bricks for the redox cascade.

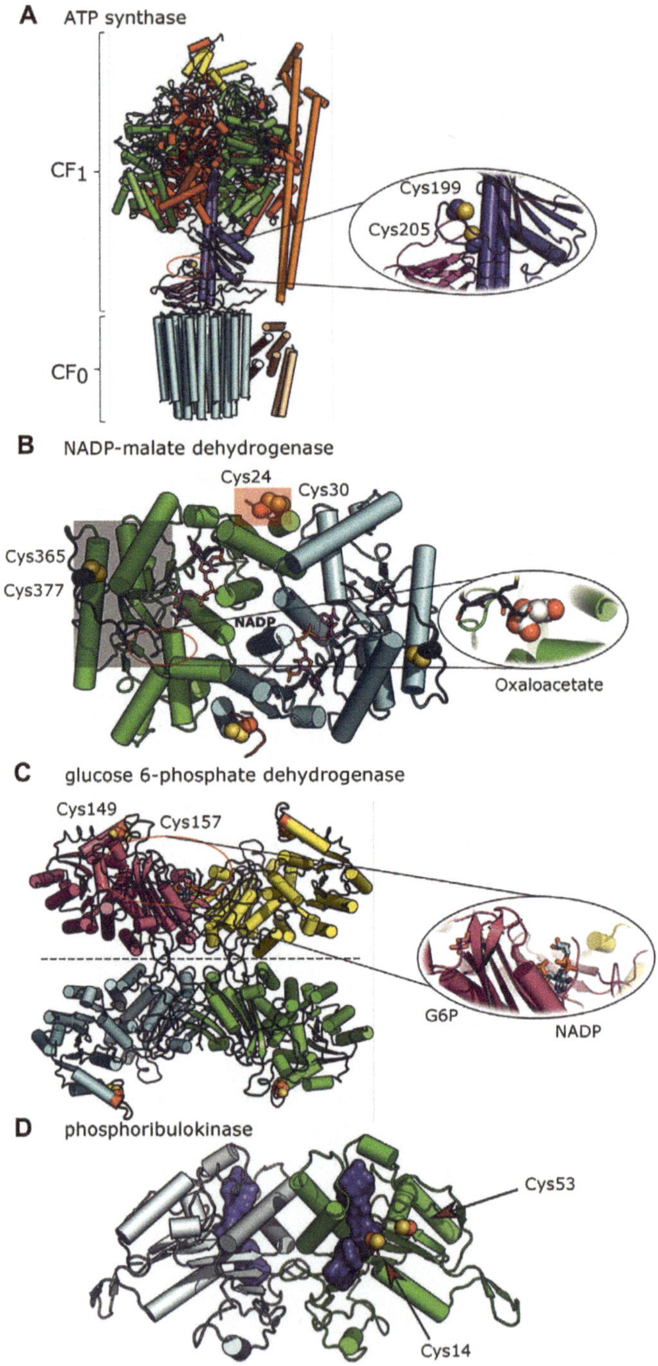

Fig. 1 3D structures and models of redox regulated chloroplastic enzymes. (A) CF1 ATPase, (B) NADP-malate dehydrogenase, (C) glucose 6 phosphate dehydrogenase, and (D) phosphoribulokinase. Cys labeled 30 of NADP-MDH is, in fact, Cys 29. *From Gütle, D.D., Roret, T., Hecker, A., Reski, R., & Jacquot, J.P. (2017). Dithiol disulphide exchange in redox regulation of chloroplast enzymes in response to evolutionary and structural constraints. Plant Science 255, 1–11.*

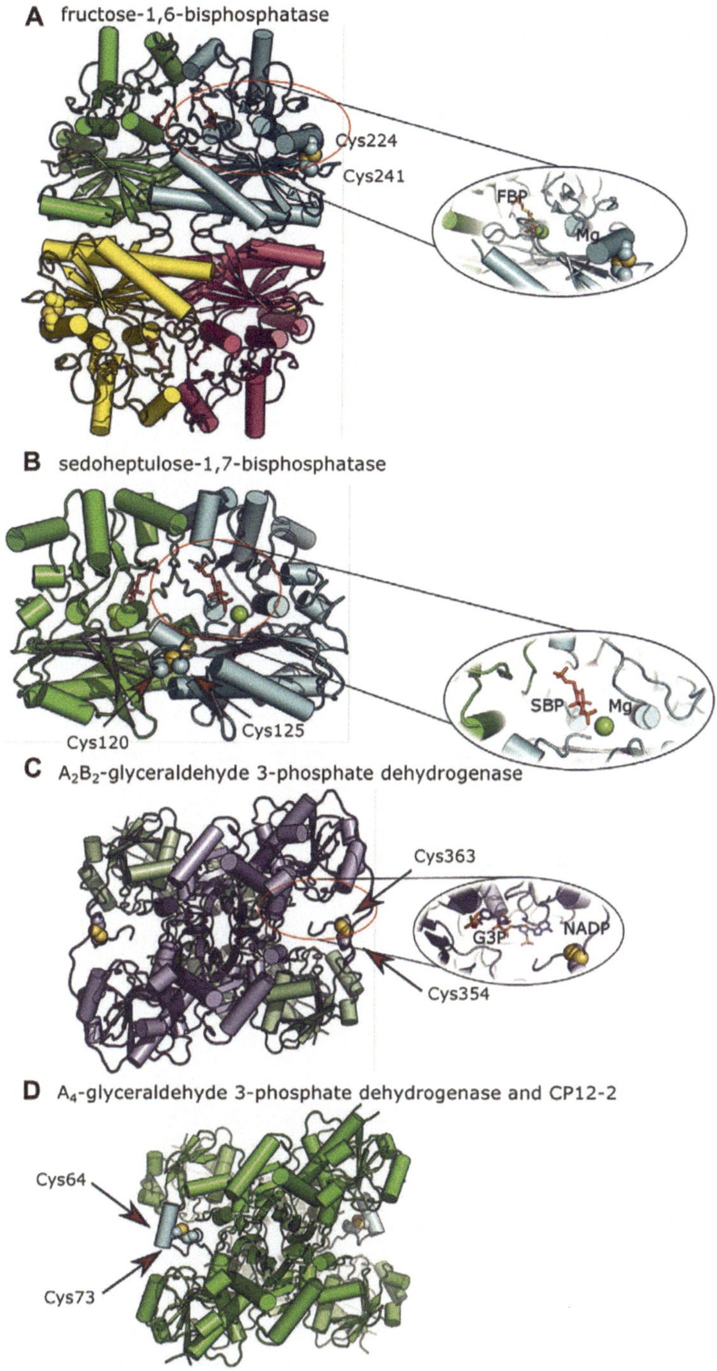

Fig. 2 3D structures and models of redox regulated chloroplastic enzymes. (A) FBPase, (B) SBPase, (C) GAPDH A2B2, and (D) GAPDH A4 with CP12. *From Gütle, D.D., Roret, T., Hecker, A., Reski, R., & Jacquot, J.P. (2017). Dithiol disulphide exchange in redox regulation of chloroplast enzymes in response to evolutionary and structural constraints. Plant Science 255, 1–11.*

My first research work was to set up a culture system and thanks to J-P. ingenuity, my naive enthusiasm and the benevolence of P. Gadal, we were able together with Heather Huppe to grow *Chlamydomonas* in big 25 L glass bottles, with a large magnetic bar rotating at the bottom, pressurized air bubbling from the top and neon light shining from the side, sitting on the floor of the main lab. This particular noise and light would accompany me throughout my thesis.

The second part of my work was to learn how to harvest and concentrate the cells, then extract and purify the proteins. Everything went fine thanks to the strong support and expertise of Paulette and Myroslawa. Ferredoxin was swiftly purified and sequenced. We were not able to obtain a highly purified form of the MDH but the fraction was active enough to conduct light activation experiments Two thioredoxins were identified as well as the FTR. The sequence of the two thioredoxins was determined by Paulette. One was clearly an m type and the other one was later identified as an h type. These first results were published in 1988 (Schmitter et al., 1988) and 1990 (Huppe et al., 1990). This study made clear that the FTR system was ubiquitous in all photosynthetic organisms, reinforcing its cornerstone pôsition and the interest of its study. At the time, I wasn't really surprised at how easy it was, being successful after just one or two attempts; it was only later that I realized how lucky I had been.

During my PhD work, we then focused on thioredoxins. In the redox cascade transmitting electrons from the photosystems to the activation site of the target enzymes, thioredoxins exhibit heterogeneity with the m, f and h types known at the time. We wanted to understand what carries the specificity of each of these proteins. The sequence of the Chlamydomonas type m showed quite a lot of similarity with the thioredoxin from *E. coli* (TrxA). As, unlike the ones of our green algae, the *trxA* gene was available, we decided to use the bacterial protein for this study. Here again, I was lucky as, in Janine Yon-Khan lab, next door, Philippe Minard was just coming back from his post doc in a UK lab where he worked on the first site-directed mutagenesis experiments. This was actually initiated after Pierre Gadal's suggestion who came back from a meeting thrilled by a presentation and telling us: you have to do site-directed mutagenesis, it's the future.

As enthusiastic and intrepid as ever, we have embarked on the development of an overproduction system for thioredoxin in a bacterial system. We cloned the *trxA* gene into a high copy number plasmid and rapidly set up conditions for the production and the purification of more than 60 mg of purified thioredoxin par liter of culture. That yield was considered as gigantic at the time. The purified protein was able to mimic the plant thioredoxin m behavior in the photomodulation system.

The next step was closer to a giant leap rather than a regular step as it was to make site-directed mutagenesis in an era before PCR was available... The first oligonucleotides carrying the mismatch were made on a (very) first generation of oligonucleotides synthesizer which would require to push four syringes (for the four nucleotides) in the correct sequence. The purified oligos were hybridized on a single stranded DNA (we needed to purify it from phage M13mp19, and that was very difficult, it worked only when the phage was willing) and the double stranded DNA was synthesized in vitro. The yield of this method (5%) was a real drawback until Amersham and Bio-Rad came with some kits. The price was extremely expensive but, once again, P. Gadal provided us a strong support, pushing forward. This new kit allowed me to produce my first mutants. The choice of the amino acids to be modified was made on the basis of the available biochemical data and a comparison of the different published primary structures. The mutants fell into two categories corresponding to the folding of the protein and the putative interactions with the target proteins. It should be noted that it was very advantageous to have the 3D structure of the *E. coli* TRX in order to localize each amino acid, and thus be able to refine the choices.

The first mutant was named D61N, meaning that Aspartic acid number 61 was replaced by its neutral equivalent Asparagine. With the exception of the thioredoxin f, all thioredoxins known at the time had a charged residue at position 61. Interestingly, thioredoxin f is the only plant thioredoxin that forms a stable complex with its target enzyme. Would it be possible to shift the behavior of the thioredoxin from the m type to the f type by changing just one residue?

We purified the modified protein and its wild-type counterpart. The recombinant proteins were analyzed with a whole set of tests, physicochemically, biophysically and catalytically. HPLC, IEF and native electrophoresis confirmed the charge difference compared to the wild type. Fluorescence spectra and circular dichroism detected no differences, as expected.

Three reduction sources were used: DTT, NADPH with *E. coli* NADPH-TRX reductase (NTR), and spinach FTR. The reduced thioredoxins were used to reduce four targets: DTNB (dithiobis nitrobenzoic acid), insulin, NADPH with NADPH-TRX reductase from corn (thioredoxin m target) and FBPase from spinach (thioredoxin f target). These tests revealed differences in efficiency in favor of the mutant for FBPase activation when DTT provides the reducing power. This confirms the hypothesis that the charge carried by

Asp61 plays a detrimental role in the interaction with FBPase. A less predictable result was the greater efficiency of the mutant in activating chloroplastic enzymes (NADP-MDH and FBPase) in the reconstituted system. This improved efficacy is explained by an improvement in the interaction between TRX and FTR. This work was published in 1991 (de Lamotte-Guéry et al., 1991).

The second one was named P40A, in which the Proline was replaced by an Alanine. This proline, found in all known thioredoxins, is located in the middle of the α2 helix. This position can have a significant impact on the structure of the protein and modify its thermostability.

The enzymatic behavior of P40A and the wild protein measured in the same conditions appeared identical. With the same tests used for D61N, only the fluorescence analysis showed a difference. At that time, it was impossible for us to interpret these results. Later, I moved to Montpellier, where I had the opportunity to collaborate with the Centre de Biochimie Structurale for the study of wheat thioredoxins. We built a molecular model of P40 A in order to interpret my observations.

P40A slightly differs from WT, the RMSD is 1.5 Å. The major difference is the helix α2 which is less bent, the angle formed by this helix shifting from 59° to 13°. This straightening induces a displacement of the active site by 6 Å without modifying its structure. The hydrophobic contacts involving TRP31 and the loop between ß2 and α3 have been removed. W31 becomes accessible to the solvent, which explains the fluorometry results. These calculations also predicted that P40A should be less stable and more hydrophobic than WT. This was verified by hydrophobic interaction chromatography and denaturation with Guanidine Chloride followed by fluorescence at 340 nm; ΔGo went from 8.9 down to 6 kcal/mol.

Thus, contrary to our hypothesis, the proline that bends the helix α2 has no overall destabilizing effect. This can be explained by an unfavorable local effect on the helix and a more important favorable effect by optimization of the hydrophobic contacts. This was finally published in 1997 (de Lamotte et al., 1997). We can see here that a more precise and better argued answer has been brought by a multidisciplinary work. This kind of approach has been, since then, my routine.

On a more personal ground, I recall the following. By the end of my PhD work, I applied for a Post-Doctoral fellowship with a research project to EMBO. I was asked by EMBO to visit a member of their scientific board for an "informal discussion" about my project. The person I should discuss with was based in Warwick and the only practical way to get there was by airplane. The round-trip ticket was pretty expensive (about 1 month of my graduate student salary). As soon as Jean-Pierre informed Pierre Gadal, he offered to cover the cost of the ticket with the lab's budget. I was touched by this act of kindness, all the more so as it was about my own future and the repercussions of this event were very limited for Pierre Gadal and his laboratory. This is a fine example of the altruism that Pr. Gadal could show.

There is a common metaphor about scientists being dwarfs standing on the shoulder of a giant. Pierre Gadal was such a giant, a friendly giant. His shoulders were pretty high but hopefully I was helped to reach them by (in order of size more than in order of importance...) Paulette, Myroslawa and Jean-Pierre. My time spent in P. Gadal's lab had a deep and long-lasting influence on what I have become.

Without ever approaching his scientific radiance, I tried to copy his benevolence and love of people. In this, one can say that Pierre Gadal has had a major influence on my life as a researcher.

Side note. I would like to stress one point. I won, without a possible doubt, the Nelson contest with Jean-Pierre...

2.2.3 Stéphane Lemaire's recollections

In the 90s, Pierre Gadal was the director of the master school (at that time named DEA) dedicated to *Plant Molecular and Cellular Biology* in Orsay. I had studied genetics and molecular biology during my undergraduate studies and I was fascinated by plant systems. It rapidly became clear to me that integrating Pierre's master school was the most interesting path for me. As this path was very competitive, I thought that I would increase my chances of being selected if I could join the group of Pierre Gadal for a short internship during the summer of 1994. I therefore contacted J-P. Jacquot and I was very lucky to be selected for this internship during which I discovered the true essence of scientific research. I recall the stimulating and friendly environment of the Gadal lab with the mentoring of Jean-Pierre. The aim of this internship was to perform site-directed mutagenesis on FBPase to confirm the exact cysteines involved in its thioredoxin-dependent activation. Although M13-based site-directed mutagenesis had been established in the group, Jean-Pierre wanted to test a new method based on PCR. These experiments were successful although I did not realize at that time that it was the first PCR-based mutagenesis performed in the lab. I clearly remember reading and writing by hand with Jean-Pierre the autoradiography of the 35S Edman sequencing gel (on a dedicated illuminated desk located in Jean Vidal's office) and our joint excitement when we confirmed that the mutagenesis had perfectly worked as expected. We could then describe the molecular mechanism of activation of FBPase by thioredoxin (Jacquot et al., 1995, 1997). After this internship I could join the master school program where I embraced molecular biology through the guidance of Pierre Gadal who was advocating, as mentioned by Jean-Pierre, the power of molecular biology for plant biology research. I joined again the lab for my Master internship where I studied the structural determinants of thioredoxin thermal stability with Jean-Pierre taking advantage of PCR site-directed mutagenesis. I could not immediately start a PhD since I had to join the military forces for 10 months, as mandatory at that time in France. I was very worried that being away from science and the lab for nearly 1 year would compromise my chances to be granted a PhD fellowship. Pierre Gadal had a decisive role for me during this period since he allowed me, thanks to his network, to obtain a military position near the lab to be able to keep in touch. He also played a decisive role to allow me to finally obtain a fellowship to initiate a PhD under the supervision of Jean-Pierre Jacquot in 1996. My work was the direct continuation of the work initiated in the Gadal group on thioredoxins and was aimed at pursuing the biochemical and structural analyses on thioredoxins but also growing Chlamydomonas under different light regimes to monitor the changes in expression of thioredoxin and related genes (Lemaire, Hours, et al., 1999; Lemaire, Keryer, et al., 1999; Lemaire et al., 2000). After my postdoc, I was hired by the CNRS in the same group, then led by Myroslawa Miginiac-Maslow. In 2010, I created my own group in Sorbonne University. This group is like many others a heritage of the Gadal school of thinking. Pierre has followed my career with a strong interest and was always very keen to discuss science, science policy and to provide his advices and recommendations. Today, we are still exploring redox regulation and the Calvin-Benson cycle in Chlamydomonas (Michelet et al., 2013). Recently, we embraced the emerging field of synthetic biology (Crozet et al., 2018; Vavitsas et al., 2019) and I am advocating to my colleagues and students the importance of these new approaches. This reminds me the pioneering view of Pierre who was advocating the importance to develop molecular biology in the plant research community. Throughout all these years under the guidance of Pierre and his mentees, Jean-Pierre and Myroslawa, I learned not only the passion of scientific research but maybe more importantly the human adventure of the scientific enterprise. Beyond his enthusiasm for science, Pierre is probably one of the most kind, gentle and caring scientist I met. He always took time to discuss and listen to the people around him, whatever their social status could be. Finally, Pierre Gadal has clearly shaped the scientist I am today. I will continue to humbly walk in his footsteps to perpetuate his heritage and I am very proud to be standing on the shoulders of this giant.

2.3 Light activation via ferredoxin-thioredoxin reductase: The NADP-malate dehydrogenase model

2.3.1 Myroslawa Miginiac-Maslow: Recollection from 1981 until Pierre Gadal's retirement in 2003 and beyond

In 1981, Pierre Gadal obtained a full professor position in Orsay, in replacement of Alexis Moyse who was retiring. Remarkably, almost all the members of the Nancy team followed him, despite family ties, witnessing the strong charisma of their leader. Several researchers from the former Moyse lab moved to other labs, but a few (including myself) remained and joined the new lab. At that time, many studies dealt with the ideal ATP/NADPH ratio necessary for optimal photosynthetic CO_2 fixation. My own results showed that this ratio could vary according to the ambient conditions, suggesting that linear and cyclic electron flow could occur simultaneously. Another intriguing observation was the stimulation of CO_2 fixation activities in some intact chloroplast preparations by Antimycin A, an inhibitor of cyclic electron flow, but also by some uncouplers, activating the electron flow (Miginiac-Maslow & Hoarau, 1975). In line with these results, evidence was obtained in the lab (Champigny & Bismuth, 1976) that these inhibitors enhanced the activity of several Calvin cycle enzymes measured in chloroplast stroma, suggesting the involvement of the reducing power in these phenomena. The arrival of persons working on enzyme redox regulation reactivated my interest in the problem, suggesting that the ideal ATP/NADPH ratio should take into account the amount of reducing power needed for enzyme activation. Then we set up a collaboration with J-P. Jacquot, with the blessing of Pierre Gadal, to determine the light saturation values of the activation of NADP-MDH in a reconstituted light activation system. This was made possible by the availability, in my former lab, of a Warburg apparatus I had equipped with strong light bulbs, and a series of neutral filters allowing to establish light saturation curves. As a prerequisite to these experiments, we optimized a reconstituted light activation system, by purifying all of its components (ferredoxin, ferredoxin-thioredoxin reductase and the target enzyme) to homogeneity, and using thylakoids isolated from pea leaves, much more stable than those from spinach, which could be stored in a frozen form (Jacquot, Droux, Miginiac-Maslow, Joly, & Gadal, 1984). This rather trivial modification allowed ending a controversy between Bob Buchanan (using thylakoids from spinach) and Louise Anderson (using thylakoids from peas), the former presenting Trx as the activating factor, and the latter proposing a membrane-bound activating factor named LEM. In fact, with once washed pea thylakoids, we could perform enzyme activation without adding Trx, whereas, with twice washed pea thylakoids (one of the wash steps including NaCl) the addition of Trx was absolutely necessary. Thus, apparently, in peas, part of the Trx was membrane-bound. The purification of the protein components benefitted from the input of Michel Droux, the first of a series of clever and very efficient PhD students selected by the DEA committee headed by Pierre Gadal, who joined our group and achieved the full purification and characterization of FTR (Droux et al., 1987). At that time, there was some controversy concerning the chemical structure of FTR, in particular the presence of an iron-sulfur cluster. Using a final purification step with an affinity chromatography on a ferredoxin Sepharose column (and starting with 10 kg of spinach leaves) we could obtain a pure FTR preparation displaying a 4-iron-4-sulfur cluster but also a disulfide bridge, a feature that had never been detected before. We established also the sequence of events leading to the activation of NADP-MDH, namely sequential reduction of ferredoxin, FTR,Trx, and finally target enzyme (Droux et al., 1987). This important contribution, started during Michel's PhD, was finalized during Michel's post-doctoral stay in the lab of Bob Buchanan, in Berkeley, where the thiol groups of the different proteins involved in the activation process could be visualized using a new fluorescent agent, monobromo bimane (mBBr). The molecular mechanism of action of FTR was later elucidated in the laboratory of Peter Schürmann, who solved also its 3D structure in collaboration with the laboratory of Hans Eklund in Uppsala, Sweden (Dai, Schwendtmayer, Schürmann, Ramaswamy, & Eklund, 2000).

The light activation curves showed that enzyme activation was saturated at low light intensity (10 $W\,m^{-2}$ under nitrogen, 50 under air, where some reoxidation certainly occurred, vs 240 $W\,m^{-2}$ for reduction of NADP). Thus this mechanism was not very energy-demanding (Miginiac-Maslow, Jacquot, & Droux, 1985). Having in hand the purified components of the system, we could titrate the cysteine thiols appearing on the activated enzymes, by using a C14-labeled alkylating agent, generally iodoacetamide. Then we could compare the reduction efficiencies of NADPH and Trx and conclude that NADP acted like a competitive inhibitor of Trx reduction, i.e. was favored in the system. We could also observe that the reduction of NADP-MDH was in fact very fast, as fast as the reduction of Trx, but that the activation of the enzyme lagged behind, taking about 10 min to reach full activity, suggesting that a slow conformational change followed the reduction step (Miginiac-Maslow, Decottignies, Jacquot, & Gadal, 1990).

Meanwhile, Gadal's lab attracted new people that joined our group: Paulette Decottignies and her husband, Pierre Le Maréchal, organic chemists working previously in a neighboring biochemistry lab. Pierre Gadal considered that their integration would bring new technical skills to the lab. Indeed their presence prompted us to try to analyze further, at the molecular level, the modifications occurring on NADP-MDH upon activation. The enzyme was incubated in a reconstituted light activation system under saturating white light. Then, the cysteines were alkylated with radioactive iodoacetate, the components separated by HPLC and the N-terminus of NADP-MDH sequenced in collaboration with J-M. Schmitter (Biochemistry lab of Ecole Polytechnique). Two labeled cysteines were found in the N-terminal part of the enzyme. None was found upon incubation in the dark. Then we considered that we had found the regulatory disulfide of NADP-MDH (Decottignies et al., 1988). In parallel with their contribution to the study of NADP-MDH, Paulette Decottignies and Pierre Le Maréchal determined the complete amino acid sequences of several proteins involved in redox metabolism, such as ferredoxin-NADP reductase and ferredoxin from *Chlamydomonas* (Decottignies et al., 1995; Rogers et al., 1992). The subsequent acquisition of a protein sequencer by the lab, thanks to funds obtained by Pierre Gadal, helped them to develop this important analytical tool for further studies.

When sorghum NADP-MDH was cloned and its cDNA sequenced (Crétin et al., 1988, 1990) we realized that each monomer of this dimeric enzyme possessed eight cysteine residues, strictly conserved in all the light-activatable NADP-MDH, but also two conserved sequence extensions compared to permanently active NAD-MDH. The two cysteines putatively involved in the activation process were located in the N-terminal extension (Cys 64 and 69 in the full-length sequence, comprising the transit peptide of this nuclear-encoded chloroplastic protein, corresponding to Cys 24 and 29 of the mature protein). There was also one Cys located in a C-terminal extension, with another cysteine located nearby. An expression and recombinant protein production in *E. coli* having been set up, and after showing that the recombinant protein displayed the same regulatory properties as the enzyme isolated from sorghum leaves (Jacquot et al., 1991), we decided to try to confirm the results obtained by chemical analyses by using site-directed mutagenesis. Emmanuelle Issakidis, a new PhD student, who acquired a strong expertise in molecular biology, compared various recombinant protein expression systems and developed an efficient site-directed mutagenesis protocol using a pET-type plasmid. The first mutations she created targeted the two N-terminal cysteines previously identified, which were substituted individually or simultaneously for either alanines or serines. We were expecting to obtain fully active, redox-insensitive enzymes. The first activity tests brought a full disappointment: our mutated NADP-MDHs still required reduced Trx to reach full activity (Issakidis et al., 1992). However, looking more precisely at the activation kinetics, we noticed that enzyme activation was very fast, as fast as the reduction of Trx we had observed before. Clearly, there was no lag anymore between reduction and activation: the enzyme was ready to function upon reduction. Then, we turned to the substitution of the two most C-terminal cysteines, still individually or together. The resulting mutant proteins exhibited the same slow activation kinetics as the WT enzyme, but, while the WT enzyme was totally inactive without reduction, the mutants displayed a discrete activity with an increased Km for their substrate, oxaloacetate. This activity could reach 20% of the activity of the fully activated enzyme, suggesting that the active site became at least partially functional. Combination of mutations at both ends, either triple (double N-terminal associated to one of the C-terminal cysteines), or quadruple, yielded fully active enzymes, insensitive to reducing treatments (Issakidis et al., 1994). These permanently active forms were strongly retained on affinity columns, requiring very high salt concentrations to be eluted, and, once purified, were rather unstable.

Meanwhile, another PhD student, Martine Lemaire, developed, under the supervision of Paulette Decottignies, a chemical derivatization approach targeting the putative active site histidine (Lemaire et al., 1994), and identified this active site histidine by sequencing the corresponding derivatized peptide. Later, she confirmed, using the same approach, that the active site of the enzyme comprised a His-Asp pair, like all the other malate dehydrogenases (Lemaire, Miginiac-Maslow, & Decottignies, 1996). This approach became very helpful for the confirmation of site-directed mutagenesis results. Indeed, it provided a simple means to test the accessibility of the active site of the enzyme. Thus, a pretreatment of the non-activated WT or N-terminal mutant enzymes with the derivatizing agent DEPC (which reacts with His) did not impair their subsequent activation by reduced Trx. In contrast, the pretreatment of the C-terminal mutants, or the activated WT enzyme prevented their subsequent activation by reduced Trx

(Issakidis, Lemaire, Decottignies, Jacquot, & Miginiac-Maslow, 1996). Clearly, the active site was accessible in the C-terminal mutants and in all the activated enzymes but not in the N-terminal mutated and WT non-activated enzymes. This led us to propose that the formation of the C-terminal disulfide bridge shielded the access to the active site, and that the reduction of the N-terminal disulfide triggered a slow conformational change allowing the enzyme to reach an optimal activity conformation.

Started with PhD students from the Orsay University, specialized in Plant Biology, the NADP-MDH program continued with PhD students coming from other universities and specialized in biochemistry. They benefited from the know-how of Emmanuelle concerning site-directed mutagenesis. A hint about the nature of the conformational change of NADP-MDH upon reduction of its N-terminal disulfide bridge was brought about by Eric Ruelland, who obtained his master degree in Biochemistry at the "Ecole Normale Supérieure". Erics' PhD work led to propose a putative role of a conserved Cys situated in the middle of the primary sequence (Cys 207) in this conformational change. Indeed, a Trx-reducible disulfide bridge could be formed between the most N-terminal Cys (Cys24) and Cys 207 when the second N-terminal Cys (Cys29) was mutated. It is thus highly probable that an isomerization of the Cys24-Cys 29 disulfide into a Cys24-Cys207 disulfide occurs during the process of activation (Ruelland et al., 1997). Concerning the interaction of the C-terminal disulfide with the active site of the enzyme, a discussion with Kenth Johansson, a PhD student from the crystallography lab of Hans Eklund (Uppsala) attracted our attention to the fact that the two most C-terminal residues of the C-terminal extension were acidic residues (glutamate and valine) and that they could mimic the substrate oxaloacetate, a dicarboxylic acid. Thus, we deleted these two residues, or replaced them by neutral residues (e.g. glutamine for glutamate). We also combined these mutations with the elimination of one or both of the possible regulatory N-terminal disulfides (Cys24-Cys29 or Cys24-Cys207) by mutating the corresponding cysteines. Proteins mutated at the most C-terminal residues had no activity in the oxidized form but were partially inhibited when pretreated with the histidine-specific reagent DEPC, showing that their active site was partially accessible. Proteins missing the N-terminal regulatory disulfides reached almost full activity without activation upon elimination of the negative charge of the penultimate Glu. These results strongly support a model where the C-terminal extension is docked into the active site through a negatively charged residue, acting as an internal inhibitor. They show also that the reduction of both possible N-terminal bridges is necessary to completely release the C-terminal extension from the active site (Ruelland, Johansson, Decottignies, Djukic, & Miginiac-Maslow, 1998). This interpretation was fully confirmed later, when the 3D structure of oxidized NADP-MDH was solved, in collaboration with the laboratory of Hans Eklund (Johansson et al., 1999). This took 7 years, and many purification campaigns of NADP-MDH. Indeed, it was very difficult to get a fully homogeneous preparation, the enzyme being prone to partial proteolysis at its N-terminal end, as verified by N-terminal sequencing. Finally, a bright and simple idea of Martine Lemaire, who spent a year as a post doc in the lab after her PhD, helped solve the problem: in the oxidized enzyme, the spontaneous proteolysis stopped after the deletion of the 15 most N-terminal amino acids. Thus, we prepared a recombinant MDH deprived of these 15 amino acids, a deletion that did not modify its regulatory properties, and obtained a very homogeneous preparation that yielded useful crystals. The structure showed that the N-terminal disulfide motif (Cys24-Cys29) was inserted between the two subunits of the dimer, thereby locking the domains of each subunit. The C-terminal extension was kept tight to the enzyme surface and shielded the access to the active site. Reduction of the N-terminal disulfide would loosen the interaction between the domains and give the enzyme the necessary flexibility. The reduction of the C-terminal disulfide would free the C-terminal residues from binding to the enzyme and make the active site accessible. As no structure of the reduced enzyme could be solved, this form being unstable, we turned to a collaboration with Jean-Marc Lancelin (University of Lyon) who was able to show, using two-dimensional NMR in solution, that in the reduced MDH, the C-terminal extension moves like a free peptide, allowing the direct sequence-specific assignment of its structure, while it is not visible in the oxidized enzyme, being probably stuck in the active site. Thus, it is obviously released from the active site upon reduction (Krimm, Goyer, Issakidis-Bourguet, Miginiac-Maslow, & Lancelin, 1999). The availability of the structure allowed to design useful mutations that led to the identification of active site arginines interacting with the acidic C-terminal residues, thereby preventing the access of the substrate, and of residues involved in cofactor specificity, unique to thioredoxin-dependent malate dehydrogenases (the permanently active forms are NAD-dependent), as well as of residues important for dimerization (Schepens et al., 2000; Schepens, Decottignies, Ruelland, Johansson, & Miginiac-Maslow, 2000; Schepens, Ruelland, Miginiac-Maslow, Le Maréchal, & Decottignies, 2000). The PhD work of Isabelle Schepens, after her master degree in structural biology, showed also that the specificity for

NADPH could be changed for NAD by site-directed replacement of selected residues, without changing the thioredoxin dependency, and that the inhibition of activation by the oxidized form of the cofactor (first observed by Scheibe & Jacquot, 1983) was retained in the NAD-specific mutants, suggesting that the cofactor interacted, through its positive charge, with the negatively charged C-terminal residues, preventing the release of the C-terminal extension from the active site.

An attempt to confer redox regulation on the thermostable NAD-dependent MDH from *Thermus flavus* by grafting the extensions of sorghum NADP-MDH showed that a successful transfer of redox regulation properties requires the grafting of both extensions, but also the introduction of specific hydrophobic residues in the core part of the protein. These residues are very likely involved in the interaction between monomers, and structural changes at the active site (Issakidis-Bourguet, Lavergne, Trivelli, Decottignies, & Miginiac-Maslow, 2006). The MDH work was completed by the demonstration, by Stéphane Lemaire, who came back to Orsay as a permanent CNRS researcher, after a post-doctoral stay in Geneva, that the complex activation mechanism of this enzyme was specific to higher plants. In the green alga *Chlamydomonas* only the C-terminal disulfide is present and full activation is reached within seconds (Lemaire et al., 2005). Then, obviously the activation mechanism evolved toward higher complexity when the role of NADP-MDH in exporting reducing equivalents (the so-called "malate valve": Scheibe, 2004) became more prominent.

The sequential model of activation of NADP-MDH was also supported by the determination of the redox potentials of the different disulfide bridges of NADP-MDH (Hirasawa et al., 2000), in collaboration with the late David Knaff. The N-terminal disulfide was shown to be less negative, i.e. more easily reducible than the C-terminal disulfide. The latter was more negative than thioredoxins, hence required an excess of reducing power to be reduced. It can thus be hypothesized that depending on the ambient reducing state of the chloroplast, NADP-MDH might exist in a partially reduced form which would then be prone to faster activation when sufficient reducing power becomes more available.

The occurrence of the Cys24-Cys 207 disulfide raised the problem of the accessibility to thioredoxin of Cys 207, located in the middle of the primary sequence. The problem was tackled in the PhD work of Aymeric Goyer, holding a master degree in chemistry and biochemistry from the University of Rennes, who studied the interaction of Trx with NADP-MDH through the creation of mixed disulfides between a monocysteinic mutant of Trx and specific cysteines of its target enzyme (Goyer et al., 1999; Goyer, Decottignies, Issakidis-Bourguet, & Miginiac-Maslow, 2001). Indeed, inspired by the pioneering work of Fred Hartmann (Brandes, Larimer, & Hartman, 1996), this approach allowed trapping the cysteines targeted by the reductant. Aymeric was able to show that Cys 207 was the only, among the four internal Cys, that could form a mixed-disulfide-bound complex with monocysteinic Trx, hence was accessible to Trx. Having established efficient protocols for generation of mixed disulfides, Aymeric applied this technology to set up a monocysteinic Trx affinity column for trapping putative Trx targets. This allowed us to develop proteomic studies aimed at identifying Trx targets at a whole cell level, with the mass spectrometry facilities developed by Paulette Decottignies and Pierre le Maréchel at the Orsay Institute of Biochemistry (Goyer et al., 2002; Lemaire et al., 2004; Marchand et al., 2004, 2010). This now widely used technology allowed the identification of a large number of new Trx targets. This global proteomic approach was further developed by Stephane Lemaire to study redox-dependent post-translational modifications of proteins, such as glutathionylation (Chardonnet et al., 2015; Michelet et al., 2005; Zaffagnini et al., 2012) or nitrosylation (Morisse, Zaffagnini, Gao, Lemaire, & Marchand, 2014).

The early work on NADP-MDH activation, started in Nancy, had revealed the existence of two different thioredoxins in chloroplasts: Trx m and Trx f, the former seeming to activate specifically NADP-MDH (Jacquot et al., 1976, 1978). However, at the turn of last century, the Arabidopsis whole genome sequencing revealed that the plastidial Trxs of this organism comprised 10 isoforms: 4 *m* type, 2 *f* type, 1 *x*, 2 *y* and 1 *z*. This feature prompted us to reinvestigate the specificity of these various isoforms in a more physiological context. First, we tested, in vivo, the ability of these isoforms to compensate for TRX deficiency of a *Saccharomyces cerevisiae* mutant strain, in collaboration with Yves Meyer (University of Perpignan). Seven Arabidopsis plastidial Trx (four *m*, two *f* and a newly discovered TRX *x*) were expressed in a *Trx*-deficient strain of yeast in a putative mature form. None of these heterologous *Trx* was able to restore growth on sulfate or methionine sulfoxide of the mutant cells. When we tested their ability to rescue the oxidant-hypersensitive phenotype of this strain, we found that Trx *m* and Trx *x*, but not Trx *f*, modified the tolerance to oxidative stress induced by either hydrogen peroxide or an alkyl hydroperoxide. Arabidopsis Trxs *m*1, *m*2, *m*4 and *x* induced hydrogen peroxide tolerance, like the endogenous yeast thioredoxins. Trx *m*3, in contrast, rendered yeast more susceptible (Issakidis-Bourguet, Mouaheb, Meyer, & Miginiac-Maslow, 2001).

In parallel, we produced all the plastidial Trx as recombinant proteins and checked their specificity in vitro, toward known targets, or putative targets involved in oxidative stress resistance. This research was led essentially by Emmanuelle Issakidis-Bourguet, who supervised the PhD work of Valérie Collin. The results confirmed the specificity of the 2 Trx f for FBPase. MDH appeared to be activatable by Trx m1, 2 and 4, but also by Trx f that seemed to be even more efficient than Trx m. In contrast, neither Trx x, nor Trx y were able to activate these two enzymes, but were efficient in donating their reducing equivalents to 2Cys-Prx (for Trx x) and Prx Q (for Trx y), two thioredoxin-dependent plastidial peroxidases (Collin et al., 2003, 2004; Lamkemeyer et al., 2006). These new thioredoxins had also a less negative redox potential. This was the first direct evidence for the role of specific chloroplastic Trx as electron donors to thiol-dependent peroxidases. This finding opened the way to a new research program, further developed by Emmanuelle Issakidis-Bourguet, and aimed at deciphering the role of Trx in oxidative stress responses in vivo in Arabidopsis, through the use of specific knockout mutant plants now freely available in libraries (Keryer, Collin, Lavergne, Lemaire, & Issakidis-Bourguet, 2004; Laugier et al., 2013; Rahantaniaina et al., 2017; Thormählen et al., 2013, 2017; Vanacker, Guichard, Bohrer, & Issakidis-Bourguet, 2018, among others).

Thus, the redox program first launched by Pierre Gadal in mid-70s developed considerably during the following decades, integrating a net of fruitful national and international collaborations. Without all these collaborations, we would have never been able to solve the complex mechanism of activation of NADP-MDH, the first example in the field of redox regulation of the so-called "intrasteric inhibition," a concept developed by Bruce Kemp for protein kinases (Kobe & Kemp, 1999). In the lab, this very active and enthusiastic research work was done in a friendly and relaxed atmosphere, with many interactions with the other members of the lab, from the morning coffee prepared by a devoted Hungarian lady, Margareth Nagy, and attended by everybody, including Pierre Gadal, to birthday celebrations with a home-made cake and champagne. Sometimes we also enjoyed excellent cakes coming from the local bakery that Bob Buchanan brought to the lab each time he was visiting us.

This research program allowed also most of our PhD students to get permanent positions either in French institutions (CNRS for Michel Droux, Emmanuelle Issakidis-Bourguet, Eric Ruelland and Stéphane Lemaire, INRA for Frederic de Lamotte-Guéry and Martine Lemaire) or abroad (a biotechnology company in Switzerland for Isabelle Schepens, Oregon University, USA, for Aymeric Goyer). Two of them (Emmanuelle and Stéphane) are still active in the redox domain and continue to develop our knowledge about the fascinating topic of redox regulation.

Pierre Gadal officially retired in 2003, but was regularly present at the Institute during several following years, as an emeritus. During these years he acted as an advisor, following the research progress of PhD students belonging to the Orsay Doctoral School of Plant Sciences and helping them find interesting post-doctoral stays. His advice was highly appreciated.

2.3.2 Michel Droux recollections from 1980 to 1988

My first memories of Professor Pierre Gadal go back to the years 1980–1981, when I was taking his courses in plant physiology at the University of Paris XI in Orsay. Exciting classes where he told us stories about the life of the plant. Always smiling, and in constant dialog with his students. At that time, I was finishing my studies (now M1) for a master's degree in Ecology of organisms and populations. In line with this aspect, I continued my studies within the framework of a DEA (now M2) in the doctoral school of Professor Miginiac (University of Paris VI), with for training course a purely biochemical subject presented by Professor Pierre Gadal. I thus followed his teaching in the field of thioredoxins (TRX) under the supervision of the future Doctor Jean-Pierre Jacquot and in collaboration with Myroslawa Miginiac-Maslow. I must confess that I had no notion of protein biochemistry and that I learned everything at the bench with the encouragement of Pierre Gadal, a human and respectful person. On the other hand, my two supervisors had to tear their hair out and had headaches. What a bad experience for Jean-Pierre, I was his first PhD student.

This scientific project focused on the study of the enzymatic mechanisms involved in the photoregulation of chloroplastic enzymes, the Ferredoxin-ThioRedoXin (FdTRX) system. My aim was to purify and characterize the ferredoxin-thioredoxin reductase (FTR). This approach was then pursued during a post-doctoral internship in the laboratory of Professor Bob B. Buchanan at the University of California, Berkeley that I cannot dissociate from my Pierre Gadal experience.

In photosynthetic organisms, light plays an essential role in the regulation of metabolism (Buchanan, 1980). This role is reflected by post-translational modifications affecting the redox state of several target enzymes, which are inactive in the dark and become active in the light. At the beginning of this work, several hypotheses were proposed to account for these photoregulatory phenomena which are dependent on photosynthetic activity and which are mimicked in the dark by the addition of a chemical reducing agent, dithiothreitol (DTT) (Jacquot et al., 1976). The three proposed systems used either redox components associated with the thylakoid membranes of the chloroplast (the Light Effect Mediator (LEM) system) (Anderson, Nehrlich, & Champigny, 1978) or located in the stroma (the Ferredoxin-ThioredoXin (FdTRX) and the Ferralterin systems) (Lara, De la Torre, & Buchanan, 1980; Wolosiuk, Buchanan, & Crawford, 1977). My thesis and post-doctoral work demonstrated that the ferredoxin-thioredoxin system is the unique physiological system that efficiently couples light energy to the photoactivation of several chloroplastic target enzymes. This system is composed of ferredoxin (Fd), ferredoxin-thioredoxin reductase (FTR) and thioredoxins (TRX m and f). At that time, thioredoxins were classified into two functional categories characterized by their target specificity, thioredoxin m associated with the photomodulation of NADP-malate dehydrogenase (NADP-MDH) and thioredoxin f modulating the activity of four of the Calvin cycle enzymes (including fructose-1,6-bisphosphatase, FBPase) (Droux, Crawford, & Buchanan, 1987; Droux, Jacquot, et al., 1987).

The study of the FdTRX system was made possible by the development of a rapid assay method for ferredoxin-thioredoxin reductase (FTR) activity. This method, called the reconstituted chloroplastic system, requires thylakoids (or electron transporters), purified protein components (Fd, FTR, TRX m or f) and a target enzyme (NADP-MDH or FBPase) (Jacquot, Droux, et al., 1984). The determination of the activity of FTR was performed by measuring its ability to promote activation of the target enzyme via TRX. The degree of activation of the target enzyme is finally assessed by measuring its catalytic activity.

The implementation of this system has led to a fine characterization of the FTR of plants of C3 type (spinach), C4 type (corn), and a cyanobacterium (*Nostoc muscorum*) (Droux, Jacquot, et al., 1987). At this stage, in the absence of molecular biology tools, FTR characterization used conventional protein purification techniques from kilograms of plants leaves (e.g. during my thesis I may have ground more than 100 k of spinach leaves). FTR is a protein that is poorly represented in the chloroplast and rather unstable. Therefore, a major advance in its purification has been achieved using affinity chromatography. This step, based on the coupling of ferredoxin (Fd) to a solid support (Shin & Oshino, 1978), allowed, in addition to the purification of FTR, the isolation of several chloroplastic proteins using ferredoxin as an electron donor for their activity (e.g. ferredoxin-NADP reductase, nitrite reductase, glutamate synthase, and sulfite reductase) (Knaff & Hirasawa, 1991). This method has thus led to the concept of the protein-protein complex (Fd-FTR) within the FdTRX system. These protein-protein interactions involve negative charges located within a domain of Fd and positive charges at the FTR level (Hirasawa et al., 1988).

FTR is a heterodimeric protein, consisting of two different polypeptides, and carrying a 4 Iron-4 Sulfur center. A comparison of the masses of the polypeptides of the protein in different species revealed the presence of one conserved polypeptide in mass and another of variable mass (Droux, Jacquot, et al., 1987). From pure proteins we prepared antibodies in rabbits. One special memory was working with Pierre Gadal on Sundays taking care of the rabbits, our antibody factories. Yes, we had to clean the cages, give them food. So there, my surprise was to see the University professor turn into a farmer wearing wooden clogs or sabots as they are known in France. No matter, we had to get down to work, always with the same joviality. My "anti-FTR" rabbits responded well, because antibodies were the basis for the notion of similar and variable subunits among three organisms analyzed (Droux, Jacquot, et al., 1987).

Furthermore, analysis of the cysteine content of these polypeptides indicated the presence of a disulfide bridge within the conserved polypeptide. This important result introduced the concept of an active subunit in the transfer of the reducing power from photoreduced Fd to TRX, through thiol-disulfide exchanges. Following the physicochemical characterization of FTR, the project consisted in elucidating, using biochemical approaches, the photoactivation mechanism of NADP-MDH and FBPase. These approaches, developed in the reconstituted chloroplast system (Droux, Miginiac-Maslow, et al., 1987), were complemented by a study in the intact chloroplast. For this purpose, a specific radioactive probe (iodoacetamide) of the SH groups was first used, but a great progress was made thanks to the introduction of thiolyte (or monobromobimane) (Crawford, Droux, Kosower, & Buchanan, 1989). This probe has the advantage of rapidly crossing cell membranes and allowing direct measurement, by fluorescence, of the redox status of FdTRX system proteins isolated from intact chloroplasts by immunoprecipitation. These experiments showed the reduction of the disulfide bridge of the conserved FTR subunit, of the canonical site of TRX, and of disulfide bridges on the target enzymes during incubation in the reconstituted chloroplast system, or of the intact chloroplasts, in light, thus clarifying the concept of photoactivation of chloroplastic enzymes (Crawford et al., 1989) (Fig. 3).

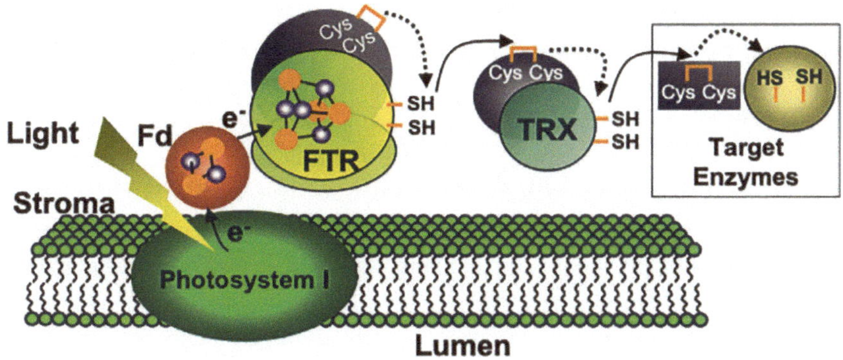

Fig. 3 Schematic representation of photoactivation of target enzymes in chloroplasts. Electron transfer from the photoreduced ferredoxin to the Fe-S center of FTR achieves light activation. FTR thus converts the electron flow into a thiol signal, which, through a thiol-disulfide exchange, reduces the active site of the thioredoxins (Dai et al., 2000; Walters et al., 2005). The thioredoxins, by a thiol-disulfide exchange, will then reduce the regulatory site of the target enzymes. This oxidation-reduction cascade of disulfide bridges leads to the transformation of an inactive target enzyme present in the dark into an active target enzyme in the light.

Beyond science, Pierre Gadal helped me a lot during my thesis, and even beyond since he visited me in Berkeley and then we attended a Gordon conference together. He introduced me there to another University Professor, Roland Douce and his colleagues, a move that proved determinant for obtaining my future CNRS position and thus the rest of my career.

2.3.3 Emmanuelle Issakidis-Bourguet: Recollections from 1986 to 2003 and beyond

I met Pierre Gadal in 1986 as my Professor in Plant Physiology at Orsay University (Paris-Sud 11). I don't have a clear memory of his lectures, but for sure they confirmed my choice to study biology at the molecular level. I have been attracted to plants since my childhood, and studying plant molecular functions has proven to be really exciting to me. I also don't remember exactly how I managed to get a 6-week laboratory internship in Prof. Gadal's lab, but what is certain is that it determined the outcome of my studies and my entire subsequent scientific career.

During this internship, mentored by Pierre Gadal, I was supervised by his PhD student Ridong Chen. This was a very nice scientific and human experience. I loved the experimental action and the intellectual activity which came with it. The subject of the internship was to measure and compare NADP-dependent isocitrate dehydrogenase activities and immuno-reactivities with an antibody obtained by Ridong, from different biological materials. This work involved not only plant extracts but also extracts from bacteria and animal organs. I remember Ridong arriving at the lab in the morning with bloody animal samples in his ice box, which he had retrieved from colleagues in the neighboring Animal Physiology laboratory. It was quite difficult for me to start the working day with this type of material. I think that this practical experience definitely convinced me that I would be working on plants and it made me want to make a career out of it later on.

We, with Ridong and Claire, another undergraduate student I was paired with, did a job good enough allowing our data to be published 2 years later (Chen, Bismuth, Champigny, & Gadal, 1989; Chen et al., 1989). How proud I was of this very first personal contribution to science! I wanted many more to come afterwards... As I mentioned, this first internship was also a very positive experience on a human level. I immediately liked the working atmosphere of a research laboratory. Its international dimension, Ridong was Chinese, the visits of collaborators from abroad. I really liked this multicultural context. In Pierre Gadal's lab, there was a very good working atmosphere, lots of exchanges over coffee, celebrating personal events (birthdays, baby births and weddings...). Prof. Gadal was attentive to everyone, always kind and caring with all lab members.

I very much wanted to come back to work in his laboratory, but in order to be sure about my career choice for the future, for the next 2 years I pursued my studies at Orsay University and did voluntary internships in laboratories where the research was more applied (at INRA and in a private company), to finally confirm my interest for fundamental research.

I chose specializing in biochemistry and plant physiology and once more had Prof. Gadal as teacher (course on plant metabolic pathways). Logically, in 1989, I applied for the DEA (Diplôme d'Etudes Approfondies) which, I think, P. Gadal had created and co-directed with Prof. Francis Quétier. This training course was the equivalent of a Master's degree specialized in *Plant Molecular and Cellular Biology* and was ending with a 6-month full-time internship. My plan was to be back working in P. Gadal's lab on the isocitrate dehydrogenase project. But it happened that P. Gadal proposed an internship to one of my classmates who came from a collaborating Spanish team and he reassured me that the research project he directed could give rise to two internship subjects. I remember that Francis Quétier, concerned about the continuation of my Master's degree and the possibility of pursuing a PhD thesis, strongly advised me to consider another internship subject proposed in the same laboratory. I followed his advice and this is how I started working on NADP-dependent malate dehydrogenase (NADP-MDH), initially under the direction of Jean-Pierre Jacquot and Claude Crétin. Looking back, this change of plan, which at the time had frustrated me, was a great opportunity. Claude had isolated the cDNA of NADP-MDH from sorghum and Jean-Pierre cloned it in a vector for expression in *E. coli*. My mission was to increase the production rate of the recombinant enzyme in order to facilitate its purification and obtain it in quantities large enough to conduct the structure-function study of its regulatory mechanism by a site-directed mutagenesis approach. This was done and it paved the way for my thesis project...

At that time, obtaining a site-specific modified protein was a long-lasting and hard task. PCR was not readily available, synthetic DNA production was just starting and not commercially available. Notably, automated DNA sequencing was not accessible yet. Following the method described by Kunkel in 1985, to introduce a point mutation in a cDNA, it was necessary, first, to obtain a single stranded DNA version using bacteriophage technology, second, to perform its in vitro copy using a mutagenic oligonucleotide (primer extension), third, to transfer it into a bacterial strain where the uracil-containing parental DNA strand (bearing the starting wild-type sequence) would be counter-selected. All this process had a limited efficiency (50–60% in my hands) and the suspense about obtaining a mutagenized DNA lasted until getting DNA sequencing results. Sequencing was another fastidious task in the process since it was manually operated. We used to perform Sanger sequencing with 35S labeled ATP and handling the thin layer acrylamide gels was always a risky step. A mutant clone bearing the desired modification once identified, the cDNA (or possibly a part of it when a judicious sub-cloning could be done) had to be entirely sequence verified prior to transfer into an expression vector for recombinant protein production. Nowadays, mutagenesis is performed using PCR and sequencing is custom realized; chemical synthesis of a site-specific modified DNA sequence encoding a whole protein has even become financially affordable and the development of the CRISPR/Cas9 technology now allows mutagenesis to be performed in vivo. In the early 90s, at the time of my PhD work, tagged protein technology to simplify purification by affinity chromatography had only just been developed and had not yet been disseminated. Purifying a mutated recombinant protein could be another fastidious task. For NADP-MDH, the purification procedure needed a number of days of experimental work including several ammonium sulfate fractionations and chromatography steps. One can then imagine the satisfaction of obtaining a purified mutated protein, when the most interesting part of the work could finally start...

The NADP-MDH project in P. Gadal's lab has been a real saga, i.e. by definition a long story, with a series of incidents and sort of heroic achievements. Myroslawa Miginiac-Maslow has described the whole project to which I had the chance to contribute. NADP-MDH is reductively activated in the light by the ferredoxin/thioredoxin system and it has the particularity, among thioredoxin-dependent enzymes, to be totally inactive in its oxidized state. Previous work from Paulette Decottignies and Myroslawa had revealed that the two cysteine residues located near the N-terminus of the enzyme were reduced upon its light activation, strongly suggesting that they could be linked in a disulfide bridge. My PhD thesis work, under the supervision of Myroslawa, started with the production of mutant proteins where these putative regulatory cysteines were substituted, separately or simultaneously. Their replacement gave enzymes that still needed a reducing treatment to be active (incident!) but we noticed that their activation kinetic was much accelerated in comparison with the wild-type enzyme. Then, I created new mutated proteins where the two conserved cysteines located in the C-terminal part of NADP-MDH, corresponding to a sequence extension specific to the redox regulated MDH isoforms, were substituted. These proteins still needed a reducing treatment to be fully active and their activation was as slow as the wild-type enzyme, but they showed a substantial activity level (ca. 20% of the maximum activity) in the oxidized state, strongly suggesting that they could be involved in the activation process, possibly in a second regulatory disulfide (Issakidis et al., 1992). Then, by combining the mutations of both the N- and C-terminal cysteines, we could obtain a thioredoxin-independent fully active NADP-MDH (achievement!) (Issakidis et al., 1994; Issakidis, Decottignies, & Miginiac-Maslow, 1993).

Then, in 1994 I went to Australia for a post-doctoral period in J. Andrews' laboratory (ANU Canberra) when the new building 630 on the Orsay university campus hosting IBP—Institut de Biotechnologie des Plantes—founded by P. Gadal and F. Quétier was completed. When I came back, I was able to get a CNRS permanent position in the team directed by Myroslawa with a research project on redox regulation by thioredoxins, including my favorite target enzyme. Since my PhD work, we had the ambition to transfer the redox control properties of NADP-MDH to a thermostable MDH isoform. To do so, I had visited Prof. T. Beppu's laboratory in 1992, at the occasion of my trip to Japan attending the IXth International Congress on Photosynthesis, which provided us with the cDNA clone for NAD-MDH from the thermophilic bacteria *Thermus flavus*. Thanks to the accumulation of data obtained over a period of almost 15 years by me and the other subsequent PhD students supervised by Myroslawa or Paulette, and the valuable help from our collaborators, experts in structural biology, we were able to reach our goal (achievement!). I personally regret that this accomplishment did not receive the interest that, in my opinion, it should have had. The corresponding paper was published in 2006 (Issakidis-Bourguet et al., 2006), after several attempts at publication in biochemistry-oriented journals where one of the reviewers even pointed out that he/she could not see the interest of our work. Pierre Gadal certainly was not one of those "pitbull reviewers." Although he could not be directly involved in all the research projects underway in his laboratory, he always kept updated about everyone's progress. His criticism was always constructive. He used to sow the seeds of doubt and reflection in your mind and would come back to you later to check whether they had germinated...

At the beginning of my PhD work, obtaining purified recombinant NADP-MDH protein in large quantities allowed us to produce antibodies. We had them made by rabbits raised at the animal house of the laboratory of Animal Physiology of the Orsay campus. Jean Vidal helped me for the injections of the rabbits, and the final step of their bleeding, which was quite disturbing, not only for the rabbits... They used to show resistance. I still have a scar on my forearm in memory of the "MDH rabbit." No one wanted to finish off the rabbits. It was said that P. Gadal was in charge of it and that the rabbits would eventually end up on his plate... If that were true, I do not think it showed his lack of sensitivity, but rather a reference to his rural origins.

Over recent years until 2017, I met him from time to time at IPS2. The Institute of Plant Sciences Paris-Saclay, housed at building 630, was created in 2015 from the combination of 3 plant research institutes—including IBP—located in the southern Paris region. P. Gadal was passing by to give and enquire as to the latest news, with his cans for softened water for his fish tanks.

In retrospect, I realize that Pierre Gadal was present at all the important steps of my student (first internship, Master, PhD: he was President of my defence jury) and professional (CNRS position) lives, and I'll always be most grateful to him for all the support and opportunities he gave me. I will warmly remember Professor Gadal as a good person and a great scientist.

2.3.4 Martine Lemaire-Chamley: Recollections from the Orsay era (1991–1997)

Was it a long time ago? Yes, it was in another century, between 1991 and 1997, when I discovered Research and Plant Biotechnologies in Orsay. At that time, plants and enzymes were my focal points. Thus in 1991, I performed my first 8-week training in the Plant Physiology laboratory headed by Pr. Pierre Gadal. I started my research activity in a cold room to purify plant enzymes and then in front of the spectrophotometer to measure their enzymatic activity, under the supervision of Paulette Decottignies. Of course, after this very interesting and successful internship, I wished to dedicate myself to scientific research and therefore I applied for the "DEA in Plant Molecular and Cellular Biology" directed by Prs Pierre Gadal and Francis Quétier and was very happy to be selected. Plant enzymology was not that much fashionable at that time, when molecular biology and bioinformatics were starting to take-off in the plant field. However, I managed to obtain a PhD grant to continue my work, still on the Orsay campus, first in the "valley" in building 430, then at the Institut de Biotechnologie des Plantes on the Saclay plateau, an institute that Pierre Gadal initiated and co-supervised the construction of. There, we benefited from unbelievable working conditions, with so much space and functionality. For a PhD student, Pierre Gadal was a very impressive person because of his responsibilities, of course: he was at the top of our hierarchical pyramid and at the interface with the rest of the scientific or administrative world... Tall and rapid in his movements, his speciality was to appear suddenly in the corridor to ask for someone or something. He scared me a couple of times, at night in the lab, when I emerged from the cold room where I was watching my chromatography column and almost bumped into him. He made me nervous by searching frantically for someone to carry important files to the ministry the day of the deadline. But he was also a very caring person, very concerned by his staff: he moved me to tears during my thesis defense, evoking a colleague who passed away way too early. I was proud that he trusted me enough to offer me a short-term teaching position as an assistant professor (ATER) after my post doc in Germany, before I applied successfully for a permanent position at the INRA. In brief, through his goodwill to everyone and his investment in research and teaching management, Pierre Gadal, together with the permanent researchers present in the lab at that time, showed me how a scientific lab could be run properly, in a fully serene working environment. I am very grateful to him for his outburst of good vibes during the first years of my scientific career.

2.3.5 *Aymeric Goyer's recollections of the Orsay era (1998–2001)*

When I think about persons who have made a difference in my life, Pierre Gadal comes immediately to my mind. I did my DEA (eq. MSc) and PhD research theses at the Institut de Biotechnologies des Plantes that Pierre Gadal founded, in the group directed by Myroslawa Miginiac-Maslow, between 1998 and 2001. My first encounter with Pierre Gadal was during my interview in the spring of 1997 to get into graduate school in the Plant Molecular and Cellular Biology program at the University of Paris XI. I was finishing my undergraduate studies at the University of Rennes 1 in Bretagne (Brittany), and I had applied to several graduate schools throughout France. The Plant Molecular and Cellular Biology program was my first choice because of the plethora of top plant science laboratories involved, including Pierre Gadal's lab. The interview was a short 10–15 min round-table with the members of the selection committee. I had never met any of those members before, but each had a nametag in front of them. When members would ask questions, I would take a pick at their nametags, and I quickly realized that many of the top plant scientists from the region were gathered right there in front of me. I still remember Pierre Gadal's question about my internship at Bretagne Biotechnologies Végétales, a question that was indirectly related to Brittany of course, a region where we both are originally from (Breizh da viken!!!). Soon after, I would learn that I had been accepted in the Plant Molecular and Cellular Biology program. So, I moved to the beautiful woody Orsay campus and started to take courses and to rotate between labs at several scientific centers throughout the Paris region, to eventually settle down in the Redox Regulation group at the Institut de Biotechnologies des Plantes where I would do my thesis research for the next 4 years. To me, this lab offered the right balance of biochemistry, molecular biology, and plant physiology that I was looking for, and the people (faculty, technicians, students) made me feel that it was the right environment for me to go through the stressful years of graduate school. There, I studied the regulation of the activity of NADP-malate dehydrogenase, a mechanism that involves molecular interactions of the enzyme with thioredoxins via the formation of specific cysteine disulfide bridges. I made my first steps into molecular biology by creating mutations in proteins coding sequences by site-directed-mutagenesis, and by verifying the mutations in the plasmid sequences by running long polyacrylamide gels, a tedious protocol that would astonish the current generation of graduate students. I would then produce the corresponding recombinant proteins in *E. coli* and purify the enzymes by using a series of size exclusion, ion exchange, and affinity chromatography columns. Once purified, I would prepare mixtures of mutated thioredoxins and NADP-malate dehydrogenase, and analyze the resulting complexes by SDS-PAGE and HPLC. The lab had become expert in all these techniques, capable of obtaining several milligrams of highly purified proteins. So, my first experiences at the bench went without major roadblocks, and in just a few months, I was lucky enough to have my research results published (I would learn later on that research is often made of seldom ups and long excruciating downs). During these years, I had the chance to get to know Pierre Gadal better. Although he was quite intimidating at first, he was somewhat the opposite of the first impression he projected, someone very approachable and always happy to converse, whether it be in the lab (unaware of how unsettling interrupting a graduate student running an experiment can be), in the hallways of the institute, or at the coffee breaks. During one of those conversations, after returning from the Plant Biology 2000 annual meeting in San Diego, CA, where he received the Corresponding Membership Award, he informed me of an open post-doctoral position at the University of Florida in the laboratory of Andrew Hanson whom he had met there, and encouraged me to apply. Which I did. I then moved to Florida in January 2002 shortly after receiving my PhD degree, a decision that drastically changed the course of my professional and personal life.

3. Nitrogen metabolism pathways and regulation

3.1 Akira Suzuki's recollections

Pierre Gadal, Professor at the University of Nancy I and later Paris XI, founded a laboratory of international reputation in plant science. I have been privileged to have worked with him in this laboratory for 3 years at the Faculty of Sciences, University of Nancy I (presently University of Lorraine) in 1978–1981 and then for 3 years at the Faculty of Sciences, University of Paris XI in 1981–1984. As initial studies are determinant for the subsequent development of research, my recollections here describe what happened in the first years-work in this lab. Besides, a review article is dedicated to Pierre's memory (see the chapter "Glutamate synthase and amino acid synthesis in higher plants" by Suzuk). The review refers to the study of the primary nitrogen metabolism in his lab in the 1970s and the early 1980s together with the outcomes of worldwide researchers working in the field of primary nitrogen metabolism.

3.1.1 Early years in Nancy

It all started by an appointment in the office of Prof Pierre Gadal in the Faculty of Sciences, University of Nancy I. After the interview, I came out of his office located at the end of fifth floor of the circular building, holding in my hands two recent master theses on nitrogen metabolism defended by two students in his lab. This happened in 1978. The study was initiated in the Plant Biology Laboratory (ERA au CNRS 799) as a DEA (today Master) student after receiving a Bachelor of Science degree in plant biology at the Faculty of Science, Tohoku University, Japan. My first laboratory notebooks delineate the daily notices of basic protocols and more advanced methods on nitrogen metabolism.

Year 1978:

- September 11: Seeding cultures of five plants (maize, rice, bean, pea and barley) on Petri dishes using a medium of Hoagland and Arnon (1938) or Lockard (1959).
- September 27: Crude enzyme extraction and assays of nitrate reducase (NR), nitrite reductase (NiR), glutamate dehydrogenase (GDH), glutamine synthetase (GS), glutamate synthase (GOGAT), glutamate-oxaloacetate transaminase (GOT), and organelle markers of triose-phosphate isomerase, fumarase and catalase.
- September 27: Spectroscopic determination of protein concentrations (Scopes, 1974).
- October 11: Organelle/cytosol localization of the enzymes by differential centrifugation.
- October 13: Biuret-Folin colorimetric assay of soluble proteins (Lowry, Rosenbrowgh, Farr, & Randall, 1951).
- October 26: Organelle localization of the enzymes by ultracentrifugation on linear or discontinuous density sucrose gradients.
- November 23: Memo "Could you prepare 20 Petri dishes and place maize seeds today? This will allow me to do experiments on roots on Saturday." Pierre Gadal sometimes tried to do experiments himself despite his numerous national and international commitments.
- November 28: Appending asparagine synthetase (AS) and carbamoylphosphate synthase (CPSase) to the localization study.

Year 1979:

- March 13: Assay of Fd-GOGAT by separating glutamate (Glu) from glutamine (Gln) by paper chromatography for 4 h (phenol/H_2O: 4/1, v/v) and measuring Glu and Gln by ninhydrin.

- March 19: Profiling of activity of NiR, Fd-GOGAT and NADH-GOGAT in greening etiolated leaves for 30 min, 1 h, 2 h up to 72 h.
- April 23: Spectrophotometric GOGAT assay by measuring reduced MV oxidation under anaerobic conditions (Asada, Tamura, & Bandurski, 1969).
- May 14: Profiling of activity of NiR and Fd-GOGAT using cycloheximide to inhibit protein synthesis on cytosolic ribosomes in greening etiolated leaves.
- May 28: Profiling of activity of NiR and Fd-GOGAT using chloramphenicol to inhibit protein synthesis on plastidial ribosomes in greening etiolated leaves.

During all this time, the lab members had the advantage of interacting with visiting scientists from France and abroad. Since the 1970s, Pierre Gadal worked with leader scientists from the UK and Canada in the field of nitrogen metabolism. He and the CNRS researchers of the lab made available a range of analytical tools, skills and competence in the study of isoenzymes of nitrogen assimilation: isolation and measurement of inorganic and organic molecules and proteins, chromatography by gel filtration, ion exchange and hydroxylapatite, electrofocalization, PAGE, etc., used for the PhD projects. In a series of experiments, we reported the localization of the key enzymes in the plastids, mitochondria, peroxisomes and cytosol of the non-photosynthetic roots of five plants (Suzuki, Gadal, & Oaks, 1981).

Meanwhile, Pierre Gadal launched a new approach of immunocytochemical localization of the enzymes implicated in the carbon metabolism, one of the two axes of the laboratory. Magnificent and specific images of the indirect immunofluorescent localization of the enzymes of CO_2 fixation by a lab mate unequivocally clarified the carbon metabolic networks in the leaf. The lab members shared this new concept and technology, and each of us carried out the purification of proteins of interest to homogeneity, immunization of rabbits by initial and booster antigen injections, blood collection and IgG isolation, etc. Years later, immunochemistry became one of the indispensable tools of plant biology to study target proteins, e.g., in screening of recombinant DNA libraries and DNA cloning, immunohistochemistry, quantification of specific proteins, identification of cross-reacting materials including recombinant proteins, immunotitration of proteins such as in vivo synthesized proteins marked with labeled amino acids, etc.

3.1.2 Later in Orsay

Following the appointment of Pierre Gadal as Professor in Orsay in 1981, most of the lab members moved with him to the Plant Molecular Physiology Laboratory (ERA au CNRS 799) at the Faculty of Sciences, University of Paris XI. After installation in Building 430, he organized a series of visits: Institute of Nuclear Physics (Building 100) and Linear Accelerator Laboratory (Building 200) located on the campus along the Yvette river, INRA Versailles (present INRAE), etc. In the subsequent years, the INRA Versailles center has become the site of my research activities after I became an INRA researcher.

The study in Orsay aimed at answering the biological questions about GOGAT and primary nitrogen metabolism.

- Since the discovery of Fd-GOGAT in pea leaves in 1974 (Lea & Miflin, 1974), the coexistence of Fd-GOGAT and NAD(*P*)H-GOGAT activities was reported in non-photosynthetic roots (Miflin & Lea, 1975). We provided evidence that Fd-GOGAT and NADH-GOGAT were distinct proteins encoded by separate genes in a variety of plant species (Suzuki, Jacquot, & Gadal, 1983; Suzuki, Vidal, & Gadal, 1982; Suzuki, Vidal, Nguyen, & Gadal, 1984).
- The structure and kinetic properties of the major Fd-GOGAT in the green leaves were characterized using the purified Fd-GOGAT protein (Suzuki & Gadal, 1982).
- The systemic in silico analysis reviewed the studies of structure, function and regulation of Fd-GOGAT and NAD(*P*)H-GOGAT in animals, archaea, bacteria, yeasts, cyanobacteria, algae and higher plants (Suzuki & Gadal, 1984).
- Other questions remained unanswered. Since the discovery of Fd-NiR activity in barley roots in 1966 (Joy & Hageman, 1966) and Fd-GOGAT activity in pea roots (Miflin & Lea, 1975), the physiological function of these Fd-dependent enzymes remained unanswered because there was no report in roots either on redox proteins analogous to leaf Fd or concerning an electron transport system independent of the photochemical reactions. We identified two novel proteins in the roots: Fd-like electron carrier (REC = root electron carrier) and ferredoxin-NADP$^+$ oxidoreductase (FNR)-like enzyme (PNR = pyridine nucleotide oxidoreductase). Without light energy in the roots, electrons are transferred, in the reverse flow compared to the situation around chloroplast PSI, from

NADPH to PNR and then to REC for the reactions of NiR and GOGAT (Suzuki, Oaks, Jacquot, Vidal, & Gadal, 1985).

- As the environmental elements, e.g., light and nutrients are key signals to regulate inorganic nitrogen assimilation into amino acids, and plant growth and productivity (Beevers, 1981), we studied the light induction of Fd-GOGAT activity in greening etiolated leaves (Suzuki et al., 1982).

In these years in Orsay, the laboratory stepped into an advanced phase of study in the redox regulation and the molecular biology and genetics of the enzymes, and these pioneer studies were led by the groups of the carbon metabolism. Over this period, several lab mates worked week-ends as well to keep ongoing experiments as part of the overall effort. Pierre Gadal passed from time to time on Saturday and/or Sunday and exchanged with us on a variety of subjects or via a brief greeting when we were doing bench work. Sometimes, he left directly to the outdoor greenhouse behind Building 430 to relax himself through gardening, a favorite practice that he missed during his busy days. It is a contagious and distinctive trait of him to be able to share with the lab members his dedication to research by making them realize that what they undertake is indispensable. Such a sense of trust created a free atmosphere in the lab, as recalled by a former Master student of Nancy in her thesis preface: "I worked with my colleagues of the laboratory in a friendly atmosphere," which sounds commonplace but is of high value (this was in one of the two master theses that Pierre Gadal handed to me in 1978).

3.1.3 In Versailles

The regulatory mechanisms in response to light and metabolites are major tools of the molecular and genetic studies in the analysis of nitrogen assimilation, biomass and plant productivity (Crawford & Arst Jr., 1993; Thomsen, Eriksson, Moller, & Schojoerring, 2014). The light and metabolite signals control the primary nitrogen metabolism throughout plant development in germinating seeds, vegetative organs, senescing and reproductive organs, and maturing seeds (Gaufichon, Reisdorf-Cren, Rothstein, Chardon, & Suzuki, 2010; Ireland & Lea, 1999). By molecular and genetic approaches, we dissected the regulation of isoenzymes of nitrogen assimilation located in the distinct cells, organelles and tissues by analyzing specific transgenic plants and mutants relative to:

- vegetative plants for the complex nitrogen flow from the NO_3^- reduction, photorespiratory NH_4^+ cycle and translocation of amino acids within the plant (see Review in the chapter "Scientific contributions

of Pierre Gadal and his Lab. Nitrogen metabolism pathways and regulation—Akira Suzuki's recollections" by Suzuki) and
- reproductive plants for the synthesis of Gln, Glu, asparagine (Asn) and aspartate (Asp) and their remobilization within the plant.

We were able to show that light induces the specific mRNA transcription and protein synthesis of Fd-GOGAT in greening etiolated leaves: the study was carried out in a Canadian lab working with Pierre Gadal since the 1970s (Suzuki, Audet, & Oaks, 1987). It was later demonstrated that phytochromes mediate the induction of mRNA and protein of Fd-GOGAT via a reversible low fluence red/far-red treatment whereas NADH-GOGAT remained at a constant level (Suzuki et al., 2001; Ziegler et al., 2003). Using two different monoclonal antibodies against tobacco Fd-GOGAT, a competitive ELISA assay showed that the Fd- and MV-interacting domain share a common epitope (Suzuki, Vergnet, Morot-Gaudry, Zehnacker, & Grosclaude, 1994).

The first molecular study of GOGAT reported in 1987 concerned the cloning of a 6.3kb NADPH-GOGAT gene of *E. coli* containing the 5′- and 3′-flanking regions and two ORFs for α large subunit (1514 amino acids of 166kDa) and β small subunit (471 amino acids of 52kDa) (Oliver et al., 1987). In the second half of the 1980s, increasing molecular and genetic studies were undertaken by different groups for enzymes of nitrogen assimilation in plants (Lam, Coschigano, Oliveira, Melo-Oliveira, & Coruzzi, 1996). Together with a Canadian group we succeeded at isolating a cDNA cloning for maize Fd-GOGAT (Suzuki, Burkhart, & Rothstein, 1996) and then shifted to that of Arabidopsis as a model plant by working in this same lab in Canada. The cDNA sequence of Arabidopsis Fd-GOGAT was determined by screening cDNA clones isolated from a lambda ZAP II library and genomic clones from a lambda GEM-11 library, and the sequencing was carried out on both strands by the dideoxy-nucleotide chain termination method (Sanger, Nicklen, & Coulson, 1977). The complete cDNA of 5178bp of *GLU1* gene, derived from a partial cDNA (4458bp) complemented by the 5′ sequence of a partial gene (1216bp), encodes a single peptide of 1648 amino acid-protein (180kDa) including the 131-amino acid transit peptide (14.6kDa) (Suzuki & Rothstein, 1997). The structural comparison indicated that *GLU1* Fd-GOGAT may have originated from the bacterial α large subunit (Suzuki & Rothstein, 1997). *GLU1* Fd-GOGAT shares the N-terminal class II glutamine-aminotransferase (GAT) domain with *E. coli* NADPH-GOGAT (Oliver et al., 1987; Suzuki & Rothstein, 1997). Analysis of the Arabidopsis Genome Initiative (2000) indicated a 32 exon-intron structure of 9.57Kb for the *GLU1* gene. Another Fd-GOGAT gene was suggested by our Southern blots of the restriction fragments of genome (Suzuki & Rothstein, 1997). Afterwards, the second *GLU2* gene of Arabidopsis Fd-GOGAT was characterized (Coschigano, Melo-Oliveira, Lim, & Coruzzi, 1998).

The cellular functions are tightly related to the cell-specific location. We dissected the spatial expression profile of Fd-GOGAT in the leaves by analyzing the Arabidopsis transgenic plants expressing a β-glucuronidase (*GUS*) reporter gene. Transgenic plants were generated by Gateway approach. The recombinant constructs of *GUS* fused with either *GLU1* promoter (2362 bp) or *GLU2* promoter (1580 bp) were electroporated into *Agrobacteria*, which were then inserted into the Arabidopsis genome by dipping floral tissues (Potel et al., 2009). *GUS* expression in mesophyll is consistent with the role of Fd-GOGAT in the primary NH_4^+ assimilation (Ireland & Lea, 1999) and photorespiratory NH_4^+ re-assimilation (Feraud et al., 2005; Somerville & Ogren, 1980). Moreover, *GUS* expression occurred in the phloem companion cell-sieve element (CC-SE) complex of minor leaf veins where signals of in situ mRNA hybridization visualized by a microscopy and the specific immunofluorescence by a confocal laser-scanning microscopy overlapped (Potel et al., 2009).

In the senescing and reproductive organs, the GS1/Fd-GOGAT cycle plays the primary role to maintain the Glu level in the phloem for nitrogen remobilization to the sinks (Masclaux-Daubresse et al., 2006). Four amino acids, Gln, Glu, Asn and Asp account for more than 50% of the total amino acids transported in the phloem (Gaufichon et al., 2013; Ireland & Lea, 1999; Valadier et al., 2008). Although the general view is that Asn is a long-distance nitrogen carrier, little is known about where Asn is synthesized, metabolized and translocated during the seed nitrogen filling. Asparagine synthase (AS) is encoded by a small multigene family in plants (Gaufichon et al., 2010), and *ASN1* encoding AS was more highly expressed in the reproductive organs of Arabidopsis than the other two genes (*ASN2* and *ASN3*) (Gaufichon et al., 2013).

Our molecular and genetic studies provided evidence for the roles of Asn in the reproductive organs during seed nitrogen filling. Quantitative changes of mRNA and metabolites were analyzed in the floral organ at three embryonic stages from ovule (stage 0), globular (stage 1) and heart (stage 3) stages, and in the silique from heart (stage 4), torpedo (stage 7) and cotyledon and mature (stage 11) stages of Arabidopsis (Gaufichon et al., 2017).

- The profiling by HCA (hierarchical cluster analysis) and PCA (principal component analysis) was carried out by measuring mRNA by real-time

RT-PCR for 16 selected genes of nitrogen and carbon metabolism such as *GLN2*, *GLU1*, *ASN1/ASN2/ASN3*, ASPGA/ASPGB for cytosolic asparaginase (ASPG) releasing the Asn-amide group, and *AGT1* for peroxisomal asparagine aminotransferase (AsnAT) using the Asn-amino group (Gaufichon, Rothestein, & Suzuki, 2016). The onset of pollination caused a distinct mRNA clustering at stage 0 compared to the other stages due to the enhancement of mRNAs for *GLU1*, *ASN1*, *AspAT2* for aspartate aminotransferase (AspAT), *ASPGA1*, *ASPGB1* and to a lesser extent of *AGT1* (Gaufichon et al., 2017).

- The profiling by HCA of 137 metabolites annotated by a gas chromatography-mass spectroscopy (GC–MS) differentiated stage 0 from the other stages according to sample maps using a distance linkage in the dendrogram and color density. This specificity was accompanied by the accumulation of Asn and the decline of Gln, Glu and Asp toward stage 1 (Gaufichon et al., 2017).

The spatial distribution of *ASN1* AS in the CC-SE complex detected by confocal microscopy in the silique cell wall was found to extend over the septum of the ovary along the central vascular strand, which branched to the ovule via the funiculus as well as the outer and inner seed integuments (Gaufichon et al., 2017). These results provided evidence that the post-phloem Asn synthesis in the CC-SE complex in the silique and septum meets a high demand for Asn for transit to its symplastic extension of seed integument during ovule differentiation into the embryo. Besides, the induction of *ASPGA1*, *ASPGB1* and to a lesser extent *AGT* at the ovule stage implies an efficient flux of the amide and amino groups of Asn for Asp synthesis during the seed nitrogen filling. This Asn turnover is consistent with the phenotype of promoter *Napin2S::ASN1* lines expressing *ASN1* later up to the desiccating stage of seed maturation (*Napin 2S* for storage 2S albumin) (De Silva Conceição & Krebbers, 1994). The wild-type seed yields of the promoter *Napin2S::ASN1* lines (seed weight and seed number per silique) indicated that the promoter *Napin2S::ASN1* lines were impaired at the timing of Asn synthesis in the embryonic cells for flux of its amide and amino groups, while the *CaMV35S::ASN1* overexpressors had 5–9% higher seed yields (Gaufichon et al., 2017).

The other enzymes of NH_4^+ assimilation are co-localized in the companion cells including GS1, Fd-GOGAT, NADH-GOGAT, NADH-GDH and CPSase (Carvalho, Pereira, Sunkel, & Salema, 1992; Potel et al., 2009). As the companion cells in the photoheterotrophic reproductive organs contain a heterogenous mixture of photosynthetic chloroplasts and non-photosynthetic plastids at different sites of the same organ (DeWitt & Sussman, 1995), the amino acid synthesis in these cells requires an electron transfer system via NADPH/FNR/Fd in contrast to the chloroplast PSI driven as demonstrated in the roots (Suzuki & Knaff, 2005; Yoneyama & Suzuki, 2019).

An impact of Asn deficiency of the *ASN1* T-DNA mutant lines (*asn1*) was found in the defective differentiation of embryos from the images of whole-mount *asn1* embryonic cells, stained by pseudo-Schiff propidium iodide and analyzed by confocal microscopy according to Truernit et al. (2008). The aberrant embryo cell divisions in the upper suspensor cell layers (diagonal/additional divisions) and hypophysis surrounded by embryo proper and suspensor (transversal division) from globular to heart could be related to a deficiency of Asp as its entry into carbamoylphosphate is a limiting step in the uridine nucleotide synthesis (Gaufichon et al., 2017; Schröder, Giermann, & Zrenner, 2005).

Over this period, Pierre Gadal and I had many occasions to meet in Orsay, Versailles, Gif-sur-Yvette, Toulouse, Paris, Nara, Brittany, Nancy, etc. It was always a pleasure to witness his enthusiasm for research and to follow his comments and discussions. In the meantime, he accepted the demand of our scientific committee of France-Japan symposium organization, and coordinated the French delegation to the bilateral symposium at Nara, Japan (October 29–31, 2002). More than 100 attendees joined the meeting along with presentations of 18 and 16 leader scientists from France and Japan, respectively. He and his colleagues from longstanding collaborations confirmed that the two scientific communities share complementary research and methodologies and a common interest in certain model plants and crop species (CNRS-IRN France-Japan Frontiers in Plant Biology program 2020–2024). His advices to our scientific committee served for the subsequent symposia in Toulouse in 2005 (October 25–29, 2005) and Tsukuba, Japan in 2008 (October 19–24, 2008). He had already been in Tsukuba in the early 1980s and briefly shared a bit of springtime with my family on the way to Tsukuba to visit the newly opened research center on the campus. He paid attention to details for lab members such as birthday celebrations, afternoon coffee breaks, numerous defense parties in the lab, etc. As far as I am concerned, his influence was determinant for the renewals of fellowship from the French government and short-term CNRS position, planning of postdoc stay, unwavering support for my application for an INRA position, elaboration of the symposium frames, etc. More recently, his dedication to research was still apparent as he agreed to be the Series Co-Editor of *Advances in Botanical Research* during 8 years (2012–2019) despite his long disease.

In September 2019, my wife and I went to visit him in the Cochin Hospital at Port-Royal. The day before, Mrs. Françoise Gadal confirmed that we could visit him. At the RER railway station, we found Mrs. Gadal in front of us. Her train from La Hacquinère should have coincided with the arrival of our train from Saint Michel-Notre Dame. She showed us the way to the Hospital that she took every day. In his room, he was checking the agricultural training report of his grandson. By staying with him and Mrs. Gadal for a while, we remembered about our walk with them on the Bayard street of Paris in the spring of 2017. They wished to visit the neighboring museum and its surroundings by extending the walk after lunch. We were very pleased to observe his large strides of walk, despite his battle against the disease, similar to some time ago at Foesnant, Beg Meil and Ile Tudy in Finistère. Also, we called to mind a remembrance that he enjoyed a laborious pruning of the tall trees in his vast nursery close to the home in Brittany.

3.2 Bertrand Hirel's recollections

In 1974, the chloroplast enzyme Ferredoxin-dependent glutamate synthase (Fd-GOGAT; E.C. 1.4.7.1) was identified in plant tissues by Peter Lea and Benjamin Miflin, two researchers based at Rothamsted Experimental Station in Harpenden, UK. This discovery drastically changed the existing ideas regarding the pathway of ammonium assimilation in plants (Lea & Miflin, 1974). Before this discovery, it was generally considered that glutamate dehydrogenase (GDH, E.C. 1.4.1.3), an enzyme that had been known in plants since 1938 as reported by Smith, Austen, Blumenthal, and Nyc (1975), was chiefly responsible for the incorporation of nitrogen (N) into the α-amino group of amino acids using NADH as a cofactor (Bulen, 1956). In addition, Lea and Miflin's research demonstrated that most of the ammonium that is metabolized by higher plants is assimilated via the glutamine synthetase/GOGAT (GS/GOGAT) cycle. This cycle carries out the incorporation of inorganic N into glutamate and glutamine, the two amino acids that are the fundamental molecules used for the synthesis of virtually all the compounds in plants that contain N. An important role for the enzyme GS was also proposed in the ammonia assimilatory pathway (E.C. 6.3.1.2), (Elliott, 1953). Later on, evidence for the key role of the GS enzyme demonstrated that irrespective of their origin, (nitrate reduction, dinitrogen fixation, photorespiration, protein and amino acid N recycling) ammonium ions are incorporated into the amide position of the amino acid glutamine in the presence of glutamate, via an ATP-dependent reaction catalyzed by the enzyme. This reaction, in conjunction with that catalyzed by GOGAT, is thus the major if not the only route facilitating the incorporation of inorganic N into organic molecules. GOGAT recycles glutamate and incorporates carbon skeletons into the GS/GOGAT cycle using the organic acid 2-oxoglutarate as a substrate (Hirel & Lea, 2001; Lea & Miflin, 2018). In this coupled reaction, 2-oxoglutarate originates from photosynthetic CO_2 assimilation, via the glycolysis pathway and the Krebs cycle (Hodges, 2002), which represents a direct interface between C- and N-metabolism.

Following the discovery of the GS/GOGAT ammonium assimilatory pathway, it was thought that the GS enzyme was present as a single protein species located in the plastids, although enzyme activity was also found to occur in non-photosynthetic tissues such as roots (Kanamori & Mastumoto, 1972), thus indicating that there could be other forms of the enzyme, likely to be present in the cytosol (Miflin & Lea, 1977).

Even following the discovery of the GS/GOGAT cycle, the possible role of mitochondrial NADH-dependent glutamate dehydrogenase in the process of ammonia assimilation was still put forward on a regular basis. This was due to the ability of glutamate dehydrogenase to fix ammonium in vitro utilizing the organic acid 2-oxoglutarate to synthesize glutamate. Some authors vehemently argued that GDH could be involved in the assimilation of inorganic N in plants, even though its K_m for ammonium was at least 10 times higher than that of GS (Oaks, 1995). It was also proposed that under certain physiological conditions to avoid a build-up of ammonium ions, the enzyme is able to synthesize glutamate (Skopelitis et al., 2006). However, so far, all the experiments performed *in planta*, using 15N-labeled techniques, have demonstrated that GDH operates in the direction of glutamate deamination (Stewart, Shatilov, Turnbull, Robinson, & Goodall, 1995), even when the enzyme is overexpressed several times in transgenic plants (Labboun et al., 2009). It was therefore concluded that GDH is involved in the recycling of carbon molecules by supplying 2-oxoglutarate to tissues that are carbon limited, rather than in the assimilation of ammonium when carbon is in excess, a finding which was largely confirmed using Arabidopsis mutants totally devoid of GDH activity (Fontaine et al., 2012).

I joined Pierre Gadal's group as a Masters student at the University of Nancy, France, in 1979. Following the Rothamsted group in the UK, Pierre Gadal set up one of the pioneer teams working on plant

N-metabolism in France. The idea behind the development of such research was to improve N use efficiency (NUE) in plants both for agronomic and ecological purposes (Hirel & Krapp, 2020).

Pierre Gadal first asked me to demonstrate that non-symbiotic N_2-fixing bacteria such as *Azospirillum* could provide N to plants, in a similar manner that can occur in legumes. Some of these bacteria were isolated by a group located nearby at the Centre de Pédologie Biologique of the CNRS (Bally, Thomas-Bauzon, Balandreau, Richard, & De Rey, 1983). As my investigations were not really conclusive, this diminished my motivation to pursue further research on the topic. Pierre Gadal then asked me to move my research to the biochemical characterization of GS. Later on, other members of Pierre Gadal's group, using a sophisticated gnotobiotic system, managed to show that part of the bacterial N originating from atmospheric N_2 acted as N fertilizers, thus allowing the demonstration of a complete plant developmental cycle (Maudinas, Chemardin, Yovanovitch, & Gadal, 1981). Interestingly, a simplified version of this gnotobiotic system is currently used by my group to study the interaction between non-symbiotic N_2-fixing bacteria and maize (Brusamarello-Santos et al., 2017). Ironically, this topic which did not excite me so much at the early stage of my career, has become one of the most extensively studied fields of plant research, principally due to the promising applications in sustainable agriculture to limit the overuse of N fertilizers that are detrimental to the environment (Dellagi, Quilleré, & Hirel, 2020).

Following the demonstration that in higher plants, ammonia is incorporated into organic molecules via the GS/GOGAT pathway, two groups discovered more or less simultaneously, the occurrence of two types of glutamine synthetase isoenzymes, one located in the cytosol and the other in the plastids. The two GS isoenzymes were originally identified by two PhD students (Caroline Guiz in Pierre Gadal's group and Fred Mann in George Stewart's group) using ion exchange chromatography experiments in which two peaks of enzyme activity could be separated in rice and barley leaves (Guiz, Hirel, Schedlofsky, & Gadal, 1979; Mann, Fentem, & Steward, 1979). The first one eluted from the column was thus called GS1 and the second one GS2. Detailed biochemical analyses, the kinetic properties of the two peaks of enzyme activity (Hirel & Gadal, 1980a) and identification of their subcellular localization revealed that GS1 and GS2 were different proteins, the first one located in the cytosol and the second one in the plastids (Hirel & Gadal, 1980b; Wallsgrove, Lea, & Miflin, 1979). Spinach leaves were first used to perform cellular fractionation in order to locate the two

GS isoenzymes, because it was relatively easy to isolate intact chloroplasts in this plant using gradients of aqueous solutions containing increasing concentration of sucrose. However, following ion exchange chromatography, only one peak of GS activity could be detected in this species, which started to worry us about the real occurrence of two isoenzymes of the GS enzyme in plants. However, more difficult to handle, non-aqueous fractionation techniques were successfully used to demonstrate that GS1 was located in the cytosol while GS2 was present in the plastids (Hirel & Gadal, 1980a), a result confirmed later in pea which also contained the two forms of the enzyme in leaves (Hirel & Gadal, 1981).

The puzzling differences observed between spinach and rice with respect to their GS isoenzyme content, prompted me to examine other plant species including C_4 plants such as sorghum. Sorghum was at that time one of the favorite model species used to study the enzyme phopho*enol*pyruvate carboxylase by Pierre Gadal's team (Vidal, Cavalie, & Gadal, 1976). This led Pierre Gadal to make the comment, "this is horizontal research, you should concentrate on the characterization and on deciphering the physiological role of GS1 and GS2." However, we soon realized that looking at different plant species was the right thing to do first, as we found that in sorghum the relative proportions of GS1 and GS2 were similar (Hirel & Gadal, 1982). This led to the conclusion that the GS isoenzyme complement was species-specific and to the hypothesis that their physiological role was also different, depending on whether they were C_3 or C_4 plants.

In 1981 at a meeting of the Society for Experimental Biology in Leicester, our group (from the University of Paris-Orsay, France) and that of George Stewart (from Birkbeck College, University of London, UK) recognized the similarity of their studies, and decided to publish a joint paper (McNally, Hirel, Gadal, Mann, & Stewart, 1983). This also marked the beginning of the most productive period of research I can remember, leading to the publication of seven joint papers from 1983 to 1984 which also led to the development of many close friendships. Certainly, the subsequent collaborative work that we pursued proved no less rewarding-scientifically, gastronomically, and enologically.

Together with the discovery of GOGAT, this new finding strongly rekindled the interest of plant biologists to reconsider the importance of inorganic N assimilation in plants and its regulation considering both the tissue and subcellular localization of the different GS isoenzymes. This also transformed glutamine synthetase from a rather mundane enzyme, previously considered as being mostly involved in the synthesis of the storage compound glutamine, into a key enzyme of plant N-metabolism. We were able to show that plant species differed with respect to the occurrence and relative amounts of glutamine synthetase isoenzymes. We found that four groups of plants could be recognized on the basis of differences in the stoichiometry of the isoenzymes in leaf tissue. One of the most important findings was that many plants with C_3 photosynthesis lacked the cytosolic isoenzyme of glutamine synthetase. In contrast, most of the C_4 species that we examined had a high content of the cytosolic isoenzyme. The importance of this lay in the proposal from the Rothamsted group that photorespiratory ammonium was reassimilated via the cytosolic isoenzyme (Keys et al., 1978). Our results ruled this out for many species, leaving a role in the photorespiratory N cycle for chloroplastic glutamine synthetase because of the lack of GS1 activity in the cytosol of leaf parenchyma cells in a number of C_3 plant species (McNally et al., 1983). This finding was further confirmed by immunocytochemical studies using antibodies raised against the GS protein that recognized both GS1 and GS2 (Hirel, Perrot-Rechenmann, Suzuki, Vidal, & Gadal, 1982). Ironically, the most decisive evidence confirming our hypothesis came from mutants of barley isolated by both the Rothamsted and Lancaster groups that lacked chloroplastic glutamine synthetase and could only grow under conditions that suppressed photorespiration (Blackwell, Murray, & Lea, 1987; Wallsgrove, Turner, Hall, Kendall, & Bright, 1987).

Later on, a research group from Portugal using immunogold transmission electron microscopy was able to show that a small proportion of GS1 was present in the cytosol of the phloem companion cells (Carvalho et al., 1992), a finding which turned out to be a general feature in all higher plants examined so far. The involvement of GS1 in the phloem in the control of long-distance transport of N-containing molecules such as proline was further deciphered by a number of groups for example using transgenic plants impaired in cytosolic GS activity in this tissue (Brugière et al., 1999; Caputo, Victoria Criado, Roberts, Geslo, & Barneix, 2009; Kamachi, Yamaya, Hayakawa, Mae, & Ojima, 1992).

The original work by McNally et al. (1983) has been frequently cited over the last 3 decades because it provides a comprehensive account of the occurrence of glutamine synthetase isoenzymes in plants having different physiological and ecological characteristics in terms of plant adaptation to a particular habitat, to various environmental conditions, and in different tissues during the plant developmental cycle from germination to senescence (Cren & Hirel, 1999).

Later on, the main reason for the success concerning the discovery of the GS isoenzymes was the rise of plant molecular biology. In the two decades since the original biochemical findings of McNally et al. (1983), an increasing number of studies carried out on different legume and non-legume species have been developed at the molecular level by a number of groups including ours own. The aim of these studies was to further decipher the role of the different GS isoenzymes and the regulation of their synthesis and the assembly of their different subunits during N assimilation and N remobilization (Masclaux, Quilleré, Gallais, & Hirel, 2001; Yamaya & Kusano, 2014). One of the main milestones in these studies was the cloning of a cDNA encoding the cytosolic form of the enzyme present in large amounts in nodules of legume plants. Such cloning was facilitated by the fact that the transcripts encoding GS1 were strongly induced in N_2-fixing nodules. This is due to the large amount of ammonium produced by the enzyme nitrogenase that is first incorporated into glutamine to allow the synthesis of N-containing molecules in the form of ureides that are further exported to other parts of the plant (Cullimore et al., 1984). Homologies of sequences between the different cDNAs encoding GS1 and GS2 then allowed the cloning of the corresponding transcripts from different plant species (Cren & Hirel, 1999).

Other remarkable pieces of work at the molecular level were the use of a reporter gene to show the tissue-specific synthesis of the GS isoenzymes in organs such as nodules, leaves or roots of transgenic plants (Miao, Hirel, Marsolier, Ridge, & Verma, 1991). Strong evidence for the putative role of GS in plant productivity was first demonstrated following one of the pioneer studies based on the exploitation of maize genetic diversity and the detection of QTLs. Rather unexpectedly, QTLs for grain yield were found to colocalize with QTLs for leaf GS activity and structural genes encoding the cytosolic form of the enzyme (Hirel et al., 2001). This led to the proposal that GS1 could be a major driver involved in the control of grain yield, which prompted the development of numerous genetic manipulations in a variety of plant species to increase the activity of the enzyme in order to boost plant performance. As such, GS became one of the most intensively studied enzymes in terms of NUE enhancement both in monocot and dicot plants. In species such as rice, maize, barley, pea and poplar, transgenic plants overexpressing GS1 or GS2 displayed improvement in NUE, increased biomass and higher grain yield (James, Borphukan, Fartyal, Achary, & Reddy, 2018; Plett, Garnett, & Okamoto, 2017). One of the most remarkable results was the increase in grain yield obtained in maize plants overexpressing GS1 and a reduction in yield observed in mutants deficient for two forms of the cytosolic enzyme (Martin et al., 2006). This study was another example of another remarkable collaboration established between my group and two other groups based in the UK, namely those of Keith Edwards at the University of Bristol and Peter J. Lea at Lancaster University. As a conclusion, my personal feeling about the pioneer work conducted when I was in Pierre Gadal's laboratory on the process of ammonia assimilation involving the discovery of the different GS isoenzymes, is that it illustrates how rewarding collaborative science can be. Such a collaboration has involved the synthesis of the findings of several colleagues in the UK and has proved far more scientifically significant than if the research had been published independently.

Acknowledgements: We are particularly grateful to Professor Peter J. Lea, who was one of the pioneer scientists in the field of nitrogen assimilation as well as an outstanding collaborator for more than 20 years, for critically reviewing the manuscript.

4. Carbon-nitrogen interactions and isocitrate dehydrogenases

4.1 Ridong Chen's recollections

After having graduated with a bachelor from a Chinese University in the summer of 1984, I was chosen out of many talented students to gain the privilege for studying abroad. In the spring of the following year, I traveled to France, carrying with me poor French but a big dream. That dream was to convert the infinite resource of sunshine into an abundance of electrical energy. I was initially accepted in the Master program of biophysics at the Center of Atomic Energy. However, an insufficient education in advanced mathematics and quantum mechanics compelled me to switch to the Master program in plant molecular biology and biotechnology, led by Pierre and Francis Quétier. I met Mr. Gadal in his office after arriving in Paris from Language School of Montpellier. In my first impression, he was a tall, energetic gentleman. He told me he had checked my background with full satisfaction by writing to my Chinese professor of plant biochemistry. After spending several dull months applying 2D gel electrophoresis to study proteomic changes during light-induced greening of plants with Jean Vidal and an INRA team, I wanted to change direction and pursued research on restriction enzymes for my doctoral degree. In the summer of 1986, Pierre invited me several times to discuss about my thesis work in his lab.

I showed no interest, as at the time, protein research was no longer considered advanced. In other words, it was unfashionable. With his persistence, and his recruitment of one friend to convince me, we finally met in his office. My eyes glazed over as the conversation quickly turned to isocitrate dehydrogenase, purification, and enzymes. Bluntly, I asked him why I should work on this. He told me that this enzyme might be the key to improving nitrogen assimilation and crop yield. My interest piqued, and as he described to me the possibilities, my imagination began to run wild with images of important discovery, perhaps even a new metabolic pathway in a textbook.

In the late summer of 1986, I had naught on my mind but the purification of the protein. After 3 months of hard work, I managed to purify small quantity of protein to homogeneity, made possible by the guidance of Pierre and the help of certain publications. I was so excited that I enthusiastically invited my graduate student friends to view the "beauty" of a faint blue band on the gel. I then spent 6 months working day and night in the cold room of the basement, often staying until the wee hours of the morning, monitoring chromatographic processes. With the precious 200 μg protein purified from tons of fresh pea leaves, Jean Vidal helped me immunize three rabbits, with one producing good titer of polyclonal antibodies. I compared the kinetic and immunological properties of the enzymes purified from roots and leaves. Pierre and I worked with great joy and together published our very first paper in *European Journal of Biochemistry* in 1988 (1). With the help from a kind and competent technician named Evelyne Bismuth and chromatographic and immunological tools, we discovered that a distinct isozyme was present in the isolated chloroplast (2). Evelyne and I managed to purify a sufficient quantity of the chloroplast isozyme for analysis and a manuscript was ready for submission. However, due to the presence of a band of 80 kDa on the gel, we could not be certain whether chloroplastic isozyme was a monomer of 80 kDa or a dimer of 2 × 40 kDa. For the sake of avoiding the possibility of reporting false findings and for peace of mind, the manuscript was never published. With the help from Evelyne and two motivated undergraduate students, Emmanuelle and Claire, I went out to the forest near Orsay, collected plant leaves, and then identified species with the generous help from a Professor of the fourth floor. We showed that NADP-isocitrate dehydrogenase was present and immunologically related in all the lower and higher plants we surveyed (3). I also initiated determination of the ratio of chloroplastic vs cytosolic isozymes and eventually completed the work when I became a professor of Biochemistry in Canada (4).

With the help of Claude Crétin, I isolated a partial cDNA and concluded the experimental work of my 3-year doctoral research.

At the time I started to write up my thesis, an aconitase isozyme was also reported in the plant cytosol. This, along with the presence of citrate in the cytosol, citrate transporter on the mitochondrial membrane, and lack of α-ketoglutarate-degrading enzyme in cytosol, inspired me to propose a novel pathway in the cytosol. With Pierre's encouragement and enthusiastic participation, we published the new concept of "division of labor": NADP-isocitrate dehydrogenase and the cytosolic pathway provide NADPH and α-ketoglutarate for biosynthesis, while NAD-isocitrate dehydrogenase and Krebs cycle produces NADH and α-ketoglutarate for ATP generation (5, and see Fig. 4). Pierre and I also co-authored a review paper, the very first systemic review of this enzyme in plants and other species (6). Seven years later, our cytosolic pathway was confirmed with NMR by a German team and adopted by the textbook of *Plant Biochemistry and Molecular Biology* (7). The dream had come true!

Pierre was a kind, joyful, and sometimes humorous man. Over the 3 years of interaction, we occasionally talked about personal life. He shared about how he had met his wife. He had spotted her in her father's dentist office, and immediately fell in love. He was very proud of being granted the title of professor at the age of 31, a rare achievement in the French education

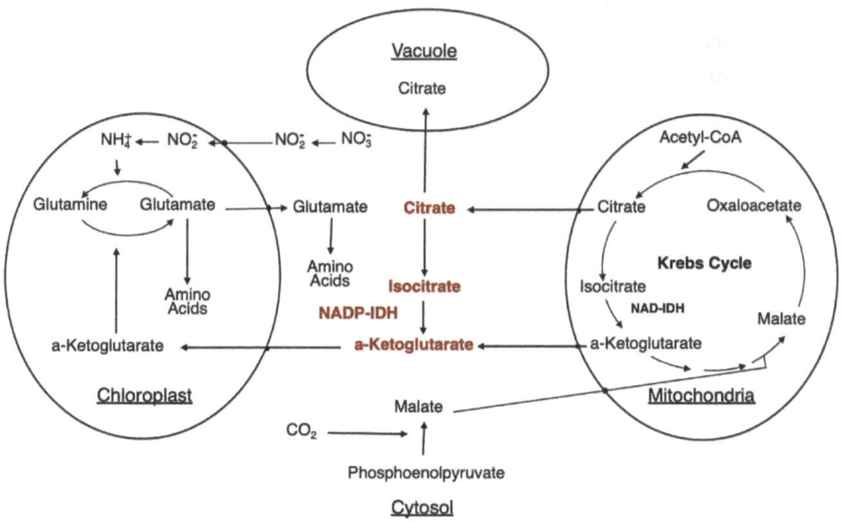

Fig. 4 Critical role of cytosolic NADP-dependent isocitrate dehydrogenase in providing α-ketoglutarate needed for plant nitrogen assimilation.

system. I feel a sense of warmth, as I think back on the good times we shared as friends during my time at Orsay. When Pierre left for sabbatical leave at the Washington University, I felt lonely as it was often difficult to get in touch with him. I was also mildly frustrated, since sometimes I was unable to order research reagents without his approval. Nevertheless, it worked out in the end, as this forced me to read more which was very helpful for the review article later on. I remember that every time I asked him when he would retire. He said loudly "Chen, I am only fifty something."

After I left the lab, Pierre remained a good friend and provided unconditional support for my faculty applications in the USA and Canada. We exchanged greetings and family pictures during Christmas time. I visited him in 2002 when I had business in France. He came to pick my family up from the train station and proudly showed me the new Plant Biotech Institute he and Francis established. My wife and I were fortunate enough to see him again in his house during the Summer of 2018 and I will remember fondly our 90 min conversation in rusty French. Despite fighting cancer for some time and recent surgery, he remained joyful and optimistic. He was very curious about what I had been doing these years. I shared with him how much I was blessed and how much I enjoyed my biomedical discovery. He emphasized that there was still so much to do for plant scientists in order to secure food safety for humanity. We left with repeated promise of "seeing each other again" while he was standing at the entrance of the house.

A few months before he passed away, he informed me of his new email account and comforted me by saying "he is doing great." In spite of challenges of both language and culture, I consider my Orsay journey with Pierre as both unexpected and amazing. This allowed me to enter the fascinating enzyme world and find something to build on. I will dearly miss and warmly remember Pierre as a great mentor and a very good friend.

4.2 Michael Hodges's recollections

In 1988, I, Michael Hodges that is, was working at the CNRS campus in Gif-sur-Yvette and already collaborating occasionally with Myroslawa Miginac-Maslow when my laboratory was closed for renovation as part of the creation of a new Institute to be called the ISV (Institut des sciences du végétal). It was under these circumstances that Pierre Gadal allowed me to join his laboratory at the Université Paris-Sud in Orsay to work on the purification of LHC kinase with Myroslawa. I then went on to investigate FNR (ferredoxin-NADP reductase) phosphorylation before deciding to move on to thioredoxins with Jean-Pierre Jacquot where I notably worked on a thioredoxin f from pea (Hodges et al., 1994) using a cDNA isolated by Loïc Lepiniec when he was screening for PEPC! I was a biophysicist by trade and joining the laboratory of Pierre Gadal gave me the opportunity to discover and to learn some biochemistry and protein purification techniques. This was then further extended to molecular biology, in the days when the laboratory had a couple of classic MacIntosh computers with Tetris inside and PCR was done using water baths before the first PCR machine was bought!

I ended up working at a laboratory bench in a room at the end of the corridor where I found myself sharing bench space with Susana Galvez Garcia, a Spanish student who had come from the laboratory of Francisco Canovas in Malaga to do her PhD with Pierre. She was given the task of characterizing NADP-dependent isocitrate dehydrogenases (ICDHs) after R-D. Chen had successfully finished his PhD and left. This enzyme had become a favorite of Pierre Gadal because it produced 2-oxoglutarate (2OG) and he assumed it to be the missing link between carbon metabolism and nitrogen metabolism since ammonium assimilation by the GS/GOGAT cycle required this keto-acid to make glutamate. At the time, Pierre Gadal was preoccupied with creating the IBP thus leaving Susana just a little bit under-supervised and so I decided to help her out with discussions and advice. Who would have thought that this was to be the start of a long story leading to us getting married and sharing more than just memories about a common research goal?

Before my involvement, Susana had already purified and characterized cytosolic and chloroplastic ICDH isoforms from mixotrophic tobacco cell suspensions (Galvez, Bismuth, Sarda, & Gadal, 1994). The idea of using this biological material was the brain child of Evelyne Bismuth, who had worked with Ridong Chen and now she was giving Susana some technical support. It was then decided by Pierre Gadal that Susana should develop a more modern molecular biology strategy. The first aims were to identify and isolate ICDH cDNA clones. Having no experience, Susana was lucky to be taken under the wings of Claude Crétin who helped her to make a cDNA library using mRNA isolated from tobacco cell suspensions. Claude became a dear friend but this was cut short by his untimely death. It was around this time that Susana moved from the basement to the laboratory bench in the room at the end of the corridor on the first floor of the building number 430 that I was using. First, Susana screened her cDNA library using antibodies raised against her purified ICDH proteins from cell suspensions but this strategy was unsuccessful. Eventually, several ICDH cDNA clones were isolated using a probe made from a partial soybean ICDH cDNA generously sent by Dr. M. Khan of Washington State University. Full-length cDNAs were then generated using PCR fragments amplified from the cDNA library. Based on the sequencing of tryptic peptides from purified tobacco ICDH isoforms a correspondence was made between cytosolic ICDH and an isolated cDNA (Gâlvez et al., 1996). This allowed the production, purification and characterization of recombinant tobacco cytosolic ICDH (Gâlvez et al., 1995). Although, my failing memory does not allow me to say what my actual contributions to this work were, I ended up as co-author of both papers cited above. During this period, Susana had met Sabine Fieuw at a meeting where she learnt that Sabine was working on ICDH in the group of Lothar Willmitzer at the MPIMP Gölm Germany. Together, they isolated a cytosolic ICDH cDNA from potato using our tobacco iCDH cDNA as a probe (Fieuw, Müller-Röber, Gâlvez, & Willmitzer, 1995).

During our time in the building number 430 at the Université Paris-Sud in Orsay, we were also involved with some of Pierre Gadals friends: Francis Martin at the University of Nancy sent Vincent Boiffin to the laboratory to clone ICDH from Eucalyptus (Boiffin et al., 1998) and Rajbir Sangwan at the University of Picardie in Amiens helped us to immuno-localize ICDH by electron microscopy (Gâlvez et al., 1996). Rajbir remained a friend over the years with several invitations to join PhD juries and recruitment committees as well as to correct the English of his numerous publications.

Anyway, back to my story. We could be proud of our ICDH discoveries and the first part of my ICDH story ended in 1994 when Susana obtained her PhD in the new amphitheater of the IBP that Pierre Gadal had convinced influential people to build. I can remember that I made some Indian food for the laboratory celebration and the day finished with an enjoyable typically French meal with Pierre, Susana, and Francisco Canovas at a nearby restaurant at Villiers-le-Bâcle that Pierre Gadal had chosen. It was the last PhD supervised by Pierre Gadal and after it was defended, Susana stayed in the team as a post doc paid for by an EU grant obtained by Pierre Gadal where we found ourselves in a consortium composed mainly of people working on fructans. This money allowed us to carry on working on ICDHs, to visit Copenhagen for a consortium meeting and to help organize the following one at the IBP, and to learn about fructans. During this period, a major aim was to show that one of our isolated tobacco ICDH cDNAs encoded the chloroplastic isoform. This involved developing in vitro translation and import of the resulting ICDH protein into isolated chloroplasts and eventually transforming tobacco to express an ICDH-GFP fusion protein. Together, these experiments strongly suggested that our cDNA encoded a mitochondrial ICDH and not the chloroplastic isoform but at the end of the day we were over the moon because our work was published in the prestigious journal, *Proc. Natl. Acad. Sci. U.S.A.* (Gálvez et al., 1998). That said, it is now admitted that the ICDH gene we were working on actually encodes an ICDH co-localized to both chloroplasts and mitochondria. Some unpublished work by Valérie Flesch at the IBP indicated that this was probably due to two translation start sites that defined the specificity of the transit peptide involved in addressing the preprotein to the organelles. We also adventured into making transgenic tobacco plants with highly reduced cytosolic ICDH activity but they did not show any growth phenotype! This made us think that perhaps it was mitochondrial NAD-dependent isocitrate dehydrogenase (IDH) that was producing the 2OG required for ammonium assimilation. This became the focus of the thesis of Muriel Lancien, and we were the first to identify IDH cDNAs and to show by the complementation of yeast *idh* mutants that tobacco IDH was heteromeric and required at least two different subunits to be active (Lancien, Gadal, & Hodges, 1998). In collaboration with Bertrand Hirel, a former member of the Gadal laboratory, who was working at the "Institut Jean-Pierre Bourgin" (IJPB) in Versailles, we investigated the nitrate-response of IDH genes (Lancien et al., 1999).

At the end of Muriel's PhD thesis, I was given the opportunity to manage my own research team at the IBP. After a 4-year break to investigate mitochondrial organic acid transporters, work on IDH continued during the thesis of Thomas Lemaitre and by this time it had been decided to move away from tobacco and to use Arabidopsis to benefit from mutant libraries and the genome sequence that had indicated the presence of five functional IDH genes. Again, single IDH mutants did not show any growth phenotype under standard laboratory growth conditions (Lemaitre et al., 2007). Although the isocitrate dehydrogenase topic at Orsay was initiated by Pierre Gadal, the Arabidopsis IDH work was no longer associated with him. However, we still interacted up to and beyond his retirement with friendly discussions in the corridors of the IBP, and every time Pierre crossed my path he would enquire about the latest news concerning me and Susana. I am most grateful to Pierre Gadal because he welcomed me into his laboratory, he gave me the liberty to decide which research areas to pursue and finally he allowed me to take over his original ICDH story and to go on to develop my personal research interests. This started at the end of the 1990s and the identification and characterization of mitochondrial organic acid transporters with Nathalie Picault, a Master then PhD student, and our Italian collaborators, Fernando and Luigi Palmieri (Picault, Palmieri, Pisano, Hodges, & Palmieri, 2002). It continued with further studies on 2OG but this time as a possible signal modulating N-metabolism via the PII protein. This work was done in collaboration with Sylvie Ferrario-Méry and Sebastien Baud of the IJPB in Versailles and by Ana-Belén Feria Bourrellier, a Spanish Post doc whom Jean Vidal knew. The PII story and my love for 2OG ended with me spending 10 months in hospital and two major discoveries: arginine and fatty acid biosynthesis are regulated by a 2OG-dependent interaction of PII with *N*-acetyl glutamate and acetyl-CoA carboxylase, respectively (Feria Bourrellier et al., 2010; Feria Bourrellier, Ferrario-Méry, Vidal, & Hodges, 2009). If it was not for Pierre Gadal, I would have never become acquainted with this keto-acid and I would have never pursued so many happy years of my research career trying to elucidate how it was made, how it was transported and how it could be a signal to regulate plant metabolism.

5. Concluding remarks

Shortly after this series of recollections was initiated many of the former students and colleagues of Pierre Gadal have expressed interest in participating. All of them could not do it for a variety of reasons but there is no doubt that their recollections would have expressed the same feeling. First, we thought that the best way to honor Pierre was to report the science that has been done in his lab as he was foremost a researcher. Hopefully the readers will get more than a glimpse of how it's been done, why it's been done that way and of the connections between the many researchers involved. In many ways Pierre has altered our scientific paths but also more broadly our lives. The readers will have noticed similar terms coming all over again, benevolence, simplicity, tenacity and unwaivering support for his students. Following the example given by Pierre large freedom of expression has been given to all contributors and the resulting article is inherently rather heterogenous. Still all of us have written it with their soul and guts and in the end, we hope that we have paid to Pierre the tribute that was so well deserved.

References

Anderson, L. E., Nehrlich, S. C., & Champigny, M. L. (1978). Light modulation of enzyme activity. Activation of the light effect mediators by reduction and modulation of enzyme activity by thiol disulfide exchange. *Plant Physiology*, *61*, 601–605.

Arabidopsis Genome Initiative. (2000). Analysis of the genome sequence of the flowering plant *Arabidopsis thaliana*. *Nature*, *408*, 220–234.

Asada, K., Tamura, G., & Bandurski, R. S. (1969). Methyl viologen-linked sulfite reductase from spinach leaves. *Journal of Biological Chemistry*, *244*, 4904–4915.

Bally, R., Thomas-Bauzon, D., Balandreau, J., Richard, C., & De Rey, J. (1983). Determination of the most frequent N_2-fixing bacteria in a rice rhizosphere. *Revue Canadienne de Microbiologie*, *29*, 881–887.

Beevers, L. (1981). Nitrogen metabolism in plants. In H. Bothe, & A. Trebst (Eds.), *Proceedings in life sciences. Biology of inorganic nitrogen and sulfur* (pp. 15–29). Berlin, Heidelberg: Springer.

Blackwell, R. D., Murray, A. J. S., & Lea, P. J. (1987). Inhibition of photosynthesis in barley with decreased levels of chloroplastic glutamine synthetase. *Journal of Experimental Botany*, *38*, 1799–1809.

Boiffin, V., Hodges, M., Gálvez, S., Balestrini, R., Bonfante, P., Gadal, P., et al. (1998). Eucalypt NADP-dependent isocitrate dehydrogenase: cDNA cloning and expression in ectomycorrhiza. *Plant Physiology*, *117*, 939–948.

Brandes, H. K., Larimer, F. W., & Hartman, F. C. (1996). The molecular pathway for the regulation of phosphoribulokinase by thioredoxin f. *The Journal of Biological Chemistry*, *271*, 3333–3335.

Brugière, N., Dubois, F., Limami, A., Lelandais, M., Roux, Y., Sangwan, R., et al. (1999). Glutamine synthetase in the phloem plays a major role in controlling proline production. *The Plant Cell*, *11*, 1995–2011.

Brusamarello-Santos, L. C., Gilard, F., Brulé, L., Quilleré, I., Gourion, B., Ratet, P., et al. (2017). Metabolic profiling of two maize (*Zea mays* L.) inbred lines inoculated with the nitrogen fixing plant-interacting bacteria *Herbaspirillum seropedicae* and *Azospirillum brasilense*. *PLoS One*, *12(3)*, e0174576.

Buchanan, B. B. (1980). Role of light in the regulation of chloroplast enzymes. *Annual Review of Plant Physiology*, *31*, 341–374.

Buchanan, B. B., Schürmann, P., & Kalberer, P. P. (1971). Ferredoxin-activated fructose diphosphatase of spinach chloroplasts. Resolution of the system, properties of the alkaline fructose diphosphatase component, and physiological significance of the ferredoxin-linked activation. *The Journal of Biological Chemistry*, *246*, 5952–5959.

Bulen, W. A. (1956). The isolation and characterization of glutamic dehydrogenase from corn leaves. *Archives of Biochemistry and Biophysics*, *62*, 173–183.

Caputo, C., Victoria Criado, M., Roberts, I. N., Geslo, M. A., & Barneix, A. J. (2009). Regulation of glutamine synthetase 1 and amino acid transport in the phloem of young wheat plants. *Plant Physiology and Biochemistry*, *47*, 335–342.

Carvalho, H., Pereira, S., Sunkel, C., & Salema, R. (1992). Detection of a cytosolic glutamine synthetase in leaves of *Nicotiana tabacum* L. by immunocytochemical methods. *Plant Physiology*, *100*, 1591–1594.

Champigny, M.-L., & Bismuth, E. (1976). Role of photosynthetic electron transfer in light activation of Calvin cycle enzymes. *Physiologia Plantarum*, *36*, 95–100.

Chardonnet, S., Sakr, S., Cassier-Chauvat, C., Le Maréchal, P., Chauvat, F., Lemaire, S. D., et al. (2015). First proteomic study of S-glutathionylation in cyanobacteria. *Journal of Proteome Research*, *2*, 59–71.

Chen, R. D., Bismuth, E., Champigny, M. L., & Gadal, P. (1989). Chromatographic and immunological evidence that chloroplastic and cytosolic pea leaf NADP-isocitrate dehydrogenases are distinct isoenzymes. *Planta*, *178*, 157–163.

Chen, R. D., Bismuth, E., Issakidis, E., Pacot, C., Champigny, M. L., & Gadal, P. (1989). Comparative immunochemistry of NADP-isocitrate dehydrogenase: A highly conserved cytosolic enzyme in higher plants. *Comptes Rendus de l'Académie des Sciences*, *308*, 459–465.

Chiadmi, M., Navaza, A., Miginiac-Maslow, M., Jacquot, J. P., & Cherfils, J. (1999). Redox signalling in the chloroplast: Structure of oxidized pea fructose-1,6-bisphosphatase. *EMBO Journal, 18*, 6809–6815.

Chollet, R., Vidal, J., & O'Leary, M. H. (1996). Phosphoenolpyruvate carboxylase: A ubiquitous, highly regulated enzyme in plants. *Annual Review of Plant Physiology and Plant Molecular Biology, 47*, 273–298.

Collin, V., Issakidis-Bourguet, E., Marchand, C., Hirasawa, M., Lancelin, J.-M., Knaff, D. B., et al. (2003). The Arabidopsis plastidial thioredoxins: New functions and new insights into specificity. *The Journal of Biological Chemistry, 278*, 23747–23752.

Collin, V., Lamkemeyer, P., Miginiac-Maslow, M., Hirasawa, M., Knaff, D. B., Dietz, K.-J., et al. (2004). Characterization of plastidial thioredoxins from Arabidopsis belonging to the new y-type. *Plant Physiology, 136*, 4088–4095.

Coschigano, K. T., Melo-Oliveira, R., Lim, J., & Coruzzi, G. M. (1998). Arabidopsis *gls* mutants and distinct Fd-GOGAT genes: Implications for photorespiration and primary nitrogen metabolism. *The Plant Cell, 10*, 741–752.

Crawford, N. M., & Arst, N. H., Jr. (1993). The molecular genetics of nitrate assimilation in fungi and Plants. *Annual Review of Genetics, 27*, 115–146.

Crawford, N. A., Droux, M., Kosower, N. S., & Buchanan, B. B. (1989). Evidence for function of the ferredoxin thioredoxin system in the reductive activation of target enzymes of isolated intact chloroplasts. *Archives of Biochemistry and Biophysics, 271*, 223–239.

Cren, M., & Hirel, B. (1999). Glutamine synthetase in higher plants: Regulation of gene and protein expression from the organ to the cell. *Plant Cell Physiology, 40*, 1187–1193.

Crétin, C., Luchetta, P., Joly, C., Decottignies, P., Lepiniec, L., Gadal, P., et al. (1990). Primary structure of sorghum malate dehydrogenase (NADP) deduced from cDNA sequence. Homology with malate dehydrogenase (NAD). *European Journal of Biochemistry, 192*, 299–303.

Crétin, C., Luchetta, P., Joly, C., Miginiac-Maslow, M., Decottignies, P., Jacquot, J.-P., et al. (1988). Identification of a cDNA clone for sorghum leaf malate dehydrogenase (NADP). Light-dependent mRNA accumulation. *European Journal of Biochemistry, 174*, 497–501.

Crozet, P., Navarro, F. J., Willmund, F., Mehrshahi, P., Bakowski, K., Lauersen, K. J., et al. (2018). Birth of a photosynthetic chassis: A MoClo toolkit enabling synthetic biology in the microalga *Chlamydomonas reinhardtii*. *ACS Synthetic Biology, 7*, 2074–2086.

Cullimore, J. V., Gebhardt, C., Saarelainen, R., Miflin, B. J., Idler, K. B., & Barker, R. F. (1984). Glutamine synthetase of *Phaseolus vulgaris* L.: Organ-specific expression of a multigene family. *Journal of Molecular and Applied Genetics, 2*, 589–599.

Dai, S., Saarinen, M., Ramaswamy, S., Meyer, Y., Jacquot, J. P., & Eklund, H. (1996). Crystal structure of *Arabidopsis thaliana thaliana* NADPH dependent thioredoxin reductase at 2.5A° resolution. *Journal of Molecular Biology, 264*, 1044–1057.

Dai, S., Schwendtmayer, C., Schürmann, P., Ramaswamy, S., & Eklund, H. (2000). Redox signaling in chloroplasts: Cleavage of disulfides by an iron-sulfur cluster. *Science, 287*, 655–658.

de Lamotte, F., Pruvost, C., Minard, P., Delsuc, M. A., Miginiac-Maslow, M., Schmitter, J. M., et al. (1997). Structural and functional roles of a conserved proline residue in the alpha-2 helix of *Escherichia coli* thioredoxin. *Protein Engineering, 10*, 1425–1432.

de Lamotte-Guéry, F., Miginiac-Maslow, M., Decottignies, P., Stein, M., Minard, P., & Jacquot, J. P. (1991). Mutation a negatively charged amino acid in of thioredoxin modifies its reactivity with chloroplastic enzymes. *European Journal of Biochemistry, 196*, 287–294.

De Silva Conceição, A., & Krebbers, E. (1994). A cotyledon regulatory region is responsible for the different spatial expression patterns of Arabidopsis 2S albumin genes. *The Plant Journal, 5*, 493–505.

Decottignies, P., Le Maréchal, P., Jacquot, J. P., Schmitter, J. M., & Gadal, P. (1995). Primary structure and post-translational modification of ferredoxin-NADP reductase from *Chlamydomonas reinhardtii*. *Archives of Biochemistry and Biophysics*, *316*, 249–259.

Decottignies, P., Schmitter, J. M., Dutka, S., Jacquot, J. P., & Miginiac-Maslow, M. (1991). Characterization and primary structure of a second thioredoxin from the green alga, *Chlamydomonas reinhardtii*. *European Journal of Biochemistry*, *198*, 505–512.

Decottignies, P., Schmitter, J. M., Jacquot, J. P., Dutka, S., Picaud, A., & Gadal, P. (1990). Purification, characterization and complete amino acid sequence of a thioredoxin from a green alga *Chlamydomonas reinhardtii*. *Archives of Biochemistry and Biophysics*, *280*, 112–121.

Decottignies, P., Schmitter, J.-M., Miginiac-Maslow, M., Le Maréchal, P., Jacquot, J.-P., & Gadal, P. (1988). Primary structure of the light-dependent regulatory site of corn NADP-malate dehydrogenase. *The Journal of Biological Chemistry*, *263*, 11780–11785.

Dellagi, A., Quilleré, I., & Hirel, B. (2020). Beneficial soil borne bacteria and fungi: A promising way to improve plant nitrogen acquisition. *Journal of Experimental Botany*, *71*, 4469–4479. https://doi.org/10.1093/jxb/eraa112.

DeWitt, N. D., & Sussman, M. R. (1995). Immunological localization of an epitope-targeged plasma membranes pump (H^+-ATPase) in phloem companion cells. *The Plant Cell*, *7*, 2053–2076.

Droux, M., Crawford, N. A., & Buchanan, B. B. (1987). Mechanism of thioredoxin-linked activation of chloroplast fructose-1,6-bisphophatase. *Comptes Rendus de l'Academie des Sciences. Serie III, Sciences de la Vie*, *305*, 335–341.

Droux, M., Jacquot, J.-P., Miginiac-Maslow, M., Gadal, P., Huet, J.-C., Crawford, N. A., et al. (1987). Ferredoxin-thioredoxin reductase, an iron-sulfur enzyme linking light to enzyme regulation in oxygenic photosynthesis: Purification and properties of the enzyme from C3, C4, and cyanobacterial species. *Archives of Biochemistry and Biophysics*, *252*, 426–439.

Droux, M., Miginiac-Maslow, M., Jacquot, J.-P., Gadal, P., Crawford, N. A., Kosower, N. S., et al. (1987). Ferredoxin-thioredoxin reductase: A catalytically active dithiol group links photoreduced ferredoxin to thioredoxin functional in photosynthetic enzyme regulation. *Archives of Biochemistry and Biophysics*, *256*, 372–380.

Echevarria, C., & Vidal, J. (2003). The unique phosphoenolpyruvate carboxylase kinase. *Plant Physiology and Biochemistry*, *41*, 541–547.

Elliott, W. H. (1953). Isolation of glutamine synthetase and glutamotransferase from green pea. *Journal of Biological Chemistry*, *201*, 661–672.

Feraud, M., Masclaux-Daubresse, C., Ferrario-Méry, S., Pageau, K., Lelandais, M., Ziegler, C., et al. (2005). Expression of a ferredoxin-dependent glutamate synthase gene in mesophyll and vascular cells and functions of the enzyme in ammonium assimilation in *Nicotiana tabacum* (L.). *Planta*, *222*, 667–677.

Feria Bourrellier, A. B., Ferrario-Méry, S., Vidal, J., & Hodges, M. (2009). Metabolite regulation of the interaction between *Arabidopsis thaliana* PII and N-acetyl-l-glutamate kinase. *Biochemical and Biophysical Research Communications*, *387*, 700–704.

Feria Bourrellier, A. B., Valot, B., Guillot, A., Ambard-Bretteville, F., Vidal, J., & Hodges, M. (2010). Chloroplast acetyl-CoA carboxylase activity is 2- oxoglutarate-regulated by interaction of PII with the biotin carboxyl carrier subunit. *Proceedings of the National Academy of Sciences of the United States of America*, *107*, 502–507.

Fieuw, S., Müller-Röber, B., Gálvez, S., & Willmitzer, L. (1995). Cloning and expression analysis of the cytosolic NADP(+)-dependent isocitrate dehydrogenase from potato. Implications for nitrogen metabolism. *Plant Physiology*, *107*, 905–913.

Fischer, N., Hippler, M., Sétif, M., Jacquot, J. P., & Rochaix, J. D. (1998). The PsaC subunit of photosystem I provides an essential residue for docking of ferredoxin. *EMBO Journal*, *17*, 849–858.

Fontaine, J. X., Tercé-Laforgue, T., Armengaud, P., Clément, G., Renou, J. P., Pelletier, S., et al. (2012). Characterization of a NADH-dependent glutamate dehydrogenase mutant of Arabidopsis demonstrates the key role of this enzyme in root carbon and nitrogen metabolism. *The Plant Cell, 24*, 4044–4065.
Galvez, S., Bismuth, E., Sarda, C., & Gadal, P. (1994). Purification and characterization of chloroplastic NADP-isocitrate dehydrogenase from mixotrophic tobacco cells (comparison with the cytosolic isoenzyme). *Plant Physiology, 105*, 593–600.
Gálvez, S., Hodges, M., Bismuth, E., Samson, I., Teller, S., & Gadal, P. (1995). Purification and characterization of a fully active recombinant tobacco cytosolic NADP-dependent isocitrate dehydrogenase in *Escherichia coli*: Evidence for a role for the N-terminal region in enzyme activity. *Archives of Biochemistry and Biophysics, 323*, 164–168.
Gálvez, S., Hodges, M., Decottignies, P., Bismuth, E., Lancien, M., Sangwan, R. S., et al. (1996). Identification of a tobacco cDNA encoding for a cytosolic NADP-isocitrate dehydrogenase. *Plant Molecular Biology, 30*, 307–320.
Gálvez, S., Roche, O., Bismuth, E., Brown, S., Gadal, P., & Hodges, M. (1998). Mitochondrial localization of a NADP-dependent isocitrate dehydrogenase isoenzyme by using the green fluorescent protein as a marker. *Proceedings of the National Academy of Sciences of the United States of America, 95*, 7813–7818.
García-Sánchez, M. I., Gotor, C., Jacquot, J. P., & Vega, J. M. (2000). Glutamate synthase from *Chlamydomonas reinhardtii*: Interaction studies with its substrate ferredoxin and molecular clonin. *Plant and Soil, 221*, 59–65. https://doi.org/10.1023/A:1004715504065.
Gaufichon, L., Marmagne, A., Belcram, K., Yoneyama, T., Sakakibara, Y., Hase, T., et al. (2017). *ASN1*-encoded asparagine synthetase in floral organs contributes to nitrogen filling in Arabidopsis seeds. *The Plant Journal, 91*, 371–393.
Gaufichon, L., Masclaux-Daubresse, C., Tcherkez, G., Reisdorf-Cren, M., Sakakibara, Y., Hase, T., et al. (2013). *Arabidopsis thaliana ASN2* encoding asparagine synthetase is involved in the control of nitrogen assimilation and export during vegetative growth. *Plant, Cell & Environment, 36*, 328–342.
Gaufichon, L., Reisdorf-Cren, M., Rothstein, S. J., Chardon, F., & Suzuki, A. (2010). Biological functions of asparagine synthetase in plants. *Plant Science, 179*, 141–153.
Gaufichon, L., Rothestein, S., & Suzuki, A. (2016). Asparagine metabolic pathways in Arabidopsis. *Plant and Cell Physiology, 57*, 675–689.
Goyer, A., Decottignies, P., Issakidis-Bourguet, E., & Miginiac-Maslow, M. (2001). Sites of interaction of thioredoxin with sorghum NADP-malate dehydrogenase. *FEBS Letters, 505*, 405–408.
Goyer, A., Decottignies, P., Lemaire, S., Ruelland, E., Issakidis-Bourguet, E., Jacquot, J.-P., et al. (1999). The internal Cys-207 of sorghum leaf NADP-malate dehydrogenase can form mixed disulphides with thioredoxin. *FEBS Letters, 444*, 165–169.
Goyer, A., Haslekås, C., Miginiac-Maslow, M., Klein, U., Le Maréchal, P., Jacquot, J.-P., et al. (2002). Isolation and characterization of a thioredoxin-dependent peroxidase from *Chlamydomonas reinhardtii*. *European Journal of Biochemistry, 269*, 272–282.
Guiz, C., Hirel, B., Schedlofsky, G., & Gadal, P. (1979). Occurrence and influence of light on the relative proportions of two glutamine synthetases in rice leaves. *Plant Science Letters, 15*, 272–277.
Gütle, D. D., Roret, T., Hecker, A., Reski, R., & Jacquot, J. P. (2017). Dithiol disulphide exchange in redox regulation of chloroplast enzymes in response to evolutionary and structural constraints. *Plant Science, 255*, 1–11.
Hirasawa, M., Droux, M., Gray, K. A., Boyer, J. M., Davis, D. J., Buchanan, B. B., et al. (1988). Ferredoxin-thioredoxin reductase: Properties of its complex with ferredoxine. *Biochimica et Biophysica Acta, 935*, 1–8.
Hirasawa, M., Ruelland, E., Schepens, I., Issakidis-Bourguet, E., Miginiac-Maslow, M., & Knaff, D. B. (2000). Oxidation-reduction properties of the regulatory disulfides of

sorghum chloroplast nicotinamide adenine dinucleotide phosphate-malate dehydrogenase. *Biochemistry*, *39*, 3344–3350.

Hirel, B., Bertin, P., Quillere, I., Bourdoncle, W., Attagnant, C., Dellay, C., et al. (2001). Towards a better understanding of the genetic and physiological basis for nitrogen use efficiency in maize. *Plant Physiology*, *125*, 1258–1270.

Hirel, B., & Gadal, P. (1980a). Glutamine synthetase in rice. A comparative study of the enzymes from roots and leaves. *Plant Physiology*, *66*, 619–623.

Hirel, B., & Gadal, P. (1980b). Sur la localisation intracellulaire de deux formes isofonctionnelles de la glutamine synthétase chez le Riz. *Comptes Rendus de l'Académie des Sciences*, *291 D*, 441–444.

Hirel, B., & Gadal, P. (1981). Glutamine synthetase isoenzymes in pea leaves. *Zeitschrift für Pflanzenphysiologie*, *102*, 315–319.

Hirel, B., & Gadal, P. (1982). Glutamine synthetase isoenzymes in leaves of a C_4 plant: *Sorghum vulgare* L. *Physiologia Plantarum*, *54*, 69–74.

Hirel, B., & Lea, P. J. (2001). Ammonia assimilation. In P. J. Lea, & J. F. Morot-Gaudry (Eds.), *Plant nitrogen* (pp. 79–99). Berlin, Heidelberg: Springer.

Hirel, B., & Krapp, A. (2020). Nitrogen utilization in plants I biological and agronomic importance. In *Encyclopedia of biochemistry* (3rd ed.). Elsevier.

Hirel, B., Perrot-Rechenmann, C., Suzuki, A., Vidal, J., & Gadal, P. (1982). Glutamine synthetase in spinach leaves. Immunological studies and immuno-cytochemical localization. *Plant Physiology*, *69*, 983–987.

Hoagland, D. R., & Arnon, D. I. (1938). In Circular California Agricultural Experiment Station (Ed.), *The water-culture method for growing plants without soil* (p. 347). Berkeley, CA: University of California, College of Agriculture, Agricultural Experiment Station.

Hodges, M. (2002). Enzyme redundancy and the importance of 2-oxoglutarate in plant ammonium assimilation. *Journal of Experimental Botany*, *53*, 905–916.

Hodges, M., Miginiac-Maslow, M., Decottignies, P., Jacquot, J.-P., Stein, M., Lepiniec, L., et al. (1994). Purification and characterisation of pea thioredoxin f expressed in *Escherichia coli*. *Plant Molecular Biology*, *26*, 225–234.

Holmgren, A., Buchanan, B. B., & Wolosiuk, R. A. (1977). Photosynthetic regulatory protein from rabbit liver is identical with thioredoxin. *FEBS Letters*, *82*, 351–354.

Huppe, H., de Lamotte, F., Jacquot, J. P., & Buchanan, B. B. (1990). The ferredoxin-thioredoxin system of a green alga, Chlamydomonas reinhardtii. *Planta*, *180*, 341–351.

Ireland, R. J., & Lea, P. J. (1999). The enzymes of glutamine, glutamate, asparagine and aspartate metabolism. In B. K. Singh (Ed.), *Plant amino acids* (pp. 49–109). New York: Marcel Dekker, Inc.

Issakidis, E., Decottignies, P., & Miginiac-Maslow, M. (1993). A thioredoxin-independent fully active NADP-MDH obtained by site-directed mutagenesis. *FEBS Letters*, *321*, 55–58.

Issakidis, E., Lemaire, M., Decottignies, P., Jacquot, J.-P., & Miginiac-Maslow, M. (1996). Direct evidence for the different roles of the N- and C-terminal regulatory disulfides of sorghum leaf NADP-malate dehydrogenase in its activation by reduced thioredoxin. *FEBS Letters*, *26*, 121–124.

Issakidis, E., Miginiac-Maslow, M., Decottignies, P., Jacquot, J.-P., Crétin, C., & Gadal, P. (1992). Site-directed mutagenesis reveals the involvement of an additional thioredoxin-dependent regulatory site in the activation of recombinant sorghum leaf NADP-malate dehydrogenase. *The Journal of Biological Chemistry*, *267*, 21577–21583.

Issakidis, E., Saarinen, M., Decottignies, P., Jacquot, J.-P., Crétin, C., Gadal, P., et al. (1994). Identification and characterization of the second regulatory disulfide bridge of recombinant sorghum leaf NADP-malate dehydrogenase. *The Journal of Biological Chemistry*, *269*, 3511–3517.

Issakidis-Bourguet, E., Lavergne, D., Trivelli, X., Decottignies, P., & Miginiac-Maslow, M. (2006). Transferring redox regulation properties from sorghum NADP-malate dehydrogenase to *Thermus* NAD-malate dehydrogenase. *Photosynthesis Research*, *89*, 213–223.

Issakidis-Bourguet, E., Mouaheb, N., Meyer, Y., & Miginiac-Maslow, M. (2001). Heterologous complementation of yeast reveals a new putative function for chloroplast m-type thioredoxin. *The Plant Journal*, *25*, 127–135.

Jacquot, J. P., Buchanan, B. B., Martin, F., & Vidal, J. (1981). Enzyme regulation in C4 photosynthesis: Purification and properties of thioredoxin linked NADP malate dehydrogenase from corn leaves. *Plant Physiology*, *68*, 300–304.

Jacquot, J. P., de Lamotte-Guéry, F., Fontecave, M., Schürmann, P., Decottignies, P., Miginiac-Maslow, M., et al. (1990). Human thioredoxin reactivity: Structure-function relationship. *Biochemical and Biophysical Research Communications*, *173*, 1375–1381.

Jacquot, J.-P., Droux, M., Miginiac-Maslow, M., Joly, C., & Gadal, P. (1984). Frozen thylakoids—An improvement for reconstituted chloroplast enzyme light-activation systems. *Plant Science Letters*, *35*, 181–185.

Jacquot, J.-P., Keryer, E., Issakidis, E., Decottignies, P., Miginiac-Maslow, M., Schmitter, J.-M., et al. (1991). Properties of recombinant NADP-malate dehydrogenases from *Sorghum vulgare* leaves expressed in *Escherichia coli* cells. *European Journal of Biochemistry*, *199*, 47–51.

Jacquot, J. P., Lopez-Jaramillo, J., Chueca, A., Cherfils, J., Lemaire, S., Chedozeau, B., et al. (1995). High level expression of recombinant pea chloroplast fructose-1,6-bisphosphatase and mutagenesis of its regulatory site. *European Journal of Biochemistry*, *229*, 675–681.

Jacquot, J. P., Lopez-Jaramillo, J., Miginiac-Maslow, M., Lemaire, S., Cherfils, J., Chueca, A., et al. (1997). Cysteine 153 is required for redox regulation of pea chloroplast fructose-1,6-bisphosphatase. *FEBS Letters*, *401*, 143–147.

Jacquot, J. P., Maudinas, B., & Gadal, P. (1979). Occurence of thioredoxins m in the photosynthetic bacteria *Rhodopseudomonas capsulata*. *Biochemical and Biophysical Research Communications*, *91*, 1371–1376.

Jacquot, J. P., Rivera-Madrid, R., Marinho, P., Kollarova, M., Le Maréchal, P., Miginiac-Maslow, M., et al. (1994). *Arabidopsis thaliana thaliana* NADPH thioredoxin reductase. *Journal of Molecular Biology*, *235*, 1357–1363.

Jacquot, J. P., Stein, M., Hodges, M., & Miginiac-Maslow, M. (1992). PCR cloning of a nucleotidic sequence coding for the mature part of *Chlamydomonas reinhradtii* thioredoxin Ch2. *Nucleic Acids Research*, *20*, 617.

Jacquot, J. P., Stein, M., Suzuki, A., Liottet, S., Sandoz, G., & Miginiac-Maslow, M. (1997). Residue Glu 91 of *Chlamydomonas reinhardtii* ferredoxin is essential for electron transfer to ferredoxin thioredoxin reductase. *FEBS Letters*, *400*, 293–296.

Jacquot, J. P., Vidal, J., & Gadal, P. (1976). Identification of a protein factor involved in dithiothreitol activation of NADP-malate dehydrogenase from French bean leaves. *FEBS Letters*, *71*, 223–227.

Jacquot, J. P., Vidal, J., & Gadal, P. (1977). Evidence for chloroplastic localization of spinach leaf NADP-malate dehydrogenase activating factors. *Planta*, *137*, 89–90.

Jacquot, J.-P., Vidal, J., Gadal, P., & Schürmann, P. (1978). Evidence for the existence of several enzyme-specific thioredoxins in plants. *FEBS Letters*, *96*, 243–246.

James, D., Borphukan, B., Fartyal, D., Achary, V. M. M., & Reddy, M. K. (2018). Transgenic manipulation of glutamine synthetase: a target with untapped potential in various aspects of crop improvement. In S. S. Gosal, & S. H. Wani (Eds.), *Biotechnology of crop improvement* (pp. 367–416). Springer International Publishing AG.

Johansson, K., Ramaswamy, S., Saarinen, M., Lemaire-Chamley, M., Issakidis-Bourguet, E., Miginiac-Maslow, M., et al. (1999). Structural basis for light activation of a chloroplast enzyme: The structure of sorghum NADP-malate dehydrogenase in its oxidized form. *Biochemistry*, *38*, 4319–4326.

Johnson, H. S., & Hatch, M. D. (1970). Properties and regulation of leaf nicotinamide-adenine dinucleotide phosphate-malate dehydrogenase and 'malic' enzyme in plants with the C4-dicarboxylic acid pathway of photosynthesis. *Biochemical Journal, 119*, 273–280.

Joy, K. W., & Hageman, R. H. (1966). The purification and properties of nitrite reductase from higher plants, and its dependence on ferredoxin. *Biochemical Journal, 100*, 263–273.

Kagawa, T., & Hatch, M. D. (1977). Regulation of C4 photosynthesis—characterization of a protein factor mediating activation and inactivation of NADP-malate dehydrogenase. *Archives of Biochemistry and Biophysics, 184*, 290–297.

Kamachi, K., Yamaya, T., Hayakawa, T., Mae, T., & Ojima, K. (1992). Vascular bundle-specific localization of cytosolic glutamine synthetase in rice leaves. *Plant Physiology, 99*, 1481–1486.

Kanamori, T., & Mastumoto, H. (1972). Glutamine synthetase from rice plants. *Achives of Biochemistry and Biophysics, 125*, 404–412.

Keryer, E., Collin, V., Lavergne, D., Lemaire, S., & Issakidis-Bourguet, E. (2004). Characterization of Arabidopsis mutants for the variable subunit of ferredoxin: thioredoxin reductase. *Photosynthesis Research, 79*, 265–274.

Keys, A. F., Bird, I. F., Cornelius, M. J., Lea, P. J., Wallsgrove, R. M., & Miflin, B. J. (1978). Photorespiratory nitrogen cycle. *Nature, 275*, 741–742.

Knaff, D. B., & Hirasawa, M. (1991). Ferredoxin-dependent chloroplast enzymes. *Biochimica et Biophysica Acta, 1056*, 93–125.

Kobe, B., & Kemp, B. E. (1999). Active site-directed protein regulation. *Nature, 402*, 373–376.

Krimm, I., Goyer, A., Issakidis-Bourguet, E., Miginiac-Maslow, M., & Lancelin, J.-M. (1999). Direct NMR observation of the thioredoxin-mediated reduction of the chloroplast NADP-malate dehydrogenase provides a structural basis for the relief of autoinhibition. *The Journal of Biological Chemistry, 274*, 34539–34542.

Labboun, S., Tercé-Laforgue, T., Roscher, A., Bedu, M., Restivo, F. M., Velanis, C. N., et al. (2009). Resolving the role of plant glutamate dehydrogenase: I. *In vivo* real time nuclear magnetic resonance spectroscopy experiments. *Plant Cell Physiology, 50*, 1761–1773.

Lam, H.-M., Coschigano, K. T., Oliveira, I. C., Melo-Oliveira, R., & Coruzzi, G. M. (1996). The molecular genetics of nitrogen assimilation into amino acids in higher plants. *Annual Review of Plant Physiology and Plant Molecular Biology, 47*, 569–593.

Lamkemeyer, P., Laxa, M., Collin, V., Li, W., Finkemeier, I., Schöttler, M. A., et al. (2006). Peroxiredoxin Q of *Arabidopsis thaliana* is attached to the thylakoids and functions in context of photosynthesis. *The Plant Journal, 45*, 968–981.

Lancelin, J. M., Bally, I., Arlaud, G. J., Blackledge, M., Gans, P., Stein, M., et al. (1994). NMR structures of ferredoxin chloroplastic transit peptide from Chlamydomonas reinhardtii promoted by trifluoroethanol in aqueous solution. *FEBS Letters, 343*, 261–266.

Lancelin, J. M., Stein, M., & Jacquot, J. P. (1993). Secondary structure and protein folding of recombinant chloroplastic thioredoxin Ch2 from the green alga *Chlamydomonas reinhardtii* as determined by ^{1}H NMR. *Journal of Biochemistry, 114*, 421–431.

Lancien, M., Ferrario-Méry, S., Roux, Y., Bismuth, E., Masclaux, C., Hirel, B., et al. (1999). Simultaneous expression of NAD-dependent isocitrate dehydrogenase and other Krebs cycle genes after nitrate resupply to short-term nitrogen starved *Nicotiana tabacum*. *Plant Physiology, 120*, 717–725.

Lancien, M., Gadal, P., & Hodges, M. (1998). Molecular charcterization of higher plant NAD-dependent isocitrate dehydrogenase: Evidence for a heteromeric structure by the complementation of yeast mutants. *Plant Journal, 16*, 325–333.

Lara, C., De la Torre, A., & Buchanan, B. B. (1980). Ferralterin: An iron-sulfur protein functional in enzyme regulation in photosynthesis. *Biochemical and Biophysical Research Communications, 94*, 1337–1344.

Laugier, E., Tarrago, L., Courteille, A., Innocenti, G., Eymery, F., Rumeau, D., et al. (2013). Involvement of thioredoxin y2 in the preservation of leaf methionine sulfoxide reductase capacity and growth under high light. *Plant, Cell & Environment, 36*, 670–682.

Lea, P. J., & Miflin, B. J. (1974). Alternative route for nitrogen assimilation in higher plants. *Nature, 251*, 614–616.

Lea, P. J., & Miflin, B. J. (2018). Nitrogen assimilation and its relevance to crop improvement. In C. H. Foyer, & H. Zhang (Eds.), *Nitrogen metabolism in plants in the post-genomic era: Vol. 42. Annual plant reviews* (pp. 1–40). Chichester, UK: Wiley-Blackwell.

Lemaire, S. D., Guillon, B., Le Maréchal, P., Keryer, E., Miginiac-Maslow, M., & Decottignies, P. (2004). New thioredoxin targets in the unicellular photosynthetic eukaryote *Chlamydomonas reinhardtii. Proceedings of the National Academy of Sciences of the United States of America, 101*, 7475–7480.

Lemaire, S. D., Hours, M., Gérard-Hirne, C., Trouabal, A., Roche, O., & Jacquot, J.-P. (1999). Analysis of light/dark synchronization of cell-wall-less *Chlamydomonas reinhardtii* (Chlorophyta) cells by flow cytometry. *European Journal of Phycology, 34*, 279–286.

Lemaire, S., Keryer, E., Stein, M., Schepens, I., Issakidis-Bourguet, E., Gérard-Hirne, C., et al. (1999). Heavy-metal regulation of thioredoxin gene expression in *Chlamydomonas reinhardtii. Plant Physiology, 120*, 773–778.

Lemaire, M., Miginiac-Maslow, M., & Decottignies, P. (1996). The catalytic site of chloroplastic NADP-dependent malate dehydrogenase contains a His/Asp pair. *European Journal of Biochemistry, 236*, 947–952.

Lemaire, S. D., Quesada, A., Merchan, F., Corral, J. M., Igeno, M. I., Keryer, E., et al. (2005). NADP-malate dehydrogenase from unicellular green alga *Chlamydomonas reinhardtii.* A first step toward redox regulation? *Plant Physiology, 137*, 514–521.

Lemaire, S. D., Richardson, J. M., Goyer, A., Keryer, E., Lancelin, J.-M., Makhatadze, G. I., et al. (2000). Primary structure determinants of the pH- and temperature-dependent aggregation of thioredoxin. *Biochimica et Biophysica Acta, 1476*, 311–323.

Lemaire, M., Schmitter, J.-M., Issakidis, E., Miginiac-Maslow, M., Gadal, P., & Decottignies, P. (1994). Essential histidine at the active site of sorghum leaf NADP-dependent malate dehydrogenase. *The Journal of Biological Chemistry, 269*, 27291–27296.

Lemaitre, T., Urbanczyk-Wochniak, E., Flesch, V., Bismuth, E., Fernie, A. R., & Hodges, M. (2007). NAD-dependent isocitrate dehydrogenase mutants of *Arabidopsis* suggest the enzyme is not limiting for nitrogen assimilation. *Plant Physiology, 144*, 1546–1558.

Lepiniec, L., Thomas, M., & Vidal, J. (2003). From enzyme activity to plant biotechnology: 30 years of research on phosphoenolpyruvate carboxylase. *Plant Physiology and Biochemistry, 41*, 533–539.

Lillig, C. H., Prior, A., Schwenn, J. D., Aslund, F., Ritz, D., Vlamis-Gardikas, A., et al. (1999). New thioredoxins and glutaredoxins as electron donors of 3'-phosphoadenylylsulfate reductase. *The Journal of Biological Chemistry, 274*, 7695–7698.

Lockard, R. G. (1959). In E. J. Hewitt (Ed.), *Sand and water culture methods* (p. 547). Bucks, England: The Commonwealth Agricultural Bureau.

Lopez-Jaramillo, J., Chueca, A., Jacquot, J. P., Hermoso, R., Lazaro, J. J., Sahrawy, M., et al. (1997). High-yield expression of pea thioredoxin m and assessment of its efficiency in chloroplast fructose-1,6-bisphosphatase activation. *Plant Physiology, 114*, 1169–1175.

Lowry, O. H., Rosenbrowgh, N. J., Farr, A. L., & Randall, R. J. (1951). Protein measurement with folin phenol reagent. *Journal of Biological Chemistry, 193*, 265–275.

Mann, A. F., Fentem, P. A., & Steward, G. R. (1979). Identification of two forms of glutamine synthetase in barley (*Hordeum vulgare* L.). *Biochemical and Biophysical Research Communications, 88*, 512–521.

Marchand, C., Le Maréchal, P., Meyer, Y., Miginiac-Maslow, M., Issakidis-Bourguet, E., & Decottignies, P. (2004). New targets of Arabidopsis thioredoxins revealed by proteomic analysis. *Proteomics, 4*, 2696–2706.

Marchand, C. H., Vanacker, H., Collin, V., Issakidis-Bourguet, E., Le Maréchal, P., & Decottignies, P. (2010). Thioredoxin targets in Arabidopsis roots. *Proteomics, 10*, 2418–2428.

Martin, A., Lee, J., Kichey, T., Gerentes, D., Zivy, M., Tatou, C., et al. (2006). Two cytosolic glutamine synthetase isoforms of maize (*Zea mays* L.) are specifically involved in the control of grain production. *The Plant Cell, 18*, 3252–3274.

Masclaux, C., Quilleré, I., Gallais, A., & Hirel, B. (2001). The challenge of remobilisation in plant nitrogen economy. A survey of physio-agronomic and molecular approaches. *Annals of Applied Biology, 138*, 69–81.

Masclaux-Daubresse, C., Reisdorf-Cren, M., Pageau, K., Lelandais, M., Grandjean, O., Kronenberger, J., et al. (2006). Glutamine synthetase/glutamate synthase pathway and glutamate dehydrogenase play distinct roles for sink/source nitrogen cycle in tobacco (*Nicotiana tabacum* L.). *Plant Physiology, 140*, 444–456.

Maudinas, B., Chemardin, M., Yovanovitch, E., & Gadal, P. (1981). Gnotobiotic cultures of rice plants up to ear stage in the absence of combined nitrogen source but in the presence of free-living nitrogen fixing bacteria *Azotobacter vinelandii* and *Rhodopseudomonas capsulata*. *Plant and Soil, 60*, 85–97.

McNally, S. F., Hirel, B., Gadal, P., Mann, A. F., & Stewart, G. R. (1983). Glutamine synthetase of higher plants: Evidence for a specific isoenzyme content related to their possible physiological role and their compartmentation within the leaf. *Plant Physiology, 72*, 22–25.

Miao, G. H., Hirel, B., Marsolier, M. C., Ridge, R. W., & Verma, D. P. S. (1991). Ammonia regulated expression of a soybean gene encoding cytosolic glutamine synthetase in transgenic *Lotus corniculatus*. *The Plant Cell, 3*, 11–22.

Michelet, L., Zaffagnini, M., Marchand, C., Collin, V., Decottignies, P., Tsan, P., et al. (2005). Glutathionylation of chloroplast thioredoxin f is a redox signaling mechanism in plants. *Proceedings of the National Academy of Sciences of the United States of America, 102*, 16478–16483.

Michelet, L., Zaffagnini, M., Morisse, S., Sparla, F., Perez-Perez, M.-E., Francia, F., et al. (2013). Redox regulation of the Calvin-Benson cycle: Something old, something new. *Frontiers in Plant Science, 4*, 470.

Miflin, B. J., & Lea, P. J. (1975). Glutamine and asparagine as nitrogen donors for reductant-dependent glutamate synthesis in roots. *Biochemical Journal, 149*, 403–409.

Miflin, B. J., & Lea, P. J. (1977). Amino acid metabolism. *Annual Review of Plant Physiology, 28*, 299–329.

Miginiac-Maslow, M., Decottignies, P., Jacquot, J.-P., & Gadal, P. (1990). Regulation of corn leaf NADP-malate dehydrogenase light-activation by the photosynthetic electron flow. Effect of photoinhibition studied in a reconstituted system. *Biochimica et Biophysica Acta, 1017*, 273–279.

Miginiac-Maslow, M., & Hoarau, A. (1975). The reducing power and the CO_2 fixation activity of isolated choroplasts. *Physiologia Plantarum, 35*, 186–190.

Miginiac-Maslow, M., Jacquot, J.-P., & Droux, M. (1985). Energetic aspects of the light activation of two chloroplast enzymes: Fructose-1,6-bisphosphatase and NADP-malate dehydrogenase. *Photosynthesis Research, 6*, 201–213.

Miranda-Vizuete, A., Damdimopoulos, A. E., Gustafsson, J., & Spyrou, G. (1997). Cloning, expression, and characterization of a novel *Escherichia coli* thioredoxin. *Journal of Biological Chemistry, 49*, 30841–30847. https://doi.org/10.1074/jbc.272.49.30841. PMID: 9388228.

Mittard, V., Morelle, N., Brutscher, B., Simorre, J. P., Marion, D., Stein, M., et al. (1995). 1H, 13C, 15N-NMR resonance assignments of oxidized thioredoxin h from the eukaryotic green alga *Chlamydomonas reinhardtii* using new methods based on two-dimensional triple resonance NMR spectroscopy and computer assisted backbone assignment. *European Journal of Biochemistry, 229*, 473–485.

Morisse, S., Zaffagnini, M., Gao, X. H., Lemaire, S. D., & Marchand, C. H. (2014). Insight into protein S-nitrosylation in *Chlamydomonas reinhardtii. Antioxidants & Redox Signaling, 20*, 1271–1284.

Oaks, A. (1995). Evidence for deamination by glutamate dehydrogenase in higher plants: Reply. *Canadian Journal of Botany, 73*, 1116–1117.

Oliver, G., Gosset, G., Sanchez-Pescador, R., Lozoya, E., Ku, L. M., Flores, N., et al. (1987). Determination of the nucleotide sequence for the glutamate synthase structural genes of *Escherichia coli* K-12. *Gene, 60*, 1–11.

Picault, N., Palmieri, L., Pisano, M., Hodges, M., & Palmieri, F. (2002). Identification of a novel transporter for dicarboxylates and tricarboxylates in plant mitochondria. Bacterial expression, reconstitution, functional characterization, and tissue distribution. *Journal of Biological Chemistry, 227*, 24204–24211. https://doi.org/10.1074/jbc.M202702200.

Plett, D., Garnett, T. P., & Okamoto, M. (2017). Molecular genetics to discover and improve nitrogen use efficiency in crop plants. In M. Anwar Hossain, T. Kamiya, D. Buritt, L. S. Phan Tran, & T. Fujiwara (Eds.), *Plant macronutrient use efficiency* (pp. 93–122). Academic Press.

Potel, F., Valadier, M.-H., Ferrario-Méry, S., Grandjean, O., Marin, H., Gaufichon, L., et al. (2009). Assimilation of excess ammonium into amino acids and nitrogen translocation in Arabidopsis thaliana—Roles of glutamate synthases and carbamoylphosphate synthetase in leaves. *The FEBS Journal, 276*, 4061–4076.

Rahantaniaina, M.-S., Li, S., Chatel-Innocenti, G., Tuzet, A., Issakidis-Bourguet, E., Mhamdi, A., et al. (2017). Cytosolic and chloroplastic DHARs cooperate in oxidative stress-driven activation of the salicylic acid pathway. *Plant Physiology, 174*, 956–971.

Rivera-Madrid, R., Mestres, D., Marinho, P., Jacquot, J. P., Decottignies, P., Miginiac-Maslow, M., et al. (1995). Evidence for five divergent thioredoxin h sequences in *Arabidopsis thaliana. Proceedings of the National Academy of Sciences of the United States of America, 92*, 5620–5624.

Rogers, W. J., Hodges, M., Decottignies, P., Schmitter, J.-M., Gadal, P., & Jacquot, J.-P. (1992). Isolation of a cDNA fragment coding for *Chlamydomonas reinhardtii* ferredoxin and expression of the recombinant protein in *Escherichia coli. FEBS Letters, 310*, 240–245.

Ruelland, E., Johansson, K., Decottignies, P., Djukic, N., & Miginiac-Maslow, M. (1998). The autoinhibition of sorghum NADP malate dehydrogenase is mediated by a C-terminal negative charge. *The Journal of Biological Chemistry, 273*, 33482–33488.

Ruelland, E., Lemaire-Chamley, M., Le Maréchal, P., Issakidis-Bourguet, E., Djukic, N., & Miginiac-Maslow, M. (1997). An internal cysteine is involved in the thioredoxin-dependent activation of sorghum leaf NADP-malate dehydrogenase. *The Journal of Biological Chemistry, 272*, 19851–19857.

Sanger, F., Nicklen, S., & Coulson, A. R. (1977). DNA sequencing with chain-terminating inhibitors. *Proceedings of the National Academy of Sciences of the United States of America, 74*, 5463–5467.

Scheibe, R. (2004). Malate valves to balance cellular energy supply. *Physiologia Plantarum, 120*, 24–29.

Scheibe, R., & Jacquot, J.-P. (1983). NADP regulates the light activation of NADP-dependent malate dehydrogenase. *Planta, 157*, 548–553.

Schepens, I., Decottignies, P., Ruelland, E., Johansson, K., & Miginiac-Maslow, M. (2000). The dimer contact area of sorghum NADP-malate dehydrogenase: Role of aspartate 101 in dimer stability and catalytic activity. *FEBS Letters, 471*, 240–244.

Schepens, I., Johansson, K., Decottignies, P., Gillibert, M., Hirasawa, M., Knaff, D. B., et al. (2000). Inhibition of the thioredoxin-dependent activation of the NADP-malate dehydrogenase and cofactor specificity. *The Journal of Biological Chemistry*, *275*, 20996–21001.

Schepens, I., Ruelland, E., Miginiac-Maslow, M., Le Maréchal, P., & Decottignies, P. (2000). The role of active site arginines of sorghum NADP-malate dehydrogenase in thioredoxin-dependent activation and activity. *The Journal of Biological Chemistry*, *275*, 35792–35798.

Schmitter, J. M., Jacquot, J. P., de Lamotte-Guéry, F., Beauvallet, C., Dutka, S., Gadal, P., et al. (1988). Purification, properties and complete amino-acid sequence of the ferredoxin from a green alga, *Chlamydomonas reinhardtii*. *European Journal of Biochemistry*, *172*, 405–412.

Schröder, M., Giermann, N., & Zrenner, R. (2005). Functional analysis of the pyrimidine de novo synthesis pathway in *Solanaceous* spiecies. *Plant Physiology*, *138*, 1926–1938.

Schürmann, P., & Jacquot, J. P. (1979). Improved in vitro light activation and assay systems for two spinach chloroplast enzymes. *Biochimica et Biophysica Acta*, *569*, 309–312.

Schürmann, P., Wolosiuk, R. A., Breazeale, V. D., & Buchanan, B. B. (1976). Two proteins function in the regulation of photosynthetic CO2 assimilation in chloroplasts. *Nature*, *263*, 257–258.

Scopes, R. K. (1974). Measurement of protein by spectrophotometry at 205 nm. *Analytical Biochemistry*, *59*, 277–282.

Shin, M., & Oshino, R. J. (1978). Ferredoxin-sepharose 4B as a tool for the purification of ferredoxin-NADP reductase. *Biochemical Journal*, *83*, 357–361.

Skopelitis, D., Paranychiankis, N. V., Paschalidis, K. A., Plianokis, E. D., Delis, I. D., Yakoumakis, D., et al. (2006). Abiotic stress generates ROS that signal expression of anionic glutamate dehydrogenase to form glutamate for proline synthesis in tobacco and grapevine. *Plant Cell*, *18*, 2767–2781.

Smith, E. L., Austen, B. M., Blumenthal, K. M., & Nyc, J. F. (1975). 5 Glutamate dehydrogenases. *The Enzymes*, *11*, 293–367.

Somerville, C. R., & Ogren, W. L. (1980). Inhibition of photosynthesis in Arabidopsis mutants lacking in leaf glutamate synthase activity. *Nature*, *286*, 257–259.

Stein, M., Chedozeau, B., & Jacquot, J. P. (1996). Cloning and sequencing of a ferredoxin gene (Genbank U29516) from *Chlamydomonas reinhardtii*. *Plant Physiology*, *109*, 721.

Stein, M., Jacquot, J. P., Jeannette, E., Decottignies, P., Hodges, M., Lancelin, J., et al. (1995). *Chlamydomonas reinhardtii* thioredoxins: Structure of the genes coding for the chloroplastic m and cytosolic h isoforms; expression in Escherichia coli of the recombinant proteins, purification and biochemical properties. *Plant Molecular Biology*, *28*, 487–503.

Stein, M., Jacquot, J. P., & Miginiac-Maslow, M. (1993). A cDNA clone encoding *Chlamydomonas reinhardtii* preferredoxin. *Plant Physiology*, *102*, 1349–1350.

Stewart, G. R., Shatilov, V. R., Turnbull, M. H., Robinson, S. A., & Goodall, R. (1995). Evidence that glutamate dehydrogenase plays a role in oxidative deamination of glutamate in seedlings of *Zea mays*. *Australian Journal of Plant Physiology*, *22*, 805–809.

Suzuki, A., Audet, C., & Oaks, A. (1987). Influence of light on the ferredoxin-dependent glutamate synthase in maize leaves. *Plant Physiology*, *84*, 578–581.

Suzuki, A., Burkhart, W., & Rothstein, S. (1996). Nitrogen effects on the induction of ferredoxin-dependent glutamate synthase and its mRNA in maize leaves under the light. *Plant Science*, *114*, 83–91.

Suzuki, A., & Gadal, P. (1982). Glutamate synthase from rice leaves. *Plant Physiology*, *69*, 848–852.

Suzuki, A., & Gadal, P. (1984). Glutamate synthase. Physicochemical and functional properties of different forms in higher plants and in other organisms. *Physiologie Végétale*, *22*, 471–486.

Suzuki, A., Gadal, P., & Oaks, A. (1981). Intracellular distribution of enzymes associated with nitrogen assimilation in roots. *Planta*, *151*, 457–461.

Suzuki, A., Jacquot, J. P., & Gadal, P. (1983). Glutamate synthase in rice roots. Studies on the electron donor specificity. *Phytochemistry*, *22*, 1543–1546.

Suzuki, A., & Knaff, D. B. (2005). Glutamate synthase: Structural, mechanistic and regulatory properties, and role in the amino acid metabolism. Review. *Photosynthesis Research*, *83*, 191–217.

Suzuki, A., Oaks, A., Jacquot, J. P., Vidal, J., & Gadal, P. (1985). An electron donor system in maize roots for reaction of glutamate synthase and nitrite reductase. Physiological and immunological properties of the electron carrier and pyridine nucleotide reductase. *Plant Physiology*, *78*, 374–378.

Suzuki, A., Rioual, S., Lemarchand, S., Godfroy, N., Roux, Y., Boutin, J. P., et al. (2001). Regulation by light and metabolite of ferredoxin-dependent glutamate synthase in maize. *Physiologia Plantarum*, *112*, 524–530.

Suzuki, A., & Rothstein, S. (1997). Structure and regulation of ferredoxin-dependent glutamate synthase in *Arabidopsis thaliana*. Cloning of cDNA, expression of different tissues of wild-type and *gltS* mutant strains, and light induction. *European Journal of Biochemistry*, *243*, 708–718.

Suzuki, A., Vergnet, C., Morot-Gaudry, J.-F., Zehnacker, C., & Grosclaude, J. (1994). Immunological characterization of ferredoxin and methyl viologen interacting domains of glutamate synthase using monoclonal antibodies. *Plant Physiology and Biochemistry*, *32*, 619–626.

Suzuki, A., Vidal, J., & Gadal, P. (1982). Glutamate synthase isoforms. Immunological studies of enzymes in green leaf, etiolated leaf and root tissues. *Plant Physiology*, *70*, 827–832.

Suzuki, A., Vidal, J., Nguyen, J., & Gadal, P. (1984). Occurrence of ferredoxin-dependent glutamate synthase in plant cell fraction of soybean nodules (*Glycine max*). *FEBS Letters*, *173*, 204–208.

Thomsen, H. C., Eriksson, D., Moller, I. S., & Schojoerring, J. K. (2014). Cytosolic glutamine synthetase: A target for improvement of crop nitrogen use efficiency? *Trends in Plant Science*, *19*, 656–663.

Thormählen, I., Ruber, J., von Roepenack-Lahaye, E., Ehrlich, S. M., Massot, V., Hümmer, C., et al. (2013). Inactivation of thioredoxin f1 leads to decreased light activation of ADP-glucose pyrophosphorylase and altered diurnal starch turnover in leaves of Arabidopsis plants. *Plant, Cell & Environment*, *36*, 16–29.

Thormählen, I., Zupok, A., Rescher, J., Leger, J., Weissenberger, S., Groysman, J., et al. (2017). Thioredoxins play a crucial role in dynamic acclimation of photosynthesis in fluctuating light. *Molecular Plant*, *10*, 168–182.

Truernit, E., Bauby, H., Dubreucq, B., Grandjean, O., Runions, J., Barathélémy, J., et al. (2008). High-resolution whole-mount imaging of three-dimensional tissue organisation and gene expression enables the study of phloem development and structure in Arabidopsis. *The Plant Cell*, *20*, 1494–1503.

Valadier, M.-H., Yoshida, A., Grandjean, O., Morin, H., Kronenberger, J., Boutet, S., et al. (2008). Implication of the glutamine synthetase-glutamate synthase pathway in conditioning the amino acid metabolism in bundle sheath and mesophyll cells of maize leaves. *FEBS Journal*, *275*, 3193–3206.

Vanacker, H., Guichard, M., Bohrer, A.-S., & Issakidis-Bourguet, E. (2018). Redox regulation of monodehydroascorbate reductase by thioredoxin y in plastids revealed in the context of water stress. *Antioxidants*, *7*, E183.

Vavitsas, K., Crozet, P., Vinde, M. H., Davies, F., Lemaire, S. D., & Vickers, C. E. (2019). The synthetic biology toolkit for photosynthetic microorganisms. *Plant Physiology*, *181*, 14–27.

Vidal, J., Cavalie, G., & Gadal, P. (1976). Studies on bean and sorghum phosph*oenol*pyruvate carboxylase with polyacrylamide gel electrophoresis. *Plant Science Letters*, 7, 265–270.
Vidal, J., & Chollet, R. (1997). Regulatory phosphorylation of C_4 PEPC carboxylase. *Trends in Plant Science*, *2*, 230–237.
Vidal, J., Jacquot, J. P., & Gadal, P. (1980). Light activation of NADP malate dehydrogenase in a reconstituted chloroplastic system. *Phytochemistry*, *19*, 1919–1924.
Vidal, J., Jacquot, J. P., Gadal, P., & Vidal, D. (1978). Influence of protein factors on activation of NADP-malate dehydrogenase by dithiothreitol. *Physiologia Plantarum*, *42*, 307–314.
Vidal, J., Jacquot, J. P., Membre, H., & Gadal, P. (1978). Detection and study of protein factors involved in dithiothreitol activation of NADP-malate dehydrogenase from a C4 plant. *Plant Science Letters*, *11*, 305–310.
Wallsgrove, R. M., Lea, P. J., & Miflin, B. J. (1979). Distribution of the enzymes of nitrogen assimilation within the pea leaf cell. *Plant Physiology*, *63*, 232–236.
Wallsgrove, R. M., Turner, J. C., Hall, N. P., Kendall, A. C., & Bright, S. W. J. (1987). Barley mutants lacking chloroplast glutamine synthetase. Biochemical and genetic analysis. *Plant Physiology*, *83*, 155–158.
Walters, E. M., Garcia-Serres, R., Jameson, G. N., Glauser, D. A., Bourquin, F., Manieri, W., et al. (2005). Spectroscopic characterization of site-specific [Fe(4)S(4)] cluster chemistry in ferredoxin:thioredoxin reductase: Implications for the catalytic mechanism. *Journal of the American Chemical Society*, *127*, 9612–9624.
Wetterauer, B., Jacquot, J. P., & Véron, M. (1992). Thioredoxins from *Dictyostelium discoideum* are a developmentally regulated multigene family. *The Journal of Biological Chemistry*, *267*, 9895–9904.
Wildner, G. F., & Criddle, R. S. (1969). Ribulose diphosphate carboxylase I. A factor involved in light activation of the enzyme. *Biochemical and Biophysical Research Communications*, *37*, 952–956.
Wollmann, E., d'Auriol, L., Rimsky, L., Shaw, A., Jacquot, J. P., Wingfield, P., et al. (1988). Cloning and expression of a cDNA for human thioredoxin. *The Journal of Biological Chemistry*, *263*, 15506–15512.
Wolosiuk, R. A., & Buchanan, B. B. (1977). Thioredoxin and glutathione regulate photosynthesis in chloroplasts. *Nature*, *266*, 565–567.
Wolosiuk, R. A., Buchanan, B. B., & Crawford, N. A. (1977). Regulation of NADP-malate dehydrogenase by the light-activated ferredoxin/thioredoxin system of chloroplasts. *FEBS Letters*, *81*, 253–258.
Wolosiuk, R. A., Crawford, N. A., & Buchanan, B. B. (1979). Isolation of three thioredoxins from spinach leaves. *The Journal of Biological Chemistry*, *254*, 1627–1632.
Yamaya, T., & Kusano, M. (2014). Evidence supporting distinct functions of three cytosolic glutamine synthetases and two NADH-glutamate synthases. *Journal of Experimental Botany*, *19*, 5519–5525.
Yoneyama, T., & Suzuki, A. (2019). Exploration of nitrate-to-glutamate assimilation in non-photosynthetic roots of higher plants by studies of ^{15}N-tracing, enzymes involved, reductant supply, and nitrate signaling: A review and synthesis. *Plant Physiology and Biochemistry*, *136*, 245–254.
Zaffagnini, M., Bedhomme, M., Groni, H., Marchand, C. H., Puppo, C., Gontero, B., et al. (2012). Glutathionylation in the photosynthetic model organism *Chlamydomonas reinhardtii*: a proteomic survey. *Molecular and Cellular Proteomics*, *11*, M111.014142.
Ziegler, C., Feraud, M., Jouglet, T., Viret, L., Spampinato, A., Paganelli, V., et al. (2003). Regulation of promoter activity of ferredoxin-dependent glutamate synthase. *Plant Physiology and Biochemistry*, *41*, 649–655.

CHAPTER FOUR

Glutamate synthase and amino acid synthesis in higher plants

Akira Suzuki*
INRAE, Institut Jean-Pierre Bourgin, UMR1318, AgroParisTech, Versailles Cedex, France
*Corresponding author: e-mail address: akirasuzuki004@gmail.com

Contents

Abstract

The discovery of glutamate synthase (NADPH-GOGAT, EC 1.4.1.13 and Fd-GOGAT, EC 1.4.7.1) in the 1970s gave rise to a new pathway of NH_4^+ assimilation into amino acids in concert with glutamine synthetase (GS, EC 6.3.1.2) (GS/GOGAT pathway). The subsequent studies led to an overall reassessment of the inorganic nitrogen assimilation into amino acids. This review dissects the research in Prof Pierre Gadal's laboratory in the 1970s and early 1980s, and the subsequent studies that contributed to our understanding of the primary nitrogen metabolism in plants.

Abbreviations

AS asparagine synthetase
AsnAT asparagine aminotransferase
AspAT aspartate aminotransferase
Fd ferredoxin
GDH glutamate dehydrogenase
GOGAT glutamate synthase
GS glutamine synthetase
CPSase carbamoylphospahte synthase

Advances in Botanical Research, Volume 100
ISSN 0065-2296
https://doi.org/10.1016/bs.abr.2021.01.005
Copyright © 2021 Elsevier Ltd
All rights reserved.

1. Introduction

Plants utilize inorganic nitrogen mainly as nitrate (NO_3^-) and ammonium (NH_4^+) absorbed in roots and transported in part to aerial tissues (Fig. 1). The inorganic nitrogen is assimilated into amino acids at different stages of life cycle from seed germination, vegetative growth, flowering and senescence to seed filling and maturation. Unraveling the molecular mechanism of inorganic nitrogen assimilation into amino acids and enabling an improved utilization of nitrogen have a strong impact on the yield and quality of plants because nitrogen is a rate-limiting element for plant growth.

Feeding studies of $^{15}NO_3^-$ or $^{15}NH_4^+$ to bacteria, yeasts and plants in the 1960s have assessed ^{15}N assimilatory pathways into amino acids (see Yoneyama & Suzuki, 2019 for review). The discovery of NADPH-glutamate synthase in the 1970s in *Aerobacter aerogenes* (Tempest, Meers, & Brown, 1970) and ferredoxin (Fd)-glutamate synthase in pea chloroplasts (Lea & Miflin, 1974) revealed a new pathway of NH_4^+ assimilation into Gln and Glu coupled with glutamine synthetase (GS, EC 6.3.1.2). A rapid increase in [^{15}N]Gln at the amide group in leaves and roots fed with $^{15}NO_3^-$ or $^{15}NH_4^+$ correlated with a reaction catalyzed by GS. Subsequent labeling of [^{15}N]Glu matched the GOGAT reaction, which transfers the [^{15}N]amide group of Gln to the carbon-2 of 2-oxoglutarate (2-OG), yielding one [2–^{15} N]Glu and one Glu (Fig. 1).

Glutamate synthase, designated by the acronym GOGAT (glutamine:2-oxoglutarate aminotransferase), occurs as Fd-GOGAT isoform (EC 1.4.7.1) and NADH-GOGAT isoform (EC 1.4.1.14), which are both located in the chloroplasts or plastids in plants (Suzuki & Gadal, 1984). The functional GS/GOGAT pathway led to an overall re-review of the traditional route of glutamate formation *via* the 2-OG amination with NH_4^+ by mitochondrial NADH-glutamate dehydrogenase (NADH-GDH, EC 1.4.1.2). *In vitro*, GDH catalyzes the reversible deamination of Glu to NH_4^+, 2-OG forming NADH.

Nitrogen assimilation into amino acids is preconditioned by the GS/GOGAT cycle. In an ATP-dependent reaction, cytosolic asparagine synthetase (AS, EC 6.3.5.4) transfers the Gln-amide group to Asp generating Asn and Glu. Glu can be transaminated back to Asp *via* aspartate aminotransferase (AspAT, EC 2.6.1.1) (Fig. 1). These four amino acids (Gln, Glu, Asn and Asp) present the major nitrogen transport compounds of the plant (Fig. 1). Also, carbamoylphosphate synthetase (CPSase, EC 6.3.5.5) uses the Gln-amide

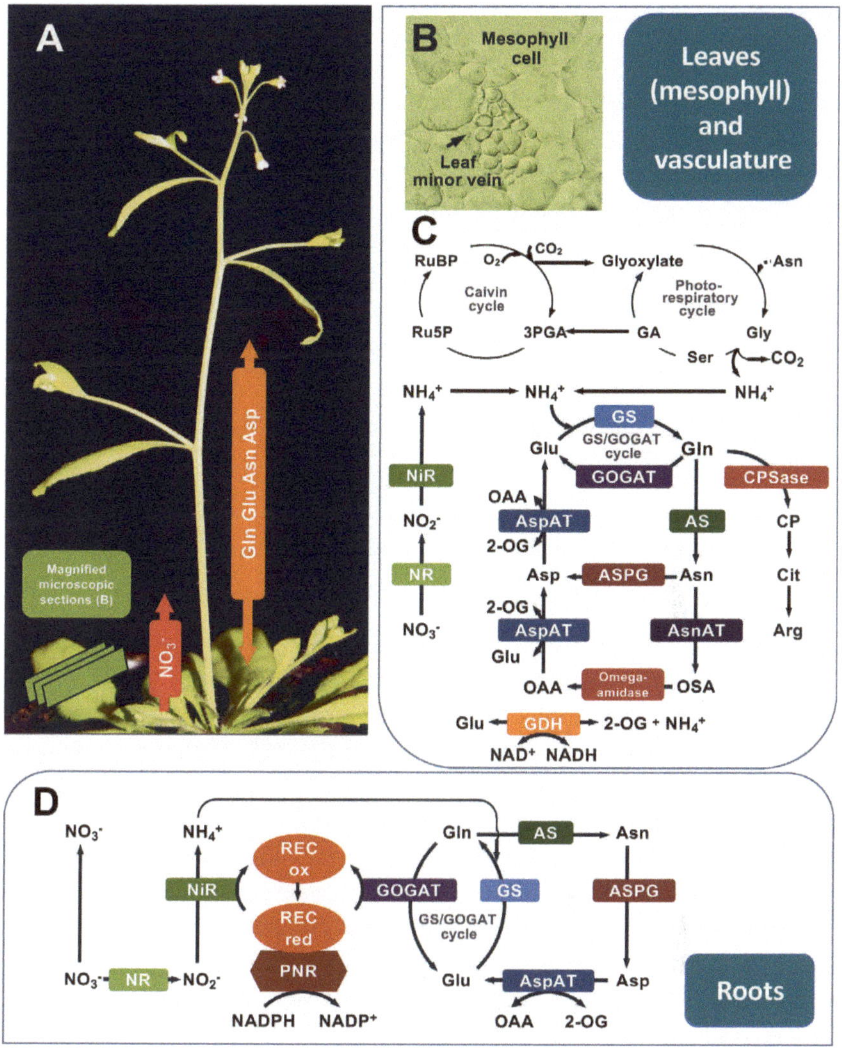

Fig. 1 Schematic pathways of nitrogen distribution during primary assimilation, photorespiratory re-assimilation and translocation in Arabidopsis. (A) Transport of nitrate in the xylem and amino acids in the phloem, and cauline leaf sectioning for microscopic observation, (B) microscopic image of cauline leaf section, (C) simplified primary nitrogen metabolism in leaf mesophyll and leaf vascular bundle, and (D) selected primary nitrogen metabolism in roots. Amino acids are shown as the three-letter symbols. 2-OG, 2-oxo-glutarate; 3PGA, 3-phosphoglycerate; AS, asparagine synthetase (EC, 6.3.5.4); AsnAT, asparagine aminotransferase (EC 2.6.1.45); AspAT, aspartate aminotransferase (EC, 2.6.1.1); ASPG, asparaginase (EC 3.5.1.1); CP, carbamoyl phosphate; CPSase, carbamoylphosphate synthetase (glutamine hydrolyzing CPSase, EC 6.3.3.5); GA, glycerate; GDH, NADH-glutamate dehydrogenase (NADH-GDH, EC 1.4.1.2); GOGAT, Fd-glutamate synthase (Fd-GOGAT, EC 1.4.7.1)/NADH-glutamate synthase (NADH-GOGAT, EC 1.4.1.14); GS, glutamine synthetase (GS, EC6.3.1.2); NH_4^+, ammonium; NiR, Fd-nitrite reductase (Fd-NiR, EC 1.6.6.4); NO_2^-, nitrite; NO_3^-, nitrate; NR, NAD(P)H-nitrate reductase (NAD(P)H-NR, EC 1.6.6.1/EC 1.6.6.2/EC 1.6.6.3); OAA, oxaloacetate; Omega-amidase (EC 3.5.1.3); OSA, 2-oxosuccinamate; PNR, pyridine nucleotide oxidoreductase – FNR-like root protein (EC 1.18.1.2); REC, root electron carrier; Ru5P, ribulose-5-bisphosphate; RuDP, ribulose-1,5-bisphosphate.

group, bicarbonate (HCO_3^-) and ATP to form carbamoylphosphate (CP), a precursor of arginine and pyrimidine synthesis (Slocum, 2005) (Fig. 1).

The research in Prof Pierre Gadal's laboratory in the 1970s and early 1980s contributed to our understanding of the inorganic nitrogen assimilation into amino acids in plants.[a] This review covers the molecular mechanisms of nitrogen assimilation revealed by the research in the Prof Gadal's laboratory together with the structure and reaction mechanism of GOGATs. This article also assesses the specific and/or redundant functions of GOGAT isoforms mainly in *Arabidopsis thaliana* demonstrated by the current genetic engineering and phenotyping, imaging of cells and organelles, databases of -omics and high throughput technologies.

2. Structure and reaction mechanism of GOGAT

2.1 Structural relationships of three forms

Three forms of GOGAT are deduced by the DNA sequencing of genomes, genes and cDNA in a wide variety of organisms including bacteria, yeasts, fungi, algae and higher plants. Bacterial GOGAT is an NADPH-dependent iron-sulfur flavoprotein composed of a dissimilar $(\alpha\beta)_x$ structure. Two genes *gltB* and *gltD* encode the α subunit (≈160 kDa) and β subunit (≈52 kDa), respectively. Archaeal genomes contain an ORF (open reading frame) similar to the α subunit active with bacterial-type ferredoxin (Fd), while a sequence similar to the β subunit is either absent or not detected (Suzuki & Knaff, 2005). The second form of NADH-GOGAT is a single long polypeptide (≈220 kDa) homologous to a fused bacterial α and β subunits specific to NADH in yeasts (encoded by *gltBD*), fungi (*gltBD*) and higher plants (*GLT*), while, cyanobacterial NADH-GOGAT is an (αβ) heterodimer. The α subunit harbors one FMN cofactor and one $[3Fe—4S]^{0,+1}$ cluster and the β subunit contains one FAD, whereas two additional $[4Fe—4S]^{+1,+2}$ appear to be located at the interface between the two subunits. The third form of Fd-GOGAT in plants is a single flavoprotein similar both in size (≈165 kDa) and amino acid sequence up to 40% to the bacterial α subunit, but it lacks a region similar to the prokaryotic β subunit (Suzuki & Knaff, 2005). Fd-GOGAT has one FMN and one $[3Fe—4S]^{0,+1}$ in cyanobacteria (encoded by *glsF* or *gltS*), algae (*gltB* or *glsF*) and higher plants (*GLU, glu, gluS* or *gltS*).

[a] Remarks on the research in the Prof Gadal's laboratory: Prof Gadal headed a team to investigate the carbon and nitrogen metabolism of the plant. The research was carried out successively at Univesité de Nancy (actual Université de Loraine) then at Université de Paris XI following his mutation to Orsay in 1981.

Provided that the sequence similarity of Fd-GOGAT to the *gltB*-encoded α subunit is correlated with evolution, prokaryotic *gltB* genes may have given rise to the cyanobacterial Fd-GOGAT gene by endosymbiosis, followed by a plastid gene transfer to plants. On the other hand, *gltB* and *gltD* may have been fused to a single gene for NADH-GOGAT in yeasts, fungi and plants (Suzuki & Knaff, 2005; Suzuki & Rothstein, 1997).

2.2 Spatial properties and reaction mechanism

X-ray crystallographic study determined the tertiary structure of α subunit of NADPH-GOGAT from *Azospririllum brasilense* (Binda et al., 2000) and Fd-GOGAT from *Synechocystis* sp. (Van den Heuvel, Curti, Vanoni, & Mattevi, 2004). A high similarity of the bacterial α subunit *vs* the amino acid sequence of *Synechocystis* Fd-GOGAT (45%) and plant Fd-GOGAT such as Arabidopsis *GLU1*-encoded Fd-GOGAT (43%) (Suzuki & Rothstein, 1997) indicates a similar and ordered reaction mechanism of these GOGATs in the synthase domain, NH_2-terminal Gln amidotransferase (GAT) domain and ammonia tunnel to synthase domain.

(1) Synthase domain: Fd-GOGAT reaction is initiated by the binding to the synthase domain of 2-OG and Fd reduced though Photosystem I (PSI) (Fig. 2). This domain is composed of Pro927 to Leu1378 on the mature Arabidopsis *GLU1* Fd-GOGAT (At-*GLU1* Fd-GOGAT) or Lys829 to Gln1265 on *Synechosystis* Fd-GOGAT (Sy-Fd-GOGAT) and displays a $(\beta/\alpha)_8$ barrel associated with the FMN and the $[3Fe{-}4S]^{0,+1}$ cluster (Fig. 2).

(2) N-terminal GAT domain: Binding of Gln to a class II (or *purF* type) N-terminal glutamine amidotransferase (GAT) domain forms a cysteinyl-glutamine tetrahedral intermediate with Cys (Cys132 of At-*GLU1* Fd-GOGAT or Cys43 of Sy-Fd-GOGAT) (Suzuki & Rothstein, 1997; Fig. 2). Its hydrolysis to NH_3 and Glu occurs in the GAT domain, which has a topology of a four-layer α/β/β/α (Cys132-Lys556 of At-*GLU1* Fd-GOGAT or Cys43-Leu464 of Sy-Fd-GOGAT). Removal of NH_3 forms λ-glutamyl thioester intermediate and its hydrolysis releases the first molecule of Glu (van den Heuvel et al., 2004) (Fig. 2). Contrary to the GDH reaction, the GAT reaction is irreversible, and the absence of an ordered reaction prevents the overall reaction (Yoneyama, Fujimori, Yanagisawa, Hase, & SuzukI, 2015).

(3) Ammonia tunnel to synthase domain: Free NH_3 is channeled by fusion through the ammonia tunnel to the synthase domain (Fig. 2). It is

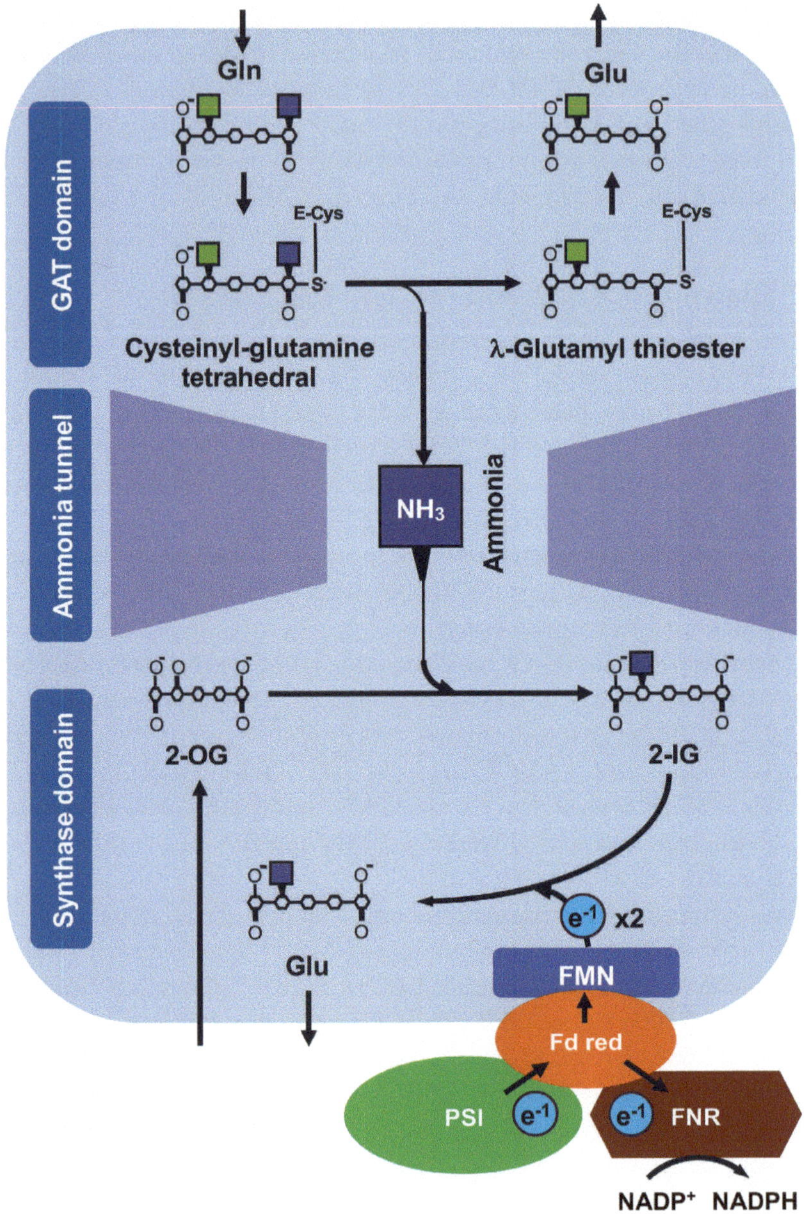

Fig. 2 Reaction model of Fd-glutamate synthase in photosynthetic tissues. 2-IG, 2-iminoglutarate; Cys, enzyme bound cysteine residue; E, enzyme polypeptide; Fd_{red}, reduced ferredoxin; FMN, enzyme bound flavin mononucleotide; FNR, ferredoxin $NADP^+$ oxidoreductase; GAT domain, glutamine amidotransferase domain; PSI, Photosystem I.

composed of an extended half $(\beta/\alpha)_8$ barrel (Glu557-Arg929 of At-*GLU1* Fd-GOGAT or Ala465-Arg828 of Sy-Fd-GOGAT). Transfer of NH_3 on the C2 carbon of 2-OG forms a 2-iminoglutarate intermediate (2-IG). Fd delivers the reducing equivalents presumably *via* the $[3Fe—4S]^{0,+1}$ to FMN, and the reduction of 2-IG by the FMN N5 atom produces the second Glu (Fig. 2).

3. Nitrogen assimilation into amino acids during vegetative growth

Extensive studies have been carried out on the assimilation, cycling and translocation of nitrogen vital for vegetative growth of plants.

3.1 Primary nitrogen assimilation and amino acid translocation

Arabidopsis prefers NO_3^- as an inorganic nitrogen source, which is absorbed by the roots and transported in part to leaves *via* the xylem (Fig. 1A). Both roots and leaves are the site of primary NO_3^- assimilation into amino acids (Fig. 1C and D). Nitrate is a substrate of nitrate reductase (NR, EC 1.6.6.1) to form nitrite (NO_2^-) and it also up-regulates *NIA1* (At1g77760) and *NIA2* (At1g37130) encoding NR. Moreover, light and sucrose induce *NIA1* expression in leaves (Cheng, Acedo, Cristinsin, & Conkling, 1992). Nitrite is reduced to NH_4^+ by plastidial nitrite reductase (NiR, EC 1.7.7.1) followed by the NH_4^+ assimilation through the GS/GOGAT cycle (Fig. 1C and D).

Fd-GOGAT is the major activity in both leaves (97%) and roots (68%), and the rest being catalyzed by NADH-GOGAT in Arabidopsis (Suzuki & Rothstein, 1997). Two genes code for Fd-GOGAT on the Arabidopsis genome: *GLU1* (At5g04140, mapped to the top of chromosome 5: 1129344..1138917, 9.57 Kb) and *GLU2* (At2g41220, Chr2: 1717894..1718917, 12.43 kb) (Arabidopsis Genome Initiative, 2000). *GLU1* is highly expressed in leaves and to a lesser extent in roots (Potel et al., 2009; Suzuki & Rothstein, 1997). Spatial expression profile of *GLU1* Fd-GOGAT supports its essential role in the primary nitrogen assimilation and amino acid translocation. The signals of *GLU1* promoter::*GUS* expression and *in situ GLU1* mRNA hybridisation were localized by microscopy to the mesophyll where the reduction of NO_3^- to NH_4^+ mainly occurs (Potel et al., 2009) (Fig. 1B). With magnified leaf phloem minor veins, the signals of these constructs were found in a cell layer of the companion cell-sieve element (CC-SE) complex: cellular compartment where the transported solutes (NO_3^-, NH_4^+, amino acids, signaling molecules etc.) first become available (Potel et al., 2009) (Fig. 1B). The immuno-fluorescence specific to Fd-GOGAT was confined to these mesophyll and CC-SE complex by a confocal laser-scanning microscopy (Potel et al., 2009). The primary NH_4^+ assimilation *via* the GS2/*GLU1* Fd-GOGAT cycle is in line with the impaired phenotypes of *GLU1* T-DNA mutants in non-photorespiratory high CO_2 (350 μL^{-1}). These mutants were deficient in Glu (63% of the wild-type level accompanied by 50% increase in Gln) (Potel et al., 2009) and chlorophylls (65% of the wild-type level) (Lancien et al., 2002). The co-expression of GS1 (*Gln1;2*, At1g66200, Chr1: 24654985.. 24657831, 2.85 kb) and AS (*ASN1*, At3g47340) in the phloem CC-SE complex shown by the *in situ* mRNA hybridisation and immuno-localization (Gaufichon et al., 2017) pointed out that the GS1/*GLU1* Fd-GOGAT cycle and *ASN1*-encoded AS in the phloem synthesize Gln, Glu and Asn which are transported throughout the plant (Gaufichon et al., 2017) (Fig. 1A).

Much less is known about the function of *GLU2* Fd-GOGAT and NADH-GOGAT encoded by a single *GLT* (At5g53460, Chr5: 21699681.. 21710800, 11.12 kb). *GLU2* Fd-GOGAT shares a high sequence similarity with *GLU1* Fd-GOGAT (80%) and *GLT* NADH-GOGAT (α subunit part) (40%). *GLU2* and *GLT* were expressed at higher levels in roots than in leaves where their mRNAs were expressed in the phloem (Potel et al., 2009). In contrast to *GLU1* induced in leaves by light through phytochromes or by sucrose, light signals did not alter the mRNA level of *GLU2* and *GLT* (Coschigano, Melo-Oliveira, Lam, & Coruzzi, 1998; Potel et al., 2009; Suzuki & Rothstein, 1997). This led to less than 5% of the total activity of *GLU2* Fd-GOGAT plus *GLT* NADH-GOGAT in green leaves (Potel et al., 2009; Suzuki & Rothstein, 1997). *GLU2* T-DNA mutant leaves had a wild-type level of NH_4^+ in ambient air and of Glu/Gln ratio in non-photorespiratory high CO_2 conditions, indicating a minor role for *GLU2* Fd-GOGAT in the primary NO_3^- assimilation and in the photorespiratory NH_3 cycling (Potel et al., 2009). It was proposed that *GLT* NADH-GOGAT is involved in Glu translocation in leaf vasculature (Lancien et al., 2002). This is consistent with the expression profile of *GLT* NADH-GOGAT mRNAs in the phloem (Potel et al., 2009), and defective fresh weight (reduced by 30% of the wild-type level), chlorophylls (20%) (Lancien et al., 2002) and leaf Glu (30–40%) in non-photorespiratory high CO_2 conditions (Lancien et al., 2002; Potel et al., 2009).

3.2 Photorespiratory nitrogen pathway

3.2.1 GS/GOGAT cycle

Photorespiratory screening of EMS-mutagenized Arabidopsis allowed the isolation of *gltS* mutants impaired on a single nuclear gene (Somerville & Ogren, 1980). These mutants, containing <5% of the wild-type Fd-GOGAT activity, became chlorotic due to NH_4^+ accumulation in ambient air under illumination. However, they were viable in enriched CO_2 (350 μL^{-1}), which represses the photorespiration (Somerville & Ogren, 1980). These conditional phenotypes demonstrated the re-assimilation through the GS1/Fd-GOGAT cycle of photorespiratory NH_4^+ produced by the mitochondrial glycine decarboxylase complex (GDC, EC 1.4.4.2/2.1.2.10) and transported to the cytosol (Fig. 1C). Keys et al. (1978) estimated that the photorespiratory NH_4^+ release was an order of magnitude greater than the primary NH_4^+ production by means of the [^{15}N, 1-^{14}C]Gly tracing with the isolated organelles (Fig. 1C). The stoichiometry of Glu formed by Fd-GOGAT in the photorespiratory nitrogen cycle depicted in (Keys et al., 1978) was found to correlate with the Glu flux calculated from the *in vitro* kinetic parameters of the purified GS1, GS2 and Fd-GOGAT (specific activity, Km, etc.) (Suzuki & Gadal, 1982). This photorespiratory Glu flux reflects prevailing GS over GOGAT in both molecular numbers *in situ* (GS1:GS2:Fd-GOGAT ≈ 1:11:2 in rice) and total activity (Suzuki & Gadal, 1982).[b]

The single gene of the *gltS* locus was later identified to be analogous to *GLU1* because of the similar conditional phenotypes shared with the *GLU1* T-DNA mutants (Coschigano et al., 1998). The photorespiratory NH_4^+ re-assimilation *via* the GS1/*GLU1* Fd-GOGAT cycle was reinforced by a spatial microscopic imaging of *GLU1* Fd-GOGAT in the mesophyll cells where photorespiration takes place (Fig. 1B). These cells were overlaid with the specific signals of the *GLU1* promoter::*GUS* construct, *in situ GLU1* mRNA hybridisation and immuno-fluorescence of Fd-GOGAT (Potel et al., 2009). Furthermore, the reversible lethal phenotype of *gltS* mutants revealed that mitochondrial NADH-GDH was unable to re-assimilate photorespiratory NH_3 produced within the same organelle although its amination activity *in vitro* is several-fold higher than the Fd-GOGAT activity (Somerville & Ogren, 1980; Suzuki & Rothstein, 1997). The *in vivo* direction of reversible NADH-GDH reaction is controversial. NADH-GDH can exist as two homohexamers of either α subunit encoded by *GDH1*

[b] Remarks on the research in the Prof Gadal's laboratory: Graduate students and the CNRS researchers, assisted by Dr. Jean Vidal, shared in a collegial manner the experimental designs, techniques and data analysis in view of the research instructions and practices for young researchers.

(At5g18170, Chr5: 6005899..6008858, 2.96 kb) or β subunit encoded by *GDH2* (At5g07440, Chr5/2355633..2358383, 2.73 kb) and five hetero-hexamers containing both α and β subunits in Arabidopsis as observed in other plants (Skopelitis et al., 2007). Studies of simple and/or double mutants of α and β subunit coupled with the ^{15}N NMR microscopic approaches support a role for GDH in energy generation and synthesis of nitrogen transport compounds in the vasculature (Fig. 1C).

3.2.2 AS

Expression studies provided evidence that Asn synthesized by *ASN2*-encoded AS served as a nitrogen storage compound to remove excess NH_4^+ produced under stress (Wong, Chan, Coruzzi, & Lam, 2004). In growing leaves, the nitrogen flow from Asn in pea plants shifted from the amide group deamidated by cytosolic asparaginase (ASPG, EC 3.5.1.1) to the amino group transaminated by asparagine aminotransferase (AsnAT, EC 2.6.1.45) (Ta, Joy, & Ireland, 1985) (Fig. 1C). *ASN2*, one of three genes for AS in *Arabidopsis* (Arabidopsis Genome Initiative, 2000). *ASN2* (At5g65010, Chr5: 25968856-25972750, 3.9 kb) groups with the dicot-subclass II together with *ASN3* (At5g10240, Chr5:3212458-3216748, 4.29 kb) (Gaufichon, Reisdorf- Cren, Rothstein, Chardon, & Suzuki, 2010), and it is up-regulated by light and sugar in green leaves (Wong et al., 2004). *ASN1* (At3g47340, Chr3: 17437659..17441608, 3.95 kb) clusters with the dicot-subclass I and its expression in heterotrophic tissues contributes to the seed nitrogen filling in the floral organs (Gaufichon et al., 2017). Among four 2-oxo acids as substrates, AsnAT aminates glyoxylate by Asn and forms glycine and 2-oxosuccinamate in the peroxisome (Gaufichon, Rothstein, & Suzuki, 2016). To test the entry of the Asn-amino nitrogen into Gly, [2–^{15} N]Asn was fed to pea leaves, and a significant part of [^{15}N] (17%) was recovered in Gly during photorespiration (Ta et al., 1985) (Fig. 1C). To evaluate the Asn flow, the photorespiratory activity was assayed by gas exchange measurements of Arabidopsis *asn2–1* mutants containing Asn <5% of the wild-type level in leaves at 47 DAS (days after sawing) (Gaufichon et al., 2013). Photosynthetic net CO_2 fixation in the *asn2* transgenic plants was identical to the wild-type plants showing similar Warburg effects, and the oxygenase activity of ribulose 1,5-biphosphate carboxylase/oxygenase (Rubisco, EC 4.1.1.39) was not changed (Gaufichon et al., 2013). The high level [^{15}N]Gly reported by Ta et al. (1985) may result from the deamidation of [2–^{15} N]Asn to [2–^{15} N] Asp by ASPG and the subsequent AspAT-catalyzed transamination of [2–^{15} N]Asp to [2–^{15} N]Glu, which served for Gly synthesis during a long-term incubation of [2–^{15} N]Asn for 1 h.

3.2.3 CPSase

The phenotyping of *GLU1* T-DNA mutants provided evidence that CPSase catalyzes a committed step to recover photorespiratory NH_4^+ into Arg (Potel et al., 2009). The *GLU1* T-DNA mutants accumulated Arg up to 5.8-fold in atmospheric air, whereas Arg remained at the wild-type level in non-photorespiratory high CO_2 conditions, indicating that NH_4^+ was transferred to Arg *via* Gln and CP generated by CPSase in the chloroplasts (Potel et al., 2009) (Fig. 1C). CPSase is a single heterodimer in Arabidopsis. The small subunit is encoded by *carA* (AT3G27740, Chr3: 10281094.. 10284089, 2.98 kb) and the large subunit by *carB* (AT1G29900, Chr1: 10467745..10472424, 4.68 kb). The expression of *carB* promoter::*GUS* fusion was found in mesophyll chloroplasts but also in the CC-SE complex of phloem minor veins, suggesting a role of CPSase in the Arg synthesis for translocation (Potel et al., 2009).

3.3 Nitrogen assimilation in nonchlorophyllous tissues

Joy and Hageman (1966) reported NiR activity in root extracts of barley and tomato using dithionite-reduced Fd, methyl viologen or benzyl viologen. Furthermore, it was surprising that GOGAT in pea root extracts was active with chemically reduced sugar-beet Fd (Miflin & Lea, 1975).[c] Fd was isolated in 1962 from *Clostridium pasteurianum* (Carnahan, Mortenson, Valentine, & Carnahan, 1962) and spinach chloroplasts (Tagawa & Arnon, 1962). Once reduced by PSI, Fd donates electrons to enzymes of carbon, nitrogen and sulfur metabolism, and redox reactions. Among them, Fd-$NADP^+$ oxidoreductase (FNR, EC 1.6.7.1) produces NADPH for use to cellular reactions. This ability of photoreduction of $NADP^+$ is used as one of the tests to characterize the Fd-like electron carriers. In plants, no Fd-like electron carrier was identified in the nonchlorophyllous roots and no information was available on a reducing system of electron carrier without photochemical reaction.[d] Meanwhile, several data supported a physiological role of Fd-GOGAT in the nonchlorophyllous tissues:

[c] Remarks on the research in the Prof Gadal's laboratory: Prof Gadal visited the Institute of Prof Peter Lea and Prof Ben Miflin, Rothamsted, United Kingdom and carried out experiments on NADH-GOGAT assays in the summer of 1976.

[d] Remarks on the research in the Prof Gadal's laboratory: Prof Ann Oaks at the McMaster University, Hamilton, Canada invited Prof Gadal, and he stayed in her laboratory late in the 1970s. Prof Oaks studied the nitrogen metabolisms in maize roots (Oaks et al., 1980). She also visited the laboratory of Prof Gadal at Université de Nancy and reviewed the current opinions on the intriguing nitrogen assimilation in roots in her seminars.

Fd-GOGAT activity was associated with the root plastids from five plants (Suzuki, Gadal, & Oaks, 1981); the antibodies against the purified Fd-GOGAT from rice cross-reacted with Fd-GOGAT protein in roots but did not recognize NADH-GOGAT activity; co-existence of high activity of Fd-GOGAT and NADH-GOGAT in a range of heterotrophic tissues of etiolated leaves, roots, root nodules (see Suzuki & Gadal, 1984 for review). Using *Zea mays*, Suzuki, Oaks, Jacquot, Vidal, and Gadal (1985) reported the electron transport system for the reactions of Fd-NiR and Fd-GOGAT in the roots. This root system relied on the discovery of two novel proteins: Fd-like electron carrier (REC = root electron carrier) and FNR-like enzyme (PNR = pyridine nucleotide oxidoreductase), which share the antigenic epitopes with Fd and FNR, respectively in spinach chloroplasts (Suzuki et al., 1985).[e] In the reverse electron flow in the chloroplasts, electrons are transferred from NADPH to PNR and then to REC for the reactions of NiR and GOGAT (Figs. 1 and 2).[f] In the following decade (1985–1995), the Fd-like electron carriers were identified in barley root plastids and radish roots, and the FNR-like enzymes in radish root plastids, rice and maize (see Okutani et al., 2005). In maize, Fd and FNR are encoded by a multigene family, and at least six Fd isoproteins and three FNR isoenzymes were identified and characterized (Okutani et al., 2005). The NADPH/PNR/REC system and its physiological role were further supported by the evidence that the oxidative pentose phosphate pathway (OPPP) supplied NADPH as an initial reductant (Bowsher et al., 2007). In the root plastids, glucose 6-phosphate dehydrogenase (G6PDH, EC 1.1.1.49) or 6-phosphogluconate dehydrogenase (6PGDH, EC 1.1.1.44) forms NADPH using glucose 6-phoshate (G6P) imported from the cytosol (Bowsher et al., 2007). Using the recombinant proteins and NADPH in the reconstituted system (NADPH/FNR/Fd/Fd-GOGAT) coupled with [5-^{15}N]Gln tracing, the GOGAT assay demonstrated an ordered reaction mechanism (Yoneyama et al., 2015). As a direct GOGAT assay, the reaction products of Glu and [^{15}N]compounds (NH_3, Gln, Glu) were measured using HPLC and capillary electrophoresis-mass

[e] Remarks on the research in the Prof Gadal's laboratory: Immunochemical studies in the Prof Gadal's laboratory covered multiple approaches. They turned out to be efficient tools in the field of current molecular physiology by screening of recombinant DNA library, immunochemical localisation and quantification of target-proteins *etc.* by using antibody probes.

[f] Remarks on the research in the Prof Gadal's laboratory: The ability of REC to mediate the photoreduction of $NADP^+$ was demonstrated by using the reconstituted chloroplast system (Suzuki et al., 1985) developed by Dr. Jean-Pierre Jacquot, Université de Paris XI, Orsay.

spectroscopy, respectively (Yoneyama et al., 2015).[g] Higher GOGAT activity was found in the root-specific system (NADPH/R-FNR/FdIII) than in the leaf-specific system (NADPH/L-FNR/FdI) (880 *vs.* 710 nmol Glu formed mL^{-1} 30 min^{-1}). This efficient root system may result from a favored redox potential between NADPH ($Em = -320$ mV) and FdIII ($Em = -34$ mV) compared with FdI ($Em = -423$ mV) (Yoneyama et al., 2015).[h] Furthermore, the [^{15}N] tracing revealed a fine regulation by NADPH, essential for the Fd_{red} binding to the synthase domain to initiate the reaction (Fig. 2). The absence of NADPH formed Glu and NH_3 (lower than Glu due to gaseous loss), products of a glutaminase in the GAT domain (Fig. 2). The generation of NADPH (supply of G6P/G6PDH) substantially increased [^{15}N]Glu (20-fold) and Glu with no NH_3 release compared with a supply of 140 nmol mL^{-1} NADPH (Yoneyama et al., 2015). The results underlined the coordinated reactions in the three domains (Fig. 2).

4. Concluding remarks

This review article was focused on the primary nitrogen metabolism during the vegetative growth of plants studied in the Prof Gadal's laboratory and the groups working in the nitrogen fields. However, the list of relevant studies discussed in this review is not exhaustive as a wider range of powerful multidisciplinary investigations is available. Nevertheless, the extensive physiological and biochemical studies in the 1970s and 1980s as discussed here realized in the Prof Gadal's laboratory have exerted a primary impact on the current investigations using molecular-genetic and transgenic tools available for model plants such as Arabidopsis. Our studies have been extended to other related biological materials in the plant life cycle namely

[g] Remarks on the research in the Prof Gadal's laboratory: GOGAT assay by measuring Glu by HPLC was improved into a highly sensitive method together with Dr. Francis Martin, Université de Nancy.

[h] Remarks on the research in the Prof Gadal's laboratory: Prof Gadal contributed to the collaboration between France and Japan in the plant science. He visited the laboratory of Prof Tadakatsu Yoneyama, National Institute of Agrobiological Resources, Tsukuba, Japan in early 1980s. Prof Gadal was coordinator of the French delegation to the bilateral symposium from October 29 to 31, 2002 at Nara, Japan. This meeting followed the previous workshop in INRA Versailles with the Japanese delegation organized by Prof Tatsuo Sugiyama, RIKEN, for the Japanese side (September 29–October 2, 1999). Funded by CNRS, INRA, French Embassy in Japan, JSPS, RIKEN and Nara Women's University, the meeting at Nara assembled more than 100 attendees including 18 and 16 invited speakers from France and Japan, respectively. The bilateral symposium takes place alternatively in France and in Japan, and the 11th meeting was organized from October 23 to 25, 2017 in Saint-Foy-lès-Lyon.

the floral organs, developing seeds, and germinating seeds by means of molecular and transgenic approaches, physiological phenotyping, systematic metabolite profiling, cell/organ-imaging technologies. The results of these studies support the view that the GS/GOGAT cycle plays the primary role for the NH_4^+ assimilation into amino acid and the nitrogen translocation within the photo-autotrophic and heterotrophic organs of the plants. The basic research on the genes and isoforms involved in NH_4^+ assimilation, light and metabolic signals and sensing systems, and networks of carbon and nitrogen metabolites may improve the nitrogen assimilation and translocation within the plants in response to environmental cues.

References

Arabidopsis Genome Initiative. (2000). Analysis of the genome sequence of the flowering plant *Arabidopsis thaliana*. *Nature, 408*, 220–234.

Binda, C., Bossi, R. T., Wakatsuki, S., Arzt, S., Coda, A., Curti, B., et al. (2000). Cross-talk and ammonium channeling between active centers in the unexpected domain rearrangement of glutamate synthase. *Structure, 8*, 1299–1308.

Bowsher, C. G., Lacey, A. E., Hanke, G. T., Clarkson, D. T., Saker, L. R., Stulen, I., et al. (2007). The effect of Glc6P uptake and its subsequent oxidation within pea root plastids on nitrite reduction and glutamate formation. *Journal of Experimental Botany, 58*, 1109–1118.

Carnahan, J., Mortenson, L. E., Valentine, R. C., & Carnahan, J. E. (1962). An electron transport factor from from *Clostridium pasteurianum*. *Biochemical and Biophysical Research Communications*, 7, 448–452.

Cheng, C.-L., Acedo, G. N., Cristinsin, M., & Conkling, M. A. (1992). Sucrose mimics the light induction of Arabidopsis nitrate reductase gene transcription. *Proceedings of the National Academy of Sciences of the United States of America, 89*, 1861–1864.

Coschigano, K. T., Melo-Oliveira, R., Lam, J., & Coruzzi, G. M. (1998). Arabidopsis *gls* mutants and distinct Fd-GOGAT genes: Implication for photorespiration and primary nitrogen assimilation. *The Plant Cell, 10*, 1–12.

Gaufichon, L., Marmagne, A., Belcram, K., Yoneyama, T., Sakakibara, Y., Hase, T., et al. (2017). *ASN1*-encoded asparagine synthetase in floral organs contributes to nitrogen filling in Arabidopsis seeds. *The Plant Journal, 91*, 371–393.

Gaufichon, L., Masclaux-Daubresse, C., Tcherkez, G., Reisdorf-Cren, M., Sakakibara, Y., Hase, T., et al. (2013). *Arabidopsis thaliana ASN2* encoding asparagine synthetase is involved in the control of nitrogen assimilation and export during vegetative growth. *Plant, Cell & Environment, 36*, 328–342.

Gaufichon, L., Reisdorf-Cren, M., Rothstein, S. J., Chardon, F., & Suzuki, A. (2010). Biological functions of asparagine synthetase in plants. *Plant Science, 179*, 141–153.

Gaufichon, L., Rothstein, S. J., & Suzuki, A. (2016). Asparagine metabolic pathways in Arabidopsis. *Plant and Cell Physiology, 57*, 675–689.

Joy, K. W., & Hageman, R. H. (1966). The purification and properties of nitrite reductase from higher plants, and its dependence on ferredoxin. *Biochemical Journal, 100*, 263–273.

Keys, A. J., Bird, I. F., Cornelius, M. J., Lea, P. J., Wallsgrove, R. M., & Miflin, B. J. (1978). Photorespiratory nitrogen cycle. *Nature, 275*, 741–743.

Lancien, M., Martin, M., Hsieh, M.-H., Leustek, T., Goodman, H., & Coruzzi, G. K. (2002). Arabidopsis *glt-1* mutant defines a role for NADH-GOGAT in the non-photorespiratory ammonium assimilatory pathway. *The Plant Journal, 29*, 347–358.

Lea, P. J., & Miflin, B. J. (1974). Alternative route for nitrogen assimilation in higher plants. *Nature, 251*, 614–616.

Miflin, B. J., & Lea, P. J. (1975). Glutamine and asparagine as nitrogen donors for reductant-dependent glutamate synthesis in roots. *Biochemical Journal, 149*, 403–409.

Oaks, A., Stulen, I., Jones, K., Winsper, S., Misra, I., & Boesel, I. (1980). Enzymes of nitrogen assimilation in maize roots. *Planta, 148*, 477–484.

Okutani, S., Hanke, G. T., Satomi, Y., Takao, T., Kurisu, G., Suzuki, A., et al. (2005). Differential distribution of three maize ferredoxin:NADP(H) oxidoreductases between stroma and thylakoid membranes and isoenzyme specific roles in chloroplast redox cascades. *Plant Physiology, 139*, 1451–1459.

Potel, F., Valadier, M.-H., Ferrario-Méry, S., Grandjean, O., Marin, H., Gaufichon, L., et al. (2009). Assimilation of excess ammonium into amino acids and nitrogen translocation in *Arabidopsis thaliana*—Roles of glutamate synthases and carbamoylphosphate synthetase in leaves. *The FEBS Journal, 276*, 4061–4076.

Skopelitis, D. S., Paranychianakis, N. V., Kouvarakis, A., Spyros, A., Stephanou, E. G., Kalliopi, A., et al. (2007). The isoenzyme 7 of tobacco NAD(H)-dependent glutamate dehydrogenase exhibits high deaminating and low aminating activities in vivo. *Plant Physiology, 145*, 1726–1734.

Slocum, R. D. (2005). Genes, enzymes and regulation of arginine biosynthesis in plants. *Plant Physiology and Biochemistry, 43*, 729–745.

Somerville, C. R., & Ogren, W. L. (1980). Inhibition of photosynthesis in Arabidopsis mutants lacking in leaf glutamate synthase activity. *Nature, 286*, 257–259.

Suzuki, A., & Gadal, P. (1982). Glutamate synthase from rice leaves. *Plant Physiology, 69*, 848–852.

Suzuki, A., & Gadal, P. (1984). Glutamate synthase. Physicochemical and functional properties of different forms in higher plants and in other organisms. *Physiologie Végétale, 22*, 471–486.

Suzuki, A., Gadal, P., & Oaks, A. (1981). Intracellular distribution of enzymes associated with nitrogen assimilation in roots. *Planta, 151*, 457–461.

Suzuki, A., & Knaff, D. B. (2005). Glutamate synthase: Structural, mechanistic and regulatory properties, and role in the amino acid metabolism. Review. *Photosynthesis Research, 83*, 191–217.

Suzuki, A., Oaks, A., Jacquot, J. P., Vidal, J., & Gadal, P. (1985). An electron donor system in maize roots for reaction of glutamate synthase and nitrite reductase. Physiological and immunological properties of the electron carrier and pyridine nucleotide reductase. *Plant Physiology, 78*, 374–378.

Suzuki, A., & Rothstein, S. (1997). Structure and regulation of ferredoxin-dependent glutamate synthase in *Arabidopsis thaliana*. Cloning of cDNA, expression of different tissues of wild-type and *gltS* mutant strains, and light induction. *European Journal of Biochemistry, 243*, 708–718.

Ta, T. C., Joy, K. W., & Ireland, R. J. (1985). Role of asparagine in the photorespiratory nitrogen metabolism in pea leaves. *Plant Physiology, 78*, 334–337.

Tagawa, K., & Arnon, D. I. (1962). Ferredoxins as electron carriers in photosynthesis and in the biological production and consumption of hydrogen gaz. *Nature, 195*, 537–543.

Tempest, D. D., Meers, C. M., & Brown, C. M. (1970). Synthesis of glutamate in *Aerobacter aerogenes* by hitherto unknown route. *Biochemical Journal, 117*, 405–407.

Van den Heuvel, R. H. H., Curti, B., Vanoni, M. A., & Mattevi, A. (2004). Glutamate synthase: A fascinating pathway from L-glutamine to L-glutamate. *Cellular and Molecular Life Sciences, 61*, 669–681.

Wong, H.-K., Chan, H.-K., Coruzzi, G. M., & Lam, H.-M. (2004). Correlation of *ASN2* gene expression with ammonium metabolism in Arabidopsis. *Plant Physiology, 134*, 332–338.

Yoneyama, T., Fujimori, T., Yanagisawa, S., Hase, T., & SuzukI, A. (2015). ^{15}N tracing studies on *in vitro* reactions of ferredoxin-dependent nitrite reductase and glutamate synthase using reconstituted electron donation systems. *Plant and Cell Physiology, 56*, 1154–1161.

Yoneyama, T., & Suzuki, A. (2019). Exploration of nitrate-to-glutamate assimilation in non-photosynthetic roots of higher plants by studies of ^{15}N-tracing, enzymes involved, reductant supply, and nitrate signaling: A review and synthesis. *Plant Physiology and Biochemistry*, *136*, 245–254.

CHAPTER FIVE

Algae for global sustainability?

Nathalie Bourgougnon[a,*], Anne-Sophie Burlot[b], and Anne-Gaëlle Jacquin[c]

[a]Laboratoire de Biotechnologie et Chimie Marines, Université Bretagne Sud, EA3884, UBS, IUEM, Vannes, France
[b]Olmix SA • ZA du Haut du Bois, Bréhan, France
[c]The Algonauts Project, www.algonauts.org, San Francisco, CA, United States
*Corresponding author: e-mail address: nathalie.bourgougnon@univ-ubs.fr; asburlot@olmix.com; ag.jacquin@gmail.com

Contents

Advances in Botanical Research, Volume 100
ISSN 0065-2296
https://doi.org/10.1016/bs.abr.2021.01.003
Copyright © 2021 Elsevier Ltd
All rights reserved.

Abstract

Our world has always been transforming, but even more so today with almost 7.8 billion people living on Earth. Since the beginning of the century, the world population has increased by 27%. Since 2000, Africa has shown the most significant inhabitant number increase with a positive growth rate of 65% compared to Oceania (36%) Latin America and the Caribbean (25%), Asia (24%), North America (18%) and Europe (3%) (United Nations, 2020a). Considering that our blue planet has a finite space, the role of humanity is to understand its ecosystem in order to anticipate the impact of this increase of human beings on the flora, the fauna and on the environment and in order to limit the imbalance of the ecosystem, and risk the extinction of certain species and the loss of biodiversity. In this context, in 2015, 193 countries, which are members of the United Nations Organization, adopted a universal program of sustainable development for the 2030 horizon, called the "2030 Agenda for Sustainable Development." It is focused on population numbers, the planet prosperity and on peace thanks to its partnerships. Seventeen sustainable development goals were defined. Algae (macro and microalgae) have the potential to respond to some of the world's most pressing challenges. In this chapter, the sustainable development goals will be approached by proposing solutions based on algae and on key players who utilize them in a healthy functioning of the global ecosystem. We propose a vision of a responsible seaweed industry, playing a globally significant role in food security, climate change mitigation, and supporting the marine ecosystem, as well as contributing to job-creation and economic growth. This chapter is a follow up of volume 71 sea plants and volume 95 seaweeds around the world: State of art and perspectives.

1. Introduction

The Oceans are the world's largest ecosystem, covering more than 70% of the earth's surface and they support floating forests with animals and sea plants. Microalgae and macroalgae are important marine resources playing a major role in supporting the rich biodiversity of the ocean. They constitute primary producers in the ecosystems at the base of the marine food chain, are food for many invertebrates and fish and support for detritus feeders and decomposers. Macroalgae or seaweeds form the most extensive and productive benthic marine vegetated habitats. Along the global coastlines of the oceans (intertidal rocky shores, coral reefs or seagrass meadows), its habitat was estimated to cover about 3.4 million km^2 and support a global net primary production of oxygen emissions of about 1.5 PgC $year^{-1}$ (Krause-Jensen et al., 2018). These ecosystems also provide nursery grounds, coastline protection from erosion and nutrient fixation (Gattuso et al., 2006; Keith, Kerswell, & Connolly, 2014).

The use of macroalgae dates back to Shen Nung, the father of husbandry and medicine, in approximately 3000 BC. Seaweeds were reported to be utilized in Iceland in 960 BC. The *Chinese Book of Poetry* (800–600 BC) praised housewives for cooking with algae, and the *Chinese Materia Medica* (600 BC) refers to algae as follows: "*Some algae are a delicacy fit for the most honorable guest, even for the King himself*" (Levine, 2016; Pérez-Lloréns, Mouritsen, Rhatigan, Cornish, & Critchley, 2020).

Seaweeds were seen as a fundamental component of the Ocean as well as the beings associated with it, they also held an essential connection to the land and the people who dwelt there (Pérez-Lloréns et al., 2020). For many centuries, coastal populations have harvested a wide variety of seaweeds within all algal clades: red (*Rhodophyta*), brown (*Phaeophyceae*), and green (*Chlorophyta*). They were most often first used for domestic purposes, such as for human consumption as vegetables in salads, soups, and main dishes, including sushi and as flavorings in chips and snacks. The industrial and commercial use of algae was marked successively in the 17th century by the production of glass and soap (France and Norway). Since the 19th century macroalgal biotechnology has included the production of iodine and phycocolloids (alginate, agar-agar, carrageenan) from *Phaeophycea* and *Rhodophyta.*

Because of their biodiversity and expansion in various living environments, these organisms contain many interesting metabolites for the production of horticultural, cosmetics, and animal welfare products (Fleurence & Levine, 2016; Leandro, Pereira, & Gonçalves, 2020; Levine & Fleurence, 2018). At present, the main directions in algal biotechnology are biofuels, agricultural biostimulants for crop plants, probiotics for aquaculture, wastewater treatment, and biomedical applications of extracted compounds (pigments, polyphenols, polysaccharides) (Leandro et al., 2020).

In 2013, the value of the global harvest of seaweeds was estimated at US$6.7 billion, and over 95% was produced in aquaculture, with China and Indonesia being the top producers (FAO, 2015). In addition to macroalgae, some microalgae are cultivated for food and food additives (FAO, 2016). The annual macroalgae harvest from wild and cultivated crops was 28.4 million tons in 2014 (FAO, 2016), destined for human consumption and the production of hydrocolloids used as thickening agents in foods and beverages. As their promising interest, cultivation and applications increase, their value in the market rises too. It is estimated that in 2025 this value will exceed the figure achieved in 2019 by twofold (Fig. 1).

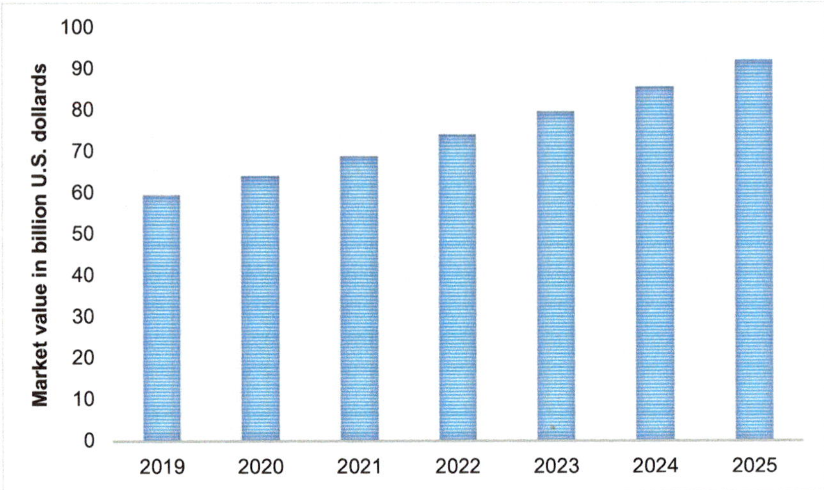

Fig. 1 Value of the seaweed market worldwide from 2019 to 2025 (in billion US dollars)—https://www.statista.com/statistics/603851/value-commercial-seaweed-market-worldwide/.

In 2015, 193 countries came together to put out a plan to improve the planet's health. At a time of global climate change we are faced with record high temperatures; cold episodes in Europe, the disruption of the Gulf Stream, the multiplication of forest fires, the migration of populations, the increase in threatened populations, the migrations of populations into coastal areas, or the resurgence of diseases and the reduction of biodiversity. These countries made a list of 17 sustainable development goals (SDG) to achieve a better and more sustainable future for all. They addressed the global challenges we face, including those related to poverty, inequality, climate change, environmental degradation, peace and justice. The 17 goals are all interconnected, and in order to leave none behind, it is important that we achieve them all by 2030.[a] Some institutions, NGO's, associations, companies, research laboratories, participatory science, as well as international and European projects are involved around Ocean challenges and support these initiatives. For example, the Intergovernmental Oceanographic Commission of UNESCO[b] proposes an ambitious program: The United Nations Decade

[a] https://www.un.org/sustainabledevelopment/sustainable-development-goals/

[b] https://en.unesco.org/

of Ocean Science for Sustainable Development (2021–2030). The European Marine Board[c] (EMB) the leading European think tank in marine science policy provides a platform to advance marine research and to bridge the gap between science and policy in order to meet future marine science challenges and opportunities.

The ocean is intrinsically linked to many of the global challenges facing the world, including climate change, food and water security and health. In this context, we propose to explore the biological, biotechnological, ecological, societal and economic solutions to climate issues that algae might contribute to, to make our planet feel great again, paraphrasing a favorite Donald trump citation who is apparently interested in making only the USA great again.

2. Algae, a long history

The term "algae" designates predominantly aquatic photosynthetic eukaryotes chlorophyll "a"-containing what stretches from unicells of a few microns in diameter, to complex multicellular forms such as the giant kelps that can reach more than 30 m in length (de Reviers, 2003). Most algae are photosynthetic organisms that acquired the trait some 1.8 billion years ago after an unknown non-photosynthetic unicellular eukaryote engulfed, or was invaded by a photosynthetic cyanobacterium close to *Gloeomargarita*. This event ultimately resulted in a cell containing a photosynthetic plastid surrounded by two membranes with a highly reduced cyanobacterially derived genome. This event, known as the primary endosymbiosis, gave rise to the extant *Archaeplastida*, which includes the *Glaucophyta*, *Rhodophyta* (red algae) and *Chloroplastida* (green algae and land plants) (Tirichine & Bowler, 2011). Brown algae have had a very different evolutionary history, as they have acquired the ability to achieve photosynthesis through a secondary endosymbiosis event. The other lines such as Cryptophytes, Haptophytes, Golden Brown Algae and Dinophytes, are therefore eukaryotes with no direct relationship with terrestrial plants. The lines resulting from secondary or even tertiary endosymbiosis between a primitive non-photosynthetic eukaryote and unicellular red algae include the Ochrophytes within Stramenopiles, Cryptophytes and Haptophytes lines.

[c] https://www.marineboard.eu/

Derived from endosymbiosis with a green unicellular alga, the lines are grouped within Euglenophyceae and Rhizaria (Niklas & Kutschera, 2010). Such events involved a second heterotrophic eukaryote that engulfed not a *cyanobacterium* but a green or red photosynthetic eukaryote and resulted in plastids with three or four membranes around them. Both the primary and secondary endosymbiosis caused massive losses of genes from the engulfed genome, many of which have been transferred to the host genome. This process of endosymbiosis gene transfer has contributed widely to the evolution of algae (Adl et al., 2012; Boudouresque, 2015; Brown et al., 2013; Burki et al., 2016; Leliaert et al., 2012; Tirichine & Bowler, 2011).

3. Sustainable development goals and algae-based solutions

3.1 No poverty: End poverty in all its forms everywhere (Goal 1)

End poverty in all its forms everywhere is the first sustainable development objective of the 2030 Agenda. According to the United Nations, more than 700 million people, 10% of the world's population, still live in extreme poverty. Consequences of poverty for the populations are the lack of the most basic needs like food supply, health, education, access to water, to electricity and sanitation. Its causes include unemployment, social exclusion, and high vulnerability of certain populations to disasters, diseases and other phenomena which prevent them from being productive (United Nations, 2020a, 2020b). The majority of people living on less than the poverty headcount ratio of USD 1.90 a day, fixed by the World Bank, live in Sub-Saharan Africa (Fig. 2).

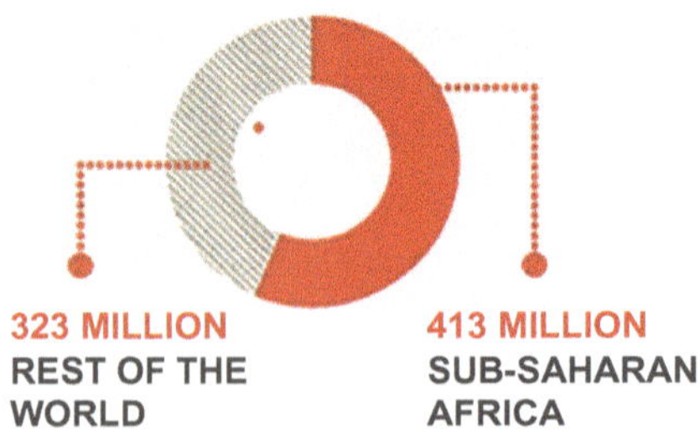

Fig. 2 736 million people lived in extreme poverty in 2015, 413 million in Sub-Saharan Africa (United Nations, 2020a).

Poverty affects populations, prosperity and peace. SDG number one is linked with all the other SDGs. Algae can bring solutions to the causes and also to the consequences of poverty (Fig. 3).

To limit poverty, its causes have to be avoided. Besides algae providing food (see Sustainable Development Goal 2), they can also provide clean water (see SDG 6), green energy (see SDG 7) improve health (SDG 3), they also represent a biomass for people:

1. to work, to liberate themselves and to flourish personally,
2. to have social interactions, health and social protection, security and,
3. to be less vulnerable in case of environmental disasters and diseases.

For coastal countries, fishery is usually the main economic activity, surpassing all other agricultural activities. Fishing escalates on a daily basis due to the involvement of a heavily populated coastal community. However, such operations are becoming more unsustainable with destructive fishing practices coupled with illegal fishery issues and communal clashes. In order to create a sustainable coastal community, especially at geographically isolated, marginalized, and vulnerable locations, the creation of alternative or sustainable livelihood options capable of unlocking the potential of the oceans seems vital. In this way, seaweed production and their use are becoming increasingly attractive to coastal communities for their socio-economic and environmental effects. Depending on seaweed cultivation and harvesting seasons, adjustments of the fishing industries can be made and offer a new and supplementary revenue. A study led in Sri Lanka has concluded that seaweed farming is reasonably profitable and generates considerable additional employment opportunities, thus is a financially profitable venture. The economic viability indicators, estimated for an average of 25 seaweed rafts, have shown an annual net income including moderate cash returns for an investment of USD 253 and a benefit-cost ratio (BCR) equal to 1.19 (Ginigaddara & Lankapura, 2018). The technology for seaweed harvesting both from the wild and farmed is relatively simple and requires low initial capital investment. The crop can be harvested in about 6 weeks and it can be dried on mats, grass, or on the sand on the beach. Seaweed farming is a system that can be replicated in other potential locations in the world. In South America, or in Brazil for example, seaweed farming is also identified as a significant source of income for individual households. Beyond the favorable economic effects on individual farm families, seaweed cultivation has extended beneficial effects to the entire economy of certain countries. For instance, it has contributed significantly to the economy of the Zanzibar Islands of Tanzania by becoming a leading foreign exchange earner that accounts for more than 90% of

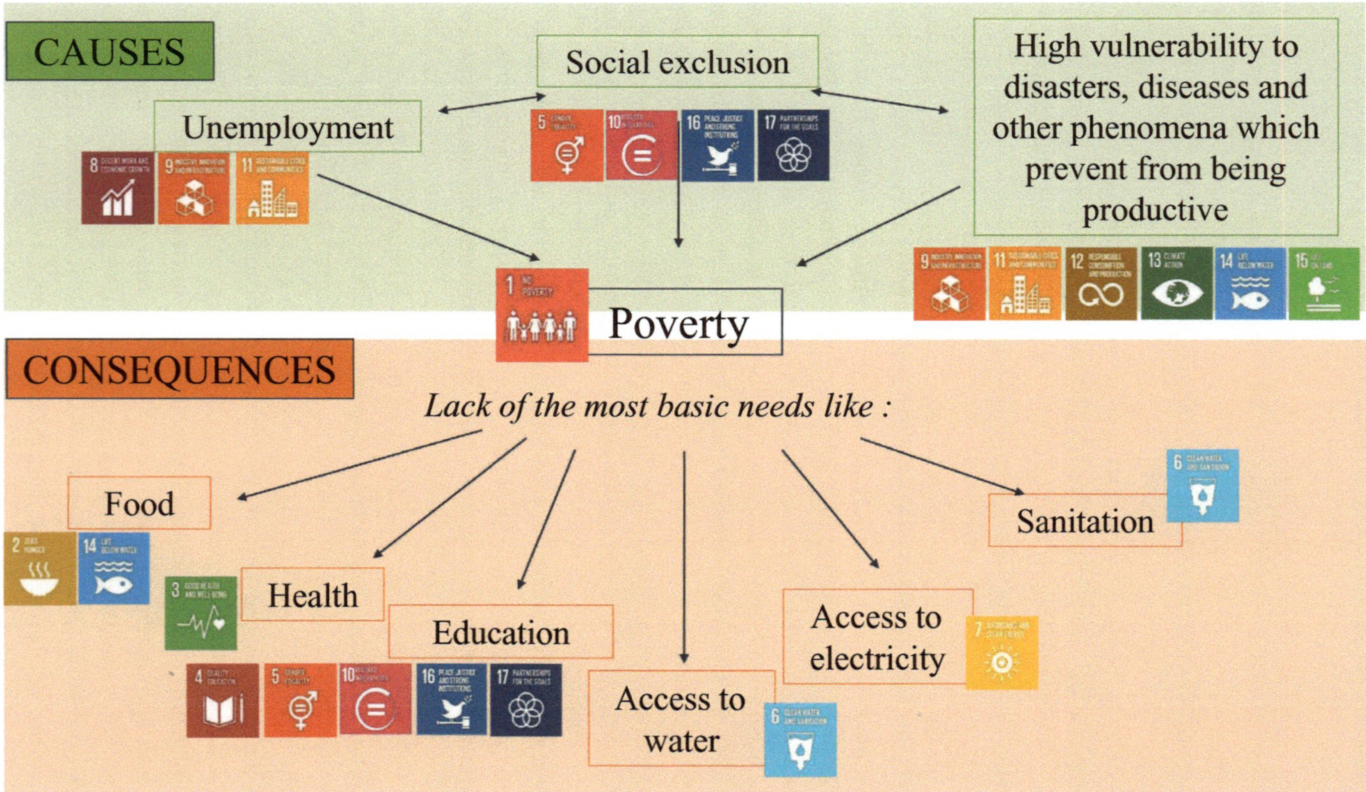

Fig. 3 Poverty affects populations, prosperity and peace. Sustainable Development Goal (SDG) number one is linked with all the other SDGs. Algae can bring solutions to the causes and also to the consequences of poverty.

Zanzibar's marine export products (Rebours et al., 2014). Nevertheless, sometimes, having a job does not guarantee a decent living. In fact, 8% of the employed workers and their families worldwide lived in extreme poverty in 2018 (United Nations, 2020a). To decrease this percentage, solutions exist. As previously mentioned, the production and transformation of algae can represent a supplementary revenue. While fishing, fishermen can also harvest or cultivate seaweeds and then sell them both. Furthermore, the development of integrated multitrophic aquaculture (IMTA) is being increasingly considered. Land based and nearshore aquaculture systems which combine fed aquaculture species (e.g. finfish), with inorganic extractive aquaculture species (e.g. seaweeds) and organic extractive species (e.g. suspension and deposit-feeders) cultivated in proximity are both systems described as IMTA. These systems should significantly increase the sustainability of aquaculture, based on a number of potential economic, societal and environmental benefits, including the recycling of waste nutrients from higher trophic-level species into the production of lower trophic-level crops with commercial value. Today, some marine IMTA systems, primarily in Asia (China), have been commercially successful at an industrial scale, while experimental projects are now scaling up toward commercialization in Canada, Chile, the USA and in some European countries. Another example is the villagers in the peninsula of Placencia in Belize, in Central America. During the sunny season, they work for the tourism industry and later in the low seasons they turn to the sea and its bountiful provision of seafood (Placencia Seaweed Farmers, 2016).

As the farming areas are generally in intertidal zones, men and women can safely access the seaweed plots. These factors allow women, in particular, an important opportunity to earn some income for themselves and their families while men work as fishermen and in other sectors. In the study of Ginigaddara et al., the employment generation of seaweed farming in the study area, in Sri Lanka, has been estimated at 3392 man days (1280 man days plus 2112 woman days) per annum. The employment potential of seaweed farming, especially as a means of female employment has been proved (Ginigaddara & Lankapura, 2018; Rebours et al., 2014). Several studies have also mentioned that women tend to be more patient and more willing to learn about sustainably managing local seaweed resources. This trend is seen in every region, particularly in Asia (China, India Indonesia, Malaysia, the Philippines) and in Africa (Ghana, Morocco, Zanzibar) (FAO, 2018; Msuya & Hurtado, 2017).

By 2030, the first objective of "No poverty" is to build up the resilience of the poor and those in vulnerable situations and reduce their exposure and vulnerability to climate-related extreme events and other economic, social and environmental shocks and disasters. This objective is to ensure the significant mobilization of resources from a variety of sources, including through enhanced development cooperation. This in order to provide adequate and predictable means for developing countries, in particular the least developed countries and to implement programs and policies to end poverty in all its dimensions. A remarkable example of sustainable development success and of international cooperation is the Red Seaweed Promise™, a program launched by the international company Cargill, that produces colloids to help ensure a long-term sustainable red seaweed supply chain. The program is specifically designed to address key sustainability challenges for the harvesting and cultivation of red seaweed, while enhancing producer livelihoods, supporting local communities and conserving the marine environment. Red seaweed is the raw material needed to produce carrageenan, a texturizer used in various applications such as dairy, confectionary and personal care products. The program also answers food industry and consumers' needs for more sustainable food ingredients, with a commitment to sourcing 60% of sustainable red seaweed by 2025. Cargill utilizes red seaweeds from four continents (South America, Asia, Africa and Europe). Furthermore, for this kind of project or program to work well and to be beneficial for both providers and users of genetic resources like algae, the Nagoya Protocol was adopted in 2010 during the 10th meeting of the Conference of the Parties to the Convention on Biological Diversity (COP 10) of United Nations, in Nagoya, Japan. It became implemented in 2014. Its objective is the fair and equitable sharing of benefits arising from the utilization of genetic resources, thereby contributing to the conservation and sustainable use of biodiversity. By helping to ensure benefit-sharing, the Nagoya Protocol creates incentives to conserve and sustainably use genetic resources, and therefore enhances the contribution of biodiversity to development and human well-being (Convention on Biological Diversity, 2015). The GlobalSeaweed STAR program has accomplished the creation of sound policy frameworks applied to the seaweed industry at national, regional and international levels, based on pro-poor and gender-sensitive development strategies, and supports accelerated investment in poverty eradication actions. It is funded to improve the seaweed industry, the fastest-growing of all aquaculture sectors, with an annual growth rate of 10% and a value in excess of USD 5 billion. The main challenges are to prevent the vulnerability of crops to disease outbreaks and pest infestation and to fill the lack of biosecurity measures and legislation governing the movement of seaweeds between regions and continents. The key ecological and socio-economic challenges preventing the sustainable economic growth of this industry were identified in a United Nation University (UNU) Policy Brief led by Dr Elizabeth Cottier-Cook from the Scottish Association for Marine Science (Cottier-Cook et al., 2016). In some cases, operations in the seaweed industry can have a negative impact on the environment and therefore they can increase the vulnerability of certain populations to disasters, diseases and other phenomena which prevent them from being productive. For example, harvesting of wild brown seaweed *Laminaria hyperborea*, using specifically designed trawling equipment, remains somewhat controversial as the removal of, and interference with natural habitats (specifically kelp forests) has the potential to affect local biodiversity and ecosystem integrity, and may contribute to coastal erosion (Stévant, Rebours, & Chapman, 2017). Nevertheless, projects like the SLAMIR project led by a natural park in Brittany, France are in partnership with national research centers and French companies' union working on seaweeds. They exist to facilitate access to data on kelp forests to maintain their ecophysiology and to sustainably manage their exploitation (Parc Naturel Marin Iroise, 2018).

Though major constraints, such as unfavorable weather and uncertain market conditions have negatively affected the progression of seaweed cultivation, many countries have illustrated the potential of seaweed farming as a profitable commercial enterprise adding opportunities for added value, integration, and earning much needed foreign exchange through exporting as well as for local development (Rebours et al., 2014). Before starting a new seaweed farm, it is essential to know that poor quality of planting materials, distortions in the market, improper aquatic environments, and poor post-harvest handling have been identified as major constraints. This is why it is important to protect our environment through understanding its serviceability. In several countries around the world, the perceived importance of algae as a vital livelihood option has proved that the system is socially and economically acceptable among coastal communities. Seaweeds have already been identified as a catalyst of social progression in coastal communities, providing substantial income while rendering extensive employment opportunities to the farming households.

3.2 Zero hunger: End hunger, achieve food security and improved nutrition and promote sustainable agriculture (Goal 2)

An estimated 821 million people were undernourished in 2017. The prevalence of undernourishment has remained virtually unchanged in the past 3 years at a level slightly below 11%. This progress sends a clear warning that more must be done, and urgently if the Sustainable Development Goal of Zero Hunger is to be achieved by 2030. The global demand for meat from livestock (excluding fish) is projected to increase from approximately 313 million tons in 2014–2016 to 455 million tons by 2050, along with a projected increase in the global demand for plant protein (i.e. vegetable and forage) for the production of livestock feed (Magnussona et al., 2019). Animal proteins come at a very high environmental cost. The largest agricultural global greenhouse gas contribution is from cattle and sheep production systems that are responsible for up to 18% of total global greenhouse gas emissions and are mainly in the form of enteric methane. These animal production systems also require more arable land (factor of deforestation) (Herrero & Thornton, 2013; Kinley et al., 2020). Population growth is driving the need to find new protein and fiber sources to feed future populations and fight food insecurity. Algae are traditionally consumed as sea vegetables in Asian countries. In Japan, the daily amount of seaweed eaten is estimated at about 4–8 g per capita (dry weight). The main species used as food are two brown seaweeds, kombu (*Laminaria japonica*) and wakame (*Undaria pinnatifida*) and the red seaweeds nori (*Porphyra/Pyropia tenera*, *Porphyra/Pyropia yezoensis*). In Europe, seaweeds are not considered as traditional foods. However, in 1990 the French National Academy of Medicine and the High Council for Public Hygiene authorized the use of 26 species of seaweeds in human food as sea vegetables. Algae are frequently sold in organic and health food shops. Marine algae contain significant quantities of vitamins, minerals, dietary fibers, proteins, polysaccharides, and various micro- and macroelements (Fig. 4). Seaweeds contain a wide diversity of minerals and trace elements: sodium, calcium, chlorine, magnesium, zinc,

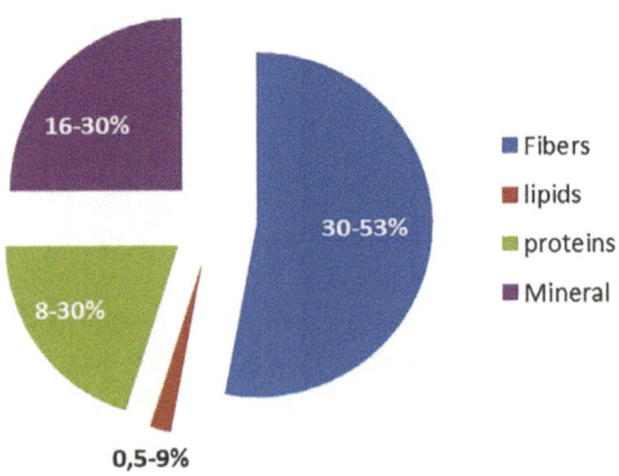

Fig. 4 Global composition of seaweed.

copper (i.e., *Palmaria palmata, Fucus vesiculosus, Laminaria* sp.). These compounds are directly implicated in the development, growth, or reproduction conditions to perform physiological functions. Nutrient contents can vary with species, geographical location, season, temperature, culture conditions and processes or biomass storage (Wells et al., 2017).

The protein content is generally high in algae, especially in red algae where it can account for up to 47% of the dry matter (red algae such as *Pyropia tenera*) (Harnedy & FitzGerald, 2011; MacArtain, Gill, Brooks, Campbell, & Rowland, 2007; Rupérez, 2002; Wells et al., 2017). Algal proteins are of importance as a source of dietary protein, essential amino acids, as well as for the bioactive potentials of specific lectins, enzymes and protein derivatives such as peptides (Stengel, Connan, & Popper, 2011). Some examples of red and green seaweeds *Pyropia* spp. (nori), *Palmaria palmata* (dulse), *Ulva ohnoi* (sea lettuce) often contain high levels of protein (Table 1). The Spirulina (Cyanobacterium *Arthrospira platensis*) and various commercial species of the unicellular *Chlorella vulgaris* (Chlorophyta) contain up to 70% dry weight protein. Proteins extracted from seaweeds have a high degree of *in vitro* digestibility. These algae also have an amino acid profile that compares well with eggs, notably containing all of the essential amino acids (EAA) that humans cannot synthesize and must obtain from foods (Fleurence &

Table 1 Main seaweeds used as foods in Japan and in Europe.

Species	Japanese name	European name	Food use
Monostroma,	Aonori	Aonori	Dried powder for cooking
Enteromorpha,			preparation
Ulva		Sea lettuce	
Porphyra tenera	Nori		Dried ingredients (makis)
Porphyra yezoensis			Salted salads
Gracilaria	Ogonori		Salted salads
Grateloupia filicina	Mukadenori		
Palmaria palmata	Dulse	Dulse	Dried ingredients
Ecklonia cava	Arame		Dried boiled with sauce
Undaria pinnatifida	Wakame	Wakame	Dried or salted soups, salads
Laminaria japonica	Kombu		Dried or salted soups, boiled with
Hizikia fusiformis	Hiziki		sauce
Laminaria digitata			Dried
Himanthalia		Kombu	Dried or salted soups, salads
elongata		Sea spaghetti	Dried Boiled with sauce

Levine, 2016). Aspartic and glutamic acids constitute a large part of the amino acid composition of these seaweed proteins and are responsible for the special flavor. Original hydrocolloids (alginates, carrageenan, agars) are the main additives used by the food industry in many dairy products (e.g. yogurts, flavored milkshakes, flans, jellies, ice creams, and beers) and meat products (e.g. hams), as a thickening, emulsifier or stabilizing agent. Agars could be a vegetarian substitute for gelatine. Lipids and more particularly polyunsaturated fatty acids, sterols, vitamins such as vitamins C or B12 with antioxidant properties are also present in macroalgae (Holdt & Kraan, 2011). Consequently, algae are low-calorie foods and already commercialized under multiple brands, and labeled with "fat-free," "gluten-free," "mineral rich" or "low in carbohydrates." Algae represent natural and healthier substitutes to pasta or to bacon (i.e. *Himanthalia elongata*, as spaghetti, and *Palmaria palmata*, as sea bacon). Other algae are used as food items such as the well-known nori sheets (*Pyropia*) used to prepare sushi rolls or crispy thin snacks. There are many other recipes such as wraps with *Undaria pinnatifida* (wakame) and *H. elongata*, or laverbread, a paste prepared with boiled nori (also recognized as laver) and in desserts such as innovative Spanish Turróns with crushed nori algae (Leandro et al., 2020) (Fig. 5).

The rise of TV cooking shows, blogs and online recipes hosted by several renowned creative chefs in different countries, has increased new food trends and encourage the consumption of algae. There are also trends for: eating less meat, eating a vegan diet and eating less dairy products in favor of vegetable proteins, with growing concerns about all aspects of sustainable and clean food production and "snacking" (Rioux, Beaulieu, & Turgeon, 2017). With these changes in attitude, the integration of algae as an alternative source in the daily diet has a great potential.

3.3 Good health and well-being: Ensure healthy lives and promote well-being for all at all ages (Goal 3)

Ensuring healthy lives and promoting well-being at all ages is essential for sustainable development. Before the COVID-19 pandemic, major progress was being made in improving the health of millions of people. Significant strides had been made in increasing life expectancy and reducing some of the common killers associated with child and maternal mortality. But more efforts are needed to fully eradicate a wide range of diseases and to address many different persisting and emerging health issues.

Back in the 1750s, doctors observed that being near the Ocean appeared to have a positive effect on human health. This thought was later scientifically

Fig. 5 Sea bacon by Seamore food company (https://www.seamorefood.com/)—Pasta by MorbiAlgues (© Nathalie Bourgougnon)—Nougat Alga Nori by Vincens (https://www.vicens.com).

proven, concluding that living (or being) next to the Ocean entailed many health benefits. Even today, foods are not only to satisfy hunger and to provide necessary nutrients but also to improve the physical and mental welfare of consumers and to prevent nutrition related diseases such as cardiovascular disease and osteoporosis (Hamed, Özogul, Özogul, & Regenstein, 2015; Levine, 2016). In addition to their nutritional value, algae are thus increasingly being marketed as "functional foods" or "nutraceuticals." These terms have no legal status in many nations but describe foods that contain bioactive compounds, or phytochemicals, that may benefit health beyond the role of basic nutrition (Holdt & Kraan, 2011; Wells et al., 2017). Indeed, seaweeds have developed effective mechanisms to survive many biotic threats, like bacteria, virus, or fungal infections. Because they are sessile organisms, seaweeds have evolved to live in variable, extreme, and hostile abiotic environmental and stressful conditions, tolerating temperature changes, salinity, environmental pollutants, or UV radiation exposure (Lalegerie, Gager, Stiger-Pouvreau, & Connan, 2020). These conditions cause these seaweeds to be able to produce a wide range of compounds called "secondary metabolites," like pigments, vitamins, phenolic compounds, sterols, polysaccharides, other bioactive agents and some extremely relevant biochemical mechanisms with biomedical applications (Fleurence & Levine, 2016; Ioannou & Roussis, 2009; Kornprobst, 2005).

These compounds relate to a vast spectrum of biological activities such as antibacterial, antifungal, antiprotozoal, anticoagulant, antithrombotic, or antiviral activities (Cardozo et al., 2007; Déléris, Nazih, & Bard, 2016; El Gamal, 2010; Schaeffer & Krylov, 2000). Fibers can increase feelings of satiety and aid the digestive transit through their bulking capacity and therefore may be present in sufficient amounts when included in the diet to prevent metabolic syndrome associated with obesity, type 2 diabetes, and cardiovascular diseases (Jakobsdottir, Nyman, & Fak, 2014). Epidemiological evidence comparing Japanese and Western diets have correlated seaweed consumption (5.3 g/day in Japan) with a decreased incidence of chronic disease (De Jesus Raposo, de Morais, & de Morais, 2016). In 2020, Cherry et al. (2019) provided an innovative approach to consider how the gut microbiota may utilize seaweed phytochemicals, such as polyphenols, polyunsaturated fatty acids, and carotenoids. They provided an updated discussion regarding the catabolism of seaweed-derived complex polysaccharides with potential prebiotic activity. As suggested by Cian, Drago, de Medina, and Martinez-Augustin (2015), algal proteins and carbohydrates that escape complete digestion in the small intestine may benefit humans by stimulating an immune response indirectly via the promotion of microbial responses. Dietary modulation of the colonic flora and the impact of bacterial fermentation products on human health are rapidly evolving areas of research and are likely to be especially important considerations in assessing the health benefits of algal-derived foods (Wells et al., 2017).

In the literature, sulfated polysaccharides rich in alginate, fucose, carrageenan and carotenoid are always cited to be seaweed's most important active metabolites as potential chemotherapeutic or prebiotic/chemo preventive agents. Two examples of sulfated polysaccharide antiviral properties against anti-HIV are respectively the gel named Carraguard used as vaginal microbicide that blocks HIV and other STD (Sexually Transmitted Diseases) or the nasal or throat spray Carragelose® used against respiratory viruses (Fig. 6).

In the presence of calcium, alginates form an irreversible matrix well adapted for tissue engineering, as entrapment and/or a delivery vector of

Fig. 6 Example of commercial products. Carragelose® marinomed (https://www.carragelose.com/en).

a variety of drugs and for use in cell, bacteria, fungi or enzyme entrapment. These biomaterials are known to be biocompatible, nonedible, and immunogenic. Alginates can be use as composite materials to elaborate hemostatic wound dressings. AlgiDERM and Sorbsan are calcium alginate dressings made of sterile purified alginate fibers. Up to 20 times their weight of exudate is absorbed into a web-like gel. They are easily removed without damaging healthy tissues (Rinaudo, 2008). A hydroxyapatite putty or self-curing resin was made by mixing sodium alginate in a non-decay-type fast-setting calcium phosphate cement, which could be very useful for certain surgical procedures. This hydroxyapatite/alginate mixture is potentially a valuable new biomaterial for use in plastic and reconstructive surgery, as well as in oral and maxillofacial surgery. Alginate also offers a mucoadhesive property which could serve as a potential advantage in mucosal drug delivery through conjugation by the formation of covalent bonds with thiol-bearing compounds (Kassem et al., 2015). These authors conjugated sodium alginate with thioglycolic acid to increase the mucoadhesive potential of resveratrol (a polyphenolic phytoalexin) microspheres. This complex shows a promising therapeutic efficiency toward the treatment of periodontal disease in vitro. In fact, resveratrol was reported to possess significant antimicrobial properties on periodontal pathogens in vitro and to stimulate osteoblastic differentiation and bone formation in vitro. However, the therapeutic applications of this drug are strongly limited because of its poor oral bioavailability due to rapid metabolism in the liver together with the entero-hepatic cycle. Clinical evaluation is still in progress (Kassem et al., 2015).

Non-polar or neutral lipids are an important lipid class in seaweeds known to include molecules with attractive bioactivities such as sterols, hydrocarbons, and carotenoids. Phytosterols have been reported to be efficient in cardiovascular diseases and as anti-inflammatory agents. Fucoxanthin, a carotenoid derivative found in brown seaweeds, is known for its anti-obesity effect and the ability to reduce blood glucose and plasma insulin levels (Terme, Chénais, Fournière, Bourgougnon, & Bedoux, 2019). Fucoxanthin and its deacetylated metabolite also exhibit antiproliferative and cancer-preventing influences (Déléris et al., 2016). Phlorotannins have been extensively investigated for various biological activities such as antioxidant, anticancer, anti-inflammatory, anti-allergic, antidiabetic, antihypertensive, and neuroprotective effects (Thomas & Kim, 2011). The anti-dyslipidemia activity of dieckol extracted from brown seaweeds *Eisenia*, *Ecklonia*, and *Ishige* genera must be highlighted because it was more effective than lovastatin in an in vivo model. The lectin griffthsin, first isolated from aqueous extracts of *Griffthsia* sp. showed excellent antiviral (Japanese encephalitis virus, SARs-) results over a variety of animal models and will probably move into clinical trials soon (Rosa et al., 2019).

In the pursuit of our well-being, marine cosmetics are also an economic reality in France, the USA or Brazil (Couteau & Coiffard, 2016; Wang, Chen, Huynh, & Chang, 2015). The global beauty market stated that the cosmetics industry will continue to develop due to the growth of the middle class in many developing countries. The advantages of natural ingredients are their environmental friendliness, fewer side effects, and safe use (Jesumani, Du, Pei, Aslam, & Huang, 2020). Cosmetic products not only have to comply with strict rules concerning the use of chemical substances (EU regulation No. 1223/2009 on cosmetic products) but must also satisfy consumer demands for products containing effective natural and non-toxic ingredients. Seaweed is used as a plant extract and there are no restrictions for cosmetic use (Bedoux, Hardouin, Burlot, & Bourgougnon, 2014). The main classes of actives in cosmetic care products are used for anti-aging care, anti-photoaging, slimming care, photoprotection, moisturizing and skin-whitening. Seaweed is of interest, because it contains active molecules also for cosmetics. In addition, the production of these molecules can be optimized: by selecting a season when the algae synthesize a lot of these compounds, by setting culture conditions that promote the production of these active molecules by the algae, and by developing a transformation process which isolates them by eliminating the others. It is therefore concentrated. HELIONORI® (Gelyma, France) supplies natural skin UVA protection against sun burn that burns human cells, as well as the protection of cells and their components such as membrane lipids and DNA. HELIONORI® patented concentrated active ingredients prepared from *Pyropia umbilicalis* (L.) Kützing contains special marine compounds named mycosporine-like amino acids (MAAs) that act as natural sunscreens in some marine organisms. Heliogard 365 (Mibelle Biochemistry) is a natural UVA-screening compound from sea algae used to protect the skin against photoaging.

However, advances in the understanding of the composition of algae according to their environment, of interactions between the compounds of algae in the metabolism of humans with their microbiome as well as the bio accessibility of algae compounds for human or animal nutrition and health must continue to be explored (Finley, Finley, Ellwood, & Hoadley, 2014; Wells et al., 2017).

3.4 Decent work and economic growth: Promote inclusive and sustainable economic growth, employment and decent work for all (Goal 8), quality education: Ensure inclusive and quality education for all and promote lifelong learning (Goal 4), partnerships for the goals: Revitalize the global partnership for sustainable development (Goal 17)

Sustained and inclusive economic growth can drive progress, create decent jobs for all and improve living standards. This part is highly linked to the first SDG. Seaweed farming is a global business and was carried out in at least 56 countries worldwide in 2018 (12 in Africa, 12 in the Americas, 14 in Asia, 11 in Europe and 7 in Oceania) (FAO, 2020). Farms vary from large industrial enterprises to smaller family-run businesses (Hoegh-Guldberg et al., 2019). More than 99% of the production is found in Asian countries. In 2018, OECD countries accounted for over 60% of the total export value (close to USD 1 billion) of seaweeds traded for direct human consumption or as raw materials to produce other food or non-food products. Europe accounted for nearly half of the world import value (USD 1.3 billion) of seaweed-based thickeners (Giercksky & Doumeizel, 2020; United Nations, 2020a, 2020b). In 2018, global seaweed production was 33 million tons (wet weight), of which 97% was farmed, and 3% came from wild seaweed. Global seaweed production accounted for around 28% of the total aquaculture tonnage. By focusing on marine seaweed resources, developing countries provided an opportunity to access an emerging market, propelled by a diversification of the demand for seaweed products from traditional uses, to bioenergy, cosmetics, and biomedicine applications. Many governments, IGOs, NGOs and private initiatives have initialized ambitious projects to scale up seaweed production outside the major producing areas in Eastern and South-Eastern Asia. The industry doubled in size between 2005 and 2015 (FAO, 2018a, 2018b). In Europe and in Norway several food and food supplement products directly prepared from algae are already certified as organic food for example, and this has given opportunities for producers to access a niche market. The interest in algae creates new markets in different fields and new jobs. In addition to the value chain, sustainable management of coastal resources creates new economic activities based on the exploitation of a raw material, and participates in the local socio-economic development in coastal areas and communities in the World. In Latin America or European coastal communities, fishermen harvesting seaweed from boats and with diving gear complement their income by participating in other fishing such as abalone, sea urchin or fish. Although the seaweed fishery is economically and environmentally sustainable, this livelihood for the fishermen is not always attractive. As explained in Section 3.1., it is therefore important to look at the fishery in a larger context, and the cultivation of seaweeds (as opposed to their harvesting) may actually offer a better alternative to coastal communities, as it can give them the opportunity to increase production and improve productivity and quality. Local farmers sell the seaweeds directly to the processing companies or companies with concessions which directly employ their own workers for harvesting during the year and contracted divers in the summer.

In Norway, the first European producer of the raw materials harvested annually using paddlewheel cutters and seaweed trawlers, corresponds to nearly 200,000 tons of *Laminaria hyperborea* and *Ascophyllum nodosum*. This activity employs 250 people during harvesting and processing for diversified markets (alginate industry), health (additives), agricultural supplies (biostimulants, fertilizers, feed), cosmetics, health and well-being.

In Brittany (France), almost 70.000 tons of fresh seaweed are produced annually (except *Gelidium* which is gathered in the Basque country). In Brittany there is a traditional shoreline seaweed harvest. In 2014, it represented about 6.000 tons. In 2017, it amounted to around 7.700 tons. Nearly 600 tons of macroalgae were used for the decoration of fish stands. Half of the 7.700 tons was mainly *Ascophyllum nodosum*, which is still the most demanded alga. It is used in fertilizers, animal meal and the alginate industry. Only 50–350 tons came from aquaculture. Today, with a growing demand for raw materials and increased pressure on the foreshore due to the coveted coastal areas for marine energy production, port activities, industries and aquaculture, the sector is becoming more structured and is gradually acquiring more efficient resource management tools. Shoreline seaweed harvesting is supervised by the Regional Committee for Sea Fisheries and Marine Farming in Brittany (CRPMEM), a commission made up of fishermen, seaweed harvesters, processing companies and scientists. The Chambre syndicale des algues et des végétaux marins and the syndicat des récoltants d'algues de rives professionnels (The Syndicate of Seaweed and Marine Plants) and the Syndicate of Professional Shoreline Seaweed Harvesters (SRPAR) is strongly involved in the organization of the harvest, the recognition of jobs and the sustainable exploitation of local resources (NETALGAE, 2012). In 2017, more than 60 structures have been identified. Within these structures, 180 people have been identified as working all or part of their time on the seasonal collection of algae (Figs. 7 and 8).

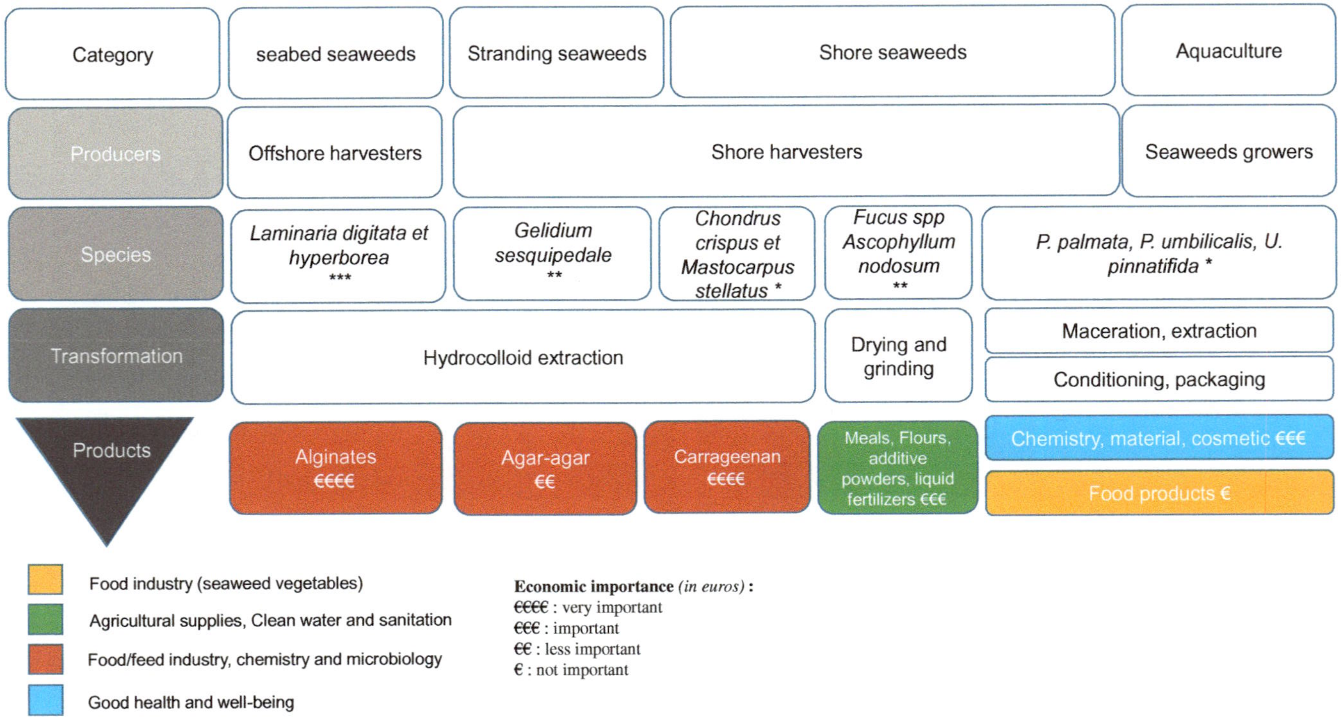

Fig. 7 Organization of macroalgae industry, from seaweeds to uses. Degree of importance in harvested tonnage *(wet tons)* : **** : ≤ 100,000; *** : 50,000; ** : 5000; * : ≤ 1000 (NETALGAE, 2012).

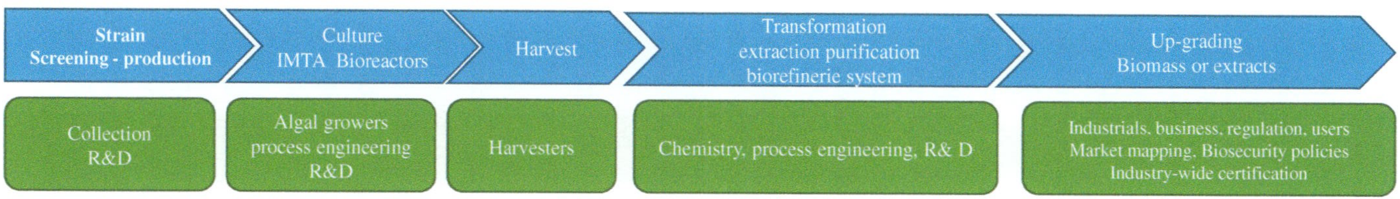

Fig. 8 Blue value chain.

To conclude, seaweed farming could create jobs and improve livelihoods in coastal communities. In low- and middle-income countries, there is a need for increased research on local seaweed communities, the strengthening of value chains and improved resilience in these areas, as well as local capacity building for safe and sustainable production and effective marketing. Development of the industry could be a new source of revenue, especially for emerging economies and vulnerable people. A blue-growth farming coalition platform might also become an anchor for "market mapping initiatives," which are necessary tools when providing directions for sustainable private/public investments (Giercksky & Doumeizel, 2020).

Only a deep understanding and education of the complexity of the marine ecosystem will enable human beings to protect the oceans and organisms populating them, and pave the way for sustainable exploitation of marine resources with decent jobs. The training of young adults in environmental, technological, societal and economic issues is the key to building another form of development of a solidary and united society, to learn to live with dignity, to reduce inequalities, to reduce vulnerabilities. Trained human resources are required in order to provide education to coastal communities, based on best practices for harvesting and cultivation, in order to establish profitable businesses which could provide socio-economic development leading to better living conditions for the rural coastal communities. Several actions are already being carried out around the world in engineering schools, universities and high schools. A first action of knowledge transfer, research and technology development (RTD) by small and medium enterprises (SMEs) is required. Representatives from Norway, Argentina, Brazil, Canada, Chile, Mexico, Peru, and Portugal have established a common working framework in order to support the development of the Latin American seaweed sector. These industry/academia networks encourage cooperation among the seaweed stakeholders; across the project areas; in all aspects related to seaweed production, research, ecosystem services, and management of artisanal and small-scale aquaculture; and traditional and alternative markets and economics (Cottier-Cook et al., 2016; Rebours et al., 2014). The concept of Planetary Health, recently formalized by the launch of a new scientific journal The Lancet Planetary Health, offers a powerful bridge between very different disciplines: it considers that the health of human populations is connected to the health of the environment and that the health of one cannot be examined without examining the other. Planetary Health is defined as the achievement of the highest standard of health, well-being and equity around the world

(Whitmee, Haines, & Beyrer, 2015). The encompassed broad spectrum of disciplines enables the investigation, not only of the effects of environmental changes on human health, but also of human systems (political, economic, social) that govern these effects. The emphasis on the human health consequences of the degradation of natural systems throws into sharp relief the urgency and fragility of the current situation. The Planetary Health Alliance aims to bring together a diverse group of researchers, instructors, policy makers, students and interested individuals in order to understand and communicate the human health impact of global environmental change. MareNet, the worldwide Network of Marine Research Institutions and Documents, provides a set of information services for marine scientists. The website (http://www.marenet.de) gives a detailed description of MareNet and its different services as Marine Research Institutions worldwide or Marine Science Societies and Organizations. In Europe, the European Algae Biomass Association (EABA) promotes mutual interchange and cooperation in the field ofbiomass production and its use, including biofuels uses and all other utilizations (https://www.eaba-association.org/en). It aims at creating, developing and maintaining solidarity with links between its Members and defending their interests at European and international level. Its main target is to act as a catalyst for synergies among scientists, industrialists and decision makers in order to promote the development of research, technology and industrial capacities in the field of Algae. In France, 15 universities have created the network of Marine Universities around 86 Master's degrees (Ocean Physics Oceanography Environment Marine Chemistry Engineering Economics Marine Energy Ecology Marine Biology Geography Climate Maritime Law Earth Sciences History and Archaeology, ...) and 27 PhD degrees (see https://www.universites-marines.fr/fr/formations). To address the increasing challenges facing ocean and coastal ecosystems, the "Interdisciplinary graduate School for the blue planet" (ISblue—https://www.isblue.fr/) was created to train the next generation of ocean innovators and science leaders and place them at the forefront of research to answer the needs of the growing blue economy through top level research-based training. ISblue includes 15 research units and teams showing an internationally recognized expertise in many disciplines (e.g. physical and spatial oceanography, geosciences, biogeochemistry, ecology, microbiology, data analytics and also geography, economics, law, and engineering), all dedicated to tackle questions related to the ocean in all its dimensions and interfaces. ISblue will move interdisciplinarity to an unprecedented level to push back the frontiers in marine science and technology and ocean innovation.

The last and 17th goal of the SDGs emphasizes the need for inclusive partnerships between governments, the private sector and civil society in order to achieve a successful sustainable development agenda. It must place people and the planet at the center and be implemented at global, regional, national and local levels. It implies enhancing cooperation and strengthening domestic resources in all sectors, be it finance, technology, capacity building, trade or systemic issues. Concurrently, over the last few years in sustainability science, the need for transdisciplinarity approaches and an integration of knowledge from experience and science has become conspicuous. The separation of natural and social science in academia led to working in "silos," producing an incomplete understanding of the nature-society interactions and of the integrated dynamics of the "Earth system" as a whole. The necessity for cross disciplinary boundaries also gives rise to challenges of collaboration across complex disciplines. Development of regulations and directives enabling the sustainable exploitation of natural resources must therefore be brought to the national and international political agenda in order to ensure environmental, social, and economic values in the coastal areas around the world. Two conditions are necessary to achieve this: political will and the will of the people. The link between these challenges is solidarity. Examples of partnerships were described in Section 3.1 "No poverty." Indeed, the Red Seaweed Promise of the international company Cargill, the Nagoya protocol or the GlobalSeaweed STAR program highlighted a cooperation and a solidarity between key players from the algae industry.

3.5 Gender equality: Achieve gender equality and empower all women and girls (Goal 5), reduce inequalities: Reduce inequality within and among countries (Goal 10) and peace, justice and strong institutions: Promote just, peaceful and inclusive societies (Goal 16)

Over the last decades, the international community has made considerable progress in lifting people out of poverty. In the 25 years from 1990 to 2015, extreme global poverty has decreased rapidly: the share of the world's population living below the extreme poverty line of $1.90 per day has fallen from 35.6% in 1990 to 10.0% in 2015 (World Bank, 2018). The vulnerable, least developed countries and small island developing States continue to score points in poverty reduction. In this context, international development key players, bilateral development agencies and countries themselves, have united around the goal SDG1 of "ending" extreme poverty by

2030. Concurrently, the development policy debate has been increasingly paying attention to the level of inequality in countries around the world. Inequalities can be estimated in numerous ways, ranging from a denial of individual opportunity to a more structural arrangement of inequality of social conditions. They continue to persist and there are still wide disparities, for example in terms of access to health services and education. They threaten long-term social and economic development which in turn can impact global human health, environmental health and global peace. Thus, the Sustainable Development Goals include both a goal to end poverty (SDG1) and a goal to reduce inequality within and among countries (SDG10).

Among these challenges, gender inequality remains a major barrier toward sustainability. Deep-rooted in every society, women suffer from a lack of accessibility to decent work, education, healthcare and are under-represented among political and economic decision makers. In many situations, they are victims of violence and discrimination. Therefore, the SDG5 focuses on achieving gender equality and empowering all women and girls. According to the Food and Agriculture Organization, of the estimated 180 million people worldwide working in fisheries and aquaculture, nearly half of them are women (FAO). While this ratio tends to be balanced, the situation is far from equal in terms of rights, recognition and economic independence. All around the world, men and women have historically been involved in fisheries in coastal areas. A common pattern is observed: while men work on boats and go to sea, women are in charge of fisheries activities on the shore, in addition to their household tasks (Hussin & Khoso, 2017). These activities are numerous and crucial to the economy: some examples are the maintenance of the fishing tools, market sales or the creation of added-value products. Unfortunately, women's work is less valued and their contribution to the economy is often invisible. This informal economy creates an insecure labor market and puts women at a greater risk to fall into poverty. As the monetary economy remains associated with the men, women remain dependent on them despite their difficult and time-consuming involvement. This lack of recognition occurs at all levels (local, national, international) and must be addressed in order to achieve sustainability and gender equality (Frangoudes et al., 2019).

Neither a reduction in social inequality nor a gender equality can be advanced without peace and respect for the rule of law. The SDG16 aims to promote peaceful societies at national levels, as well as the role of cooperation at the international level. Access to justice for all and the creation of more peaceful and inclusive societies requires effective and responsible transparent institutions and regulations at all levels. Several actions have been initiated in order to manage and structure the "algae blue chain" in many countries. These actions also aim to reduce economic and social inequalities.

Seaweed farming has been introduced and presented as a way to alleviate poverty in many countries. As mentioned in Section 3.1 "No poverty," the development of this new activity has offered opportunities for the empowerment of women. The cultivation process of these species is based on vegetative propagation and is relatively simple in adequate conditions, as their growth rate can reach 3–8% per day from propagules attached on ropes. Depending on the region and the cultivation methods, such as off-bottom cultivation, rafts, or long-lines, the production cycle before harvesting lasts from 30 to 60 days. In most developing countries, the majority of people involved with seaweed farming are women. Meanwhile in Asia, the gender balance is about even among seaweed farmers (Msuya & Hurtado, 2017).

Seaweed cultivation has become a source of income and livelihood for women living in Tamil Nadu's coastal communities in India or along the coastline of Jepara in Indonesia (Fig. 9). After learning modern seaweed farming, local fisherwomen formed Seaweed Squads to harvest the marine commodity. Thanks to their new job, local women have increased their income and bargaining power, and become role models within and outside their fishing communities (World Bank, 2018). In Tanzania, seaweed farming is unique in that over 80% of its farmers are women. Valderrama, Cai, Hishamunda, & Ridler (2013) reported on studies in Indonesia,

Fig. 9 Local women along the coastline of Jepara in Indonesia. *Photo by Ita Widowati, University of Diponegoro.*

Philippines, India, Tanzania, Solomon Islands and Mexico: they highlighted that for in off-bottom cultivation in shallow water, women were more involved as they can tie and harvest the crop by themselves, while in deeper water they tend to have a smaller role as boats are necessary for raft or floating line techniques. In Tanzania, seaweed farming was introduced in 1989 using pegs and ropes. Farmers have been producing the two carrageenophyta (*Kappaphycus alvarezii* and *Euchema denticulatum*) and agarophyta (*Gracilaria* and *Hypnea*) seaweeds and selling them to buyers who then exported them to Denmark, France, the USA, and Spain. As men were often discouraged by the labor-intensiveness and relatively low profitability of seaweed farming, and hence preferred other activities such as fishing, tourism and the shell polishing trade, it became more a female-oriented activity and an important livelihood source for women. Seaweed farming successfully changed the position of women in society.

This traditional off-bottom method is carried out in shallow intertidal areas by the beach, where conditions for growth used to be optimal. Unfortunately, temperature and salinity can vary rapidly in these environmental conditions and lead to "ice-ice" disease outbreaks and infestation by epiphytes, affecting the quality of the seaweeds (Brugère, Aguilar-Manjarrez, Beveridge, & Soto, 2019). This has been occurring increasingly over the last two decades with the rise of water temperature in the Indian Ocean. In addition to the already volatile and competitive global market, women have also been suffering from the economic and sociocultural burdens of climate change. The industry and all that it has brought to rural women is under threat. In addition, Frocklin, de la Torre-Castro, Lindstrom, Jiddawi, and Msuya (2012) highlighted the potential occupational health hazards linked to seaweed farming in Tanzania: seaweed farmers reported more health issues than non-seaweed farmers such as fatigue, musculoskeletal pain, hunger, respiratory problems, allergies, eye related problems and injuries from hazardous animals and sharp shells in the water.

Initiatives and actions have been implemented in order to address these constraints. Piloted by Dr Flower Msuya, these projects on seaweed farming help fight against social, economic and gender inequalities. The SeaPoWer project started in 2016 in Zanzibar has introduced an improved seaweed farming technology through the use of tubular nets. This method has proven to have a greater tolerance to the roughness of the sea, to provide a higher yield than the off-bottom method and to be more environmentally friendly as it could stimulate fisheries productivity and increase fish catches locally. This technology is used in deeper water and therefore requires swimming and boat handling skills. Most women of Zanzibar do not have these skills and the technology needs to be adapted to the special conditions of Zanzibar as local culture and traditions tend to prevent women from performing some tasks. Women seaweed producers' experience with tubular nets has been very positive overall and presents advantages such as increased productivity, improved welfare and work conditions. While women participants are still not fully independent from men in their community, the SeaPoWer project has boosted their confidence and self-esteem though training and collaboration. The SeaPoWer project has become a concept under which seaweed farming innovation and production cannot be separated from women's empowerment (Brugere et al., 2019). In the same dynamics, the Zanzibar Seaweed Cluster Initiative (www.zasci.webs.com) corresponds to Commercial collaboration with coastal communities—primarily women—now producing, processing and selling commercial products into African and European markets.

The Seaweed Cluster Initiative stands out to be the most successful in Tanzania. The group partners include scientists from the Zanzibar Institute of Marine Science and the government through the Commission for Science and Technology (Costech). The aim of the Seaweed Cluster Initiative is to address these problems and tap the scientific information for the benefit of the farmers and the country at large. For example, the basis of the aim of the Cluster is that there is a possibility to increase seaweed production through modifying the farming technique. Thanks to this Cluster, local people produce more seaweed with the production of *Kappaphycus*, also raising the income of the farmers by enabling them to farm the *Kappaphycus* seaweed as it fetches a high price.

In the Philippines, seaweed farming is a major economic endeavor among the households in the key seaweed production areas because of its high profitability potential relative to the farming effort. Hired workers (mostly female) were paid USD 3.00–4.25/day to remove impurities from seaweed (FAO, 2014). The study of Suyo, Masson, Shaxson, and Luhan (2020) used Social Network Analysis to explore the social structure and the nature of support mechanisms in the production segment of the seaweed value chain to understand the differential in the access of women and men to information, resources, and services. The study shows that there is a strong potential for improving the production and market outcomes by addressing the gaps arising from the current social network systems which include promoting equal access to training, technology, and social relations. There is a need to promote increased knowledge of efficient farming techniques and product standards to minimize losses along the value chain. In the Philippines, and also in other countries, the growth of the seaweed industry will be facilitated if the different stakeholders in the industry are empowered and needs and access of women and men are taken into account in the planning of, and in the formulation of programs aimed at enhancing production and market expansion.

In Indonesia, female involvement is greatest in seaweed cultivation and processing. It is just as important as that of men for the improvement of the coastal economy. The processing of raw seaweed material into various types of products, including products for the food industry, can be an alternative livelihood for fishing households and can be fully managed by women (FAO, 2018a). The Indonesian National Medium-Term Development Plan 2015–2019 explicitly stipulated efforts toward equality in its principles. The main areas foreseen were women's access to higher education (notably the knowledge of seaweed processing is transferred through training activities), health care and labor force opportunities, based on gender inequalities highlighted in the Human Development Index and Gender Development Index, such as gender gaps in education levels and health status and a low representation of women in main decision-making institutions.

In response to the problems of high coordination costs among the poor, efforts are underway in many countries to organize the poor through "self-help groups" (SHGs) or cooperatives—membership-based organizations that aim to promote social cohesion through a mixture of education, access to finance, and connections to wider development programs. The seaweed sector in coastal India has all the potential to rise from the low-income conditions normally associated with basic livelihood activities to higher levels of employment-income-consumption relationships. So far, as many as 1200 families are engaged in seaweed farming, of which 60% of the farmers are women. Seaweed farming has had a remarkably positive effect on the socio-economic status of female farmers as it allows them to engage in an income-earning activity. The SHG consists of members linked by a common bond like caste, sub-caste, community, place of origin, activity, etc., in these "natural groups" or "affinity groups." The SHG provides the benefits of economies in certain areas of the production process by undertaking common action programs like a cost effective credit delivery system, generating a forum for collective learning with rural people, promoting a democratic culture, fostering an entrepreneurial culture, providing a firm base for dialog and cooperation in programs with other institutions, possessing the credibility and power to ensure participation and helping to assess the individual member's management capacity (FAO, 2014; Krishnan & Narayanakumar, 2010). The concept of the SHG was founded on the basic premise that women are more responsible and have a better disposition to work toward achieving social and economic independence.

Through their direct and indirect roles, women contribute to livelihoods, communities, and economics at large. Frangoudes, Gerrard, and Kleiber (2019) cite the example of women in seaweed farming households in Japan who create different high-quality products from seaweed, thereby increasing its value. The emphasis of the quality of their products over quantity is also branded and marketed online. This often requires new skills such as using social media to identify and reach a national market to sell their products. By occupying these positions, women contribute to other activities such as strengthening and revitalizing communities, or food security at a larger scale (Frangoudes et al., 2019).

While successful examples of women's empowerment through seaweed farming are numerous around the world, there is still a long way to go, constraints and difficulties to bypass, efforts and progresses to be made in order to achieve sustainability. Many of the women involved in seaweed farming have low-skilled, low-paid jobs, without contracts or health, safety and labor rights. These communities are particularly fragile to the effects and consequences of climate change and environmental degradation. Sometimes they suffer from the cumulative effects of environmental degradation and measures taken for protecting the environment. Frangoudes et al., (2019) reported the example of how the implementation of Marine Protected Areas, such as in the Gulf of Mannar in India, impacted marginal groups, such as women: they lost their license for seaweed harvesting to harvest seaweeds by diving on the natural wild stocks. Consequently, they lost their only source of income. Women resisted the decision by establishing new management measures (reducing the number of harvesting days and not using modern equipment for example). In addition to the implementation of Marine Protected Area, the fisheries department used to allow only male heads of families to get social benefits, thereby keeping women dependent on men. Following training in 2014, women formulated new claims and, after negotiations, the Tamil Nadu state recognized women seaweed harvesters as fishers, providing them with identity cards and protective equipment. The example shows how women resist unfair decisions and how training and support can emancipate women (Frangoudes et al., 2019).

Some other challenges lie in cultural aspects. For example, in Malaysia, in the Suluk community, women are kept away from seaweed cultivation activities because their monthly menstruations are said to have a negative supernatural effect on seaweed production. In some other areas, women face opposition from male relatives to get involved in seaweed cultivation because it causes shame for the men or because they consider that this role will prevent the women from doing household chores (Hussin & Khoso, 2017).

Unfortunately, COVID-19 has deepened existing inequalities, hitting the poorest and most vulnerable communities the hardest. It has put a spotlight on economic inequalities and fragile social safety nets that leave vulnerable communities to bear the brunt of the crisis. At the same time, social, political and economic inequalities have amplified the impacts of the pandemic. Lakner, Gerszon Mahler, Negre, & Prydz (2020) estimate through a machine-learning algorithm that the pandemic may have driven around 60 million people into extreme poverty in 2020 and, that with current growth rates and inequality index, the number of extreme poor could remain above 600 million in 2030. Since seaweed farming has potential for sustainable development of coastal areas, support and changes toward better working conditions should be encouraged. In order to achieve these SDGs by 2030, research must be transdisciplinary, include women's situations and problems and involve more gender focus by male researchers (Frangoudes et al., 2019).

3.6 Affordable and clean energy: Ensure access to affordable, reliable, sustainable and modern energy for all (Goal 7)

Energy is central to nearly every major challenge and opportunity the world faces today. Be it for jobs, security, climate change, food production or increasing incomes, access to energy for all is essential. Fossil fuels are the current major sources of energy, providing up to 80% of the annual demand. The major industrial and threshold nations, such as the USA, Germany, Japan, China and India, rely heavily on carbonaceous combustibles to supply their domestic industries and to provide for their population.

However, dependence on fossil fuels has decreased recently, dropping from 85% in 2005 to 80%–83% in 2011 due to the development of a number of alternative renewable sources of energy, such as geothermal, wave, wind, solar and biomass (Barbot, Al-Ghaili, & Benz, 2016). Due to diminishing petroleum reserves, deleterious environmental consequences of Green House Gas (GHG) emissions from fossil-based fuels, research on renewable and environmentally friendly fuels has received considerable impetus in recent years. Clean and renewable marine biomass is an excellent contributor to the choice of alternative sources of energy. To ensure access to affordable, reliable, sustainable and modern energy, several technologies have emerged: renewable energy with algae and algae biofuel, algae bio refinery, and eco-friendly innovative green-chemistry-based systems for the sustainable extraction of all major value-added components from the selected algal species.

Biodiesel is a potential fuel that has the ability to replace conventional fossil diesel fuel. Algae are a promising alternative source to the conventional feedstocks for the third-generation biofuel production. They have fast growth rates with up to 4–6 harvest cycles per year and unlike first- and second-generation biofuel feedstocks, macroalgae can be grown in the sea thus eliminating issues relating to land use and water for irrigation. They also have other advantages over a terrestrial biomass: higher carbohydrate levels and biomass yields, their widespread availability, the absence of competition with food or agricultural surfaces and the high quality of their by-products. Their use as a means to capture CO_2 and their suitability for being integrated into wastewater treatments to reduce pollution make them one of the most attractive renewable sources for a sustainable energy strategy, such as their growth in saltwater and municipal wastewater and their lack of requirements for arable land and industrial fertilizers. Macroalgae can be used in the production of biomethane generated from thermal or biological gasification. A wide variety of seaweed species can be anaerobically digested to produce energy-rich methane, which allows flexibility for the choice of biomass source. Bucholc et al. (2014) studied the potential of collecting macrophyta accumulating on the beach from the Baltic sea as a resource for biogas production. The authors suggested that collecting macroalgae removes more nitrogen than phosphorus, that would decrease the N/P ratio in the sea, but not deteriorate the environment. Macroalgae biomass contains substantial reserves of renewable energy, which could be exploited with the appropriate technology for biomass collection/preservation and third-generation gaseous biofuel production as an additive to other carbon-rich biomasses.

Stranding macroalgae from beach and marine areas (e.g. *Ulva* in France, in Venice lagoon and China, *Codium* in Florida USA, Golden tide with *Sargassum* in the Caribbean, ...), could be used and it supports beach management for local tourism and decreases seawater pollution caused by massive algal growth (Allen, Browne, Hynes, & Murphy, 2013; Barbot et al., 2016; Kraan, 2013). The process of accessing a suitable beached macroalgae feedstock could be facilitated by the use of supporting technologies, such as satellite imagery and climate simulation models, enabling the prediction and localization of the quality and quantity of eutrophic biomass sources.

Furthermore, seaweed industries generate a considerable quantity of marginal biowaste streams, which could present a potential use in biomethanation. Incorporated into the concept of a circular economy, macroalgal wastes from industry or eutrophication could find a niche in serving as substrates for biomethanation.

Different government research institutes around the world have investigated the development of biorefinery technologies for seaweed, with the objective to produce biofuel (Okoli, Adams, Brigljević, & Liu, 2016). Lehahn, Ingle, and Golberg (2016) used a modeling approach to investigate the global potential for macroalgae. They estimate that 98 Gigatons dry weight of macroalgae per year can be grown globally over a surface area of approximately 108 km^2 and conclude that with near-future aquaculture technologies, offshore cultivation of macroalgae has a huge potential to provide fuels and chemicals for humans. Currently, research efforts into biofuels suitable for gasoline replacement have shifted focus to butanol instead of ethanol because of advantages such as lower miscibility with water, Higher Heating Value (HHV), and better compatibility with existing gasoline engines and fuel pipeline infrastructure. Butanol can be produced from macroalgae using a biochemical route via the acetone, butanol and ethanol (ABE) process. Using *Clostridium beijerinckii* as the fermentation organism Van Der Wal et al. (2013) obtained butanol yields of 0.23 g butanol/g sugars from the green seaweed *Ulva lactuca*. Huesemann, Kuo, Urquhart, Gill, and Roesijadi (2012) carried out a study of butanol fermentation from brown algae (*Saccharina*), but obtained very low butanol yields of 0.12 g butanol/g sugars.

There has also been a considerable discussion over the recent years about the potential of microalgae for the production of biofuels (Suganya, Varman, Masjuki, & Renganathan, 2016). Microalgae can produce 19,000–57,000 L oil/acre in 1 year, better than any previous biodiesel source (Demirbas & Demirbas, 2011). It is converted into biodiesel through a standard transesterification process. The remaining biomass also contains valuable components such as residual lipids, proteins, and soluble polysaccharides, which can be used for the production of bio-oil, bioethanol, biohydrogen, and biogas through various thermo-chemical and biochemical processes for the improvement of an overall energy balance (Deshmukh, Kumar, & Bala, 2019). Various innovative extraction methods like microwave, ultrasound, supercritical fluids and ionic liquids for oil extraction and also the direct transesterification have been evaluated. A higher oil yield was obtained in recent extraction techniques compared to that of conventional solvent extraction techniques. However, in spite of new processes, rapid growth and high lipid content, the production of microalgal biodiesel is an energy intensive and costly affair. Developments in harvesting, drying, oil extraction and direct transesterification stages need to be studied in a global refinery concept. Macroalgae seem to be preferable to microalgae for biofuel production because its plant-like characteristics make it easier to harvest, and its high concentration of carbohydrates in comparison to microalgae make it a potentially better biofuel feedstock.

Management of environmental and health issues are key elements for competitiveness and sustainability. The main challenge, which determines the industrial utilization of algae potential, is the development of efficient extraction methods while preserving most of the intrinsic qualities of the raw material and finding a compromise between the cost of producing sufficient quantities and the quality of compounds in the shortest timeframe. Finding the optimum processing conditions and meeting the principles of green chemistry and green technology will be required for this task. Several procedures exist for extracting and separating biologically active compounds from algae biomass. Indeed, for agriculture, animal feed and health, most of them include water extraction under high pressure and/or alkaline extraction while algal fertilizers are usually prepared with a chemical method where potash lyse and high temperatures are applied. However, such conditions are quite severe and cause the decomposition of most of the biologically active compounds.

In this context, most efforts will be put on cell wall structure disruption in order to enhance the liquefaction of the seaweeds, the release of their internal components and their partial conversion. Another challenge, which will determine the industrial use of algae, is the development of eco-friendly extraction methods in order to preserve the intrinsic qualities of the raw material. Currently, Enzyme-Assisted Extraction (EAE), Microwave Assisted Extraction (MAE) and Ultrasound-Assisted Extraction (UAE) are among the five novel techniques most mentioned in the literature for natural biomass processing (Kadam, Tiwari, & O'Donnell, 2013). The processes are described as feasible and efficient alternatives to traditional extraction methods and could be transposed to a larger scale for seaweed processing. To maximize the added value of the biomass by allowing the co-extraction of valuable components, a strategy of biorefinery will be developed at laboratory and industrial levels. The production of chemicals and materials from renewable feedstocks is an important part of a biobased economy and could potentially make algal biomass and processing steps economically feasible and environmentally beneficial. In this respect it is important to introduce the biorefinery concept that integrates biomass conversion processes in value-added chemicals from biomass, analogous to today's petroleum refineries, which produce multiple fuels and products from crude petroleum (Taylor, 2008). Nowadays, the concept of biorefinery involves the development of eco-friendly processes for sequential recovery of molecules, especially bioactive molecules. Furthermore, EAE, UAE, UAEH (combination of enzymes and sonication) and may be MAEH (combining enzyme and microwave) extraction technologies have demonstrated their interest in the extraction of seaweed biomolecules. These novel extraction technologies enable reducing the use of organic solvents and the treatment time. They also help preserve the activity of target compounds leading to higher extraction yields with a lower cost in comparison to traditional extraction methods. Nevertheless, the main challenge will be transferring them to the industrial scale.

In conclusion, an important point noted by some of these review studies was that despite the potential for macroalgae-based biorefineries, technological improvements in the whole supply chain of macroalgae-based biorefineries (such as seaweed cultivation, harvesting and transporting, pre-treatment, co-digestion, eco-friendly extraction processes and fuel conversion technologies) are needed for economically feasible macroalgae fuel and chemical processes.

3.7 Industry, innovation and infrastructure: Build resilient infrastructure, promote sustainable industrialization and foster innovation (Goal 9) and sustainable cities and communities: Make cities inclusive, safe, resilient and sustainable (Goal 11)

The world is becoming increasingly urbanized. Since 2007, more than half the world's population has been living in cities, and that share is projected to rise to 60% by 2030. The world's cities occupy just 3% of the Earth's land, but account for 60–80% of energy consumption and 75% of carbon emissions. Innovation and technological progress are key to finding lasting solutions to both economic and environmental challenges, such as increased resources and energy-efficiency. Globally, investment in Research and Development (R&D) as a proportion of Gross Domestic Product (GDP) increased from 1.5% in 2000 to 1.7% in 2015 and remained almost unchanged in 2017, but was only less than 1% in developing regions. Basic infrastructures like roads, materials, sanitation, electrical power and water remain scarce in many developing countries.

The most traditional uses of seaweed are for domestic purposes such as for building materials. In some regions, where the quantity of building materials is limited (e.g. Danish islands), seaweed was traditionally used in vernacular architecture. Today, when many countries are threatened with deforestation, some companies are looking for materials which offer a cheaper alternative to wood. Widera (2014) presents some examples of creative and contemporary dwellings made of seaweed as well as brand-new construction methods. For example, seaweed pillows were used as cladding for one holiday house on the Danish island of Læsø by architecture studio Vandkunsten and the non-profit making organization Realdania Byg. Seaweed has some great potential as a building material: it offers good insulation (which is typical for mineral materials), it is non-toxic and fireproof, has great acoustics, humidity control, reduction in CO_2 emissions, low-energy and is biodegradable with a life expectancy of more than 150 years. The seaweed is covered with sea salt and is thus naturally protected against bacteria or insects that could destroy the structure. Seaweed can be visually attractive and its usage establishes the balance between traditional and modern architecture. However, it is important to note that seaweed is a good material choice when it is harvested naturally, i.e. collected on the beach, dried on the meadows (not in ovens) and transported to short distances only. Used in that way seaweed can promote the respect for the uniqueness of regional architecture but simultaneously can be easily adapted to the specific local conditions, including different cultural and climatic factors. This is in accordance with the statement of Peter Sørensen and Winnie Friis Møller that "Architecture is a connecting link between place, climate and human life" (Sørensen & Møller, 2008; Widera, 2014). The preservation initiative of Realdania Byg helped involve local communities in the process which increased the awareness both of the natural and cultural heritage of the island. Consequently, that leads to the protection of the architecture of the past and at the same time to the development of sustainable architecture of the future.

In the study of Ramasubramani, Praveen, and Sathyanarayanan (2016), marine brown algae *Ascophyllum nodosum* was used as additive material to concrete. The authors show that an 8% addition of marine algae to concrete showed an increase in strength properties. Susilorinia et al. (2014) presented an innovative green construction material with a natural algal polymer modified mortar. The authors studied the compressive strength of a natural polymer modified mortar with red seaweed gel (*Eucheuma cottonii*) and seaweed powder (*Gracilaria* sp.). They are very effective in gelling and thickening and may increase the bonding mechanism and density. The research has shown that natural polymer modified mortar with *Gracilaria* sp. powder possesses great compressive strength and splitting tensility.

Paper industries are also dependent on wood-based raw materials. However, global warming and restriction on carbon dioxide emissions make wood exploitation more difficult. The sector of paper and cardboards from non-wood plants and agricultural residues has attracted renewed interest. Some industries using wood resources have been looking for alternate innovative sources to wood with better end products and reduced paper manufacturing expenses. The strength of the material (paper, cardboard) depends on the cellulose content of raw materials. *Cladophora* sp. cosmopolitan green alga can be found in huge masses of a variety of marine and fresh waters. This specie is known to contain fibrillary cellulose, the concentration of which is in the same order of magnitude as wood (Baweja, Kumar, Sahoo, & Levine, 2016). Recently, the TMB Marine Box company (Fort de France) deposited a patent on the fabrication of a cardboard coffin for cremation, using the invasive brown seaweed *Sargassum* as raw material for this process.

In association with the challenge of materials usage, there is the question of plastics. Plastics are undeniably part of our daily lives. Their low price/cost and weight are the main attractions behind their universal use. Unfortunately, the oceans and animals in the world suffer from these huge quantities of plastics (Lavers, Dicks, Dicks, & Fin, 2019; Lebreton et al., 2018; Rochman et al., 2015). In this context, biodegradable and/or biobased polymers are the most extensively explored as new packaging materials. Biodegradable materials are made from both natural and fossil resources and are biodegraded by microorganisms in their natural environment. Biodegradable refers to the ability of materials to break down and return to nature within a short time after disposal (Nešic et al., 2019). The use of algal polysaccharide-based materials can present an eco-friendly technological solution, by reducing dependence on fossil resources and reducing a product's carbon footprint, in comparison to conventional plastic packaging materials. Marine polysaccharides are a group of polymeric carbohydrates with inherent characteristics such as biocompatibility, biodegradability and non-toxicity toward living organisms. The most famous macroalgal-derived bioplastic agents are starch and cellulose derivatives, as well as alginate from brown algae. For example, the starch content of the green algae *Ulva ohnoi* varied from 1.59% to 21.44% depending on growth conditions and seasons. In the study by Prabhu et al. (2019), the authors showed that nutrient starvation significantly increased the starch concentration up to 21.4% on a dry weight basis from *Ulva* cultivated in sea water offshore. The extracted fraction contained 75.45% starch, and the starch extraction yield from the *U. ohnoi* biomass was 50.37%. Production of starch and other nutrients (lipids, proteins, salt, Ulvan) from *U. ohnoi* by offshore farming and extraction in biorefineries, shows great promise to provide a novel and efficient way of improving future food security. Furthermore, Algopack, based in Saint Malo (France; http://www.algopack.com), carried out some assays to produce bioplastics. This company proposes a rigid material made from 100% brown algae industrial waste. However, they are darker and slightly more brittle than conventional plastics. Making plastic bags is not really a concern since they are currently produced from corn. In recent decades many applications of seaweed-based polysaccharides in food packaging and coatings were investigated using agar, carrageenan and alginates. Some works revealed that multilayer films based on κ-carrageenan/agar/clay nanocomposite mixed with polylactide bioplastic or cellulose nanocrystal improved barrier, optical, thermal stability and mechanical properties for the food packaging domain. It was also demonstrated that an alginate-calcium coating enabled the reduction of lipid oxidation, increased moisture content and enhanced the flavor, quality and texture of food. These films could be used as edible coatings for numerous foods such as vegetables and fresh fruit to simply extend shelf life. by A number of studies highlighted that the addition of bioactive agents such as antioxidant or antimicrobial (lemongrass or thyme essential oil, phenolic compounds—tocopherols, lycopene, ascorbic acid) to alginate edible films is more beneficial in comparison with the industrial process using the direct application of bioactive agents in food (Nešic et al., 2019).

Art and culture can be excellent tools to raise awareness and promote sustainable development issues and marine biodiversity aimed at adults, children and students. Many artists give prominence to seaweeds (Painters, designers, plasticians, photographers, ...). Jasmine Linington (https://www.jasminelinington.com) has been exploring the potential of seaweed within the fashion and textiles industry while studying for a Master of Fine Arts at the Edinburgh College of Art. Given the devastating environmental impacts from the textile and fashion industry, she became motivated to learn more about seaweed's potential as an alternative for fibers and dye. She makes her own textiles using technology and handmade techniques, SeaCell™, produced by smartfiber AG (https://www.smartfiber.de/en/fibers/seacelltm/) which acts as the base fabric. SeaCell is a cellulose fiber and is non-allergenic and very soft. The seaweed for SeaCell is harvested from the sparsely populated fjords of Iceland in locations where there is no waste and no pollution from shipping traffic. She also explores the extraction of pigment from seaweed to make a natural dye. Currently, she has obtained a license for small-scale harvesting through Scottish Natural Heritage and East Lothian Council. She creates supple kelp sequin jewelry which has a sort of leather-like feel to it.

Designers are pushing boundaries in their quest for sustainability with upcycling, recycling, and using unusual natural products. In the textile domain, technologies move fast, forwards into the future, looking for color from the garden and waste and from plant matter, seaweeds and mushrooms. High tech and slow craft merge effortlessly into new hybrids, able to combine additional functions. The impetuous pace of innovation deals with folding, coiling, braiding and knotting; rethinking textiles first invented in the Bronze Age. Archeology and anthropology are major sources of both knowledge and inspiration, as more and more textiles take on cultural and humanitarian endeavors. The eye-opening concept of reconciling to the impermanent rhythm of living beings invites us to draw on new ways of conceiving our relationship to our milieus, and question where and when the action of our hands can be re-aligned to our thoughts (Buet, 2020). Violaine Buet is designer graduate of the school at ENSCI (Paris, France) (https://violainebuet.com/). Combining biology, materials, cosmology and mesology, she studies in an interdisciplinary way the relationship between living organisms and their environment. Violaine Buet has now developed an innovative multidisciplinary expertise to ennoble the algae from its natural state to novel utilizations from the fine and living arts, haute couture, scenography and decoration, to visual merchandising and industry. She has exhibited her artwork in several exhibitions (Cooper Hewitt of New York, 2020; Philadelphia Museum of Art, 2019; Wawes and Milan, 2019) (Fig 10).

3.8 Responsible consumption and production: Ensure sustainable consumption and production patterns (Goal 12)

Goal number 12 of the 2030 Agenda for Sustainable Development aims to ensure sustainable consumption and production patterns. As defined by the Oslo Symposium in 1994, sustainable consumption and production concerns "the use of services and related products, which respond to basic needs and bring a better quality of life while minimizing the use of natural resources and toxic materials as well as the emissions of waste and pollutants

Fig. 10 Handmade artwork from Violaine Buet.

over the life cycle of the service or product so as not to jeopardize the needs of further generations" (United Nations, 2020b). The value chain of algae has to be considered this way, from the supply of biomass, to consumption through bioprocesses for active compounds isolation and eco-responsible applications. Fig. 11 illustrates the responsible value chain from production to the consumption of algae.

The supply of algae can be obtained by harvesting them in their natural environment and/or by algae farming. For both, a sustainable management and a rational use of the resource have to be considered. In order to sell algae directly for food or for other applications, pre-treatment of the resource and bioprocesses are needed to access active compounds, or to facilitate their use due to a standardization and a stabilization of products. The efficiency of algae-based products is complete when users have some knowledge about algae and know how to apply them. This is why collaborations, partnerships and training are essential to produce and consume resources in a responsible manner. Together and through each one of us, we can sustainably consume natural and local resources by avoiding waste and preserving ecosystem services at the same time.

In 2018, world seaweed production totaled 33 million tons of fresh seaweed with the culture and capture sectors. In 10 years, seaweed production has doubled. The global seaweed industry is worth more than USD 6 billion per annum of which some 85% comprise food products for human consumption (FAO, 2018a). In comparison, 70% of all agricultural land are dedicated to livestock production which represents 30% of the land surface of the planet. 33% of arable lands are not used for food but for feed (FAO, 2006). In accordance with sustainable management and a rational use, algae represent a natural resource with a high potential for food, feed and energy production. While representing only a small fraction of total global seaweed production, harvesting and gathering wild seaweeds has had, and continues to have, an active integral role in many coastal societies, often being linked to the cultural identity of those coastal communities. Interest and demand for algae in the world is increasing, therefore pressure on this resource is more considerable. Resource scientists, managers, conservationists, governments, and other stakeholders need to be proactive in the sustainable management of these vulnerable, yet valuable, resources (Kraan, 2020; Mac Monagail, Cornish, Morrison, Araújo, & Critchley, 2017). Three pressures on wild algae have been identified: the ownership of the resource, its over-exploitation and the impact of mechanical harvesting techniques, such as for example, the boats especially equipped with an articulated hydraulic

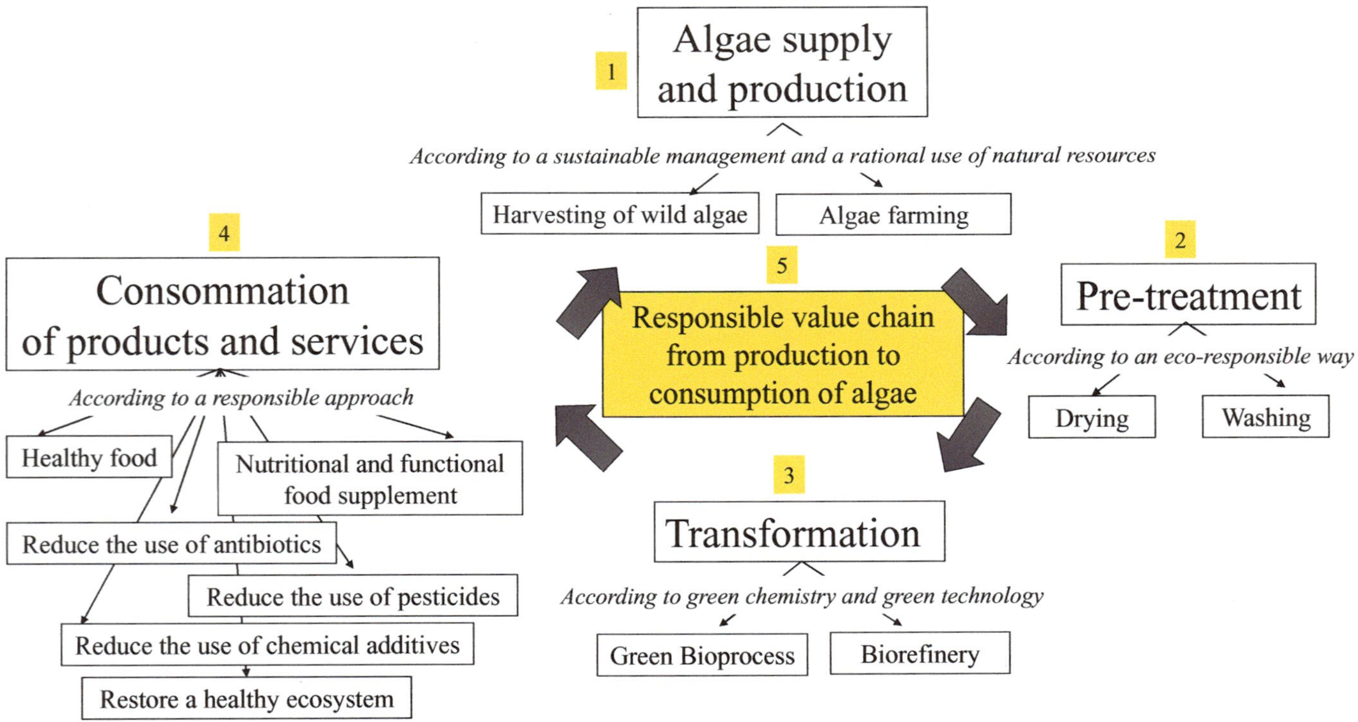

Fig. 11 Responsible value chain from production to consumption of algae. Numbers 1–5 are detailed in Section 3.8. (1) According to a sustainable management and a rational use, algae represent a natural resource with a high potential for its use in food, feed and energy production. (2) According to an eco-responsible way, algae can be prepared for compound extraction and purification. (3) According to green chemistry and green technology, compounds of interest can be isolated. (4) According to a responsible approach, people can consume healthy food, as well as nutritional and functional food supplements for well-being, by indirectly reducing the use of antibiotics in livestock, pesticides in agriculture and chemical additives. Furthermore, some companies can add value to stranded seaweed on beaches which allows the restoration of certain ecosystems. (5) The value chain of algae is responsible if short, local and circular circuits are favored. In addition, all co-products of the bioprocess for algae transformation, from the raw material to the final product, are used and consumed in order to target zero waste.

system called "Scoubidou" that collect brown seaweeds from the genus of *Laminaria* in France. To manage a wild resource, it is important to regulate the quantity and the quality of the resource. Policy regulations and programs have been created by national and international governances and are already in place (See Nagoya protocol and GlobalSeaweed STAR program in Section 3.1 for example). Furthermore, Guiry and Morrison (2013) have shown that the harvested Irish brown seaweed *Ascophyllum* populations, through using simple techniques of 4–5 years of being left fallow, and leaving sufficient material for regeneration, have produced a sustainable yearly harvest of 5.000–28.000 tons wet weight (ww) for the Irish company Arrama, from the late 1940s until the present day (Guiry & Morrison, 2013). Resource monitoring and mapping are important to regulate the growth of seaweeds and their dependent species, to maintain their biological associations which will eventually maintain the equilibrium among various species in the marine ecosystem. Regular monitoring is also important for the seaweed yield estimations and also for observations of harmful algal blooms, these can be useful for fishermen, policy makers, and companies (Beckler et al., 2019; Danish Siddiqui & Abdullah, 2016). The SLAMIR Program aims to measure the impact of the exploitation of brown seaweeds by French companies in order to establish a calendar and regulations for harvesting, in order to preserve the resource (see Section 3.1). According to the development and objectives of the PHYCOMORPH European Guidelines for the Sustainable Aquaculture of Seaweeds (PEGASUS), the development of seaweed aquaculture in Europe faces a number of challenges: the market size, potential environmental impact, preservation of local genetic diversity, regulation of food quality, heavy metals or alien species, cultivation constraints ranging from automation to issues of epiphytism, as well as the need to intensify research—both fundamental and applied. To face these challenges, a United Nations University Policy Brief specifically highlighted recommendations on how to develop this industry sustainably, and identified the key ecological and socio-economic issues preventing the sustainable economic growth of this industry. Recommendations are listed in Table 2. From these, the 4-year multi-national project—GlobalSeaweed STAR—was launched in 2017 (Barbier et al., 2019; Cottier-Cook et al., 2016; see Section 3.1).

In accordance with eco-friendly bioprocesses (Section 3.6), algae can be prepared, and active compounds of interest can be isolated. A pre-treatment of the biomass is required in order to release and dissolve the compounds of potential use. This implies sizing by milling, followed by mechanical,

Table 2 Main recommendations for sustainable aquaculture, UNU policy brief (Barbier et al., 2019; Cottier-Cook et al., 2016).

1	Establish centers of research excellence
2	Establish national seed banks
3	Maintain the genetic diversity in wild stocks
4	Exercise a precautionary approach when introducing new or nonindigenous cultivars to the marine environment
5	Focus on developing and enhancing biosecurity programs
6	Incentivise long-term investment in the industry
7	Incentivise the integration of seaweed and other extractive species with finfish in integrated multitrophic aquaculture (IMTA) systems
8	Develop assessment tools for evaluating spatial planning issues in relation to aquaculture
9	Develop and implement ecosystem-based management models and integrated coastal zone planning
10	Develop regulations and directives that enable a sustainable exploitation of the natural resource
11	Address capacity building and adaptive governance toward seaweed resources
12	Establish management regimes for the sustainable exploitation of the seaweed resources
13	Train human resources to provide education to coastal communities, based on best practices for harvesting and cultivation

chemical or enzymatic methods for the degradation of the cell walls and intercellular matrix. A specific challenge is the high water content of algae (70–90%), which implies that further water addition should be minimized. This pre-treated biomass can be subjected to enzymatic hydrolysis and fermentation for fuels or chemicals, or applied for the isolation of proteins, sulfated polysaccharides and other high-value products (de Carvalho et al., 2020; Skjermo et al., 2014). Continuous, year-round production is required for the economic feasibility of a large-scale production plant. Usually, the biomass will be harvested over a short period, seasonally, when its composition is at its optimum. This implies that the seaweed biomass will need be stored and preserved. In Norway, the brown seaweed *Laminaria hyperborea* harvested for alginate production (approx. 150.000 tons ww annually) is preserved with formaldehyde, which is not a viable option for later biochemical conversion. Other preservation methods applicable for several thousand tons have not yet been defined (Skjermo et al., 2014).

Algae are usually washed with seawater or fresh water to remove undesirable ones, like epiphytes of pieces of plastic, and then either dried or frozen to be sold as such, or prior to extracting the active compounds with higher added value. Fresh, they can also be directly transformed in a short time to keep the intrinsic properties of native compounds from algae.

Drying is a very long-established means of preservation for seaweed. Seaweed lends itself to drying because of its natural salt content. The most common methods utilize natural drying with the heat of sun, dehumidification and air movement. Then different techniques, such as mechanical, by pressing for example, and ultrasonic and microwave treatments, have been used to enhance drying rates (Kadam, Álvarez, Tiwari, & O'Donnell, 2015). Some of them are described in more detail in Section 3.6. Drying could possibly be applied for small biomass volumes intended for high-cost products, but will be too expensive for high-volume/low-cost products. Microalgal biomass pre-treatments for integrated processing into biofuels, food, and feed is an energy-costly step in microalgal biorefineries. Microalgae are sources of nutritional products and biofuels. Their economical processing is challenging, thermolysis being the most cost-efficient processes for further bio-digestion. Using an enzymatic hydrolysis and/or pulsed electric fields and autolysis also are promising steps (de Carvalho et al., 2020).

In accordance with a responsible approach, people can consume healthy food, as well as nutritional and functional food supplements for human well-being by indirectly reducing the use of antibiotics in livestock, pesticides in agriculture and chemical additives. Some companies can add value to stranded seaweed on beaches which allows the restoration of certain ecosystems. This is the case of the French group Olmix for example.

For a responsible food source, by bringing together contributors from the sea to the plate, the "Merci Les Algues, 2020"® association builds sustainable and committed sectors for future generations. The farmers at the heart of the process guarantee the quality of the food that ends up on the consumer's plate. By choosing to reduce the use of antibiotics and pesticides by using natural and local alternatives such as algae, they are committed to responsible nutrition. To encourage them in this direction, they benefit from higher remuneration (www.mercilesalgues.com).

The value chain of algae is viable if short, local and circular circuits are favored. In addition, all co-products of the bioprocess for algae transformation, from the raw material to the final product, are to be used and consumed in order to target zero waste. The biorefinery concept has been identified as the most promising way to create a biomass-based industry (Trivedi, Aila, Bangwal, Kaul, & Garg, 2015 and see Section 3.6).

The study by Peñuela et al. (2018) is a concrete example of this concept at the laboratory scale. The biomass of the carragenophyte *Solieria filiformis* cultured in IMTA system was used to obtain valuable products using a biorefinery approach. Enzymatic-assisted extraction and microwave-assisted extraction were the eco-friendly technologies used to ensure an environmentally friendly valorization of the biomass. Three valuable products were successfully recovered: some water-soluble extract rich in proteins and sulfated polysaccharides suitable as a food supplement; some lipid fraction rich in polyunsaturated fatty acids (PUFAs) with potential to be used in the nutraceutical industry; and a pure iota-carrageenan with a powerful antiviral activity against *Herpes simplex*. To make sense of the algal biorefinery concept, there is a need to establish a proper connection between the various input and output streams of the products.

3.9 Environment: Preservation of the flora, fauna and water and resilience to the climate change thanks to algae (Goals 6, 13, 14, 15)

Considering that our blue planet is composed of a determined space, the role of humanity is to understand it in order to protect and to conserve its ecosystem. Understanding enables the anticipation of the impact that the increase in human population bears on the life below water (SDG 14), on the life on land (SDG 15) and on the quality of water (SDG 6). The preservation of biodiversity and of the environment limits the imbalance of ecosystems risking species extinction and biodiversity loss. Nearly 3 decades ago, a community of eminent scientists warned that humans were on a collision course with the natural world. Twenty-five years later, Ripple et al. (2017) evaluated the human response based on their analysis of time-series data. They concluded that humanity had failed to make sufficient progress over that period in dealing with the environmental challenges. In fact, they concluded that most of these problems had worsened. The original 1992 call was supported by more than 1.700 scientists, while 25 years later over 15.000 scientists added their signatures to the recent declaration (Pyšek et al., 2020; Ripple et al., 2017). Today, it is even more essential to act and to combat climate change and its impacts for the future generations (SDG 13). Fig. 12 summarizes the causes and the consequences of anthropogenic activities on the climate of the Earth. Algae-based

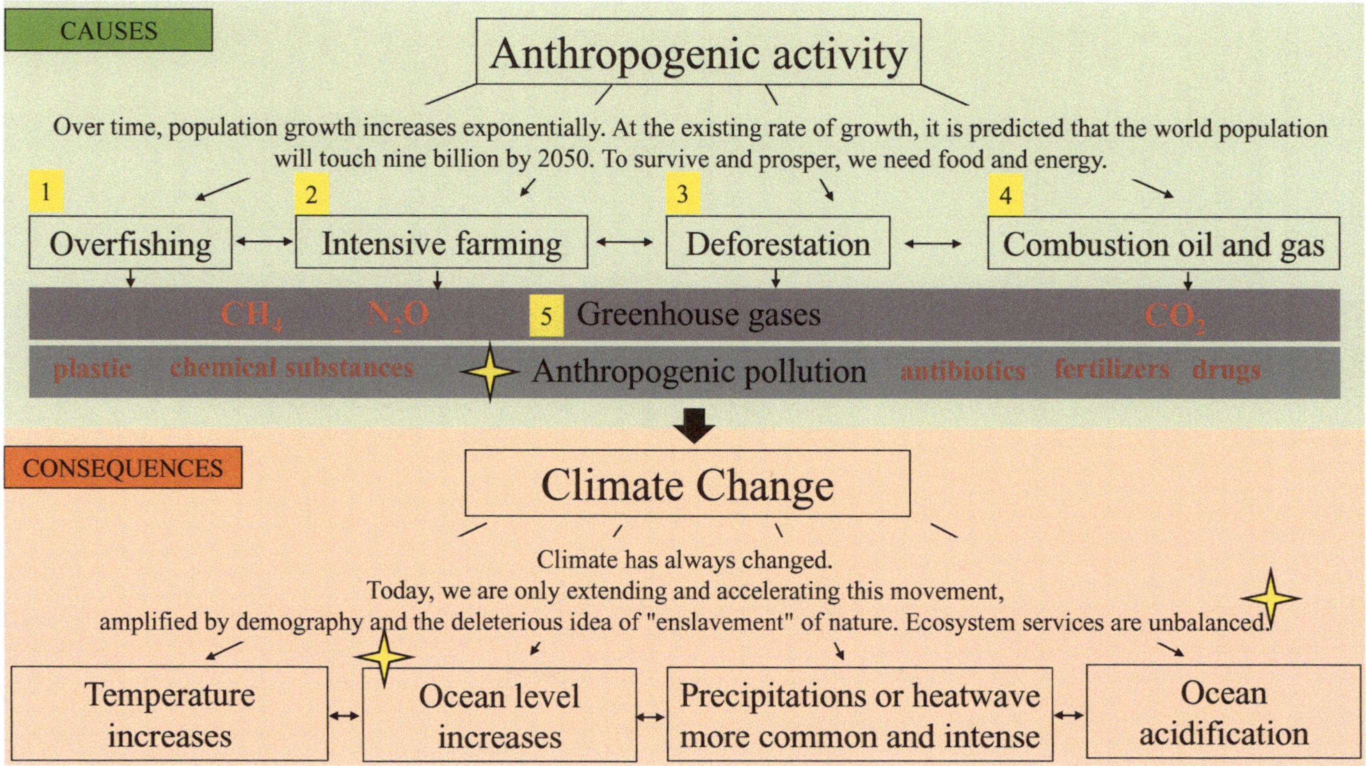

Fig. 12 See figure legend on next page.

solutions are proposed for their numerous attributes toward climate change. They are detailed in this part of the chapter.

The more humans on Earth, the greater the demand for space, food and energy. Under a globalization system, all countries depend on international markets and on a global economy, although inequalities exist and persist (see Section 3.5). Every day, to satisfy basic needs, people around the world consume products from overfishing, intensive farming using high quantities of fertilizers, pesticides, antibiotics and chemicals, deforestation and burning of fossil reserves like oil and gas. All these anthropogenic activities generate greenhouse gases, including essentially carbon dioxide (CO_2), methane (CH_4) and nitrous oxide (N_2O), and pollution on land and in water (Fig. 13).

Fig. 12 Impact of anthropogenic activities on the Climate of the Earth. Numbers 1–8 are detailed in the text of Section 3.9. For each number, algae-based solutions are proposed. (1) Algae have a real ecological role in marine and fresh water ecosystems. They provide feed and food and oxygen for animals. They neutralize carbon dioxide from the atmosphere. They promote life below water (SDG 14). (2) Algae represent a source of feed and food. To produce them, fertilizers, pesticides, antibiotics, chemicals are not needed. Also, red seaweed can be used to reduce methane emissions from intensive livestock. (3 and 4) Deforestation is caused by the need for arable fields and energy production. To produce algae for feed, food or energy, no arable land is required and life on the land is also protected and conserved (SDG 15). Seaweed farming can take place directly in the sea and Ocean. Furthermore, energy can be produced with algae as mentioned previously in Section 3.6 "Energy clean—Affordable and clean energy—Ensure access to affordable, reliable, sustainable and modern energy for all (SDG 7)." (5) Due to photosynthesis, algae, essentially under the form of phytoplankton, represent the second lung of the Earth. They produce 50% of O_2 and capture CO_2 from the atmosphere that is sequestered in the Ocean. (6) The intensive use of plastics and chemicals contaminate water sources on the Earth. Algae can be produced to filter water. They are renewable agents for bioremediation and enable access to clean water and sanitation (SDG 6). Furthermore, plastic can be made with some seaweeds which avoids polluting the ocean as mentioned in Section 3.7 "Industry, innovation and infrastructure: Build resilient infrastructure, promote sustainable industrialization and foster innovation (SDG 9) and Sustainable cities and communities—Make cities inclusive, safe, resilient and sustainable (SDG 11)." (7) Ecosystem services are unbalanced and consequently the climate changes faster. Today climate actions are essential. Citizens, researchers, politicians and businessmen must act for a responsible and sustainable production and consumption and follow a local, circular, green economy (the doughnut model), (SDG 12 and SDG 13). (8) Ocean levels are increasing which generates an erosion of the coastline, causing houses and buildings to collapse into the sea. A sustainable alternative for coastal dune restoration by sand-trapping fences was developed with algae wrack: AlgoBox®. See Section 3.1 "No poverty: End poverty in all its forms everywhere" (SDG 1).

Fig. 13 Anthropogenic pollutions. (A) Plastic residues in Saint Louis, in Senegal (April 2017) and (B) Atmosphere charged with greenhouse gas in Beijing in China (August 2009). *Photos by Burlot AS.*

In his inaugural lesson entitled "Biodiversity: from the ocean to the city" at the Collège de France in 2013, biologist Professor Gilles Boeuf explained that the climate has always changed, but the problem is that today we are extending and accelerating this movement, amplified by demography and the deleterious idea of "enslavement" of nature. According to him, the biodiversity is threatened by four major phenomena: the destruction and contamination of the environment; the predation and over-exploitation of natural resources; the anarchic introduction of species from one environment to another and, finally, global warming (Bœuf, 2014). With time, the greenhouse effect becomes more significant and the global temperature increases while the level of the oceans rises. Precipitation and heatwaves are more intense and more common and the oceans are becoming more acidic (GIEC, 2014). Ecosystems are unbalanced and some species die while others proliferate. Urban looked at over 130 studies to identify the level of risk that climate change poses to species and the specific traits and characteristics that contribute to risk. Results suggest that extinction risks will accelerate with future global temperatures, threatening up to one in six species under current policies. Extinction risks were highest in South America, Australia, and New Zealand, and risks did not vary by taxonomic group (Urban, 2015). Another known and negative impact linked with the climate change is coral bleaching. Coral reefs are the world's most diverse marine ecosystems. They provide billions of dollars in economic value through coastal protection, food, tourism, and pharmaceuticals from the sea. Rapid increases in sea surface temperatures and ocean acidification are increasing the frequency and intensity of coral bleaching events during which corals lose their endosymbiotic algae: a zooxanthella, which is a primary energy source for most reef coral. Its loss causes coral mortality that threatens coral reefs globally (Brown, 1997; Cousteau & Diolé, 1971; Sully, Burkepile, Donovan, Hodgson, & van Woesik, 2019). Biological invasions are also a global consequence of an increasingly connected world and the rise of the size of the human population. Coastal eutrophication is an increasingly significant problem worldwide, where rising nutrient levels in seawater (mainly nitrogen and phosphorus) threaten natural ecosystems. Eutrophication is often directly linked to human activities, largely as a result of fertilizers, local pollution and slurry leakage into freshwater streams eventually reaching coastal areas. Degraded water quality from increased nutrient pollution promotes the development and persistence of many harmful algal blooms and is one of the reasons for their expansion in many nations (Heisler et al., 2008). Algal blooms and their subsequent collapses lead to the development of hypoxia, causing direct effects on the trophic chain leading to community structure changes, biodiversity and habitat loss and overall impoverished ecosystems. The sea lettuce *Ulva* spp. is one of the macroalgal species most commonly linked to the formation of "green tides," where large swaths of the coastal environment are taken over by rapidly proliferating seaweed biomass. The increased occurrence of green tides since the 1970s could however be due to the amplification of fast-growing strains. Fort, Mannion, Fariñas-Franco, and Sulpice (2020) hypothesize that the selective pressure in green tide areas leads to the amplification of *Ulva* genotypes best adapted for this environment. Such selection of fast-growing strains would indicate that green tides are likely to become more prevalent and of higher magnitude over the coming years (Fort et al., 2020). The number of reports of sudden beaching of huge seaweed masses smothering the coastline has increased worldwide in recent years (Fig. 14). These "seaweed tides" can harm tourism-based economies, smother aquaculture operations or disrupt traditional artisanal fisheries (Smetacek & Zingone, 2013).

Algae can be used for the resilience of humanity and for the preservation of biodiversity. For that to happen, it is important to understand the role of each species in the global ecosystem and to change our behavior and our mentality to protect the biodiversity and the environment. Using the best scientific information available to support decision-making is fundamental to the implementation of national and international policies on the conservation of biodiversity and sustainable use of resources as well as the evaluation of ecosystemic trends in both continental and marine areas (Araújo et al., 2016). In relation to the development of public policies for the observation, the preservation and the recovery of natural

Fig. 14 The number of reports of "seaweed tides" has increased worldwide. (A) Red tide of *Solieria chordalis* in Saint-Gildas-de-Rhuys, France. (B) Green tide of *Ulva* sp. in Dakar, Senegal. *Panel A: Photo taken by Burlot AS in August 2015. Panel B: Photo taken by Burlot AS in August, 2015.*

environments, monitoring networks have been initiated to better understand the putative impact of anthropic pressures on various biotopes. Since 2000, the European Union has included in the Water Framework Directory the investigation of biological communities (phytoplankton, macroalgae, seagrasses, macroinvertebrates and fish) as bio-indicators (biological quality elements or BQEs) to evaluate the ecological status of water bodies (Ar Gall et al., 2016). The project SLAMIR (see Section 3.1) led by a natural park in France gathers data on kelp forests to sustainably manage their exploitation by local companies (Parc Naturel Marin Iroise, 2018).

Algae have a real ecological role in marine and fresh water ecosystems provided that their growth is under control and that they do not proliferate. They procure feed and food and also oxygen for animals. Due to photosynthesis, algae, essentially under the form of phytoplankton, represent the second lung of the Earth. They produce 50% of O_2 and capture CO_2 from the atmosphere that is sequestered by the Ocean.

Sequestering carbon has become part of the portfolio of solutions to mitigate carbon emissions, and thus climate change, through policy and voluntary markets, primarily by land-based reforestation or afforestation and preservation. However, land is limited, creating an interest in a rapidly growing aquatic farming sector of seaweed aquaculture (Froehlich, Afflerbach, Frazier, & Halpern, 2019). Other markets can be expanded to reduce greenhouse gas emissions, such as the use of certain seaweed strains in livestock feed, where its inclusion in cow diets may significantly (by 75%–99%) reduce ruminant methane production (Roque et al., 2019), as well as offshore seaweeds production for biofuel (Fernand et al., 2017 and see Section 3.6). The addition of the red seaweed *Asparagopsis taxiformis* (2% organic matter) or the halogenated methane analog bromoform (5 μM) reduced methane production by over 99% compared to a basal substrate-only control according to the in vitro study of Machado et al. (2018). Fernand et al. (2017) provided a review of algae-based biorefineries with offshore cultivation and consequent biomass conversion into transportation liquid biofuels. In France, by setting a target value of 60% biofuel use with a productivity of 7 g dry weight (DW) of algal biomass per square meter per day and with the possibility to convert 30% of the algal biomass into transportation biofuel, the marine area needed to produce (the 60%) of the total amount of bio-oil consumption for transportation in France is 54,795 km^2. In comparison, the total French marine area's exclusive economic zone with overseas territories included is about 11 million km^2 (Fernand et al., 2017).

Seaweed also procures marine ecosystem support by contributing to increased fish habitats and marine biodiversity, as well as ocean restoration. Seaweeds could potentially serve as the basis of an IMTA (see Section 3.1) with fish and shellfish to increase production while decreasing the environmental impact (Chopin & Robinson, 2004). They promote life under water by providing food (SDG 14). Experiments of Marzinelli, Leong, Campbell, Steinberg, and Vergés (2016) on the dominant habitat-forming macroalga *Phyllospora comosa* (Fucales) that went extinct locally along the metropolitan coastline of Sydney in the 1980s, have indicated that the restoration of key habitat-forming seaweeds not only recovers the algal species but also reduces the risks of losing habitat diversity for epifauna and their consumers. However, restoration of all the original biodiversity associated with specific seaweeds can be a difficult, complex, and long-term process (Marzinelli et al., 2016). By creating coastal habitats, seaweed farming can potentially contribute to some of the ecosystem functions that natural kelp forests and macroalgal beds support. They support high biodiversity, including calcifiers such as lobsters, crabs, mollusks, and crustaceans (Duarte, Wu, Xiao, Bruhn, & Krause-Jensen, 2017).

Furthermore, the intensive use of plastics and chemicals contaminate water sources on the Earth. Algae can also be used as biofilters. They are renewable agents for bioremediation and enable access to clean water and sanitation (SDG 6). Algae can be produced specifically for water filtration. This practice is called phycoremediation, it is the use of algae for the removal, or biotransformation of pollutants and toxic compounds from wastewater and carbon dioxide from effluent air stream. Algae also help remove the elements in water which would have led to the growth of harmful bacteria and consecutive problems. Everywhere in the world, municipal, domestic, agricultural, and industrial sources generate wastewater as an end product. For example, India produces 40 billion liters of wastewater per day (Singh, Tiwari, & Das, 2016). Wastewater is an excellent medium for algal growth, with carbon dioxide addition. Wastewater usage eliminates the need for fresh water and saves nutrient cost since nutrients are in abundance in wastewater. Nutrients such as nitrates and phosphates, help the treatment of the wastewater by assimilating organic and inorganic pollutants into their cells, and eliminate the CO_2 emissions associated with wastewater treatment. Phycoremediation reduces eutrophication. There are also huge amounts of toxic materials like chromium, zinc and lead present in industrial wastewater which are absorbed by algal cells and usually leads to their death (Azarpira, Dhumal, & Pondhe, 2014; Singh et al., 2016). Furthermore, in the context of IMTA systems, several macroalgal species, namely some species from the genus *Gracilaria*, have been shown to be efficient biofilters. Nevertheless, in general, biomass production and nutrient removal were negatively related to the cultivation densities in the system. Seaweed has proved to be an efficient component of land-based IMTA systems, with environmental and potentially economic benefits for the fish farm. Algae can reduce nitrogen and phosphorus as well as heavy metals from water and at the same time value can be added to them as resources for biofuel, active agents and biogas production (Abreu, Pereira, Yarish, Buschmann, & Sousa-Pinto, 2011; Holdt, 2015). Clean water, an efficient and less expansive aquaculture production (fish, mussels) and algal biomasses for biofuel, biogas and the production of active agents are the three major results of the use of algae as biofilters. Furthermore, to avoid plastic pollution in the Oceans; biodegradable plastic can be made with some seaweeds, as mentioned in Section 3.7.

Deforestation is caused by the need for arable fields and energy production. To produce algae for feed, food or energy, no arable land is needed. Therefore, life on land is protected and conserved (SDG 15). Seaweed farming can take place directly in the sea and the Ocean. Furthermore, energy can be produced with algae as mentioned previously in Section 3.6 "Energy clean—(SDG 7)."

In Asia, seaweeds are commonly consumed as "sea vegetables." Most of the *Saccharina japonica*, a brown seaweed, produced in China is used for food. Besides its role as a "health" vegetable, *Saccharina* is also important as raw material for its alginates (texturing and gelling compounds), mannitol and iodine. *Saccharina japonica* is the most important economic seaweed in China. Seaweed farming on artificial floating rafts started in 1952 and production increased steadily until 1980 when the production of 200.000 dry tons (about 1.500.000 wet tons) was achieved. Currently, the yield of *S. japonica* from about 200.000 acres of farms is about 250.000 dry tons from about 2.000.000 wet tons (Guiry, 2020). Some 221 species of seaweed are of commercial value. About 10 species are intensively cultivated, such as brown seaweeds (*Saccharina japonica*, *Undaria pinnatifida* and *Sargassum fusiforme*); red seaweeds (*Porphyra* spp., *Eucheuma* spp., *Kappaphycus alvarezii* and *Gracilaria* spp.); and green seaweeds (*Enteromorpha clathrata*, *Monostroma nitidum* and *Caulerpa* spp.) *S. japonica* and they account for over 33% of the global cultivated seaweed production.

Ecosystem services are unbalanced and the climate is changing faster. Today climate actions are essential. Citizens, researchers, politicians and businessmen act for a responsible and sustainable production and consumption following a local, circular and green economy (the doughnut model) (SDG 12 and SDG 13).

In 2020, a seaweed manifesto was published in open access. It has been initiated by Lloyd's Register Foundation, an independent global charity that supports research, innovation and education with a mission to make the world a safer place. The work has been actively supported by the Sustainable Ocean Business Action Platform of the United Nations Global Compact. The Action Platform is taking a comprehensive view on the role of the ocean in achieving the 17 Sustainable Development Goals. The seaweed manifesto is a visionary document for the industry exploring the opportunities and benefits, as well as outlining the challenges and barriers for the responsible development of the industry. It shows the importance of the algae sector, considered by the United Nations, playing a significant role in achieving the Global Goals by contributing to food safety and security, climate change mitigation, poverty alleviation and support to marine ecosystems. The seaweed manifesto was also elaborated to create increased interest and active contributions to the responsible development of the industry from international donors, intergovernmental organizations, nongovernmental organizations, research centers and international companies (Lloyd's Register Foundation, 2020).

The solutions based on nature are defined by the International Union for Conservation of Nature (IUCN) as actions aiming to protect, manage and restore sustainably natural or modified ecosystems, which respond effectively and adaptively to societal challenges by supplying the benefits in terms of human well-being and biodiversity (Lehmann, 2020). Until today, ecosystem services have been largely exploited by humans and some of them tend to run out. It is the case for fossil reserves for example. Ecosystem services are the benefits people obtain from ecosystems. These include provisioning services such as food, water, timber, and fiber; regulating services that affect climate, floods, disease, waste, and water quality; cultural services that provide recreational, esthetic, and spiritual benefits; and supporting services such as soil formation, photosynthesis, and nutrient cycling. The human species, while buffered against environmental changes by culture and technology, is fundamentally dependent on the flow of ecosystem services. This concept was spread by the Millennium Ecosystem Assessment (MA) that was carried out between 2001 and 2005 to assess the consequences of ecosystem change for human well-being. It also established the scientific basis for actions needed to enhance the conservation and sustainable use of ecosystems and their contributions to human well-being. The MA responds to government requests for information received through four international conventions—the Convention on Biological Diversity, the United Nations Convention to Combat Desertification, the Ramsar Convention on Wetlands, and the Convention on Migratory Species—and is also designed to meet the needs of other stakeholders, including the business community, the health sector, nongovernmental organizations, and indigenous peoples. The sub-global assessments also aimed to meet the needs of users in the regions where they were undertaken (Millennium Ecosystem Assessment, 2005). Nevertheless, the Earth is limited and these ecosystems services become unbalanced if nothing changes. Nine planetary boundaries were defined: freshwater use, biochemical flows, ocean acidification, atmospheric aerosol loading, stratospheric ozone depletion, novel entities, climate change, biosphere integrity and land-system change. If humanity does not exceed these boundaries, it will be possible to continue to develop and to prosper for the future generations. To ensure a safe and equitable space, the global economy needs to change. The British economist Kate Raworth supports the idea that we must write a novel economy manual by starting from the planetary boundaries and Human Rights. She is the author of the book titled "Doughnut Economics" (Lehmann, 2020). Today, politicians and businessmen consider this doughnut theory as explained in Section 3.8 on Responsible Production and Consumption and in Section 3.1 with the examples of the Nagoya protocol and the GlobalSeaweedSTAR program.

Other actions have taken place for the conservation, restoration of biodiversity and ecosystem services in disturbed environments. Innovated applications have been initiated by citizens around the world. In 2010, after a few years spent researching algae, the marine biologist Anne-Gaëlle Jacquin started a journey around the world, called the "Route des Algonautes" to talk to people working with algae. She interviewed more than 150 people. One of the objectives is to share with everyone, including children at school, the many virtues of algae and make them known and available. In 2016, she created the website www.algonauts.org to share this knowledge over the web. It's a place where she explores the latest developments in sustainability and algae in her on-going method (Jacquin, 2020). Another concrete action is the one of native divers in Japan for the removal of the sea urchins, the causative grazer of kelp beds. With less sea urchins, the kelp beds were restored and the fish returned. The aged and depopulated fishermen on the coast had food again thanks to these volunteer divers (Watanuki et al., 2010).

A sustainable alternative for coastal dune restoration by sand-trapping fences was developed with algae wrack: AlgoBox® (Fig. 15). Throughout

Fig. 15 The presence of (A) sustainable alternative for coastal dune restoration by sand-trapping fences (B) organic matter released thanks to Rhodophyta Solieria chordalis biomass decomposition in South Brittany (France). *"Photo by N. Bourgougnon."*

the world, numerous coastlines are subjected to severe erosion. The topography of these coasts is characterized mainly by cliffs and headlands, embayed beaches and "low-tide-terrace" pocket beaches. Wrack deposits can form significant seasonal accumulations on the beaches, thus impacting their morphology and ecosystem (Sedrati, 2018; Sedrati & Cochet, 2015).

Finally, as synthetized in Fig. 13, ocean levels are increasing, causing erosion of the coastline and resulting potentially in houses collapsing and falling into the sea. Development of seaweed farming can bring solutions to this issue. The canopies of farmed seaweeds, like those of wild seaweeds, dampen wave energy and hence, serve as live coastal protection structures buffering against coastal erosion. For example, Norwegian kelp forests dominated by the brown seaweed *Laminaria hyperborea* have been reported to reduce wave heights by up to 60%. A key difference, in terms of the capacity of farmed seaweed to reduce wave energy is that their canopies are suspended from the surface rather than being benthic. The wave-attenuating effect depends on the extent and structure of the seaweed habitat as well as the energy involved, as seaweed farms will be damaged during high-energy storms (Duarte et al., 2017).

Furthermore, algae aquaculture can be a tool for carbon fixation, reducing nutrient loads in coastal waters, and also for the conservation of ecosystem services. In 2019, the European Marine Research Network EUROMARINE supported a workshop called "RECOVER: Seaweed aquaculture: A promising tool for the restoration and sustainable development of coastal environments or an expensive end-of-pipe technology?" Twenty experts from eight EU countries, Israel and Senegal participated. The interactions of algae monocultures with their environment and their use in integrated marine aquaculture are increasingly attracting scientific interest because of their key roles in marine biodiversity, ecological processes and biogeochemical cycles. Demonstration regions included the Bay of Dakar (a heavily polluted environment with direct sewage disposal), Madeira (coastal waters described as a disturbed environment in nutrient-poor water), and the southern Baltic Sea (habitat over-fertilized by nutrient-rich surface runoff).

4. Conclusion

A holistic approach is needed for addressing complex global problems and must involve all stakeholders, researchers, policy makers, civil society, scientific and non-scientific communities. That requires efforts for finding methods and languages across and beyond each individual discipline, and covering a wide spectrum of disciplines for which education and progress must be nurtured. As described in the previous goals in this article, seaweeds withhold a tremendous potential for facing sustainability challenges. Seaweeds propose environmental benefits: renewable biomass and fast growth, CO_2 uptake and storage, increased biodiversity and dissolved oxygen in the sea, increased fish stocks in the vicinity of farms, biofilter ability (bioremediation), assistance in reducing coastal eutrophication and improve water quality. IMTA—Integrated aquaculture for cleaner waters and improved resource utilization, Seaforestation—planting artificial seaweed forests for the benefit of the environment, help mitigate climate change and ocean acidification (through large-scale cultivation). The numerous applications could also provide environmental services and an important source of livelihood in coastal regions. This potential is still underexploited. Seaweed science and industry can contribute deeply in the Planetary Health discussions.

References

Abreu, M. H., Pereira, R., Yarish, C., Buschmann, A. H., & Sousa-Pinto, I. (2011). IMTA with Gracilaria vermiculophylla: Productivity and nutrient removal performance of the seaweed in a land-based pilot scale system. *Aquaculture*, *312*, 77–87.

Adl, S. M., Simpson, A. G., Lane, C. E., Lukeš, J., Bass, D., Bowser, S. S., et al. (2012). The revised classification of eukaryotes. *The Journal of Eukaryotic Microbiology*, *59*, 429–493.

Allen, E., Browne, J., Hynes, S., & Murphy, J. D. (2013). The potential of algae blooms to produce renewable gaseous fuel. *Waste Management*, *33*, 2425–2433.

Ar Gall, E., Le Duff, M., Sauriau, P. G., De Casamajor, M. N., Gevaert, F., Poisson, E., et al. (2016). Implementation of a new index to assess intertidal seaweed communities as bioindicators for the European Water Framework Directory. *Ecological Indicators*, *60*, 162–173.

Araújo, R. M., Assis, J., Aguillar, R., Airoldi, L., Bárbara, I., Bartsch, I., et al. (2016). Status, trends and drivers of kelp forests in Europe: An expert assessment. *Biodiversity and Conservation*, *25*, 1319–1348.

Azarpira, H., Dhumal, K., & Pondhe, G. (2014). Application of phycoremediation technology in the treatment of sewage water to reduce pollution load. *Advances in Environmental Biology*, 2419–2424.

Beckler, J. S., Arutunian, E., Moore, T., Currier, B., Milbrandt, E., & Duncan, S. (2019). Coastal harmful algae bloom monitoring via a sustainable, sail-powered mobile platform. *Frontiers in Marine Science*, *6*, 587. https://doi.org/10.3389/fmars.2019.00587.

Barbier, M., Charrier, B., Araujo, R., Holdt, S. L., Jacquemin, B., & Rebours, C. (2019). In M. Barbier, & B. Charrier (Eds.), *PEGASUS—PHYCOMORPH European guidelines for a sustainable aquaculture of seaweeds, COST action FA1406*. France: Roscoff.

Barbot, Y. N., Al-Ghaili, H., & Benz, R. (2016). A review on the valorization of macroalgal wastes for biomethane production. *Marine Drugs*, *14*, 120–129.

Baweja, P., Kumar, S., Sahoo, D., & Levine, I. (2016). Macroalgae systematics. In J. Fleurence & I. Levine (Eds.), *Seaweed in health and disease prevention*. Elsevier. ISBN: 9780128027936 (Chapter 3).

Bedoux, G., Hardouin, K., Burlot, A. S., & Bourgougnon, N. (2014). Bioactive components from seaweeds: Cosmetic applications and future development. In N. Bourgougnon, & J. P. Jacquot (Eds.), *71. Advance botanical research* (pp. 345–378). London, Cambridge, Oxford: Elsevier.

Bœuf, G. (2014). *La biodiversité, de l'océan à la cité. Collège de France/Fayard Collection « Leçons inaugurales du Collège de France ». Vol. 241* (p. 85). ISBN : 978-2-213-68148-1.

Boudouresque, C. F. (2015). Taxonomy and phylogeny of unicellular eukaryotes. In J. C. Bertrand, P. Caumette, P. Lebaron, R. Matheron, P. Normand, & T. Sime-Ngando (Eds.), *Environmental microbiology: fundamentals and applications. Microbial ecology* (pp. 191–257). Springer publication.

Brown, B. E. (1997). Coral bleaching: Causes and consequences. *Coral Reefs*, *16*, 129–S138.

Brown, M. W., Sharpe, S. C., Silberman, J. D., Heiss, A. A., Lang, B. F., Simpson, A. G. B., et al. (2013). Phylogenomics demonstrates that breviate flagellates are related to opisthokonts and apusomonads. *Proceedings of the Royal Society B*, *280*, 1755–1761.

Brugère, C., Aguilar-Manjarrez, J., Beveridge, M. C. M., & Soto, D. (2019). The ecosystem approach to aquaculture 10 years on – a critical review and consideration of its future role in blue growth. *Reviews in Aquaculture*, *11*(3), 493–514. https://doi.org/10.1111/raq.12242.

Bucholc, K., Szymczak-Żyła, L., Lubecki, L., Zamojska, A., Hapter, P., Tjernström, E., et al. (2014). Nutrient content in macrophyta collected from southern Baltic Sea beaches in relation to eutrophication and biogas production. *Science of the Total Environment*, *473–474*, 298–307.

Buet, V. (2020). From seaweed to seaweeders. In *Advances botanical research: Vol. 95. Seaweeds around the world: State of art and perspectives*. ISBN: 978-0-08-102710-3 (Chapter 12).

Burki, F., Kaplan, M., Tikhonenkov, D. V., Zlatogursky, V., Minh, B. Q., Radaykina, L. V., et al. (2016). Untangling the early diversification of eukaryotes: A phylogenomic study of the evolutionary origins of Centrohelida, Haptophyta and Cryptista. *Proceedings. Biological Sciences*, *283*, 2015–2802.

Cardozo, K. H. M., Guaratini, T., Barros, M. P., Falcão, V. R., Tonon, A. P., Lopes, N. P., et al. (2007). Metabolites from algae with economical impact. *Comparative Biochemistry and Physiology Part C: Toxicology & Pharmacology*, *146*, 60–78.

Cherry, P., Yadav, S., Strain, C. R., Allsopp, P. J., McSorley, E. M., Ross, R. P., et al. (2019). Prebiotics from seaweeds: An ocean of opportunity? *Marine Drugs*, *17*, 327.

Chopin, T., & Robinson, S. (2004). Defining the appropriate regulatory and policy framework for the development of integrated multi-trophic aquaculture practices: Introduction to the workshop and positioning of the issues. *Bulletin of the Aquaculture Association of Canada*, *104*, 4–10.

Cian, R. E., Drago, S. R., de Medina, F. S., & Martinez-Augustin, O. (2015). Proteins and carbohydrates from red seaweeds: Evidence for beneficial effects on gut function and microbiota. *Marine Drugs*, *13*, 5358–5383.

Convention on Biological Diversity. (2015). https://www.cbd.int/abs/about/.

Cottier-Cook, E. J., Nagabhatla, N., Badis, Y., Campbell, M., Chopin, T., Dai, W., et al. (2016). *Safeguarding the future of the global seaweed aquaculture industry*. United Nations University (INWEH) and Scottish Association for Marine Science Policy Brief. ISBN 978-92-808-6080-1.

Cousteau, J. Y., & Diolé, P. (1971). *La vie et la mort des coraux*. France: Flammarion.

Couteau, C., & Coiffard, L. (2016). Chapter 14. Seaweed application in cosmetics. In J. Fleurence, & I. Levine (Eds.), *Seaweed in health and disease prevention* Elsevier. eBook.

Danish Siddiqui, M., & Abdullah, M. (2016). Evaluation of World View-2 satellite data for mapping seaweed beds along Karachi Coast. In *41st COSPAR Scientific Assembly, abstracts from the meeting that was to be held 30 July–7 August at the Istanbul Congress Center (ICC), Turkey, but was cancelled*. See http://cospar2016.tubitak.gov.tr/en/.

de Carvalho, J. C., Magalhães, A. I., de Melo Pereira, G. V., Medeiros, A. B. P., Sydney, E. B., Rodrigues, C., et al. (2020). Microalgal biomass pretreatment for integrated processing into biofuels, food, and feed. In *Vol. 300. Bioresource Technology* (pp. 122–719). Elsevier Ltd.

De Jesus Raposo, M. F., de Morais, A. M., & de Morais, R. M. (2016). Emergent sources of prebiotics: Seaweeds and microalgae. *Marine Drugs, 14*, 27–34.

de Reviers, B. (2003). *Biologie et phylogénie des algues, Tome 1 and 2. France : Belin sup.*

Déléris, P., Nazih, H., & Bard, J. M. (2016). Seaweeds in human health. In J. Fleurence, & I. Levine (Eds.), *Seaweed in health and disease prevention* Elsevier. eBook.

Demirbas, A., & Demirbas, F. M. (2011). Importance of algae oil as a source of biodiesel. *Energy Conversion and Management, 52*, 163–170.

Deshmukh, S., Kumar, R., & Bala, K. (2019). Microalgae biodiesel: A review on oil extraction, fatty acid composition, properties and effect on engine performance and emissions. *Fuel Processing Technology, 191*, 232–247.

Duarte, C. M., Wu, J., Xiao, X., Bruhn, A., & Krause-Jensen, D. (2017). Can seaweed farming play a role in climate change mitigation and adaptation?. *Frontiers in Marine Science, 4*, 100–108.

El Gamal, A. A. (2010). Biological importance of marine algae. *Saudi Pharmaceutical Journal, 18*, 1–25.

FAO. (2006). *Livestock long shadow*. Rome: Food and Agriculture Organisation of the United Nations.

FAO. (2014). *The state of the world fisheries and aquaculture 2014* (p. 223). Rome: FAO.

FAO. (2015). *FAO global aquaculture production database updated to 2013—Summary information*. Rome: FAO.

FAO. (2016). The state of the world fisheries and aquaculture 2016. *Contribution to food security and nutrition for all*. Rome. 200 pp.

FAO. (2018a). The global status of seaweed production, trade and utilization. In F. Ferdouse, Z. Yang, S. L. Holdt, P. Murúa, R. Smith, & FAO Consultants (Eds.), *Vol. 124. Globefish research programme*.

FAO. (2018b). *The state of world fisheries and aquaculture 2018—Meeting the sustainable development goals*. Rome. Licence: CC BY-NC-SA 3.0 IGO.

FAO. (2020). *La durabilité en action. La situation mondiale des pêches et de l'aquaculture 2020*. Rome: FAO. https://doi.org/10.4060/ca9229fr.

Fernand, F., Israel, A., Skjermo, J., Wichard, T., Timmermans, K. R., & Golberg, A. (2017). Offshore macroalgae biomass for bioenergy production: Environmental aspects, technological achievements and challenges. In *Vol. 75. Renewable and sustainable energy reviews* (pp. 35–45). Elsevier Ltd.

Finley, J. W., Finley, J. W., Ellwood, K., & Hoadley, J. (2014). Launching a new food product or dietary supplement in the United States: Industrial, regulatory, and nutritional considerations. *Annual Review of Nutrition, 34*, 421–447.

Fleurence, J., & Levine, I. (2016). In J. Fleurence, & I. Levine (Eds.), *Seaweed in health and disease prevention* Elsevier. ebook.

Fort, A., Mannion, C., Fariñas-Franco, J. M., & Sulpice, R. (2020). Green tides select for fast expanding Ulva strains. *Science of the Total Environment, 698*, 134–337.

Frocklin, S., de la Torre-Castro, M., Lindstrom, L., Jiddawi, N., & Msuya, F. (2012). Seaweed mariculture as a development project in Zanzibar, East Africa: A price too high to pay? *Aquaculture, 356*, 30–39.

Frangoudes, K., Gerrard, S., & Kleiber, D. (2019). Situated transformations of women and gender relations in small-scale fisheries and communities in a globalized world. *Maritime Studies, 18*, 241–248.

Froehlich, H. E., Afflerbach, J. C., Frazier, M., & Halpern, B. S. (2019). Blue growth potential to mitigate climate change through seaweed offsetting. *Current Biology, 29*(8), 3087–3093.

Gattuso, J. P., Gentili, B., Duarte, C. M., Kleypas, J. A., Middelburg, J. J., & Antoine, D. (2006). Light availability in the coastal ocean: Impact on the distribution of benthic photosynthetic organisms and their contribution to primary production. *Biogeosciences*, *3*, 489–513.

GIEC. (2014). *Changement climatique2014*. Rapport de synthèse.

Gierckský, E., & Doumeizel, V. (2020). *Seaweed revolution: A manifesto for a sustainable future* (p. 16). Lloyd's Register Foundation.

Ginigaddara, G. A. S., & Lankapura, A. I. Y. (2018). Farmer's perspective on importance and constraints of seaweed farming in Sri Lanka. *Current Investigations in Agriculture and Current Research*, *3*(1). https://doi.org/10.32474/CIACR.2018.03.000151. CIACR. MS.ID.000151.

Guiry. (2020). *Saccharina cultivation in China. The Seaweed Site: information on marine algae.* Site © M.D. Guiry 2000–2020. Online on July 16th 2020 http://www.seaweed.ie/aquaculture/kelp_china.php.

Guiry, M. D., & Morrison, L. (2013). The sustainable harvesting of Ascophyllum nodosum (Fucaceae, Phaeophyceae) in Ireland, with notes on the collection and use of some other brown algae. *Journal of Applied Phycology*, *25*, 1823–1830.

Hamed, I., Özogul, F., Özogul, Y., & Regenstein, J. M. (2015). Marine bioactive compounds and their health benefits: A review. *Comprehensive Reviews in Food Science and Food Safety*, *14*, 446–465.

Harnedy, P. A, & FitzGerald, R. J. (2011). Bioactive proteins, peptides, and amino acids from macroalgae. *Journal of Phycology*, *47*(2), 218–232.

Heisler, J., Glibert, P. M., Burkholder, J. M., Anderson, D. M., Cochlan, W., Dennison, W. C., et al. (2008). Eutrophication and harmful algal blooms: A scientific consensus. *Harmful Algae*, *8*, 3–13.

Herrero, M., & Thornton, P. K. (2013). Livestock and global change: Emerging issues for sustainable food systems. *Proceedings of the National Academy of Sciences of the United States of America*, *110*, 20878e20881.

Hoegh-Guldberg, O., Caldeira, K., Chopin, T., Gaines, S., Haugan, P., Hemer, M., et al. (2019). *The ocean as a solution for climate change: Five opportunities for action*. Washington, DC: World Resources Institute. Report. https://oceanpanel.org/sites/default/files/2019-10/HLP_Report_Ocean_Solution_Climate_Change_final.pdf.

Holdt, S. L. (2015). *The commercial IMTA and future seaweed biofilter potential in Denmark*. Sound/Visual Production.

Holdt, S., & Kraan, S. (2011). Bioactive compounds in seaweed: Functional food applications and legislation. *Journal of Applied Phycology*, *23*, 543–597.

Huesemann, M. H., Kuo, L.-J., Urquhart, L., Gill, G. A., & Roesijadi, G. (2012). Acetone-butanol fermentation of marine macroalgae. *Bioresource Technology*, *108*, 305–309.

Hussin, H., & Khoso, A. (2017). Seaweed cultivation and coastal communities in Malaysia: An Overview. *Asian Fisheries Science*, *30*, 87–100.

Ioannou, E., & Roussis, V. (2009). Natural products from seaweeds. In *Biomedical and life science* (pp. 51–81). Springer.

Jacquin, A. G. (2020). *The Algonauts project*. Website consulted on July 17th 2020 http://algonauts.org/.

Jakobsdottir, G., Nyman, M., & Fak, F. (2014). Designing future prebiotic fiber to target metabolic syndrome. *Nutrition*, *30*, 497–502.

Jesumani, V., Du, H., Pei, P., Aslam, M., & Huang, N. (2020). Comparative study on skin protection activity of polyphenol-rich extract and polysaccharide-rich extract from *Sargassum vachellianum*. *PLoS One*, *15*, e0227308.

Kadam, S. U., Álvarez, C., Tiwari, B. K., & O'Donnell, C. P. (2015). Processing of seaweeds. In *Seaweed sustainability: food and non-food applications* (pp. 61–78). Elsevier Inc.

Kadam, S. U., Tiwari, B. K., & O'Donnell, C. P. (2013). Application of novel extraction technologies for bioactives from marine algae. *Journal of Agricultural and Food Chemistry, 61*, 4667–4675.

Kassem, A. A., Farid, R. M., Elsayed Issa, D. A., Khalil, D. S., Abd-El-Razzak, M. Y., Saudi, H. I., et al. (2015). Development of mucoadhesive microbeads using thiolated sodium alginate for intrapocket delivery of resveratrol. *International Journal of Pharmaceutics, 487*, 305–313.

Keith, S. A., Kerswell, A. P., & Connolly, S. R. (2014). Global diversity of marine macroalgae: Environmental conditions explain less variation in the tropics. *Global Ecology and Biogeography, 23*, 517–529.

Kinley, R. D., Martinez-Fernandez, G., Matthews, M. K., de Nys, R., Magnusson, M., & Tomkins, N. W. (2020). Mitigating the carbon footprint and improving productivity of ruminant livestock agriculture using a red seaweed. *Journal of Cleaner Production, 259*, 120836.

Kornprobst, J.-M. (2005). *Substances Naturelles D'origine Marine*. Paris: Tech et Doc Lavoisier.

Kraan, S. (2013). Mass-cultivation of carbohydrate rich macroalgae, a possible solution for sustainable biofuel production. *Mitigation and Adaptation Strategies for Global Change, 18*(1), 27–46.

Kraan, S. (2020). Seaweed resources, collection, and cultivation with respect to sustainability. In *Sustainable seaweed technologies* (pp. 89–102). London: Elsevier.

Krause-Jensen, D., Lavery, P., Serrano, O., Marba, N., Masque, P., & Duarte, C. M. (2018). Sequestration of macroalgal carbon: The elephant in the Blue Carbon room. *Biology Letters, 14*, 20180236.

Krishnan, M, & Narayanakumar, R. (2010). Structure, conduct and performance of value chain in seaweed farming in India. *Agricultural Economics Research Review, 23*, 505–514.

Lakner, C., Gerszon Mahler, D., Negre, M., & Prydz, E.B. (2020). How much does reducing inequality matter for global poverty? *Global poverty monitoring technical note 13*.

Lalegerie, F., Gager, L., Stiger-Pouvreau, V., & Connan, S. (2020). The stressful life of red and brown seaweeds on the temperate intertidal zone: Effect of abiotic and biotic parameters on the physiology of macroalgae and content variability of particular metabolites. In N. Bourgougnon, & J. P. Jacquot (Eds.), *Vol. 95. Seaweeds around the world: State of art and perspectives. Advances botanical research*. London: Elsevier.

Lavers, J. L., Dicks, L., Dicks, M. R., & Fin, A. (2019). Significant plastic accumulation on the Cocos (Keeling) Islands, Australia. *Scientific Reports, 9*, 7102.

Leandro, A., Pereira, L., & Gonçalves, A. M. (2020). Diverse applications of marine macroalgae. *Marine Drugs, 18*, 17.

Lebreton, L., Slat, B., Ferrari, F., Sainte-Rose, B., Aitken, J., Marthouse, R., et al. (2018). Evidence that the Great Pacific Garbage Patch is rapidly accumulating plastic. *Scientific Reports, 8*, 4666.

Lehahn, Y., Ingle, K. N., & Golberg, A. (2016). Global potential of offshore and shallow waters macroalgal biorefineries to provide for food, chemicals and energy: Feasibility and sustainability. *Algal Research, 17*, 150–160.

Lehmann, A. (2020). *Des services écosystémiques au développement durable. Lecture Webinar on May 7th 2020 organized by Rovaltain Foundation*.

Leliaert, F., Smith, D., Moreau, H., Herron, M., Verbruggen, H., Delwiche, C., et al. (2012). Phylogeny and molecular evolution of the green algae. *Critical Reviews in Plant Sciences, 31*, 11–46.

Levine, I. (2016). Algae: a way of life and health. In J. Fleurence, & I. Levine (Eds.), *Seaweed in health and disease prevention*. Amsterdam Boston Heidelberg London: Elsevier, Academic Press.

Levine, I., & Fleurence, J. (2018). In J. Fleurence, & I. Levine (Eds.), *Microalgae in health and disease prevention*. Amsterdam Boston Heidelberg London: Elsevier, Academic Press.

Lloyd's Register Foundation. (2020). *Seaweed revolution: A manifesto for a sustainable future.* Rapport.
Mac Monagail, M., Cornish, L., Morrison, L., Araújo, R., & Critchley, A. T. (2017). Sustainable harvesting of wild seaweed resources. *European Journal of Phycology, 524*, 371–390.
MacArtain, P., Gill, C. I., Brooks, M., Campbell, R., & Rowland, I. R. (2007). Nutritional value of edible seaweeds. *Nutrition Reviews, 65*, 535–543.
Machado, L., Tomkins, N., Magnusson, M., Midgley, D. J., de Nys, R., & Rosewarne, C. P. (2018). In vitro response of rumen microbiota to the antimethanogenic red macroalga *Asparagopsis taxiformis. Microbial Ecology, 75*, 811–818.
Magnussona, M., Glassona, C. R. K., Matthew Vuckob, J., Angellb, A., Neohb, T. L., & de Nys, R. (2019). Enrichment processes for the production of high-protein feed from the green seaweed *Ulva ohnoi. Algal Research, 41*, 101555.
Marzinelli, E. M., Leong, M. R., Campbell, A. H., Steinberg, P. D., & Vergés, A. (2016). Does restoration of a habitat-forming seaweed restore associated faunal diversity? *Restoration Ecology, 24*, 81–90. https://doi.org/10.1111/rec.12292.
Merci Les Algues. (2020). https://mercilesalgues.com/ (online on July 7th 2020).
Millennium Ecosystem Assessment. (2005). *Ecosystems and human well-being: Synthesis.* Washington, DC: Island Press.
Msuya, F. E., & Hurtado, AQ. (2017). The role of women in seaweed aquaculture in the Western Indian Ocean and South-East Asia. *European Journal of Phycology, 52*, 482–494. https://doi.org/10.1080/09670262.2017.1357084.
Nešic, A., Cabrera-Barjas, G., Dimitrijevic-Brankovic, S., Davidović, S., Radovanović, N., & Delattre, C. (2019). Prospect of polysaccharide-based materials as advanced food packaging. *Molecules, 25*, 135.
NETALGAE. (2012). *Overview of the seaweed industry by country: Ireland, France, Norway, Portugal, Spain, United Kingdom, An output of the Netalgae project—Interreg IVB* (p. 8).
Niklas, K. J., & Kutschera, U. (2010). The evolution of the land plant life cycle. *New Phytologist, 185*, 27–41.
Okoli, C. H., Adams, T. A., Brigljević, B., & Liu, J. L. (2016). Design and economic analysis of a macroalgae-to-butanol process via a thermochemical route. *Energy Conversion and Management, 123*, 410–422.
Parc Naturel Marin Iroise. (2018). *Retour sur le conseil de gestion du Parc naturel marin d'Iroise.* Press release in November 2018 au Conquet, France.
Peñuela, A., Robledo, D., Bourgougnon, N., Bedoux, G., Hernández-Núñez, E., Freile-Pelegrín, Y., et al. (2018). Environmentally friendly valorization of *Solieria filiformis* (Gigartinales, Rhodophyta) from IMTA using a biorefinery concept. *Marine Drugs, 16*, 487.
Pérez-Lloréns, J. L., Mouritsen, O. G., Rhatigan, P., Cornish, M. L., & Critchley, A. T. (2020). Seaweeds in mythology, folklore, poetry, and life. *Journal of Applied Phycology, 32*, 3157–3182.
Placencia Seaweed Farmers. (2016). *Placencia SeaWeed Project Shamax Production.* https://www.youtube.com/watch?v=Tzusmxs8Jh8&feature=youtu.be.
Prabhu, M., Chemodanov, A., Gottlieb, R., Kazir, M., Nahord, O., Gozin, M., et al. (2019). Starch from the sea: The green macroalga *Ulva ohnoi* as a potential source for sustainable starch production in the marine biorefinery. *Algal Research, 37*, 215.
Pyšek, P., Hulme, P. E., Simberloff, D., Bacher, S., Blackburn, T. M., Carlton, J. T., et al. (2020). Scientists' warning on invasive alien species. *Biological Reviews, 95*(6), 1511–1534. brv.12627.
Ramasubramani, R., Praveen, R., & Sathyanarayanan, K. S. (2016). Study on the strength properties of marine algae concrete. *Rasayan Journal of Chemistry, 9*, 706–715.

Rebours, C., Marinho-Soriano, E., Zertuche-González, J. A., Hayashi, L., Vásquez, J. A., Kradolfer, P., et al. (2014). Seaweeds: An opportunity for wealth and sustainable livelihood for coastal communities. *Journal of Applied Phycology*, *26*, 1939–1951.

Rinaudo, M. (2008). Main properties and current applications of some polysaccharides as biomaterials. *Polymer International*, *57*, 397–430.

Rioux, L. E., Beaulieu, L., & Turgeon, S. L. (2017). Seaweeds: A traditional ingredients for new gastronomic sensation. *Food Hydrocolloids*, *68*, 255–265.

Ripple, W. J., Wolf, C., Newsome, T. M., Galetti, M., Alamgir, M., Crist, E., et al. (2017). World scientists' warning to humanity: A second notice. *Bioscience*, *67*, 1026–1028.

Rochman, C. M., Akbar Tahir, S., Williams, L., Baxa, D. V., Lam, R., Miller, J. T., et al. (2015). Anthropogenic debris in seafood: Plastic debris and fibers from textiles in fish and bivalves sold for human consumption. *Plastic Science Report*, *5*, 14340.

Roque, B. M., Brooke, C. G., Ladau, J., Polley, T., Marsh, L. J., Najafi, N., et al. (2019). Effect of the macroalgae Asparagopsis taxiformis on methane production and rumen microbiome assemblage. *Animal Microbiome*, *1*, 1–14.

Rosa, G. P., Tavares, W. R., Sousa, P. M. C., Pagès, A. K., Seca, A. M. L., & Pinto, D. C. G. A. (2019). Seaweed secondary metabolites with beneficial health E effects: An overview of successes in *in vivo studies and clinical trials. Marine Drugs*, *18*, 2–35.

Rupérez, P. (2002). Mineral content of edible marine seaweeds. *Food Chemistry*, *79*, 23–26.

Schaeffer, D. J., & Krylov, V. S. (2000). Anti-HIV activity of extracts and compounds from algae and cyanobacteria. *Ecotoxicology and Environmental Safety*, *45*, 208–227.

Sedrati, M. (2018). A sustainable alternative for coastal dune restoration by sand-trapping fences and algae wrack: AlgoBox®. In *20th EGU General Assembly, EGU2018, Proceedings from the conference held 4-13 April, 2018 in Vienna, Austria* (p. 7376).

Sedrati, M., & Cochet, J. (2015). AlgoBox®: Un outil écologique pour la régénération des pieds de dunes grâce aux échouages de macro-algues. In *Coastal and maritime Mediterranean conference*. available online http://www.paralia.fr.

Singh, V., Tiwari, A., & Das, M. (2016). Phyco-remediation of industrial waste-water and flue gases with algal-diesel engenderment from micro-algae: A review. *Fuel*, *173*, 90–97.

Skjermo, J., Marie Aasen, I., Forbord, S., Solbakken, R., Steinhovden, K. B., Handå Co-authors, A., et al. (2014). A new Norwegian bioeconomy based on cultivation and processing of seaweeds: Opportunities and R&D needs. In *SINTEF fisheries and aquaculture SINTEF fisheries and aquaculture*. In undefined www.sintef.no/fisk.

Smetacek, V., & Zingone, A. (2013). Green and golden seaweed tides on the rise. *Nature Publishing Group*, *504*(7478), 84–88. https://doi.org/10.1038/nature12860.

Sørensen, P., & Møller, W. F. (2008). Place and climate. In T. Dahl (Ed.), *Climate and architecture*. London, New York: Taylor & Francis.

Stengel, D. B., Connan, S., & Popper, Z. A. (2011). Algal chemodiversity and bioactivity: Sources of natural variability and implications for commercial application. *Biotechnology Advances*, *29*, 483–501.

Stévant, P., Rebours, C., & Chapman, A. (2017). Seaweed aquaculture in Norway: Recent industrial developments and future perspectives. *Aquaculture International*, *25*(4), 1373–1390. https://doi.org/10.1007/s10499-017-0120-7.

Suganya, T., Varman, M., Masjuki, H. H., & Renganathan, S. (2016). Macroalgae and microalgae as a potential source for commercial applications along with biofuels production: A biorefinery approach. *Renewable and Sustainable Energy Reviews*, *55*, 909–941.

Sully, S., Burkepile, D. E., Donovan, M. K., Hodgson, G., & van Woesik, R. (2019). A global analysis of coral bleaching over the past two decades. *Nature Communications*, *10*, 1–5.

Susilorinia, R., Rr, M. I., Hardjasaputrab, H., Tudjonoc, S., Hapsaria, G., Wahyu Sa, R., et al. (2014). The advantage of natural polymer modified mortar with seaweed:green

construction material innovation for sustainable concrete. In *Vol. 95. 2nd international conference on sustainable civil engineering structures and construction materials (SCESCM 2014). Procedia engineering* (pp. 419–425).
Suyo, J. G. B., Masson, V. L., Shaxson, L., & Luhan, M. R. J. (2020). A social network analysis of the Philippine seaweed farming industry: Unravelling the web. *Marine Policy, 118*, 104007. https://doi.org/10.1016/j.marpol.2020.104007.
Taylor, G. (2008). Biofuels and the biorefinery concept. *Energy Policy, 36*, 4406–4409.
Terme, N., Chénais, B., Fournière, M., Bourgougnon, N., & Bedoux, G. (2019). Chapter 15: Algal derived functional lipids and their role in promoting health recent advancements. In *Micro and macroalgal processing: food and health perspectives* Wiley.
Thomas, N. V., & Kim, S. K. (2011). Potential pharmacological applications of polyphenolic derivatives from marine brown algae. *Envrionmental Toxicology and Pharmacology, 32*, 325–335.
Tirichine, L., & Bowler, C. (2011). Decoding algal genomes: Tracing back the history of photosynthetic life on Earth. The plant genome: An evolutionary view on structure and function. *The Plant Journal, 66*, 45–57.
Trivedi, J., Aila, M., Bangwal, D. P., Kaul, S., & Garg, M. O. (2015). Algae based biorefinery—How to make sense? In *vol. 47. Renewable and sustainable energy reviews* (pp. 295–307). Elsevier Ltd.
United Nations. (2020a). *UN comtrade*. http://comtrade.un.org. accessed on 25 April 2020.
United Nations. (2020b). *Sustainable consumption and production. Topics*. https://sdgs.un.org/topics/sustainable-consumption-and-production-76 (online on July 18th 2020).
Urban, M. C. (2015). Accelerating extinction risk from climate change. *Sciences, 348*, 571–573.
Valderrama, D., Cai, J., Hishamunda, N., & Ridler, N. (Eds.). (2013). *Social and economic dimensions of carrageenan seaweed farming. Fisheries and aquaculture technical paper no.* Rome, Italy: FAO.
Van Der Wal, H., Sperber, B. L. H. M., Houweling-tan, B., Bakker, R. R. C., Brandenburg, W., & López-contreras, A. M. (2013). Production of acetone, butanol, and ethanol from biomass of the green seaweed *Ulva lactuca*. *Bioresource Technology, 128*, 431–437.
Wang, H. M. D., Chen, C. C., Huynh, P., & Chang, J. S. (2015). Exploring the potential of using algae in cosmetics. *Bioresource Technology, 184*, 355–362.
Watanuki, A., Aota, T., Otsuka, E., Kawai, T., Iwahashi, Y., Kuwahara, H., et al. (2010). Restoration of kelp beds on an urchin barren: Removal of sea urchins by citizen divers in southwestern Hokkaido. *Bulletin of Fisheries Research Agency, 32*, 83–87.
Wells, M. L., Potin, P., Craigie, J. S., Raven, J. A., Merchant, S. S., Helliwell, K. E., et al. (2017). Algae as nutritional and functional food sources: Revisiting our understanding. *Journal of Applied Phycology, 29*, 949–982.
Whitmee, S., Haines, A., Beyrer, C., et al. (2015). Safeguarding human health in the Anthropocene epoch: Report of The Rockefeller Foundation–Lancet Commission on planetary health. Lancet. Published online July 16. https://doi.org/10.1016/S0140-6736(15)60901-1.
Widera, B. (2014). Possible application of seaweed as building material in the modern seaweed house on Læsø. In *30th international plea conference 16-18 December 2014* CEPT University.
World Bank. (2018). *India's seaweed squads boost women's incomes and confidence*. http://wrld.bg/5bOb30koDqZ.

CHAPTER SIX

Genomics of cyanobacteria: New insights and lessons for shaping our future—A follow-up of volume 65: Genomics of cyanobacteria

Franck Chauvat* and Corinne Cassier-Chauvat
Laboratory of Biology and Biotechnology of Cyanobacteria, Université Paris-Saclay, CEA, CNRS, Institute for Integrative Biology of the Cell (I2BC), Gif-sur-Yvette, France
*Corresponding author: e-mail address: franck.chauvat@cea.fr

Contents

Abstract

Cyanobacteria, the only known photosynthetic prokaryotes, are ancient organisms regarded as the producers of the Earth's oxygenic atmosphere and the ancestors of plant chloroplasts. Contemporary cyanobacteria have evolved as widely-diverse organisms in colonizing most aquatic and soil biotopes, where they face various environmental challenges and competition or symbiosis with other organisms. Cyanobacteria exhibit a wide morphological diversity (unicellular/multicellular, cylindrical/spherical shapes) and many species differentiate specialized cells for growth and survival under adverse conditions. They convert captured solar energy at high efficiencies, to fix an enormous amount of carbon from CO_2 into a huge biomass that sustains a large part of the food chain, and they tolerate high CO_2 content in gas streams. They also synthesize a vast array of biologically active metabolites with great interests for human health and industries. Hence, they are viewed as promising "low-cost" cell factories for the carbon-neutral production of chemicals, thanks to their simple nutritional requirements, their metabolic robustness and plasticity, and the powerful genetics of some model strains.

In 2013 when ABR vol 65 entitled "Genomics of Cyanobacteria" was published, approximately 70 cyanobacteria had their genomes fully or partially sequenced. Since then, the number has increased up to about 2000 (https://genome.jgi.doe.gov/portal/), genome editing studies using CRISPR/Cas have been reported in several model cyanobacteria (Gale et al., 2019;

Advances in Botanical Research, Volume 100
ISSN 0065-2296
https://doi.org/10.1016/bs.abr.2021.01.007
Copyright © 2021 Elsevier Ltd
All rights reserved.

Pattharaprachayakul, Lee, Incharoensakdi, & Woo, 2020), and the knowledge that was presented in each of the eight chapters of the vol 65 book has greatly improved, as described below.

In the first chapter entitled "Structural, Mechanistic and Genomic Insights into OCP-Mediated Photoprotection," Cheryl Kerfeld and Diana Kirilovsky reported on the mechanisms used by cyanobacteria to protect themselves against high light fluence (Kerfeld & Kirilovsky, 2013). When the amount of light energy exceeds what is needed for assimilation of inorganic-nutrients (CO_2, NO_3, PO_4, SO_4, *etc.*), the photosynthetic electron transport chain becomes stalled in a reduced state, leading to the formation of reactive oxygen species (ROS) that generate severe cell damages. Cyanobacteria have evolved a rapid photoprotective mechanisms to cope with abrupt changes in the quality and intensity of light. Light activates a soluble orange carotenoid protein (OCP), which interacts with the light-harvesting antenna (phycobilisomes), to increase thermal dissipation of absorbed energy, thereby decreasing energy arriving at the reaction centers. The increase of thermal dissipation also causes a decrease of phycobilisome fluorescence creating a non-photochemical fluorescence quenching (Kerfeld & Kirilovsky, 2013).

Our knowledge of OCP-dependent process has greatly progressed as summarized in the recent reviews by Cheryl Kerfeld' and Diana Kirilovsky' laboratories (Dominguez-Martin & Kerfeld, 2019; Muzzopappa & Kirilovsky, 2020). OCP binds one ketocarotenoid, which spans both N- and C-terminal OCP domains (NTD and CTD). Upon photoactivation, the carotenoid is translocated to the NTD and the color of the protein changes to red. Then, the domains move away from one another and OCP becomes active. The red OCP binds to the phycobilisome, inducing thermal dissipation of excess energy.

OCP is composed of two globular domains, the α-helical N terminal domain (NTD; residues 18–165) and the α-helix/β-sheet C-terminal domain (CTD; residues 190–317), of the latter presenting homologies to the nuclear transport factor 2 (NTF2) superfamily, which are connected by a long, flexible (largely unstructured) loop linker (approximately 25 residues). In the dark, OCP is orange and has a compact (closed) globular conformation. This conformation is stabilized by strong interdomain interactions in the middle of the protein, and a ketocarotenoid, 3′-hydroxyechinenone (hECN), which spans both domains. The carotenoid is a sensor of light intensity. A strong blue-green light, absorbed by the carotenoid, induces conformational changes in the carotenoid (it moves by 12 Å) and in the two protein domains and the orange inactive form (OCP^O) is converted into a metastable red active form (OCP^R). The OCP^R state is characterized by large-scale structural changes that involve dissociation and separation of C-terminal and N-terminal domains accompanied by carotenoid translocation into the N-terminal domain (Konold et al., 2019). Only OCP^R can bind to PBS, while OCP^O is unable to bind.

When low-light conditions return, the interaction of OCP with the fluorescence recovery protein (FRP), a non-chromophore protein, is needed to block photoprotection under low light and increase the effective size of the antenna in cells containing OCP1 (Gupta et al., 2019). In the absence of FRP, the cells remain in a quenched state, even in the dark. FRP is an α-helix elongated protein that is stable as a dimer in solution. FRP binds to CTD of OCP^R as a dimer but then monomerizes. It has been proposed that FRP binding to OCP weakens the interaction between the monomers, which can lead to spontaneous monomerization of FRP; alternatively, the binding of a second OCP to the FRP dimer may provoke a clash between the two OCP molecules, causing the splitting of the 2OCP:2FRP complex into two OCP:FRP complexes. The monomerization is not essential for FRP activity but can improve its efficiency, especially at elevated concentrations of OCP^R under high light intensities (Muzzopappa & Kirilovsky, 2020). FRP has two different activities, being essential for the OCP^R-to-OCP^O conversion reaction and for detaching OCP^R from PBS. In cells, the concentration of FRP is much lower than that of OCP, which is important since high FRP concentrations completely inhibit the PBS fluorescence quenching. Thus, the concentration of FRP not only controls the rate of PBS fluorescence recovery but also the amplitude of PBS fluorescence quenching (Muzzopappa & Kirilovsky, 2020).

Genomic studies identified three paralog families of OCP, namely, OCP1, OCP2 and OCPX. Most OCPs (approximately 65% of available sequences) belong to the OCP1 family, to which the well-characterized *Synechocystis* PCC 6803, *Arthrospira maxima*, and *Anabaena* PCC 7120 OCPs belong. Only one OCP2 (*Tolypothrix* PCC 7601) and one OCPX (*Scytonema hofmanni* PCC 7110) have been characterized (Muzzopappa & Kirilovsky, 2020). The frp gene is absent in strains containing only OCP2 or OCPX. In these strains, OCPX or OCP2 is able to detach from PBS and allows fluorescence recovery in the absence of FRP. Moreover, the rate of OCP^R-to-OCP^O conversion of these OCPs is fast and does not need to be accelerated by FRP. The FRP dimer interacts with high affinity with CTD of $OCP1^R$ (Muzzopappa & Kirilovsky, 2020).

In addition to OCPs, many cyanobacterial strains contain homologs of the NTD and CTD domains of OCP. The family of NTD-like proteins, named helical carotenoid proteins (HCPs), contains at least nine different subclades (HCP1 to HCP9), while the family of CTD-like proteins (CTDHs) contains only two subclades (CTDH1 and CTDH2). Both HCPs and CTDHs are carotenoid proteins that preferentially bind CAN. While several HCPs from different subclades can coexist in cells of some strains, only one CTDH is present. While HCPs are monomers, CTDHs are dimers, sharing one canthaxanthin (CAN) molecule. HCPs from different subclades have different activities, at least in vitro. HCP2 proteins from *Anabaena* and *Tolypothrix* and HCP3 from *Anabaena* are good singlet oxygen quenchers, and HCP1 can be a carotenoid carrier. CTDHs are transient carotenoid proteins, which are able to transfer the carotenoid from the membrane to HCPs and OCP (Slonimskiy et al., 2019).

OCP and the paralogs of its domains, HCPs and CTDHs, are also excellent quenchers of singlet oxygen, decreasing the risks of oxidative damage. HCPs seem to help decrease the oxidative stress under specific stress conditions, such as desiccation. The principal role of CTDH is to transport the carotenoid from the membrane to HCP and, perhaps, to OCP. The study of these proteins allowed the discovery of a unique protein-to-protein carotenoid transfer mechanism, which could potentially be applicable to other processes involving water-soluble carotenoid proteins. These findings may also inspire the construction of stable carotenoid nanocarriers capable of targeted delivery of the antioxidant power to specific tissues, such as the macula lutea, in which the lutein-carrying NTF2-like StAR-related lipid transfer domain containing 3 (STARD3) protein plays a role in photoprotection (Dominguez-Martin & Kerfeld, 2019; Muzzopappa & Kirilovsky, 2020).

Although OCP is important for cyanobacterial photoprotection other processes exist since about one-third of cyanobacteria have no *ocp* genes, such as the PBS-lacking cyanobacteria (including *Prochlorococcus marinus* strains) and the light-tolerant model species *Synechococcus* PCC 7942 and *Thermosynechococcus elongatus* BP1 (Muzzopappa & Kirilovsky, 2020).

In the second chapter entitled "Genomics of Salt Acclimation: Synthesis of Compatible Solutes among Cyanobacteria" Martin Hagemann reviewed the existing mechanistic and genomic knowledges about cyanobacterial salt acclimation, an important process for future biotechnological applications, some of which could benefit from being performed in abundant marine waters (Cui, Sun, Chen, & Zhang, 2020; Hagemann, 2013).

Besides light, the availability of water and the amounts of dissolved ions (total salinity) are important environmental factors determining the occurrence of cyanobacteria in specific water and terrestrial environments. Because total salinity and water amount are closely linked, e.g. during desiccation, the amount of water is decreasing in parallel with the increase in total salt concentration, it is not surprising that salt and desiccation have common detrimental effects such as inhibition of photosynthesis (Yang, Wang, Liu, & Sui, 2020). In both cases, the maintenance of water and turgor pressure inside the cell is one of the central issues during the acclimation. Because water uptake is a passive process following the water potential gradient, growing cells need to establish a low water potential inside the cell relative to the surrounding medium. This is achieved by regulating the cellular osmotic potential via varying amounts of low molecular compounds. The main difference between pure water, or osmotic stress and salt stress is the additional direct ion effects on metabolic activities in the latter case. In nature, the large variations in the amount and composition of inorganic salts clearly affect the distribution of cyanobacteria. In addition to the problem that high total ion contents generally make it difficult to maintain water and turgor inside the cell, many ions are toxic for living cells. This direct toxicity is not only true for heavy metals, but for any ion at non-physiological high-cellular concentrations. Among cyanobacterial strains, three main salt tolerance groups can be distinguished: low and moderate halotolerant cyanobacteria as well as hypersaline strains. Regardless of the final salt tolerance, all cyanobacteria apply two basic strategies for a successful acclimation to high salt concentrations: accumulation of compatible solutes combined with active export of toxic ions, particularly Na^+ and Cl^- (Hagemann, 2013). These processes have been well-studied in the model cyanobacteria (Cui et al., 2020; Kirsch, Klähn, & Hagemann, 2019).

Maintaining the intracellular ion balance by expressing related-transporters is one important resistance strategy against salt stress. A series of salt-tolerance relevant transporters have been identified and characterized in cyanobacteria, such as different Na^+/H^+ antiporters, Mrp-system, and Na^+-pumping ATPase (Cui et al., 2020). It responds to increasing salinities mainly by accumulating glucosylglycerol (GG) and sucrose, whereas the (freshwater) model cyanobacterium *Synechococcus* PCC 7942 accumulates only sucrose (Liang et al., 2020).

Salt tolerance is also closely related to the primary compatible solutes in cyanobacteria. Early studies found that cyanobacterial strains with poor salt tolerance usually accumulate sucrose or trehalose, while strains with moderate salt tolerance accumulate glucosylglycerol (GG), and highly halotolerant strains synthesize the glycine betaine (GB), and glutamate betaine, respectively (Kirsch et al., 2019; Liang et al., 2020). High-salt induced accumulation of the main compatible solute GG is achieved by de novo synthesis. The key target of regulation is the enzyme GG-phosphate synthase (GgpS) and involves transcriptional, posttranscriptional, and biochemical mechanisms. Recently, the GG-degrading enzyme GG hydrolase A (GghA) was identified, which is particularly important for GG degradation during exposure to decreasing salinities. The inversely ion-regulated activities of GgpS and GghA could represent a main model for effectively tuning GG steady state levels according to external salinities. Similar to GG, the intracellular amount of sucrose is also salt-regulated and seems to be determined by the balance of sucrose synthesis via sucrose-phosphate synthase (Sps) and its degradation via invertase (Inv). In addition to their role as stress protectants, both compatible solutes also represent promising targets for biotechnology. Hence, the increasing knowledge on the regulation of compatible solute accumulation not only improves our understanding of the stress physiology of cyanobacteria but will also support their future biotechnological applications (Kirsch et al., 2019). Interestingly, a series of cyanobacteria are able to accumulate lipids and alkanes under salt stress. For example, lipid contents were enhanced in *Synechococcus* PCC 7942 under salt stress (Cui et al., 2020).

Another aspect of salt stress is that it increases the intracellular content of reactive oxygen species (ROS), which can damage the photosynthetic apparatus, inhibit the synthesis of photosystem-related proteins and destroy the thylakoid membrane structure, leading to inhibited photosynthesis (Yang et al., 2020). Consequently part of the cell responses to salt stress consist in an increase in the production of the superoxide dismutase (SOD) enzyme and the above-mentioned OCP (orange carotenoid) photoprotection protein (Yang et al., 2020). Furthermore, a series of secondary messengers have been recently identified, such as the cyclic-di-AMP and cyclic di-GMP, which play essential roles in cellular responses to environmental stresses (Cui et al., 2020).

In the third chapter entitled "Iron in Cyanobacteria" Nir Keren and Enrico Schleiff described the processes used by cyanobacteria to provide large amounts of the essential micronutrient iron (Fe) to their Fe-rich photosynthetic apparatus, and a wealth of other Fe-requiring enzymes crucial to cell metabolism (Kranzler, Rudolf, Keren, & Schleiff, 2013). Although Fe is the fourth most plentiful element in the Earth's crust, it is frequently a growth-limiting nutrient in large regions of the ocean and in many freshwater environments. This is due to both Fe concentration and chemistry. In aqueous solutions, iron has two environmentally-relevant oxidation states: Fe(II) and Fe(III). Prior to the evolution of oxygenic photosynthesis, reducing environmental conditions resulted in iron existing primarily in its reduced form, ferrous ions Fe(II). Fe(II) are relatively soluble and thereby readily bioavailable at the circumneutral pH range. Upon evolution of oxygenic photosynthesis, molecular oxygen accumulation in the atmosphere led to oxygenation of aquatic environments where Fe(II) species rapidly oxidized to ferric ions Fe(III). Fe(III) are poorly soluble at the circumneutral pH range and precipitate out of solution as ferric oxyhydroxides which are poorly bioavailable.

Fe, alone or complexed with sulfur under the form of iron–sulfur (Fe-S) clusters, serves as the cofactor of many (electron transfer) proteins involved in important physiological processes, including respiration, photosynthesis, nitrogen fixation, amino acid and purine metabolism, RNA modification and DNA replication, as well as repair and regulation of gene expression (Cassier-Chauvat & Chauvat, 2014; Gao, 2020). Hence, cyanobacteria in relying on photosynthesis to power up their whole metabolism have much larger iron requirements than non-photosynthetic organisms (Kranzler et al., 2013).

Cyanobacteria utilize multiple strategies to maintain Fe levels within a desired range. Depending on the chemistry and environmental bioavailability of Fe species, these processes include (i) synthesis, export and re-import of powerful ferric ion chelators called siderophores (Årstøl & Hohmann-Marriott, 2019); (ii) dedicated energy-consuming uptake systems; (iii) sequestration of temporary extra Fe atoms in storage proteins (ferritins) (Moparthi et al., 2016); (iii) replacing iron-containing enzymes with iron-free analogs; and (iv) degradation of abundant Fe-rich non-essential proteins in response to Fe starvation to release Fe atoms for subsequent incorporation into crucial Fe-requiring metabolic enzymes (Gao, 2020; Houot et al., 2007). These processes require a tight coordination with the homeostasis of all metal ions (Mg, Mn, Zn, Cu, etc.) that serve as cofactors for a wealth of metalloenzymes operating in cell metabolism, as occurs in most biological organisms (Foster, Osman, & Robinson, 2014; Salomon & Keren, 2015).

Siderophores are biologically produced low-molecular-weight (400–1000 kDA) compounds with an extremely high affinity for iron. Increased iron bioavailability can also be achieved by the complex formation of Fe(III) with common metabolites including citric acid and humic acids. Typically, siderophores form a strong hexadentate octahedral complex with ferric iron ($Fe3+$). This complex can then be transported inside the cell. Once inside the cell, iron has to be removed from siderophore. This can happen either through the reduction of iron from Fe(III) to Fe(II) by ferric siderophore reductases or by ferric siderophore hydrolases or by the photolysis of iron-bound siderophores, as described for synechobactin. Based on the primary oxygen-donating ligands that bind the iron, siderophores are divided into four different types. These are the hydroxamates, catecholates, and carboxylates, as well as siderophores with mixed types of these functional groups. All cyanobacterial siderophores that have been identified so far are either hydroxamates or catecholates, which have been named: synechobactin, schizokinen and anachelin (Årstøl & Hohmann-Marriott, 2019).

The only characterized siderophore export system in cyanobacteria is that of *Anabaena* PCC 7120. In this organism, the inner membrane MFS protein SchE (all4025) is required for the secretion of schizokinen, along with the TolC-like outer membrane protein hgdD (alr2887), with the two proteins likely working in tandem. The import of siderophore-iron complexes into the periplasmic space is enabled by receptor proteins, TonB-dependent transporters (TBDT) located in the outer membrane. In the periplasm, Fe-schizokinen is recognized and transported to the cytoplasm via the hydroxamate system FhuBCD (all0387, all0388, and all0389, respectively). TBDTs are unevenly distributed in cyanobacteria. In general, TBDTs seem to outnumber the amount of TonB-proteins, indicating some functional flexibility in TonB-complexes. The genome of *Anabaena* PCC 7120 contains 22 TBDTs, while no candidate genes were identified in the marine Prochlorococcus. *Synechococcus* PCC 7002, which synthesizes synechobactin, contains 6 TBDTs, with two exhibiting sequence similarity to resemble known schizokinen transporters, and two found more broadly homologous to hydroxamate transporters. The large number of TBDTs in some organisms is partially explained by the possibility of uptake of siderophores which are not produced by the organism (xenosiderophores), known as siderophore piracy. For example, the non-siderophore-producing *Synechocystis* PCC 6803 is capable of utilizing both schizokinen and the unnamed siderophore of *Anabaena variabilis* ATCC 29413 and can grow with iron-siderophore complexes as their only source of iron. (Årstøl & Hohmann-Marriott, 2019).

In the fourth chapter entitled "Functional Genomics of Metalloregulators in Cyanobacteria" Maria Fillat reviewed the major families of metalloregulators in cyanobacteria (Botello-Morte, González, Bes, Peleato, & Fillat, 2013). The major metal regulator system is composed of Fur-type regulators that comprise three paralogs named FurA (Fur, iron homeostasis), FurB (Zur, Zinc homeostasis) and FurC (PerR, responses to oxidative stress), which control the expression of a wealth of genes involved in exopolysaccharide biosynthesis, phycobilisome degradation, chlorophyll catabolism and nitrogen fixation (Cassier-Chauvat & Chauvat, 2016; González, Sevilla, Bes, Peleato, & Fillat, 2016; Rachedi, Foglino, & Latifi, 2020). These metal-sensing proteins, which harbor metal-binding and DNA-binding domains, are usually allosteric proteins. Their reversible interaction with the regulatory metal drives conformational changes that affect DNA binding and the subsequent regulation of various genes involved in various processes. These findings revealed interesting connections between metabolic networks and interactivity between major regulons. They provide a better understanding of cyanobacterial physiology including maintenance of metal homeostasis in a highly variable environment, and of the strategies to deal with different stress-generated cell damages, as follows. Shortly after the publication of vol 65, the group of Dr. Fillat has studied the molecular mechanism by which FurA from the cyanobacterium *Anabaena* PCC 7120 specifically binds to "iron boxes" in PfurA, FurA binds PfurA using a sequential mechanism: (i) a monomer specifically binds to an "iron box" and bends PfurA; (ii) two situations may occur, i.e. a second FurA monomer covers the free "iron box" or that joins to the previously used forming a dimer which would maintain the DNA kinked; (iii) trimerization in which the DNA is unbent; and (iv) finally undergoes a tetramerization; the next coming molecules covering the DNA strands un-specifically. In summary, the bending appears when an "iron box" is bound to one or two molecules and decreases when both "iron boxes" are covered. These results suggest that DNA bending contributes to the first steps of FurA repression promoting the recruitment of new molecules resulting in a fine regulation in the Fur-dependent cluster associated genes (Pallarés et al., 2014).

Besides the FUR regulators cyanobacteria have other metal regulator operating in the homeostasis of manganese (Eisenhut, 2020), nickel (Foster et al., 2017) or zinc (Pérez et al., 2017), or the tolerance to cadmium (Sun et al., 2017), mercury (Singh, Lingaswamy, Koduru, Nagu, & Jogadhenu, 2019), or the repair of Fe-S cluster (Vuorijoki, Tiwari, Kallio, & Aro, 2017).

In addition to transcription factors, many non-coding RNAs have been identified that play crucial roles in the acclimation responses of cyanobacteria to changes in the environment. Some of these regulatory RNAs are conserved throughout the cyanobacterial phylum, while others exist only in a few taxa (Muro-Pastor & Hess, 2020). A recent study on *Synechocystis* PCC 6803 showed that the gene expressing the small regulatory RNA Isar1 (Iron Stress-Activated RNA 1) is repressed by Fur. Isar1 accumulates in response to iron starvation and controls at the posttranscriptional level the expression of several genes involved in central cellular processes (photosynthesis, (iron–sulfur) cluster biogenesis, citrate cycle and tetrapyrrole biogenesis). Homologs of IsaR1 are largely conserved in the genomes of cyanobacteria, suggesting that the involvement of this riboregulator in the iron-dependent control of photosynthesis might be conserved in the cyanobacterial phylum (Rachedi et al., 2020).

In the fifth chapter entitled "Genomics of the Pleïotropic Glutathione System in Cyanobacteria" (Narainsamy, Marteyn, Sakr, Cassier-Chauvat, & Chauvat, 2013) we reported the variety of the processes used by cyanobacteria to protect themselves against oxidative stress, emphasizing on glutathione (the gamma-glutamyl-cysteinyl-glycine tripeptide, GSH) and glutathione-dependent enzymes, because they are well conserved during evolution (Cassier-Chauvat & Chauvat, 2016; Fahey, 2013). Because of their photoautotrophic lifestyle, cyanobacteria are continuously challenged with toxic reactive oxygen species (ROS) present in our oxygenic atmosphere (ozone, O_3), or generated by the metal-dependent cellular processes photosynthesis, respiration and cell metabolism. These oxidative agents are namely: singlet oxygen ($^1O_2$), the superoxide anion ($O2^{\cdot-}$), hydrogen peroxide (H_2O_2), and the hydroxyl radical ($OH^{\bullet}$). The ROS oxidants can be detoxified by various metabolites (ascorbate, carotenoids, glutathione, vitamins, etc.) and several enzymes, such as the superoxide dismutase (SOD), catalase and peroxidase, which sequentially convert the superoxide anion to hydrogen peroxide (SOD) and then to water (catalase and peroxidase). Glutathione occurs under two forms: the reduced (GSH) and the oxidized, dimeric (GSSG) forms. GSH, the major form, maintains the intracellular compartment in a reduced state. GSH can form a Fe(II)GSH complex, which likely supplies Fe for the synthesis of Fe cofactor or (Fe-S) clusters of metalloenzymes involved in photosynthesis, respiration or other cellular functions (Berndt & Lillig, 2017). GSSG, the oxidized form of glutathione generated upon cell detoxication by GSH, can be reduced back to GSH by various factors, such as the NADPH-using enzyme glutathione reductase

(GR) that occurs in many, but not all organisms (for example some cyanobacteria have no GR) (Narainsamy et al., 2013). In addition, GSH serves in the biosynthesis of ergothioneine (EGT; 2-mercaptohistidine trimethylbetaine), which has been proposed to function as a stable antioxidant and cytoprotectant (Narainsamy et al., 2016).

Among ROS-generated damages, cysteines can be oxidized to form sulfenic acid (-SOH) and disulfide (-S-S-) by a two-electrons transition; sulfinic acid (-SO_2H) by a four-electrons transition; and eventually sulfonic acid (-SO_3H) by a six-electrons transition. In *Synechocystis* PCC 6803, the redox state of ~2100 Cys residues from 1060 proteins were found to be affected by light/dark cycles (Guo et al., 2014). Two types of disulfides can be distinguished considering whether they link two cysteinyl residues of, the same or different, proteins (disulfide bridges); or from a protein and a molecule of GSH (glutathione-protein mix disulfide, also termed glutathionylation). The glutathionylation process is regarded as a transient protection of critical cysteines against irreversible oxidation (sulfinic and sulfonic acids) during oxidative stress, and as a post-translational regulatory modification. The reverse reaction, deglutathionylation is catalyzed by the NAD(*P*)H-dependent oxidoreductase thioredoxins (Trx), or the GSH-dependent enzymes GST or glutaredoxins (Grx). Using the well-studied unicellular strain *Synechocystis* PCC 6803 as a model, we recently identified 383 proteins that can be glutathionylated in vitro (Chardonnet et al., 2015). These glutathionylable proteins participate in a wide range of cellular processes and metabolic pathways, such as carbon and nitrogen metabolisms, cell division, stress responses and hydrogen production. The glutathionylation of several proteins, namely, the antioxidant peroxiredoxin II, the mercuric reductase, the AbrB2 transcription factor, and the 3-phosphoglycerate dehydrogenase, was confirmed by biochemical studies (Chardonnet et al., 2015).

Very little was known about GSTs in cyanobacteria, even though these prokaryotes are regarded as the inventor of the oxygen-evolving photosynthesis, as well as the GSH and the GSH-dependent enzymes, such as GST, to protect themselves against the ROS that they massively produce by their active photosynthesis (Fahey, 2013; Hamilton, Bryant, & Macalady, 2016; Narainsamy et al., 2016; Schopf, 2011). Attesting to the importance of GST in cyanobacteria, the well-studied model *Synechocystis* PCC 6803 possesses six GSTs: Sll0067, Sll1147, Sll1545, Sll1902, Slr0236, and Slr0605 (Kammerscheit, Chauvat, & Cassier-Chauvat, 2019a; Pandey, Chhetri, et al., 2015; Pandey et al., 2017; Pandey, Singh, Chhetri, Tripathi, &

Singh, 2015). We showed that both Sll1545 and Slr0236 operate in the protection against photo-oxidative stress triggered by high light or H_2O_2 (Kammerscheit et al., 2019a) and that Sll1147 and its human orthologs play a prominent role in the tolerance to membrane stresses triggered by heat, cold or lipid peroxidation (Kammerscheit, Chauvat, & Cassier-Chauvat, 2019b). We also reported that Sll0067 catalyzes the conjugation of GSH with MG, the first step in the detoxification of methylglyoxal (MG), a toxic metabolite by-product that causes diabetes in humans, by the Glx enzymes. These findings are important because the conjugation of GSH with MG is always described as spontaneous (non-enzymatic) in all organisms (Kammerscheit, Hecker, Rouhier, Chauvat, & Cassier-Chauvat, 2020).

In addition, ROS species, more particularly H_2O_2, can also operate in signaling, which is an important physiological process. Indeed, H_2O_2 possesses the required properties to be a secondary messenger in being enzymatically produced and degraded by the SOD and catalase enzymes, respectively. Furthermore, H_2O_2 oxidizes protein thiols in disulfides, which can be reduced back to thiols are thereby relevant as thiol redox switches for signaling (Mironov, Sinetova, Shumskaya, & Los, 2019).

Concerning the Grx enzymes, the analysis of 503 amino acid sequences from 167 cyanobacterial species led to the identification of four Grx classes (Mondal, Kumar, & Singh, 2020). Class I and II Grxs are single module oxidoreductase while class III and IV Grxs are absent in cyanobacteria. Class V and VI Grxs are multimodular proteins having additional modules at their C-terminal and N-terminal end, respectively. Furthermore, class VI Grxs are exclusively present in marine cyanobacteria. All four classes of cyanobacterial Grxs comprise several subgroups having variable dithiol (CxxC) and/or monothiol (Cxxx) catalytic active site motives (Mondal, Kumar, & Singh, 2020). However, class II Grxs possess the CGFS-type highly conserved monothiol catalytic active site motif (Daniel, Faruq, Laura Magdalena, Manuela, & Christopher Horst, 2020; Picciocchi, Saguez, Boussac, Cassier-Chauvat, & Chauvat, 2007).

The sixth chapter entitled "A Genomic View of Secondary Metabolite Production in Cyanobacteria" written by Annick Mejean and Olivier Ploux reviewed the connections between cyanobacterial secondary metabolites and their biosynthetic genes, with emphasis on the most significant cases like cyanotoxins, sunscreens and alkanes (Méjean & Ploux, 2013).

An important aspect of cyanobacteria is that they produce a wide range of secondary metabolites with diverse chemical structures and biological activities (vitamins, sunscreens, antibiotics, anti-cancer, toxins, *etc.*) (Cassier-Chauvat, Dive, & Chauvat, 2017; Demay, Bernard, Reinhardt, & Marie, 2019; Jeong et al., 2020; Mondal et al., 2020). For instance, at least 800 different secondary metabolites have been identified in marine cyanobacteria so far, a number likely representing a small fraction of the natural product repertoire. The functions of most of these secondary metabolites are usually unknown, but it is assumed that their production gives some advantage to the producers in complex ecosystems. It has also been proposed that these molecules might be communication molecules although there are no firm experimental data on this issue. Some of the cyanobacterial secondary metabolites appeared to be cytotoxic, neurotoxic or dermatotoxic to animals and/or humans. In the past 10 years, many biosynthesis of cyanobacterial secondary metabolites have been deciphered, at the genetic and biochemical level. Thanks to the advent of genomic data on cyanobacterial genomes and to new powerful bioinformatic tools, about 30 clusters of genes responsible for the production of cyanobacterial secondary metabolites have been identified, including the cyanotoxins: microcystin, cylindrospermopsin, saxitoxin and anatoxin-a. Almost all cyanobacterial secondary metabolites are the products of polyketide synthases, non- ribosomal peptide synthases or hybrid thereof. However, ribosomal peptides are also produced by cyanobacteria, like the cyanobactins and recent genome mining data suggest that these metabolites are more represented than initially thought in cyanobacteria.

Recently, genome sequencing and mining has revealed that natural cyanobacteria have the capacity to produce many more secondary metabolites than have been characterized (Cassier-Chauvat et al., 2017; Pearson et al., 2016). Most cyanobacterial natural products are non-ribosomal peptides (NRPS, such as lyngbyatoxin), polyketides (PKS, such as anatoxin-A), or hybrid peptide–polyketide compounds (for examples apratoxin, curacin A, microcystin), which are sequentially synthesized by NRPS and/or PKS enzymes. For example, anatoxin-A, the potent cyanobacterial neurotoxins are biosynthesized in cyanobacteria from proline and acetate by a pathway involving three polyketide synthases (Kust, Méjean, & Ploux, 2020). The corresponding assembly lines comprise several modules, each carrying out the addition of a single substrate to the lengthening molecule. Other cyanobacterial secondary metabolites are peptides synthesized by ribosomes and post-translationally modified, such as the *N*-acetylated tri- and tetradecapeptides microviridins (serine protease inhibitors) produced by *Anabaena*, *Microcystis*, *Nodularia*, *Nostoc*, and *Planktothrix* (Dittmann, Gugger, Sivonen, & Fewer, 2015; Pearson et al., 2016). These cascades of modular synthesis, assembly and modification activities produce a large array of secondary metabolites, some of which occurring in multiple variants (100 variants in the case of microcystin). Cyanobacteria use about 5–6% of their genome length to direct the production of secondary metabolites. For example, it is predicted that the genome of *Moorea producens* PAL-8-15-08-1 carries 18 NRPS Biosynthetic Gene Clusters (Jeong et al., 2020). These gene clusters are often large (Cassier-Chauvat et al., 2017; Dittmann et al., 2015; Pearson et al., 2016). For example, the size of the *Moorea producens* 3L (formerly *Lyngbya majuscula* 3L) gene clusters encoding curacin A (an anti-cancer inhibiting tubulin polymerization) and barbamide (a molluscicidal metabolite) are 64 and a 26 kb longs, respectively. The synthesis of aeruginosin (a protease inhibitor) is encoded by a 34-kb gene cluster in *Planktothrix* and *Microcystis*, while the synthesis of saxitoxin (a neurotoxic alkaloid, which blocks the sodium (Na+) channels in shellfish) is directed by a 25.7 kb gene cluster in *Raphidiopsis brookii* D9. The corresponding panoply of enzymes (polyketide synthases and non-ribosomal peptide synthases) of interest for synthetic biology can still be increased through gene manipulations with the tools available for the few genetically manipulable strains (Cassier-Chauvat et al., 2017; Jeong et al., 2020). Thus, cyanobacteria, which combine the ability to synthesize a vast array of bioactive molecules with powerful photosynthesis, radiation resistance and effective synthetic biology tools for several model strains, are well suited for future cheap production of a vast array of uniformly ^{14}C-labeled bioactive compounds, in order to screen for new drugs and analyze their absorption, distribution, metabolism, and excretion properties (Cassier-Chauvat et al., 2017).

In the seventh chapter entitled "Assembly and Export of Extracellular Polymeric Substances (EPS) in Cyanobacteria: A Phylogenomic Approach" the groups of Roberto De Philippis and Paula Tamagnini provided a first insight on the phylogenetic history of the EPS-related genes, and constitute a robust basis for subsequent studies aiming at optimizing EPS production in cyanobacteria (Pereira, Mota, Santos, De Philippis, & Tamagnini, 2013). In addition to secondary metabolites, many cyanobacteria produce extracellular polymeric substances (EPSs), mainly composed of polysaccharides, which can remain associated to the cells (capsular EPS) or be released into the surrounding environment (released EPS).

As the primary colonizers of constrained environments, such as desert soils and lithic and exposed substrates, cyanobacteria are the first contributors to the synthesis of the EPSs constituting the extracellular polymeric matrix, frequently associated with peptides, DNA and fatty acids, which favors the formation of microbial associations with varying levels of complexity called biofilms (Rossi & De Philippis, 2015). Thus, cyanobacteria play key roles in soil properties and functions. Cyanobacterial biofilms bind soil aggregates and create a stable surface layer that facilitates the path for colonization by other biocrust organisms such as lichens and mosses (Chamizo, Adessi, Torzillo, & De Philippis, 2020).

Thanks to their negatively-charged surface, cyanobacterial EPSs usually show a high affinity for metal cations and other positively-charged or polar molecules (Bhunia et al., 2018; Jittawuttipoka et al., 2013; Planchon et al., 2013b). On the other hand, the presence of deoxy-sugars (rhamnose and fucose), peptides and ester-linked acetyl groups can also confer a contemporary hydrophobic character, significantly affecting their rheological and emulsifying properties, as well as the capability to adhere to solid surfaces (Planchon et al., 2013a; Rossi & De Philippis, 2015; Vourc'H et al., 2018). Sugars commonly found in cyanobacterial EPS include the hexoses glucose, galactose, mannose, and fructose; the pentoses ribose, xylose, and arabinose; the deoxyhexoses fucose and rhamnose; the acidic hexoses glucuronic and galacturonic acids, the amino sugars glucosamine, galactosamine, *N*-acetyl galactosamine, and *N*-acetyl glucosamine (Bhunia et al., 2018). Many cyanobacterial EPS also possess two different uronic acids (from 2% up to 80% of the total EPS dry weight, commonly between 15% and 30%), which is a rare feature in microbial EPS. In addition, they usually contain sulfate groups, which are usually present in the EPS produced by archaea and eukaryotic EPS but absent in those produced by bacteria. These last two features, and the presence of numerous different monosaccharides (up to 13) contributing to the overall anionic charge of the cyanobacterial polymers, making them suitable for a vast array of biotechnological applications, such as metal chelation, flocculating, emulsifying or rheology agents with valuable biological activities (e.g., antiviral, antimicrobial, anticoagulant, antitumor) (Pereira et al., 2019; Rossi & De Philippis, 2015).

The mechanisms of EPS assembly and export are relatively conserved, and require the involvement of polysaccharide copolymerase and outer membrane polysaccharide export proteins. In cyanobacteria, the genes encoding these proteins occur in multiple copies, scattered throughout the genome, either isolated or in small clusters. It is also necessary to identify other EPS production genes, and understand their genomic distribution and regulation of expression (Allen, Rittmann, & Curtiss, 2019; Cui et al., 2018; Jittawuttipoka et al., 2013; Ohki et al., 2019). To better exploit the biotechnological potential of cyanobacterial EPS, there are still challenges to address, including (i) a better understanding of the relationship between central metabolism and the synthesis of sugar nucleotides, (ii) the identification and characterization of other key components of the EPS production machinery, and (iii) elucidation of the regulatory networks of the EPS production process (Bhunia et al., 2018; Pereira et al., 2019).

Last, but not least, many cyanobacteria display multicellularity and cell differentiation. In the eighth chapter entitled "Gene Expression during Heterocyst Differentiation," Antonia Herrero and Enrique Flores summarized the biochemical and morphological properties of the heterocysts and the program of gene expression that supports the process of differentiation and its regulation (Herrero, Picossi, & Flores, 2013). Many cyanobacteria can also fix atmospheric dinitrogen (N_2). Because the N_2-fixing enzyme nitrogenase is irreversibly inactivated by O_2, the N_2-fixing (diazotrophic) cyanobacteria developed several strategies to protect their nitrogenase from O_2. Unicellular cyanobacteria (*Chrocosphera*, *Gloeothece* and *Cyanothece*) temporally separate photosynthesis (production of O_2) during the day from N_2 fixation during the night (when O_2 is no longer produced by photosynthesis and is consumed by respiration). Many diazotrophic pluricellular (filamentous) cyanobacteria, such as *Nostoc* (*Anabaena*) sp., respond to nitrogen depletion by differentiating (irreversibly) some of their vegetative cells into heterocysts, the non-diving, photosynthetically inactive, cells that fix atmospheric (inorganic) nitrogen (N_2) (Harish & Kunal, 2020; Herrero et al., 2013). This differentiation process results in a heterocyst pattern consisting of two heterocysts separated by about 10 to 15 vegetative cells (Herrero, Stavans, & Flores, 2016). The thick-walled heterocyst cells provide a micro-oxic environment for the oxygen-sensitive nitrogenase enzyme to perform nitrogen fixation. Vegetative cells and heterocysts exchange metabolites, with heterocysts providing vegetative cells with fixed nitrogen, while vegetative cells provide heterocysts with photosynthates.

Recently, intercellular molecular exchange was found to occur by simple diffusion through proteinaceous structures, known as septal junctions, which connect the adjacent cells in the filament and traverse the septal peptidoglycan through perforations known as nanopores. Proteins that are necessary to produce, and that may be components of, the septal junctions-SepJ, FraC and FraD-have been identified in the heterocyst-forming cyanobacterium *Anabaena* PCC 7120 model. Additionally, several proteins that are necessary to produce a normal number of nanopores and functional septal junctions have been identified, including AmiC-type amidases, peptidoglycan-binding proteins and some membrane transporters.

Recently, two novel ABC transporter components involved in the formation of the glycolipid layer of the heterocyst envelope have been identified (Herrero & Flores, 2019). Available reports and re-evaluation of intercellular molecular transfer data for some mutants of *Anabaena* suggest that the septal junctions can be regulated, likely by a mechanism of gating (Flores, Nieves-Morión, & Mullineaux, 2019). The cyanobacterial septal junctions resemble the metazoan gap junctions, and more precisely their constituent connexons, which are protein complexes that allow the diffusion of small compounds and ions between cells and can be regulated by gating. Septal and gap junctions represent therefore a wonderful example of convergent evolution, in which different proteins and structures perform the same function (cell–cell communication) in extremely distant organisms (Flores, Nieves-Morión, & Mullineaux, 2019).

At the metabolite level, heterocyst differentiation starts as a response to a persistent high-cellular carbon-to-nitrogen balance signaled by the level of the 2-oxoglutarate. In turn, this results in activation of the global transcriptional CRP-family regulator NtcA followed by increased expression, mainly localized to differentiating cells, of the genes *ntcA* and *hetR*, which encodes HetR, the differentiation-specific transcription factor. The expression of genes encoding proteins that transform the vegetative cell into a heterocyst is then activated with a spatiotemporal specificity to produce a mature functional heterocyst. A few other regulators of gene expression participate at specific differentiation steps, and a specific transcription factor, CnfR, activates *nif* genes expression under the micro-oxic conditions of the heterocyst (Flores, Picossi, Valladares, & Herrero, 2019). Recently, the role of small RNA and interruption DNA elements that influence the heterocyst formation and function has also been identified (Harish & Kunal, 2020).

Heterocysts depend on a wide variety of metals to maintain their specific functions and cellular metabolism. N_2 fixation requires a significant number of metalloproteins of the photosynthetic and respiratory metabolism, which have high Fe and Cu requirements, to provide energy and electrons to this essential biological process for the nitrogen cycle. Nitrogenases contain 38 atoms of Fe and either two Molybdenum (Mo) or two vanadium (V) atoms. Other metalloenzymes are crucial to manage efficiently reactive oxygen species generated during the metabolism of heterocysts, while some metalloproteins are especially important in Fe storage, which is the most required transition metal in heterocysts. These sets of metalloproteins are aimed at creating the optimal working conditions for the central metabolic enzyme of heterocysts, the nitrogenase complex, but also associated hydrogenases required to metabolize efficiently the H_2 byproduct generated during the N_2 reduction reaction to recover reducing equivalents that, otherwise, would be wasted. Moreover, the uptake hydrogenase accounts for 12 atoms of Fe and one Ni atom, while the bidirectional hydrogenase requires 27 atoms of Fe and one Ni atom. The biological relevance of the metals Fe, Cu, Mo, Mn, Ni, V and Zn metals in heterocysts and their use by metalloproteins has been recently discussed (Pernil & Schleiff, 2019).

Finally, the oxygen-free environment offered by the heterocysts, in combination with an increasing arsenal of molecular biology tools for heterocyst-specific expression, has led some authors to propose that heterocysts could be exploited as cell factories for the production of hydrogen and/or chemicals (Avilan et al., 2018; Liu, Ni, Xu, & Tao, 2018; Magnuson, 2019).

1. Concluding remarks

Once again, we express our gratitude to all contributing authors of the previous vol 65 book for their dedicated efforts to clearly expose their fields of expertise to researchers and students. We noticed with great pleasure that these authors have made great contributions to the important advancement of the knowledge on cyanobacteria presented here. We also encourage cyanobacterial researcher to study new cyanobacteria endowed with specific morphological and/or physiological interests in comparison to model cyanobacteria, in order to better analyze the large biodiversity of cyanobacteria, and deploy them as industrial cell factories for the production of a wide range of chemicals in local industrial sites (Chenebault et al., 2020; Mukherjee, Madhu, & Wangikar, 2020). In this prospect it will be important to determine if, and when, the required genetic toolbox developed for the few well-studied cyanobacterial models can be adapted to the "new" cyanobacteria emerging from such investigations (Behler, Vijay, Hess, & Akhtar, 2018; Gale et al., 2019; Yao et al., 2020).

Acknowledgment

We thank Jean-Pierre Jacquot for critical reading of our manuscript.

References

Allen, R., Rittmann, B. E., & Curtiss, R. (2019). Axenic biofilm formation and aggregation by synechocystis sp. Strain PCC 6803 are induced by changes in nutrient concentration and require cell surface structures. *Applied and Environmental Microbiology*, *85*(7). e02192–18.

Årstøl, E., & Hohmann-Marriott, M. F. (2019). Cyanobacterial siderophores—Physiology, structure, biosynthesis, and applications. *Marine Drugs*, *17*(5), 281.

Avilan, L., Roumezi, B., Risoul, V., Bernard, C. S., Kpebe, A., Belhadjhassine, M., et al. (2018). Phototrophic hydrogen production from a clostridial [FeFe] hydrogenase expressed in the heterocysts of the cyanobacterium *Nostoc* PCC 7120. *Applied Microbiology and Biotechnology*, *102*, 5775–5783.

Behler, J., Vijay, D., Hess, W. R., & Akhtar, M. K. (2018). CRISPR-based technologies for metabolic engineering in cyanobacteria. *Trends in Biotechnology*, *36*, 996–1010.

Berndt, C., & Lillig, C. H. (2017). Glutathione, glutaredoxins, and iron. *Antioxidants and Redox Signaling*, *27*(15), 1235–1251.

Bhunia, B., Prasad Uday, U. S., Oinam, G., Mondal, A., Bandyopadhyay, T. K., & Tiwari, O. N. (2018). Characterization, genetic regulation and production of cyanobacterial exopolysaccharides and its applicability for heavy metal removal. *Carbohydrate Polymers*, *179*(1), 228–243.

Botello-Morte, L., González, A., Bes, M. T., Peleato, M. L., & Fillat, M. F. (2013). Functional genomics of metalloregulators in cyanobacteria. In *Advances in Botanical Research: vol 65. Genomics of Cyanobacteria* (pp. 107–156). Academic Press.

Cassier-Chauvat, C., & Chauvat, F. (2014). Function and regulation of ferredoxins in the cyanobacterium, *Synechocystis* PCC6803: Recent advances. *Life*, *4*(4), 666–680.

Cassier-Chauvat, C., & Chauvat, F. (2016). Genomics of the resistance to metal and oxidative stresses in cyanobacteria. In F. J. de Bruijn (Ed.), *Vol. 2. Stress and environmental regulation of gene expression and adaptation in bacteria* (pp. 1154–1164). John Wiley & Sons, Inc. Chapter 20.3.

Cassier-Chauvat, C., Dive, V., & Chauvat, F. (2017). Cyanobacteria: Photosynthetic factories combining biodiversity, radiation resistance, and genetics to facilitate drug discovery. *Applied Microbiology and Biotechnology*, *101*(4), 1359–1364.

Chamizo, S., Adessi, A., Torzillo, G., & De Philippis, R. (2020). Exopolysaccharide features influence growth success in biocrust-forming cyanobacteria, moving from liquid culture to sand microcosms. *Frontiers in Microbiology*, *11*, 568224.

Chardonnet, S., Sakr, S., Cassier-Chauvat, C., Le Maréchal, P., Chauvat, F., Lemaire, S. D., et al. (2015). First proteomic study of S-glutathionylation in cyanobacteria. *Journal of Proteome Research*, *14*(1), 59–71.

Chenebault, C., Diaz-Santos, E., Kammerscheit, X., Görgen, S., Ilioaia, C., Streckaite, S., et al. (2020). A genetic toolbox for the new model cyanobacterium Cyanothece PCC 7425: A case study for the photosynthetic production of limonene. *Frontiers in Microbiology*, *11*, 586–601.

Cui, L., Liu, Y., Yang, Y., Ye, S., Luo, H., Qiu, B., et al. (2018). The drnf1 gene from the drought-adapted cyanobacterium *Nostoc flagelliforme* improved salt tolerance in transgenic *Synechocystis* and *Arabidopsis* plant. *Genes*, *9*(9), 441.

Cui, J., Sun, T., Chen, L., & Zhang, W. (2020). Engineering salt tolerance of photosynthetic cyanobacteria for seawater utilization. *Biotechnology Advances*, *43*, 107578.

Daniel, T., Faruq, H. M., Laura Magdalena, J., Manuela, G., & Christopher Horst, L. (2020). Role of GSH and iron-sulfur glutaredoxins in iron metabolism—Review. *Molecules (Basel, Switzerland)*, *27*(17), 3860.

Demay, J., Bernard, C., Reinhardt, A., & Marie, B. (2019). Natural products from cyanobacteria: Focus on beneficial activities. *Marine Drugs*, *17*(6), 320.

Dittmann, E., Gugger, M., Sivonen, K., & Fewer, D. P. (2015). Natural product biosynthetic diversity and comparative genomics of the cyanobacteria. *Trends in Microbiology*, *23*(10), 642–652.

Dominguez-Martin, M. A., & Kerfeld, C. A. (2019). Engineering the orange carotenoid protein for applications in synthetic biology. *Current Opinion in Structural Biology*, *57*, 110–117.

Eisenhut, M. (2020). Manganese homeostasis in cyanobacteria. *Plants*, *9*(1), 18.

Fahey, R. C. (2013). Glutathione analogs in prokaryotes. *Biochimica et Biophysica Acta – General Subjects*, *1830*, 3182–3198.

Flores, E., Nieves-Morión, M., & Mullineaux, C. W. (2019). Cyanobacterial septal junctions: Properties and regulation. *Life*, *9*(1), 1.

Flores, E., Picossi, S., Valladares, A., & Herrero, A. (2019). Transcriptional regulation of development in heterocyst-forming cyanobacteria. *Biochimica et Biophysica Acta – Gene Regulatory Mechanisms*, *1862*(7), 673–684.

Foster, A. W., Osman, D., & Robinson, N. J. (2014). Metal preferences and metallation. *Journal of Biological Chemistry*, *289*(41), 28095–28103.

Foster, A. W., Pernil, R., Patterson, C. J., Scott, A. J. P., Pålsson, L. O., Pal, R., et al. (2017). A tight tunable range for Ni(II) sensing and buffering in cells. *Nature Chemical Biology*, *13*, 409–414.

Gale, G. A. R., Osorio, A. A. S., Mills, L. A., Wang, B., Lea-Smith, D. J., & McCormick, A. J. (2019). Emerging species and genome editing tools: Future prospects in cyanobacterial synthetic biology. *Microorganisms*, 7(10), 409.

Gao, F. (2020). Iron–sulfur cluster biogenesis and iron homeostasis in cyanobacteria. *Frontiers in Microbiology*, *11*, 165.

González, A., Sevilla, E., Bes, M. T., Peleato, M. L., & Fillat, M. F. (2016). Pivotal role of Iron in the regulation of cyanobacterial electron transport. *Advances in Microbial Physiology*, *68*, 169–217.

Guo, J., Nguyen, A. Y., Dai, Z., Su, D., Gaffrey, M. J., Moore, R. J., et al. (2014). Proteome-wide light/dark modulation of thiol oxidation in cyanobacteria revealed by quantitative site-specific redox proteomics. *Molecular and Cellular Proteomics*, *13*(12), 3270–3285.

Gupta, S., Sutter, M., Remesh, S. G., Dominguez-Martin, M. A., Bao, H., Feng, X. A., et al. (2019). X-ray radiolytic labeling reveals the molecular basis of orange carotenoid protein photoprotection and its interactions with fluorescence recovery protein. *Journal of Biological Chemistry*, *294*(22), 8848–8860.

Hagemann, M. (2013). Genomics of salt acclimation: Synthesis of compatible solutes among cyanobacteria. In *Advances in Botanical Research: vol 65. Genomics of Cyanobacteria* (pp. 27–55). Academic Press.

Hamilton, T. L., Bryant, D. A., & Macalady, J. L. (2016). The role of biology in planetary evolution: Cyanobacterial primary production in low-oxygen Proterozoic oceans. *Environmental Microbiology*, *18*(2), 325–340.

Harish, & Kunal, S. (2020). Molecular circuit of heterocyst differentiation in cyanobacteria. *Journal of Basic Microbiology*, *60*(9), 738–745.

Herrero, A., & Flores, E. (2019). Developmental biology in cyanobacteria. *Life*, *9*(2), 39.

Herrero, A., Picossi, S., & Flores, E. (2013). Gene expression during heterocyst differentiation. In *Advances in Botanical Research: vol 65. Genomics of Cyanobacteria* (pp. 281–329). Academic Press.

Herrero, A., Stavans, J., & Flores, E. (2016). The multicellular nature of filamentous heterocyst-forming cyanobacteria. *FEMS Microbiology Reviews*, *40*(6), 831–854.

Houot, L., Floutier, M., Marteyn, B., Michaut, M., Picciocchi, A., Legrain, P., et al. (2007). Cadmium triggers an integrated reprogramming of the metabolism of *Synechocystis* PCC6803, under the control of the Slr1738 regulator. *BMC Genomics*, *8*, 350.

Jeong, Y., Cho, S. H., Lee, H., Choi, H. K., Kim, D. M., Lee, C. G., et al. (2020). Current status and future strategies to increase secondary metabolite production from cyanobacteria. *Microorganisms*, *8*(12), 1849.

Jittawuttipoka, T., Planchon, M., Spalla, O., Benzerara, K., Guyot, F., Cassier-Chauvat, C., et al. (2013). Multidisciplinary evidences that *Synechocystis* PCC6803 exopolysaccharides operate in cell sedimentation and protection against salt and metal stresses. *PLoS One*, *8*(2), e55564.

Kammerscheit, X., Chauvat, F., & Cassier-Chauvat, C. (2019a). First in vivo evidence that glutathione-s-transferase operates in photo-oxidative stress in cyanobacteria. *Frontiers in Microbiology, 10,* 1899.

Kammerscheit, X., Chauvat, F., & Cassier-Chauvat, C. (2019b). From cyanobacteria to human, MAPEG-type glutathione-S-transferases operate in cell tolerance to heat, cold, and lipid peroxidation. *Frontiers in Microbiology, 10*(2248), 2248.

Kammerscheit, X., Hecker, A., Rouhier, N., Chauvat, F., & Cassier-Chauvat, C. (2020). Methylglyoxal detoxification revisited: Role of glutathione transferase in model cyanobacterium *Synechocystis* sp. strain PCC 6803. *MBio, 11*(4).

Kerfeld, C. A., & Kirilovsky, D. (2013). Structural, mechanistic and genomic insights into OCP-mediated photoprotection. In *Advances in Botanical Research: vol 65. Genomics of Cyanobacteria* (pp. 1–26). Academic Press.

Kirsch, F., Klähn, S., & Hagemann, M. (2019). Salt-regulated accumulation of the compatible solutes sucrose and glucosylglycerol in cyanobacteria and its biotechnological potential. *Frontiers in Microbiology, 10,* 2139.

Konold, P. E., Van Stokkum, I. H. M., Muzzopappa, F., Wilson, A., Groot, M. L., Kirilovsky, D., et al. (2019). Photoactivation mechanism, timing of protein secondary structure dynamics and carotenoid translocation in the orange carotenoid protein. *Journal of the American Chemical Society, 141*(1), 520–530.

Kranzler, C., Rudolf, M., Keren, N., & Schleiff, E. (2013). Iron in cyanobacteria. In *Advances in Botanical Research: vol 65. Genomics of Cyanobacteria* (pp. 57–105). Academic Press.

Kust, A., Méjean, A., & Ploux, O. (2020). Biosynthesis of anatoxins in cyanobacteria: Identification of the carboxy-anatoxins as the penultimate biosynthetic intermediates. *Journal of Natural Products, 83*(1), 142–151.

Liang, Y., Zhang, M., Wang, M., Zhang, W., Qiao, C., Luo, Q., et al. (2020). Freshwater cyanobacterium *Synechococcus elongatus* PCC 7942 adapts to an environment with salt stress via ion-induced enzymatic balance of compatible solutes. *Applied and Environmental Microbiology, 86*(7), e02904–e02919.

Liu, H., Ni, J., Xu, P., & Tao, F. (2018). Enhancing light-driven 1,3-propanediol production by using natural compartmentalization of differentiated cells. *ACS Synthetic Biology, 7*(10), 2436–2446.

Magnuson, A. (2019). Heterocyst thylakoid bioenergetics. *Life, 9*(1).

Méjean, A., & Ploux, O. (2013). A genomic view of secondary metabolite production in cyanobacteria. *Advances in Botanical Research,* 189–234.

Mironov, K. S., Sinetova, M. A., Shumskaya, M., & Los, D. A. (2019). Universal molecular triggers of stress responses in cyanobacterium *Synechocystis*. *Life, 9*(3), 67.

Mondal, A., Bose, S., Banerjee, S., Patra, J. K., Malik, J., Mandal, S. K., et al. (2020). Marine cyanobacteria and microalgae metabolites—A rich source of potential anticancer drugs. *Marine Drugs, 18*(9), 476.

Mondal, S., Kumar, V., & Singh, S. P. (2020). Phylogenetic distribution and structural analyses of cyanobacterial glutaredoxins (Grxs). *Computational Biology and Chemistry, 84.*

Moparthi, V. K., Li, X., Vavitsas, K., Dzhygyr, I., Sandh, G., Magnuson, A., et al. (2016). The two Dps proteins, NpDps2 and NpDps5, are involved in light-induced oxidative stress tolerance in the N2-fixing cyanobacterium *Nostoc punctiforme*. *Biochimica et Biophysica Acta – Bioenergetics, 1857*(11), 1766–1776.

Mukherjee, B., Madhu, S., & Wangikar, P. P. (2020). The role of systems biology in developing non-model cyanobacteria as hosts for chemical production. *Current Opinion in Biotechnology, 64,* 62–69.

Muro-Pastor, A. M., & Hess, W. R. (2020). Regulatory RNA at the crossroads of carbon and nitrogen metabolism in photosynthetic cyanobacteria. *Biochimica et Biophysica Acta – Gene Regulatory Mechanisms, 1863*(1), 194477.

Muzzopappa, F., & Kirilovsky, D. (2020). Changing color for photoprotection: The orange carotenoid protein. *Trends in Plant Science, 25*(1), 92–104.

Narainsamy, K., Farci, S., Braun, E., Junot, C., Cassier-Chauvat, C., & Chauvat, F. (2016). Oxidative-stress detoxification and signalling in cyanobacteria: The crucial glutathione synthesis pathway supports the production of ergothioneine and ophthalmate. *Molecular Microbiology*, *100*(1), 15–24.

Narainsamy, K., Marteyn, B., Sakr, S., Cassier-Chauvat, C., & Chauvat, F. (2013). Genomics of the pleïotropic glutathione system in cyanobacteria. In *Advances in Botanical Research: Vol. 65. Genomics of Cyanobacteria* (pp. 157–188). Academic Press.

Ohki, K., Kanesaki, Y., Suzuki, N., Okajima, M., Kaneko, T., & Yoshikawa, S. (2019). Physiological properties and genetic analysis related to exopolysaccharide (EPS) production in the fresh-water unicellular cyanobacterium aphanothece sacrum (suizenji nori). *Journal of General and Applied Microbiology*, *65*(1), 39–46.

Pallarés, M. C., Marcuello, C., Botello-Morte, L., González, A., Fillat, M. F., & Lostao, A. (2014). Sequential binding of FurA from *Anabaena* sp. PCC 7120 to iron boxes: Exploring regulation at the nanoscale. *Biochimica et Biophysica Acta – Proteins and Proteomics*, *1844*(3), 623–631.

Pandey, T., Chhetri, G., Chinta, R., Kumar, B., Singh, D. B., Tripathi, T., et al. (2015). Functional classification and biochemical characterization of a novel rho class glutathione S-transferase in *Synechocystis* PCC 6803. *FEBS Open Bio*, *5*(1), 1–7.

Pandey, T., Shukla, R., Shukla, H., Sonkar, A., Tripathi, T., & Singh, A. K. (2017). A combined biochemical and computational studies of the rho-class glutathione s-transferase sll1545 of *Synechocystis* PCC 6803. *International Journal of Biological Macromolecules*, *94*, 378–385.

Pandey, T., Singh, S. K., Chhetri, G., Tripathi, T., & Singh, A. K. (2015). Characterization of a highly pH stable Chi-class glutathione S-transferase from *Synechocystis* PCC 6803. *PLoS One*, *10*(5).

Pattharaprachayakul, N., Lee, M., Incharoensakdi, A., & Woo, H. M. (2020). Current understanding of the cyanobacterial CRISPR-Cas systems and development of the synthetic CRISPR-Cas systems for cyanobacteria. *Enzyme and Microbial Technology*, *140*, 109619.

Pearson, L. A., Dittmann, E., Mazmouz, R., Ongley, S. E., D'Agostino, P. M., & Neilan, B. A. (2016). The genetics, biosynthesis and regulation of toxic specialized metabolites of cyanobacteria. *Harmful Algae*, *54*, 98–111.

Pereira, S. B., Mota, R., Santos, C. L., De Philippis, R., & Tamagnini, P. (2013). Assembly and export of extracellular polymeric substances (EPS) in cyanobacteria. A phylogenomic approach. In *Advances in Botanical Research: vol 65. Genomics of Cyanobacteria* (pp. 235–279). Academic Press.

Pereira, S. B., Sousa, A., Santos, M., Araújo, M., Serôdio, F., Granja, P., et al. (2019). Strategies to obtain designer polymers based on cyanobacterial extracellular polymeric substances (EPS). *International Journal of Molecular Sciences*, *20*(22), 5693.

Pérez, A. A., Gajewski, J. P., Ferlez, B. H., Ludwig, M., Baker, C. S., Golbeck, J. H., et al. (2017). Zn^{2+}-inducible expression platform for *Synechococcus* sp. strain PCC 7002 based on the smtA promoter/operator and smtB repressor. *Applied and Environmental Microbiology*, *83*(3). e02491–16.

Pernil, R., & Schleiff, E. (2019). Metalloproteins in the biology of heterocysts. *Life*, *9*(2), 32.

Picciocchi, A., Saguez, C., Boussac, A., Cassier-Chauvat, C., & Chauvat, F. (2007). CGFS-type monothiol glutaredoxins from the cyanobacterium *Synechocystis* PCC6803 and other evolutionary distant model organisms possess a glutathione-ligated [2Fe-2S] cluster. *Biochemistry*, *46*(51), 15018–15026.

Planchon, M., Jittawuttipoka, T., Cassier-Chauvat, C., Guyot, F., Chauvat, F., & Spalla, O. (2013a). Influence of exopolysaccharides on the electrophoretic properties of the model cyanobacterium *Synechocystis*. *Colloids and Surfaces B: Biointerfaces*, *110*, 171–177.

Planchon, M., Jittawuttipoka, T., Cassier-Chauvat, C., Guyot, F., Gelabert, A., Benedetti, M. F., et al. (2013b). Exopolysaccharides protect *Synechocystis* against the deleterious effects of titanium dioxide nanoparticles in natural and artificial waters. *Journal of Colloid and Interface Science*, *405*, 35–43.

Rachedi, R., Foglino, M., & Latifi, A. (2020). Stress signaling in cyanobacteria: A mechanistic overview. *Life*, *10*(12), 312.

Rossi, F., & De Philippis, R. (2015). Role of cyanobacterial exopolysaccharides in phototrophic biofilms and in complex microbial mats. *Life*, *5*(2), 1218–1238.

Salomon, E., & Keren, N. (2015). Acclimation to environmentally relevant Mn concentrations rescues a cyanobacterium from the detrimental effects of iron limitation. *Environmental Microbiology*, *17*(6), 2090–2098.

Schopf, J. W. (2011). The paleobiological record of photosynthesis. *Photosynthesis Research*, *107*(1), 87–101.

Singh, D. K., Lingaswamy, B., Koduru, T. N., Nagu, P. P., & Jogadhenu, P. S. S. (2019). A putative merR family transcription factor Slr0701 regulates mercury inducible expression of MerA in the cyanobacterium *Synechocystis* sp. PCC6803. *MicrobiologyOpen*, *8*(9), e00838.

Slonimskiy, Y. B., Muzzopappa, F., Maksimov, E. G., Wilson, A., Friedrich, T., Kirilovsky, D., et al. (2019). Light-controlled carotenoid transfer between water-soluble proteins related to cyanobacterial photoprotection. *FEBS Journal*, *286*(10), 1908–1924.

Sun, T., Xu, L., Wu, L., Song, Z., Chen, L., & Zhang, W. (2017). Identification of a new target slr0946 of the response regulator Sll0649 involving cadmium tolerance in *Synechocystis* sp. PCC 6803. *Frontiers in Microbiology*, *8*, 1582.

Vourc'H, T., Peerhossaini, H., Léopoldès, J., Méjean, A., Chauvat, F., & Cassier-Chauvat, C. (2018). Slowdown of surface diffusion during early stages of bacterial colonization. *Physical Review E*, *97*(3).

Vuorijoki, L., Tiwari, A., Kallio, P., & Aro, E. M. (2017). Inactivation of iron-sulfur cluster biogenesis regulator SufR in *Synechocystis* sp. PCC 6803 induces unique iron-dependent protein-level responses. *Biochimica et Biophysica Acta – General Subjects*, *1861*(5), 1085–1098.

Yang, W., Wang, F., Liu, L. N., & Sui, N. (2020). Responses of membranes and the photosynthetic apparatus to salt stress in cyanobacteria. *Frontiers in Plant Science*, *11*, 713.

Yao, L., Shabestary, K., Björk, S. M., Asplund-Samuelsson, J., Joensson, H. N., Jahn, M., et al. (2020). Pooled CRISPRi screening of the cyanobacterium *Synechocystis* sp PCC 6803 for enhanced industrial phenotypes. *Nature Communications*, *11*(1).

CHAPTER SEVEN

Compatible interactions between plants and endoparasitic nematodes—A follow-up of ABR volume 73: Plant nematode interactions—A view on compatible interrelationships

Carolina Escobar[a,b,*] and Carmen Fenoll[a]

[a]Facultad de Ciencias Ambientales y Bioquímica, Área de Fisiología Vegetal, Universidad de Castilla-La Mancha, Toledo, Spain
[b]International Research Organization for Advanced Science and Technology (IROAST), Kumamoto University, Kumamoto, Japan
*Corresponding author: e-mail address: carolina.escobar@uclm.es

Contents

Abstract

Volume 73 of *Advances in Botanical Research* "Plant Nematode Interactions—A View on Compatible Interrelationships" edited by Carolina Escobar and Carmen Fenoll (series Editors Jean-Pierre Jacquot and Pierre Gadal) was published in 2015, and its contents are still valid in most aspects. Forty-one authors contributed to 14 chapters covering the biology of nematodes, as well as biochemical, genomic and biotechnological aspects of the plant parasitic nematodes interactions (PPNs). The book is centered in the compatible interaction with special emphasis on the characteristics and functions of the nematode-induced feeding sites, the effectors secreted by the nematodes and their role in exploiting plant-cellular processes or interfering with the plant defenses, as well as the application of biotechnology for nematode control. We believe that it is now appropriate for the anniversary volume 100 of ABR to summarize what was described in that volume but also to provide some new aspects of the more recent literature regarding epigenetic regulatory processes and future perspectives for nematode control.

Advances in Botanical Research, Volume 100
ISSN 0065-2296
https://doi.org/10.1016/bs.abr.2021.03.001
Copyright © 2021 Elsevier Ltd
All rights reserved.

1. Salient features of volume 73

The book concentrates on the interactions of sedentary plant root endoparasitic nematodes with their hosts, and particularly in the compatible interactions that lead to successful infection and nematode establishment. Once these nematodes invade the root as juveniles and induce their feeding sites inside the host roots, they become sessile and fully dependent on such sites for their development and reproduction. The damage inflicted to the plants is mainly due to the feeding behavior of the nematode that induces a nematode feeding site (NFS), a strong nutritional sink that compromises water and nutrient distribution to the rest of the plant, what makes these nematodes devastating agricultural pests. Additionally, during migration and establishment the nematodes cause distortions of the root vasculature, further impinging in water and solute movement.

The main groups of these biotrophic and obligate parasites are the root-knot (RKNs; *Meloidogyne* spp.) and the cyst (CNs; *Heterodera* and *Globodera* spp.) nematodes. While they share overall parasitism strategies, the mechanisms involved in root invasion and in the development of the feeding sites have very different features. The two opening chapters provide a general description of their body plan, their complex life cycles and their specialized feeding sites induced within the plants, namely the galls or knots enclosing the giant cells for *Meloidogyne* spp., and the syncytia induced by cyst nematodes (Bohlmann, 2015; Escobar, Barcala, Cabrera, & Fenoll, 2015). These chapters summarize the main cellular and molecular aspects of their life cycles and biology, emphasizing the function and structure of their feeding sites, as well as their physiology from a molecular genetics perspective. RKNs basically transform pre-vascular cells near the root tip into giant cells (GCs) by injecting different effectors that cause, among other changes, dramatic rearrangements of the cytoskeleton and cell cycle alterations, such as mitosis with partial cytokinesis and endoreduplication. Consequently, huge feeding cells differentiate, which function as sinks providing nutrients to the nematode throughout its successive developmental stages, until the mature female produces its offspring in the form of a gelatinous egg mass. CNs induce the formation of feeding cells by local cell wall dissolution, forming a syncytium made up from several hundreds of root cells that feeds the nematode till the formation of a female that produces hundreds of eggs within its body, which hardens and becomes a cyst to protect the offspring. Holistic approaches to tackle feeding site-specific gene expression are also described. Therefore, Chapters 1 and 2 introduce the most salient features that unfold in more detail in the following chapters throughout the volume.

Chapters 3–6 dwell on the particular aspects of how the feeding sites develop and function as nutritional sinks, a prerequisite for a successful compatible interaction. In this respect, in Chapter 3, Wieczorek (2015) reviews the elaborated cell wall alterations that take place during the invasion and migration of the juvenile stages (J2) inside the roots. Cell wall modifying and degrading proteins such as cellulases and expansins are secreted by the J2 to ease its way during the penetration and migration towards the vascular cylinder through the root tissues. In parallel, the expression of many plant genes encoding cell wall biosynthetic and modifying enzymes contribute to the structural modifications of the root cells during feeding site formation, and they are an important part of the remodeling mechanisms that contribute to shape the distinct NSF structure and function, either giant cells or syncytia. Finally, an update of the complex chemical composition of the feeding cell walls is addressed, despite the intrinsic difficulty for their study due to the very restricted amounts of material that can be sampled.

During NFS development, nematodes finely tune the plant cell division cycle, and Chapter 4 (de Almeida-Engler et al., 2015) provides some clues as to how they accomplish this. Both syncytia and giant cells are multinucleate and each nucleus displays genome amplification but through very different mechanisms. Giant cells experience repeated mitosis without or with partial cytokinesis and enter endoreplication cycles, while syncytia reunite the nuclei of several adjacent cells whose walls dissolve and protoplast fuse to form a syncytial structure, with endoreplication operating in each nucleus. Chapter 4 also presents some of the very few studies that show direct evidence of the endoreduplication process into the nuclei of the feeding cells by using flow cytometry to measure the ploidy levels of giant cells. These processes are believed to facilitate the large cell size and sink strength characteristic of NFS, and they are accompanied by distinct changes in the expression of key cell cycle genes. de Almeida-Engler et al. (2015) describe how several laboratories identified these genes and established their probable functions in NFS with a combination of promoter-reporter fusions, cell biology tools and genetic interference using mutants and transgenic lines with altered functionality of these regulators, as well as pharmacological studies with inhibitors of the cell cycle and immunocytochemistry. The NFS features more directly related to their sink behavior are treated in Chapter 5, where Siddique & Grundler (2015) review the primary metabolism and nutrient uptake associated to the nematode feeding structures. One of the first approaches to explore these metabolic aspects was by studying phloem loading via labelling and fluorescent dye experiments. More recently, transcriptomic analysis of NFS from different plant species revealed an abundance of transcripts related to primary metabolism and showed the up-regulation of sink-related genes. Analytical techniques such as HPLC and gas chromatography coupled to mass spectrometry were used to obtain metabolic profiles of GCs and syncytia. Both types of feeding cells showed accumulation of starch, sucrose and diverse sugars, amino-acids, and other metabolites at higher levels than in non-infected roots. This suggests metabolic similarities between them, despite their differences in ontogeny. Additionally, a subtle regulation of the sugar-processing enzymes in GCs and syncytia was observed, based on functional analysis of loss of function plant lines concerning key enzymes such as invertases or sucrose synthases. Chapter 5 also provides a detailed update of the mechanisms proposed for nutrient delivery into the nematode-induced feeding sites. Functional plasmodesmata seem to be present at medium-late stages of syncytia development, when they have a high demand for assimilates, while during their initial phase of development they are symplastically isolated, and enhanced activity of transporters seems to mediate sugar supply. However, little is known of the mechanisms that aid in accumulating nutrients in the GCs, as it is generally assumed that they are symplastically isolated, but contrasting results were obtained with other experiments as fluorescent dyes. Therefore, there is still fragmentary knowledge of those processes.

The development of the complex feeding sites induced by RKNs and CNs constitutes in itself a process of new organogenesis, as they become functional organs after the initial plant cells experience dramatic morphological, physiological, and molecular changes triggered by the nematode. Chapter 7 (Cabrera, Díaz-Manzano, Fenoll, & Escobar, 2015) is centered in the mechanisms used by RKNs and CNS to interfere with predefined plant developmental processes in order to differentiate their feeding sites. Some molecular components of the modified mechanisms are transducers common to lateral root (LR) and gall formation where auxins and cytokinins, together with nematode-secreted effectors such as CLE-like and CEP peptides, are major players. Analyses using genetic sensors, mutants and comparative transcriptomic analysis led to the identification of several of the regulators of these processes described in Chapter 7. In the recent years, several molecular players, mostly integrated in transduction pathways leading to callus, lateral root or adventitious root formation, and root meristem and root regeneration, were also characterized and identified, some of them regulated by auxin responsive factors (Olmo et al., 2020). Nevertheless, we are still far from understanding how the molecular components integrated in these transduction pathways operate to reprogram the initial root cells into feeding sites, and this is a relevant topic to be deciphered.

One of the most remarkable and conserved trends observed in the transcriptomes from early galls/giant cells of different plant-species is a global downregulation of gene expression. In the last decade an emerging topic is the study of putative gene silencing mechanisms that may be underlying this massive gene repression, particularly in feeding cells. Chapter 9 (Hewezi & Baum, 2015) reviews the host endogenous small RNAs and the RNA-mediated silencing pathways that appear to play key roles in regulating the gene expression changes aiding in the dramatic reprogramming of initial plant cells into new cell types forming the syncytia by CNs or giant-cells by RKNs. The application of holistic methods to analyze small RNA populations (sRNAs), such as next-generation sequencing, helped elucidate the diversity and complexity of sRNAs during plant–nematode interactions. This chapter details the function of several microRNAs, often targeting transcription factor-coding genes, through loss-of function lines of the target genes, as well as their miRNA-resistant versions and over-expression lines. sRNAs are powerful modulators of gene expression, as they can coordinately change the expression of hundreds and even thousands of genes. Finally, two artificial RNA interference systems, host-induced gene silencing and virus-induced gene silencing (VIGS) are shown as useful tools for functional genomics of the plant–nematode interactions that aid the identification of genetic determinants of parasitism. However, it is interesting to point that in the recent years several studies revealed that sRNAs-guided DNA methylation pathways (RdDM) might be important during the plant-nematode interaction. Methyl-seq approaches detected differential DNA methylation of nematode-feeding sites as compared to non-infected tissues, which could also partially explain the relevant accumulation of sRNAs in transposons and retrotransposons of different types, at least in the Arabidopsis genome (Ruiz-Ferrer et al., 2018; reviewed in Hewezi, 2010). Moreover, a novel effector, 32E03, from *Heterodera schachtii* has been recently described to interfere with the regulation of rRNA gene expression through chromatin modifications via the inhibition of *Arabidopsis thaliana* histone deacetylase activities (Vijayapalani, Hewezi, Pontvianne, & Bauma, 2018). This brings into the scenario the possibility that the nematode directly interferes with plant epigenetic mechanisms which can cause a vast impact in gene expression.

The biotrophic nematodes, and particularly those with obligate endoparasitism, maintain an intimate relationship with the plant during long periods to complete their life cycles. It is therefore reasonable to think that long-lasting close association would require the suppression and modulation of plant defenses by the nematode. Chapter 13 (Mantelin, Thorpe, & Jones, 2015) presents what is known of suppression of plant defenses mainly driven by nematode effectors. A large number of candidate effectors were identified from genomics and transcriptomics projects in the last decade, some of them interfering with the hormonal signaling pathways that promote plant defenses. This chapter is linked to Chapter 6 (Brown Miyara, Ionit, Buki, & Kolomiets, 2015) where modulation of plant defenses by fatty acids through a complex lipid-mediated signaling crosstalk between the plant and the nematode is brought into scene. Different studies have provided genetic evidence of the biological significance of lipid signals during the activation of local and systemic plant defenses in the plant–nematode interaction, as well as the prominent role of the jasmonate pathway. Additionally, the emerging role of some nematode effectors in manipulating lipid-based signaling in the feeding sites is also described, although little is still known on this relevant topic. In this respect, Chapters 11 and 12 (Gardner, Verma, & Mitchum, 2015; Truong, Nguyen, Abad, Quentin, & Favery, 2015, respectively) focus on the emerging roles of nematode effectors and their targets studied via molecular and functional characterization. Chapter 11 focuses on CNs and Chapter 12 on RKNs. Nematode effector molecules from their three esophageal gland cells that are secreted through their stylets play various roles. Basically, they function to modulate cellular processes and favor nematode development during invasion, migration and establishment. Some are known to alter phytohormone balances for their establishment, such as "plant peptide hormone mimicry stylet-secreted effectors." Others are involved in direct modulation of host hormone-transporter proteins, or in the modulation of stress homeostasis and defense responses by suppressing ROS production, or detoxifying ROS within the plant. In this respect, a relevant study confirmed the ability of *H. shachtii* to synthesize a functional cytokinin by means of an isopentenyltransferase gene and it is used to manipulate the host system by controlling cell cycle in order to establish a long-term parasitic interaction (reviewed in Siddique & Grundler, 2018). Subcellular localization of nematode-effectors in the target plant cell is also diverse, as some were detected in the apoplast, cytoplasm or nuclei. Different approaches as direct methods for the production of stylet secretions in semi-sterile conditions by incubation of J2 with neurostimulants of stylet thrusting, combined with protein microsequencing techniques, or the use of sensitive high-throughput proteomics-based liquid chromatography techniques, led to the identification of hundreds of *Meloidogyne* spp. secreted proteins. In addition, comparative genomics tools stemming from the growing data of plant-parasitic nematode genomic and transcriptomic information allowed the identification of novel effectors and helped identify other gene candidates which play different roles in the compatible and incompatible interaction. This topic is further revised in Chapter 10 (Bird, Williamson, & Opperman, 2015), where it is evident that the recent advances in high-throughput sequencing have accelerated the understanding of genomics in nematodes mainly due to whole genome sequencing of a diverse and high number of plant parasitic nematode species. Comparative genomics confirmed prior information obtained from EST sequencing and increased our understanding of horizontal gene transfer as a source of novelty during the evolution of parasitism. Processes such as horizontal gene transfer, gene family expansion, evolution of new genes that mediate interactions with the host, and parasitism-specific gene regulation allowed significant adaptations that helped the nematodes parasitize plants (reviewed in Kikuchi, Eves-van den Akker, & Jones, 2017).

Most of the research revised in these former chapters was conducted in a few plant species, particularly *Arabidopsis thaliana* as an easy and simple plant model, soybean, tomato and, in some instances, *Medicago* spp. or pea, but all of them are dicotyledonous species. However, sedentary endoparasitic nematodes are also major pests of important monocotyledonous crops, and thus the study of the compatible interactions with species such as wheat, barley, maize, oat, sugar cane, banana or rice among others is an emerging field revised in Chapter 8 (Fernández et al., 2015). Histological data and plant-genetic resistance sources are firstly presented during different plant–nematode interactions, i.e., wheat/barley/oat-*Heterodera avenae*, wheat/barley-*Meloidogyne* spp., Rice/maize-*Heterodera sacchari*, Rice/maize-*Meloidogyne* spp. Despite the efforts made in identifying resistance genes in monocots in the last decade, none of them has been cloned yet. Histological data of the feeding sites have proved useful for a characterization of the plant-nematode interaction. For example, *M. graminicola* forms a hook-like gall at the root tip in young roots of rice and the egg-laying occurs as early as 14 days por inoculation, whereas the establishment of *M. javanica* in Arabidopsis is in the elongation zone of the root and life cycle completion lasts at least 30–40 days at 26 °C. The chapter ends with a review of the still limited transcriptomic data on the plant–nematode interaction in monocots, which awaits the functional characterization of putative pathways involved. This Chapter presents also some prospects concerning the identification of putative effectors from nematode transcriptomes combined with comparative genomic approaches. In the last years the knowledge of the plant–monocot interactions opened new fields of research as those related to epigenetic processes, such as changes in DNA methylation guiding gene-regulation during *M. graminicola*-rice interactions (Atighi, Verstraeten, De Meyer, & Kyndt, 2020).

Finally, (Fosu-Nyarko & Jones, 2015) (Chapter 14) present a very comprehensive overview of the biotechnological solutions that have been explored for the control of plant parasitic nematodes. It starts with how molecular markers aided to identify natural resistance genes, making it possible to use transgenesis for transferring useful genes to different susceptible crops from wild or cultivated species. Initially undertaken with great expectations, there are only a few success stories, probably because of a lack of full understanding of the mechanisms through which these resistance genes operate in their hosts and incomplete reconstruction of essential transduction pathways in the target species. The following epigraph deals with the various transgenic strategies used to disrupt nematode feeding sites by targeting the so termed "susceptibility genes," the plant genes essential for feeding site formation and maintenance, or overexpressing genes that disrupt feeding sites. RNA interference in various formats is also reviewed, as are genome editing-based approaches that develop form the recent boom of CRISPR-Cas9 technologies. Other novel technological advances are discussed, not all of them translated to the field of nematode resistance and sometimes aimed at softening legislative constraints in some countries, including the genomics-based search for novel chemical nematicides. This chapter includes an interesting and often overseen analysis of issues related to what comes after the scientific proof of concept, such as patenting, regulations and safety, GMO acceptance, and others.

2. Conclusion and perspectives

Many of the chapters described here review how transcriptomic analyses together with molecular genetics combined also with cell biology techniques, revealed dramatic and specific changes in gene expression in nematode feeding sites. Nonetheless, to date, only a limited number of players of the complex regulatory networks having a role during feeding cells development have emerged, and the understanding of the subtle interaction with their hosts and their necessary cross talk is still fragmentary. Yet, integrative analysis of proteomic and transcriptomics together with genetics and molecular and cell biology tools are facilitating its comprehension. Some of the outgoing topics focus on epigenetic processes such as RdDM regulatory mechanisms that mediate DNA methylation, or histone modifications driving chromatin remodeling (reviewed in Hewezi, 2010). Those may coordinately regulate groups of hundreds or thousands of genes as well as chromatin structure and transposon stability. This is a relevant trending topic that will help to elucidate important regulatory mechanisms during the plant–nematode interaction in the coming years.

The strict restrictions of chemical nematicides due to environmental concerns, the occurrence of low durability of monogenic plant resistance in the field, in addition to the lack of acceptance in the EU of transgenic approaches and the European reluctance to new editing technologies such as CRISPR-Cas, has prompted interest on alternative methods for nematode control. An increasing field of interest is biological control, based on biopesticides or microorganisms (reviewed in Poveda, Abril-Urias, & Escobar, 2020). Other biotechnological approaches based on the application of exoRNAs, are still in progress. External application of in vitro synthesized or bacterially produced long dsRNAs, hairpin RNAs (hpRNAs), or small interfering RNAs (siRNAs) targeting essential genes of plant pathogens and pests, have proved to decrease mRNA levels of the target genes both in plants and pathogens (reviewed in Dubrovina & Kiselev, 2019). So far, there are not yet reports on exogenous RNA application *in planta* against nematodes. However, solutions of synthetic neurotransmitters mixed with dsRNA were successfully used for in vitro RNAi in plant parasitic nematodes. Additionally, *in planta* RNAi to deliver dsRNAs to feeding nematodes were effective to silence target genes, generating resistance (reviewed in Banerjee et al., 2017). The development of this new candidate strategy could greatly contribute to nematode control in the close future.

Acknowledgments

This work was supported by the Spanish Government (AGL2016-75287-R; PID2019-105924RB-I00 and RED2018-02407-T) and by the Castilla-La Mancha Government (SBPLY/17/180501/000287) to C.E. The laboratory received support from UCLM intramural funds and from European Cohesion Funds. We thank Virginia Ruiz-Ferrer for the micrographs at Fig. 1.

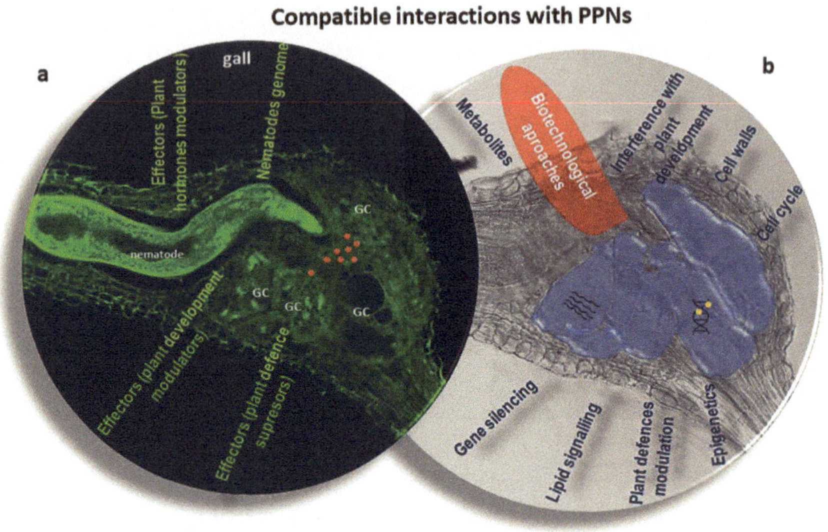

Fig. 1 Combined diagram and micrographs of an Arabidopsis root gall formed by *Meloidogyne javanica* 14 days post infection. The gall was fixed and clarified with BABB as in Cabrera et al. (2018). (A) Confocal micrography where glutaraldehyde auto-fluorescence was excited at 488 nm. The green auto-fluorescence generated by the glutaraldehyde was recovered for image capture in the range of 500–600 nm following Cabrera et al. (2018). (B) Transmission micrography from the same gall. Letterings indicate the main topics of the chapters included in the volume 73 relative to the compatible PPNs interactions, mostly focused in the feeding sites. Galls and nematode are indicated. GCs, giant cells. *The section images were kindly donated by Dr. Virginia Ruiz-Ferrer.*

References

Atighi, M. R., Verstraeten, B., De Meyer, T., & Kyndt, T. (2020). Genome-wide DNA hypomethylation shapes nematode pattern-triggered immunity in plants. *The New Phytologist*, *227*, 545–558.

Banerjee, S., Banerjee, A., Gill, S. S., Gupta, O. P., Dahuja, A., Jain, P. K., et al. (2017). RNA interference: A novel source of resistance to combat plant parasitic nematodes. *Frontiers in Plant Science*, *19*, 834.

Bird, D. M., Williamson, V. M., & Opperman, C. H. (2015). Exploiting solved genomes of plant-parasitic nematodes to understand parasitism. In C. Escobar, & C. Fenoll (Eds.). *Advances in Botanical Research. Plant nematode interactions* (pp. 241–258). Elsevier.

Bohlmann, H. (2015). Introductory chapter on the basic biology of cyst nematodes. In C. Escobar, & C. Fenoll (Eds.), *Plant nematode interactions, Advances in Botanical Research* (pp. 33–60). Elsevier.

Brown Miyara, S., Ionit, I., Buki, P., & Kolomiets, M. (2015). The role of lipid signalling in regulating plant–nematode interactions. In C. Escobar, & C. Fenoll (Eds.), *Advances in Botanical Research. Plant nematode interactions* (pp. 139–188). Elsevier.

Cabrera, J., Díaz-Manzano, F. E., Fenoll, C., & Escobar, C. (2015). Developmental pathways mediated by hormones in nematode feeding sites. In C. Escobar, & C. Fenoll (Eds.). *Advances in Botanical Research. Plant nematode interactions* (pp. 167–188). Elsevier.

Cabrera, J., Olmo, R., Ruiz-Ferrer, V., Abreu, I., Hermans, C., Martinez-Argudo, I., et al. (2018). A phenotyping method of giant cells from root-knot nematode feeding sites by confocal microscopy highlights a role for CHITINASE-LIKE 1 in Arabidopsis. *International Journal of Molecular Sciences, 19*, 429.

de Almeida-Engler, J., Vieira, P., Rodiuc, N., Grossi de Sa, & Engler, G. (2015). The plant cell cycle machinery: Usurped and modulated by plant-parasitic nematodes. *Advances in Botanical Research. Plant nematode interactions* (pp. 91–118). Elsevier.

Dubrovina, A. S., & Kiselev, K. V. (2019). Exogenous RNAs for gene regulation and plant resistance. *International Journal of Molecular Sciences, 20*, 2282.

Escobar, C, Barcala, M., Cabrera, J., & Fenoll, C. (2015). Overview of root-knot nematodes and giant cells. In C. Escobar, & C. Fenoll (Eds.), *Plant nematode interactions, Advances in Botanical Research* (pp. 1–32). Elsevier.

Fernández, D., Petitot, A-S., Grossi de Sa, M., Nguyen, V.P., de Almeida Engler, J. and Kyndt, T. (2015). Recent advances in understanding plant–nematode interactions in monocots. In: C. Escobar and C. Fenoll (Eds.), Advances in Botanical Research. Plant nematode interactions, 189-220, Elsevier. ISBN: 978-0-12-417161-9.

Fosu-Nyarko, J. and Jones, M.G.K. (2015). Application of biotechnology for nematode control in crop plants. 339-376. In: Advances in Botanical Research. Plant nematode interactions. Edited by C. Escobar and C. Fenoll (Eds.), Elsevier. ISBN: 978-0-12-417161-9.

Gardner, M., Verma, A., & Mitchum, M. G. (2015). Emerging roles of cyst nematode effectors in exploiting plant cellular processes. In C. Escobar, & C. Fenoll (Eds.), *Advances in Botanical Research, Plant nematode interactions* (pp. 259–291). Elsevier.

Hewezi, T. (2010). Epigenetic mechanisms in nematode–plant interactions. *Annual Review of Phytopathology, 58*, 119–138.

Hewezi, T., & Baum, T. J. (2015). Gene silencing in nematode feeding sites. In C. Escobar, & C. Fenoll (Eds.), *Advances in Botanical Research. Plant nematode interactions* (pp. 221–240). Elsevier.

Kikuchi, T., Eves-van den Akker, S., & Jones, J. T. (2017). Genome evolution of plant-parasitic nematodes. *Annual Review of Phytopathology, 55*, 333–354.

Mantelin, S., Thorpe, P., & Jones, J. T. (2015). Suppression of plant defences by plant-parasitic nematodes. In C. Escobar, & C. Fenoll (Eds.), *Plant nematode interactions, Advances in Botanical Research* (pp. 325–337). Elsevier.

Olmo, R., Cabrera, J., Díaz-Manzano, F. E., Ruiz-Ferrer, V., Barcala, M., Ishida, T., et al. (2020). Root-knot nematodes induce gall formation by recruiting developmental pathways of post-embryonic organogenesis and regeneration to promote transient pluripotency. *The New Phytologist, 227*, 200–215.

Poveda, J., Abril-Urias, P., & Escobar, C. (2020). Biological control of plant-parasitic nematodes by filamentous fungi inducers of resistance: Trichoderma, mycorrhizal and endophytic fungi. *Frontiers in Microbiology, 11*, 992.

Ruiz-Ferrer, V., Cabrera, J., Martinez-Argudo, I., Artaza, H., Fenoll, C., & Escobar, C. (2018). Silenced retrotransposons are major rasiRNAs targets in Arabidopsis galls induced by Meloidogyne javanica. *Molecular Plant Pathology, 19*, 2431–2445.

Siddique, S., & Grundler, F. (2015). Metabolism in nematode feeding sites. In C. Escobar, & C. Fenoll (Eds.), *Plant nematode interactions. Advances in Botanical Research* (pp. 119–138). Elsevier.

Siddique, S., & Grundler, F. M. W. (2018). Parasitic nematodes manipulate plant development to establish feeding sites. *Current Opinion in Microbiology*, *46*, 102–108.

Truong, N.M., Nguyen, C-N., Abad, P., Quentin, M. and Favery, B. (2015). Function of root-knot nematode effectors and their targets in plant parasitism. Pp. 293-324. In: Advances in Botanical Research. Plant nematode interactions. Edited by C. Escobar and C. Fenoll. Elsevier. ISBN: 978-0-12-417161-9.

Vijayapalani, P., Hewezi, T., Pontvianne, F., & Bauma, T. J. (2018). An effector from the cyst nematode *Heterodera schachtii* derepresses host rRNA genes by altering histone acetylation. *The Plant Cell*, *30*, 2795–2812.

Wieczorek, K. (2015). Cell wall alterations in nematode-infected roots. In C. Escobar, & C. Fenoll (Eds.), *Plant nematode interactions. Advances in Botanical Research* (pp. 61–90). Elsevier.

CHAPTER EIGHT

Glucosinolates revisited—A follow-up of ABR volume 80: Glucosinolates

Stanislav Kopriva*
Institute for Plant Sciences, Cluster of Excellence on Plant Sciences (CEPLAS), University of Cologne, Cologne, Germany
*Corresponding author: e-mail address: skopriva@uni-koeln.de

Contents

Abstract

Glucosinolates are an important group of plant secondary metabolites found in Brassicaceae. They are part of plant defense responses against insects, fungal and bacterial pathogens, are responsible for taste and flavor of cruciferous vegetables, and have beneficial health properties. Because of their importance, ABR Volume 80 was dedicated to these compounds, reviewing the metabolism, regulation, and function of glucosinolates. Here the content of the volume will be shortly recapitulated and the progress in various areas of glucosinolate research will be reviewed. Apart from further details on the genes and enzymes involved in glucosinolate synthesis and downstream metabolism, new emerging areas are discussed, such as the function of glucosinolates in shaping the plant microbiome and their further interactions with bacteria and other microorganisms. A number of ways for successful engineering of glucosinolate pathways toward improved yield, health, and nutritional properties of crops are discussed.

Advances in Botanical Research, Volume 100
ISSN 0065-2296
https://doi.org/10.1016/bs.abr.2021.01.008
Copyright © 2021 Elsevier Ltd
All rights reserved.

1. Glucosinolates in ABR volume 80

Glucosinolates are a fascinating class of plant secondary metabolites, that despite a narrow phylogenetic occurrence solely in the Capparales, have long been in the focus of plant biology. Glucosinolates are important for plant defense against herbivores and pathogens, are determinants of taste and smell of cruciferous vegetables, and are also responsible for many of the health benefits provided by these vegetables. As they are also formed in the model plant *Arabidopsis thaliana*, their biosynthesis and its regulation are relatively well understood. The importance of glucosinolates has been corroborated by being the sole focus of Volume 80 of Advances in Botanical Research.

In the 10 chapters of ABR Volume 80, entitled simply "Glucosinolates," researchers shaping the glucosinolate research in the last three decades summarized the state of the art knowledge. The volume started with an excellent overview of glucosinolate function, biosynthesis (Fig. 1), and storage by Halkier (2016). The chapter also pointed out the importance of inter- and intracellular transport for glucosinolate metabolism and their tissue

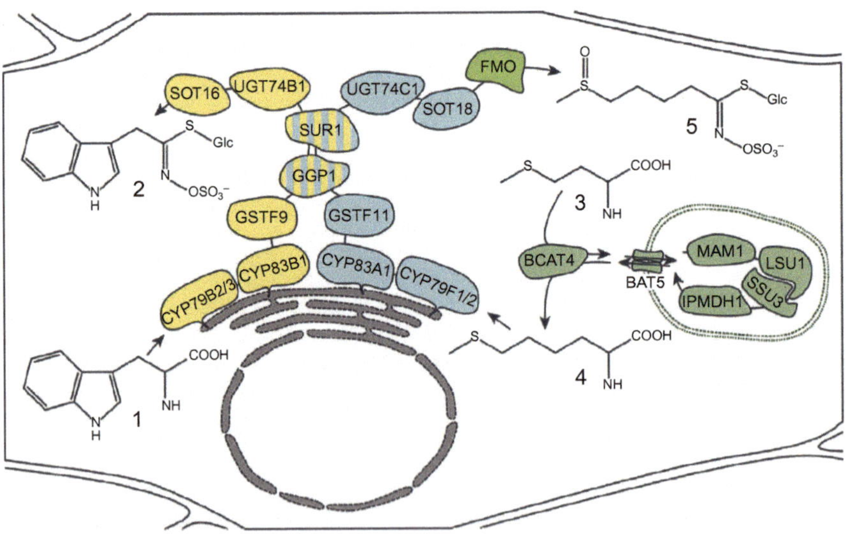

Fig. 1 The biosynthesis of glucosinolates: examples of aliphatic glucoraphanin and indolic indolyl-3-glucosinolates. *From Halkier, B. A. (2016). Chapter one—General introduction to glucosinolates. In S. Kopriva (Ed.),* Advances in botanical research *(Vol. 80, pp. 1–14). Academic Press, volume 80 ABR.*

distribution. The next chapter by Burow (2016) explained how the environment shapes glucosinolate accumulation in plants. She summarized experiments testing how different insects induce glucosinolate synthesis and concluded by stressing the important role of jasmonate signaling in this regulation. She then showed how additional abiotic stresses affect the regulation by biotic factors and pointed out to the need for integration of the signaling pathways. Two factors have been highlighted as important for the integration, the Med25 subunit of Mediator, a transcriptional complex connecting transcription factors with RNA polymerase II, and the AOP2 protein involved in glucosinolate synthesis pathway.

AOP2 was also discussed in the next chapter by Kliebenstein and Cacho (2016), as it underlies one of the major quantitative trait loci (QTL) responsible for variations in glucosinolate composition and accumulation. This chapter described the genetic variation that determines plant glucosinolate profiles not only in Arabidopsis but in the phylogenetic context of the Brassicaceae. The genetic variation has long been known and exploited in plant breeding as well as in research to identify genes regulating the pathway. This chapter described the variation in the major QTLs, GS-AOP, GS-ELONG, or GS-OX, and explained how a small number of loci determine a large variation in glucosinolate profiles. The genetic and biochemical information was compared and discussed in the frame of ecological and evolutionary observations.

The next chapter by Frerigmann (2016) addressed the transcriptional regulation of glucosinolate synthesis. The pathway is under control of a family of MYB transcription factors, three for the aliphatic glucosinolates, three for the indolic ones. It has been, however, shown that these MYB factors require a presence of bHLH factors for their activity. This chapter disentangled the network of these transcription factors, describing individual functions, their interactions, and cross-talk (Fig. 2). It also suggested a mechanism for a metabolic feedback regulation, by which glucosinolates fine tune their own synthesis.

The biological activity of glucosinolates is dependent on sulfur atoms in their structures. The chapter by Kopriva and Gigolashvili (2016) highlighted the connection between sulfur metabolism and glucosinolates. It also showed how glucosinolate synthesis is linked with the retrograde signaling molecule phosphoadenosine phosphate (PAP), important for oxidative stress signaling from chloroplasts to nucleus. In addition, the relevance of redox signaling in controlling glucosinolate synthesis was discussed.

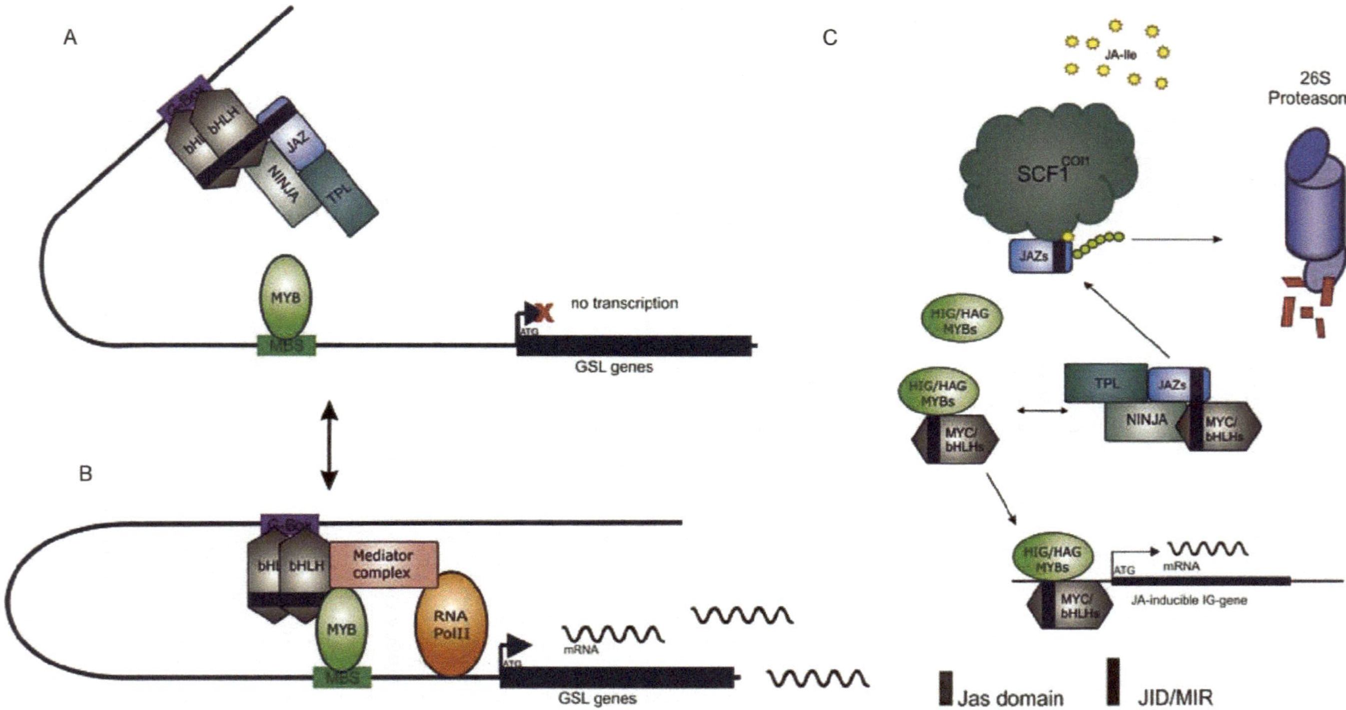

Fig. 2 Model for transcriptional regulation of glucosinolate synthesis. *From Frerigmann, H. (2016). Chapter four—Glucosinolate regulation in a complex relationship—MYC and MYB—No one can act without each other. In S. Kopriva (Ed.),* Advances in botanical research *(Vol. 80, pp. 57–97). Academic Press, volume 80 ABR.*

The next chapter of Volume 80 by Wittstock, Kurzbach, Herfurth, and Stauber (2016) focused on another process connected with biological functions of glucosinolates, their degradation. Through reaction with myrosinase, the thioglucoside bond of glucosinolates is hydrolyzed and the aglycones rearrange into volatile isothiocyanates, nitriles, thiocyanates, or epithionitriles (Fig. 3). In the classical concept of "mustard oil bomb" this happens after tissue damage, when the metabolites stored in vacuoles get into contact with cytosolic myrosinases, but formation of isothiocyanates is possible also without damage, e.g., after fungal infection, or during sulfur starvation. This chapter gave detailed information about the proteins important for glucosinolate degradation, myrosinases and specifier proteins and also identified the knowledge gaps, particularly connected to the completeness of the degradation machinery.

Where the preceding chapter ended, with the description of degradation products, the next chapter by Pastorczyk and Bednarek (2016) took over and explained the function of glucosinolate degradation products in plant immunity. This chapter pointed out the difference in response of the two glucosinolate classes to pathogenic microorganisms. For aliphatic ones the response is similar to herbivores, whereas specialized pathways evolved to

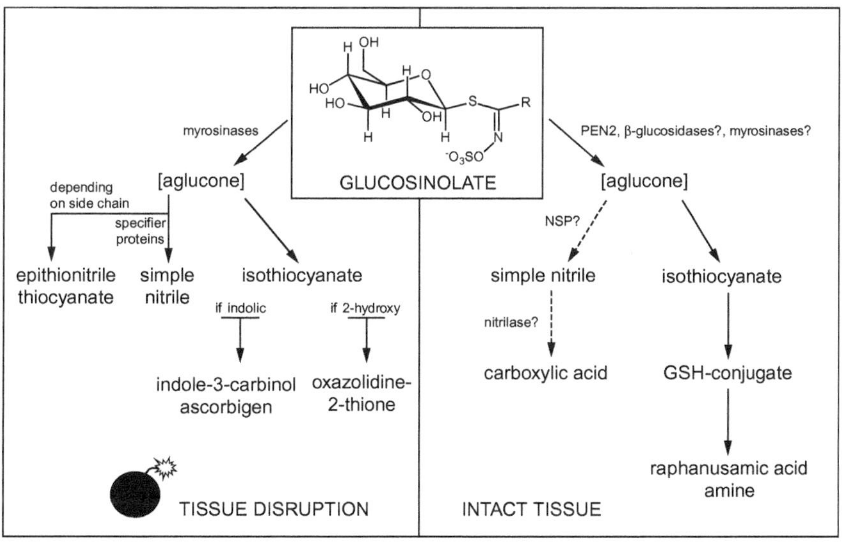

Fig. 3 Scheme of glucosinolate breakdown. *From Wittstock, U., Kurzbach, E., Herfurth, A. M., & Stauber, E. J. (2016). Chapter six—Glucosinolate breakdown. In S. Kopriva (Ed.),* Advances in botanical research *(Vol. 80, pp. 125–169). Academic Press, volume 80 ABR.*

enable the indolic glucosinolates to actively combat pathogens. The indolic glucosinolates are primarily active against fungi, they are toxins, but also play a role in signaling and through interconversion with a number of other indolic phytoalexins, such as camalexin. The variety of the indolic compounds in different Brassica species was presented and new avenues for research suggested. The need to pay more attention to studying immunity in the roots was particularly significant, as a great number of excellent papers appeared since.

The next chapter, by Jeschke, Gershenzon, and Vassao (2016), describes how insects defend themselves against glucosinolates, what mechanisms they evolved to overcome the mustard oil bomb. Two strategies have evolved, one coping with the toxic volatiles by conjugation and detoxification, the other preventing their release in the first place. This can be achieved by preventing the molecular rearrangement by hydrolysis of the sulfate residue, or by interfering with it and pushing the reaction into formation of less toxic nitriles. Finally, the strategies allowing insects to use plant glucosinolates for their own defense were explained.

The next chapter deals also with function of glucosinolates, this time one that is important beyond plants. Glucosinolates, or rather the derived isothiocyanates, have beneficial properties for human health, which is a topic of the chapter by Traka (2016). In this chapter the metabolism of the isothiocyanates and the role of glutathione transferases were explained. The chapter listed a number of clinical intervention trials examining the links between cruciferous vegetables and health benefits. It also described a number of mechanisms by which these compounds benefit humans, including their effects on metabolism of xenobiotics, regulation of expression of detoxification enzymes and influence on oxidative stress and apoptosis (Fig. 4).

The last chapter by Schnug and Haneklaus (2016) was a tour through the history of glucosinolate research for agriculture. Despite the positive properties of glucosinolates described so far, they also have a negative impact in oil-seed rape and therefore the breeding of a low glucosinolate variety was a great success and helped increase the production of this crop. The technology that enabled rapid determination of glucosinolates in seeds was described as well as the effects of sulfur deficiency on the different oil-seed rape varieties.

The final chapter of Volume 80 ends with a merger of fundamental and applied research, typical for glucosinolate research. The availability of genetic tools in Arabidopsis helped dissect many molecular mechanisms of regulation

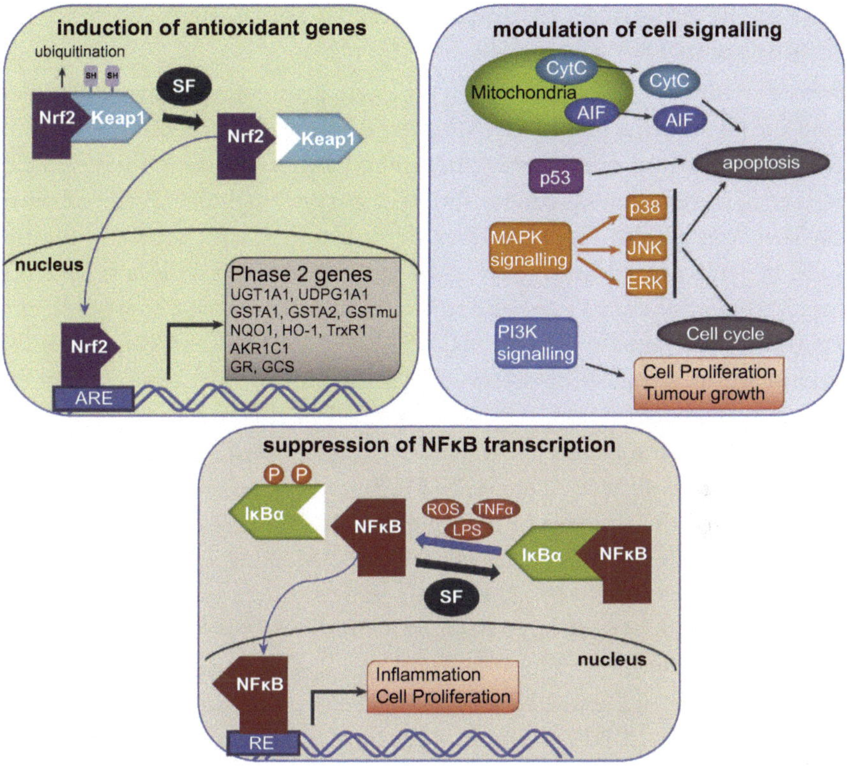

Fig. 4 Mechanisms of action of isothiocyanates in human cells. *From Traka, M. H. (2016). Chapter nine—Health benefits of glucosinolates. In S. Kopriva (Ed.),* Advances in botanical research *(Vol. 80, pp. 247–279). Academic Press, volume 80 ABR.*

of glucosinolate synthesis and made them a model class of compounds to study evolution of secondary metabolism. At the same time, glucosinolates are clearly important for brassica crops performance and quality, as well as for humans in nutrition and medicine. Volume 80 thus represented the ultimate overview of all these aspects of glucosinolate research.

2. Glucosinolate research after the ABR volume 80

ABR Volume 80 was published in 2016, with the chapters covering literature till mid-2015. Five years may seem a short time period to obtain a significant increase in knowledge, however, in the glucosinolate field this has not been the case. Almost 1000 papers have been published and several major discoveries have been made in the last 5 years. We will describe now the most significant ones.

2.1 Recent advances in understanding regulation of glucosinolate synthesis

The core pathway of glucosinolate synthesis in plants seems to be well understood and all enzymes identified (Mitreiter & Gigolashvili, 2021). For a number of modifications of the core structure in various species, however, while the class of the enzymes is obvious, the identity of the individual isoforms is not (Bell et al., 2020). This has been demonstrated, e.g., through identification of genes for synthesis of aromatic glucosinolates in *Barbara vulgaris* (Liu et al., 2016). Similarly, in *Raphanus sativus* the gene encoding a 2-oxoglutarate-dependent dioxygenase GLUCORAPHASATIN SYNTHASE 1 (GRS1) responsible for synthesis of the unique glucosinolate in radish, glucoraphasatin, was identified (Kakizaki et al., 2017). For this reason, new genome sequences from the Brassicaceae will be an important resource as demonstrated for *Eruca sativa* (Bell et al., 2020). But even in Arabidopsis, the gene responsible for the isoform of indole glucosinolate *O*-methyltransferase 5, which is responsible for synthesis of 1MOI3M, has been identified only recently (Pfalz et al., 2016). The major progress, however, has been made in characterizing several enzymes catalyzing downstream reactions from glucosinolates synthesis. These include two ß-glucosidases, BGLU28 and BGLU30, which are active in glucosinolate recycling during sulfate starvation (Zhang et al., 2020). Glucosinolates have long been considered as storage for sulfur and their levels are indeed diminishing in sulfate starved plants (Hirai et al., 2005). This has been attributed to a rapid downregulation of their synthesis due to sulfate starvation induced SDI1 (sulfur deficiency induced) and SDI2, which bind to and inhibit MYB factors activating genes for glucosinolate synthesis (Aarabi et al., 2016). Identification of the two β-glucosidases shows, that in addition to the sulfur sparing due to inhibited synthesis, the sulfur recycling is an active process (Zhang et al., 2020). In addition, a glutathione-S-transferase U13 (GSTU13) has been shown to be important for the metabolism of indolic glucosinolates in the PEN2 pathway involved in plant immunity (Pislewska-Bednarek et al., 2018). GSTU13 conjugates the unstable products of PEN2 myrosinase reaction and accordingly, a pathogen treated *gstu13* mutant lacks the end products of this pathway, such as indole-3-ylmethyl amine and raphanusamic acid. This results in increased susceptibility to several fungal pathogens (Pislewska-Bednarek et al., 2018). Even more significant is the elucidation of the pathway for indolic phytoalexins, such as brassinin or sinalbin, which have been shown to be synthesized from indolic glucosinolates (Klein & Sattely, 2017). First a new myrosinase, Brassinin-Associated β-Glucosidase (BABG), forms an indole isothiocyanate from I3M that is then methylated by Dithiocarbamate S-Methyltransferase (DTC-MT) to brassinin, a precursor of cyclobrassinin and other indolic compounds. The pathway has been reconstructed in tobacco leaves together with pathway synthesizing indolic glucosinolates showing that these phytoalexins may be introduced to other plant species than Brassicaceae (Klein & Sattely, 2017). Similarly, the mechanism by which a specifier protein, the thiocyanate-forming protein from *Thlaspi arvense*, facilitates the formation of thiocyanate over isothiocyanate has been elucidated (Eisenschmidt-Bonn, Schneegans, Backenkohler, Wittstock, & Brandt, 2019). Allylthiocyanate is formed thanks to a specific conformation of a 3L2 loop, which enables interaction between the allyl side chain of the glucosinolate and an Fe^{2+} cofactor. Since iron has been found bound also in the epithiospecifier protein and nitrile-specifier protein 3 from Arabidopsis, the specifier proteins can be seen as an iron dependent lyase enzymes (Backenkohler et al., 2018; Eisenschmidt-Bonn et al., 2019).

An interesting new feature of glucosinolate degradation is the association of glucosinolate degrading β-glucosidases with endoplasmic reticulum (ER) bodies in the roots (Nakano et al., 2017). ER bodies are derived from ER and seem to have a function in defense, based on the enzymes associated with them, which include a number of β-glucosidases, and the induction of their formation upon pathogen interaction and jasmonate signaling. The major β-glucosidase associated with ER bodies is the BGLU23 (PYK10) which was shown to possess myrosinase activity against indolic glucosinolates. Interestingly, PYK10 is more similar to the unusual myrosinase PEN2 than the canonical TGGs. Thus, contrary to a former association of ER bodies with metabolism of scopoletin, these functions of organelles in immunity seem to be in synthesis of antifungal compounds derived from indolic glucosinolates (Nakano et al., 2017). ER bodies have been found also in leaves, where they protect plants against herbivory (Nakazaki et al., 2019).

Also our knowledge on regulation of glucosinolate synthesis has substantially increased since the publication of ABR 80. The complement of transcription factors has been only marginally increased; as reviewed by Frerigmann in the ABR80, the pathway is controlled by two groups of MYB factors and one clade of bHLH factors and another 22 transcription factors contributing to regulation of individual genes for aliphatic glucosinolates. Two more factors, Far1 Related Sequence (FRS) 7 and FRS12 together with novel interactor of JAZ (NINJA) form a transcriptional repressor complex acting in a specific time of day rhythm (Fernandez-Calvo et al., 2020). This is in line with the well-established connection between glucosinolates and circadian clock (Kerwin et al., 2011) and recent observations that circadian clock-associated1 (CCA1) regulates synthesis of indolic glucosinolates (Lei et al., 2019). Not only the circadian rhythm of glucosinolate synthesis and resistance to insects is disturbed by manipulation of CCA1, its overexpression results in increased levels of indolic glucosinolates. In addition to effects of single transcription factors, a systematic analysis of multiple gene knockouts revealed a strong epistasis that affected glucosinolate synthesis at all levels (Li et al., 2018). For metabolite accumulation, most of the epistatic effects were antagonistic, i.e., the phenotypic values in double mutants were lower than the combination of single mutants, except when one of the tested partners was the key regulator, MYB28 or MYB29. The analysis also showed that the epistatic effects on the transcript levels did not predict effects on metabolites (Li et al., 2018). Many of these transcription factors showed similar epistatic effects on a number of growth and flowering phenotypes (Li et al., 2020). These analyses not only revealed the complexity of the glucosinolate regulatory network but also its connection with general plant performance.

Glucosinolate synthesis is, however, regulated also at a posttranscriptional level. A prime example is the finding of interaction of the cytoplasmic protein phosphatase 2A regulatory subunit B'γ (PP2A-B'γ) with indole glucosinolate methyltransferases (Rahikainen et al., 2017). PP2A-B'γ negatively regulates synthesis of 4MO-I3M and coordinates it with methyl utilization, possibly including DNA methylation. However, how far epigenetic mechanisms really control glucosinolate synthesis remains an open question. Analysis of recombinant inbred lines with a *ddm1* mutant in DNA methylation as one parent revealed some effect on glucosinolate variation, the effect was much lower than that of genetic variation (Aller, Jagd, Kliebenstein, & Burow, 2018). However, this may be different in a context of defense priming. Indeed, in a screen for metabolites inducing such priming, isothiocyanate sulforaphane was identified, which is derived from glucoraphanin by myrosinase action (Schillheim et al., 2018). Sulforaphane induced K4me3 methylation and K9 acetylation of histone H3 specifically in the promoter region of *WRKY6* and *PDF12,* genes important for defense against a range of pathogens. This modification correlated with chromatin unpacking in the regulatory regions of these two genes, increased WRKY6 expression, and increased resistance to *Hyaloperonospora arabidopsidis* (Schillheim et al., 2018). This is significant, because sulforaphane is produced in plants upon tissue damage. Its ability to prime defense against oomycete pathogens and its direct toxicity against fungal pathogens thus protects plants from secondary infections at the site of the damage. Thus, the progress in understanding the metabolism and regulation of glucosinolates and their downstream products has a great potential to design new ways to protect plants against a variety of pests.

2.2 Factors affecting natural variation of glucosinolates

Glucosinolates represent an excellent system to study adaptation and evolution of metabolic pathways and, on the other hand, their natural variation among closely related genotypes can be used as a potent tool to identify new genes and regulatory circuits. This has been greatly used in Arabidopsis (Chan, Rowe, Corwin, Joseph, & Kliebenstein, 2011; Kliebenstein et al., 2001; Rowe & Kliebenstein, 2008) and in the last few years increasingly in other species. For example, in Cardaminae 70 glucosinolate structures, almost the half of all known ones, have been characterized, including discovery of two new glucosinolates (Olsen et al., 2016). In several species genome or transcriptome sequencing in conjunction with glucosinolate profiling generated a knowledge base for further detailed dissection of the synthesis and regulation of the more "exotic" glucosinolates, particularly those not present in Arabidopsis. These species include *Eruca sativa* (Bell et al., 2020), *Nasturtium officinale* (Jeon et al., 2017) and *Tarenaya hassleriana* (van den Bergh, Hofberger, & Schranz, 2016). In several species of the Camelinae glucosinolates are almost not present in the leaves, and the indolic glucosinolate seem to lose their role in immunity, indicating a different evolutionary paths in different tribes of the Capparales (Czerniawski, Piasecka, & Bednarek, 2021). However, the largest progress has been made in the crop species *Brassica napus*.

Already the first genome-wide association study (GWAS) analysis through associated transcriptomics with a small set of *B. napus* varieties indicated *MYB28* as the casual gene underlying the difference between low and high glucosinolate oil-seed rape (Harper et al., 2012). This has been confirmed in a number of other studies pointing out the importance of this transcription factor for control of seed glucosinolate levels (Kittipol et al., 2019; Liu, Zhou, et al., 2020; Wang et al., 2018) and also leaf glucosinolate accumulation (Liu, Huang, et al., 2020). In addition, MYB29 was identified as a transcription factor controlling synthesis of aromatic glucosinolates in the roots and MAM3 as the enzyme responsible for elongation of aromatic glucosinolates (Kittipol et al., 2019), proving again the usefulness of GWAS in identifying new regulatory elements. Importantly, the analyses yielded a number of alleles of *MYB28* and other genes useful for breeding of crops with designed concentrations of glucosinolates.

Among other genes contributing to natural variation the methylthioalkylmalate synthase (*MAM*) seems to play the most important role. MAM is a key enzyme of the methionine elongation pathway and underlies the GS-ELONG locus (Kroymann et al., 2001). In Arabidopsis it occurs as three isoforms with different enzymatic properties and is responsible for a large part of the variation in chain length distribution of aliphatic glucosinolates (Benderoth et al., 2006; Kroymann et al., 2001; Textor, de Kraker, Hause, Gershenzon, & Tokuhisa, 2007). While MAM3 catalyzes elongation of both short and long chain substrates, MAM1 and MAM2 are responsible for the prevalence of 3C (chain length with 3 carbon atoms) or 4C glucosinolates, respectively. Arabidopsis accessions with active MAM1, either alone or alongside MAM2 accumulate prevalently 4C glucosinolates (such as Col-0), whereas those missing MAM1 produce mainly C3 molecules (Ler) (Benderoth et al., 2006). MAM1 and MAM2 are very similar and X-ray structure helped to design four amino acid mutations in the enzymes from *Brassica juncea* that altered substrate specificity (Kumar et al., 2019). Similarly, profiles of aliphatic glucosinolates can be changed by alteration of only four amino acids in Arabidopsis MAM1 (Petersen et al., 2019). The importance of MAM1 and MAM2 for the chain length distribution of glucosinolates has also been shown via a modeling approach (Sharma, Popa, Kopriva, & Ebenhoeh, 2019). Recent developments expanding knowledge on glucosinolates beyond Arabidopsis and the Brassica crops, (Abrahams, Pires, & Schranz, 2020) have demonstrated the evolution of the MAM locus in Caparales and its link to glucosinolate profiles.

2.3 Glucosinolates and their degradation products in human nutrition and health

Glucosinolates have significant health promoting effects, however, the knowledge is still largely limited to sulforaphane as a product of glucosinolate degradation and only a few other metabolites. Recent research thus focused on improving the health properties of glucosinolate containing vegetables. It was shown that the way of cooking strongly affects the balance of production between isothiocyanates and nitriles with positive effect of dilution and more acidic pH (Hanschen et al., 2017). Cruciferous vegetables are also an excellent vehicle for selenium biofortification due to synergistic effects of Se and glucosinolates (Wiesner-Reinhold et al., 2017).

Treatment with selenite or selenate not only increases the concentration of the essential micronutrient in broccoli, but it also leads to increased sulforaphane content (Tian, Xu, Liu, Xie, & Pan, 2016). It seems that at least in broccoli, the Se treatment affects glucosinolate synthesis through regulation of APS kinase and thus the availability of active sulfate (McKenzie et al., 2017).

However, the main progress over the last 5 years has been made in understanding the role of gut microbiota in facilitating the uptake and health benefits of glucosinolate derived metabolites. Since myrosinase is heat labile, the production of the active isothiocyanates, such as sulforaphane, from glucosinolates in vegetables must be catalyzed by the gut bacteria (Tian, Liu, Lei, Zhang, & Shan, 2018). Indeed, while there was evidence for an important role of gut microbiota for transformation of dietary glucosinolates, a mechanistic understanding of this process has been obtained only recently (Liou et al., 2020). Human colonic bacterium *Bacteroides thetaiotaomicron* was shown previously to metabolize sinigrin into an allyl isothiocyanate (Elfoul et al., 2001). Using a genome-wide transposon insertion screen of this strain, an operon encoding genes necessary for glucosinolate metabolism was identified (Liou et al., 2020). The four genes in this operon were active in disaccharide metabolism, however, most importantly, three genes not showing any sequence homology with known myrosinases were shown to be involved in the production of isothiocyanates from glucosinolates. Interestingly, deletion analysis revealed that all reactions are catalyzed by pairs of enzymes, but different substrates required different pairs (Liou et al., 2020). For the glucosinolates, BT2158 with either BT2156 or BT2157 represented the corresponding set of proteins. Indeed, gnotobiotic mice inoculated with *B. thetaiotaomicron* mutants in these genes metabolized significantly less glucosinolates, both pure or in context of broccoli meal, compared to WT strain (Liou et al., 2020). The knowledge on the ability of bacteria to metabolize glucosinolates can now be used in innovative therapies. For example, an *E. coli* strain was engineered to secrete myrosinase and to bind to the surface of colorectal cancer cells, which in combination with glucosinolate rich diet inhibited tumor proliferation (Ho et al., 2018). The combination of glucosinolates and microbial myrosinase was effective in a range of *in vitro* cell lines, as well as *in vivo* in murine models (Ho et al., 2018). But there is also a reverse connection between glucosinolates and the composition of the microbiome, as diets rich in glucosinolate containing vegetables, such as broccoli significantly affect the composition and function of human gut microbiome (Kaczmarek et al., 2019).

2.4 Glucosinolates and microbiota

Glucosinolates not only modulate the human microbiome, but they also affect the microbiome of plants. This is not surprising given their function in plant innate immunity, but the extent of the importance of glucosinolates in shaping the interactions of plant associated microorganisms has been revealed only recently. Glucosinolates and the derived isothiocyanates are released to soil during seed germination, root wounding, and degradation of root and shoot matter. Exogenous glucosinolates significantly affect the composition of root and soil microbial populations (Bressan et al., 2009; Siebers et al., 2018). The taxonomic diversity upon biofumigation with oil-seed rape derived isothiocyanates resulted in diminishing the diversity of bacteria and fungi. However, interestingly, the surviving bacteria had a larger proportion of strains able to promote plant growth (Siebers et al., 2018). Thus, isothiocyanates might be a mechanism by which Brassica plants shape their microbiome for a higher benefit. Similar results were obtained in field trials where introducing another Brassicaceae species, *Camelina sativa* resulted in decline in microbial abundance and changes in community structures (Hansen, Schillinger, Sullivan, & Paulitz, 2020). Interestingly, rhizobacteria may also induce resistance against insects, through induction of glucosinolate synthesis, as demonstrated with a *Pseudomonas simiae* strain WCS417r (Pangesti et al., 2016). The interaction of root associated microbes with glucosinolates is thus rather multifaceted and complex.

However, the most exciting is the new connection between glucosinolates, microbiote, and phosphate nutrition unraveled recently (Castrillo et al., 2017; Hacquard et al., 2016; Hiruma et al., 2016). Experiments with the endophytic fungus *Colletotrichum tofieldiae* revealed the importance of glucosinolates for a beneficial interaction with Arabidopsis. *C. tofieldiae* has a similar function in Arabidopsis as mycorrhiza in other plant species, the transfer of phosphate, particularly at low phosphate conditions (Hiruma et al., 2016). The growth promotion from *C. tofieldiae* colonization is limited to phosphate deficient conditions and requires intact phosphate starvation response, as a *phr1 phl1* mutant in two key transcription factors orchestrating the response were profiting much less from the colonization. In addition, the plant growth promotion (PGP) required the plants to synthesize and process indolic glucosinolates, as mutants in metabolism or regulation of their synthesis as well as the *pen2* mutant did not show any PGP (Hiruma et al., 2016). Interestingly, the fungal biomass was increased in the mutants unable to produce indolic glucosinolates, while the loss of all indolic defense metabolites in *cyp79b2 cyp79b3* mutant caused a strong inhibition of plant growth by the otherwise beneficial fungus. Accordingly, *C. tofieldiae* induces the transcription of genes for synthesis of indolic glucosinolates in high phosphate but not low phosphate (Hacquard et al., 2016). The effect of phosphate nutrition does not occur in a pathogenic relative of *C. tofieldiae, C. incanum*, pointing to the interaction of glucosinolates and phosphate nutrition as a possible evolutionary driver enabling the beneficial association with plants. These findings have been extended to the bacterial microbiome, as mutants in phosphate starvation response displayed variations in community profiles (Castrillo et al., 2017). Incubation of Arabidopsis roots with bacterial synthetic communities with different phosphate supplies resulted in transcriptional alterations of many genes connected to immunity, among them genes for glucosinolate synthesis. These genes were induced by the bacteria in low phosphate but not high phosphate conditions and in a PHR1 dependent manner (Castrillo et al., 2017). Thus, glucosinolate synthesis is under control of the phosphate starvation network and PHR1 because it is important for the interaction with fungal and bacterial endophytes in phosphate nutrition dependent manner.

Another interesting development is the identification of mechanisms by which microorganisms detoxify glucosinolates. These were previously mainly known from insects, however, recently several of those processes have been described in fungi. For example, the necrotrophic fungal pathogen *Botrytis cinerea* is variably sensitive to indolic glucosinolates and their degradation products (Buxdorf, Yaffe, Barda, & Levy, 2013). Growth of *B. cinerea* was inhibited by benzyl isothiocyanate and the treatment strongly induced expression of a *MfsG* gene (Vela-Corcia et al., 2019). Loss of this gene resulted in highly increased sensitivity to benzyl isothiocyanate and when grown on a fluorescent isothiocyanate to its increased concentration in the fungal hyphae. This and the similarity of deduced MfsG protein structure with MFS general substrate transporters indicates that *B. cinerea* uses this protein to detoxify isothiocyanates through export from the cells. Accordingly, the expression of the *MfsG* gene in *B. cinerea* growing on different Arabidopsis mutants revealed a correlation between glucosinolate levels and transcript levels and low expression on myrosinase mutants. However, loss of *MfsG* made the pathogen less virulent (Vela-Corcia et al., 2019). Another necrotrophic fungal pathogen, *Sclerotinia sclerotiorum*, is able to metabolize isothiocyanates through conjugation to glutathione and hydrolysis to amines (Chen et al., 2020). The fungus possesses an isothiocyanate hydrolase, similar to the bacterial SaxA, which converts aliphatic and aromatic isothiocyanates into corresponding non-toxic amines. Consequently, the loss of this gene significantly reduces the virulence of *S. sclerotiorum* on Arabidopsis (Chen et al., 2020).

Finally, a new mechanism of action for glucosinolates against pathogenic bacteria has been described. One of the major differences between pathogenic and commensal strains is the presence of type III secretion systems (TTSS) in the former. This system is targeted and inhibited by sulforaphane as a potent defense mechanism (Wang et al., 2020). Interestingly, the inhibition is at the transcriptional level, sulforaphane covalently modifies a cysteine residue of the transcription activator HrpS and prevents TTSS gene expression. Indeed, mutants unable to synthesize aliphatic glucosinolates were more susceptible to *P. syringae* because they were not able to inhibit TTSS gene expression (Wang et al., 2020). This finding might underpin developments of new ways to neutralize bacterial pathogens.

2.5 Glucosinolates and insects

The mechanisms by which insects cope with glucosinolates have been described already in the original chapter by Jeschke, Gershenzon, and Vassão (2016). However, since then several important and interesting findings significantly increased our knowledge on how glucosinolate affect insects and how insects handle these compounds. A direct comparative study of specialized and generalist insects showed that in contrast to the specialists that are able to strongly reduce the production of toxic isothiocyanates, these derivatives accumulate in the generalists (Jeschke et al., 2017). The generalists, such as *Spodoptera littoralis* detoxify the isothiocyanates by conjugation with glutathione, which however, leads to depletion of cysteine for protein synthesis and diminishing fitness (Jeschke, Gershenzon, & Vassao, 2016). Indeed, glucosinolates negatively affected growth of young larvae, until the last instar stage, when surprisingly the larvae grew better on WT Arabidopsis than on mutants in glucosinolate synthesis (Jeschke et al., 2017). However, plant age also affects the way insects feed (Mao et al., 2017). Larvae of both generalist *Helicoverpa armigera* and crucifer specialist *Plutella xylostella* gain less weight on older leaves. This is connected with a greater accumulation of glucosinolates and changes of defense dynamics, mainly a reduced potential for jasmonate signaling in older leaves (Mao et al., 2017).

Several new mechanisms of glucosinolate detoxification have recently been described. The phloem feeder *Bemisia tabaci*, which has been described previously as able to use sulfatase to detoxify glucosinolates (Malka et al., 2016), is also able to modify glucosinolates by additional glucosylation

(Malka et al., 2020). The addition of glucose prevents hydrolysis by myrosinase and is achieved through a transglucosidation, catalyzed by glycoside hydrolase GH13. This enzyme is found in a number of other phloem feeder insects so that this mechanism seems to be a widespread way to enable these insects to feed on plants of the Brassicaceae family (Malka et al., 2020). A myrosinase independent detoxification mechanism dependent on conjugation with aspartate was identified in larvae of leaf beetle *Phaedon cochleariae* (Friedrichs et al., 2020). Interestingly, in tritrophic interactions, manipulations of glucosinolate metabolism in the primary herbivores, affects their development and survival, but not that of their predators (Sun et al., 2019). However, even insects that do not possess these or other detoxification enzymes themselves, are able to prevent the isothiocyanate toxicity indirectly, through their gut microbiota (Shukla & Beran, 2020; Welte et al., 2016). Another recent interesting development is the finding of the mechanism by which *P. xylostella* achieves its host preference for *A. thaliana* (Liu, Zhang, et al., 2020). The diamondback moth uses glucosinolate derived isothiocyanates as signals perceived by olfactory receptors Or35 and Or49. Three specific isothiocyanates are perceived by these receptors, iberverin, 4-pentenyl ITC, and phenylethyl ITC, and the perception of these signals is necessary for correct oviposition preference (Liu, Zhang, et al., 2020).

2.6 New links of glucosinolates with other pathways

Although glucosinolates are well described as part of the plant immunity and other defenses, particularly against herbivores, recent results showed new and unexpected links with other pathways and processes. For example, exogenous sinigrin can be taken up and affect the growth of Arabidopsis, depending on the genotype (Francisco et al., 2016a). The accessions that are able to synthesize sinigrin have not been affected, those that produce primarily 3C aliphatic glucosinolates, such as Ler, gained fresh weight upon sinigrin treatment, but the growth of Col-0 which produces 4C glucosinolates has been challenged. Interestingly, the modulation of fresh weight depends on the side chain, as 4-methylthiobutyl glucosinolate inhibited Ler growth (Francisco et al., 2016a). Subsequent GWAS identified a number of genes potentially involved in the signaling pathway downstream of sinigrin (Francisco et al., 2016b). The effect of sinigrin is multiple, through various pathways, dependent on different metabolites derived from sinigrin. While raphanusamic acid and 3-butenoic acid interfere with auxin signaling, acrylic acid affects the progression of the cell cycle in the meristem

(Katz et al., 2020). This ensures an optimal and specific response for any individual environment change. Other glucosinolates use different pathways to affect plant development. For example, 3-hydroxypropylglucosinolate exploits the TOR (Target of Rapamycin) pathway, which plants and other organisms use to balance growth and development with the available energy and nutrients, to inhibit root growth and development (Malinovsky et al., 2017).

Further, a large scale metabolite analysis of the seeds focusing on glutamine metabolism was used for GWAS and revealed unexpected association of several genes of aliphatic glucosinolate synthesis (Slaten et al., 2020). Indeed, in a panel of Arabidopsis accessions with different combinations of alleles at the two major QTLs, leading to specific presence/absence of one of four aliphatic glucosinolates, presence of 2-propenyl glucosinolate correlated with low glutamine. Loss of aliphatic glucosinolates led to accumulation of Gln in seeds demonstrating the close relationship between secondary and primary metabolism (Slaten et al., 2020). The same conclusion has been made from the investigations of links between glucosinolates and ammonium nutrition. Mutants lacking aliphatic glucosinolates, *myb28 myb29* were hypersensitive to ammonium nutrition, displaying lower biomass and chlorosis (Coleto et al., 2021). However, this is not connected to glucosinolates per se, rather the MYB28 and MYB29 seem to control additional pathways to glucosinolate synthesis, as previously shown for assimilatory sulfate reduction (Yatusevich et al., 2010). Indeed, when grown on ammonium, the *myb28 myb29* plants show a strong upregulation of iron starvation response, and lower Fe content and the negative effect of ammonium on the mutant could be complemented by higher iron supply. Thus, MYB28 and MYB29 seem to be important for maintaining iron homeostasis, which is negatively affected by ammonium and may be part of the mechanism of ammonium sensitivity in Arabidopsis (Coleto et al., 2021). These reports corroborate the findings of (Marino et al., 2016) on the importance of nitrogen nutrition for glucosinolate accumulation.

Investigations of the interaction of glucosinolate metabolism with plant phytohormones focused on auxin, due to the common origin of indolic glucosinolates and IAA from tryptophan. This interaction also underlines the link between glucosinolates and drought stress tolerance (Salehin et al., 2019). Analysis of drought sensitive mutants in auxin signaling, *iaa5, 6, 19*, revealed a reduction in the expression of genes for synthesis of aliphatic glucosinolates and corresponding accumulation of aliphatic glucosinolates. This has been confirmed with mutants lacking aliphatic glucosinolates, which were not able to close stomata during drought stress (Salehin et al., 2019). Thus, aliphatic glucosinolates seem to be a new player in the network of drought stress response. Interestingly, this is a second link of these secondary metabolites to drought stress resistance, since the well described drought stress signaling compound PAP is also produced in the course of glucosinolate synthesis (Chan, Wirtz, Phua, Estavillo, & Pogson, 2013). These two signaling pathways, however, seem to be independent. Recent studies, however, showed also a link between glucosinolates and other hormones. Analysis of *flasher* mutant, which affects Arabidopsis flower architecture revealed that the underlying gene encodes a flavin monooxygenase GS-OX5, which converts methylthioalkyl glucosinolates to methylsulfinylalkyl glucosinolates (Garrido et al., 2020). The mutation causes increased cytokinin levels and reduced jasmonate-Ile conjugate accumulation, both responsible for a number of morphological alterations. How the side chain modification of glucosinolates affects the hormones, however, still needs to be established as well as whether the phenotype is retained in other Arabidopsis accessions than Ler (Garrido et al., 2020).

3. Summary and conclusions

Clearly, the 5 years from ABR80 brought a great deal of new findings and further details on the mechanisms of glucosinolate action in diverse areas of plant, microbial, and insect biology. Nevertheless, the principles and concepts described and reviewed there are still valid and an excellent reference. Glucosinolate emerged as very important players among the mechanisms by which plants shape their microbiome. Thanks to the progress in synthetic biology approaches enabling production of glucosinolates in other plants and various microorganisms they might be a viable means as bioactives controlling soil microflora. At the same time, gut microbiome is being increasingly recognized as being critical for the health benefits of glucosinolates. Thus, new approaches for increasing health benefits of cruciferous vegetables might be achieved through targeting the gut bacteria. The knowledge on new mechanisms by which insects and fungi detoxify glucosinolates and isothiocyanates might help to design new bio-pesticides. The deeper understanding of fundamental biological mechanisms of synthesis, homeostasis, and action of glucosinolates is now ready to be translated into crop improvement. The new alleles obtained from the genetic dissection of the control of glucosinolate accumulation (Kittipol et al., 2019) will be an important source of material for such approaches. The blueprint of how such translation of knowledge from Arabidopsis may look like is the successful reduction in glucosinolate content in seeds of *Brassica rapa* and *B. juncea* (Nour-Eldin et al., 2017). In Arabidopsis, loss of two glucosinolate transporters GTR1 and GTR2 prevented accumulation of glucosinolates in seeds (Nour-Eldin et al., 2012), which is a sought for agricultural trait. Already a mutation of one out of 7 *GTR* genes in *B. rapa* resulted in a 60–70% reduction of seed glucosinolates. Similarly, mutation of 4 out of 12 GTRs from *B. juncea* led to a 60% reduction in glucosinolates (Nour-Eldin et al., 2017). These results show the potential of manipulation of glucosinolate transporters to modify the distribution of glucosinolates in plant organs and in combination with, e.g., the diverse *MYB28* alleles it can be used for generation of a wide range of crops with designed glucosinolate qualitative and quantitative distribution. This is increasingly important as a part of the strategies to ensure food security for future generations.

Acknowledgments

Research in S.K.'s lab is funded by the Deutsche Forschungsgemeinschaft (DFG) under Germany's Excellence Strategy—EXC 2048/1—project 390686111.

References

Aarabi, F., Kusajima, M., Tohge, T., Konishi, T., Gigolashvili, T., Takamune, M., et al. (2016). Sulfur deficiency-induced repressor proteins optimize glucosinolate biosynthesis in plants. *Science Advances*, *2*(10), e1601087. https://doi.org/10.1126/sciadv.1601087.

Abrahams, R. S., Pires, J. C., & Schranz, M. E. (2020). Genomic origin and diversification of the glucosinolate MAM locus. *Frontiers in Plant Science*, *11*, 711. https://doi.org/10.3389/fpls.2020.00711.

Aller, E. S. T., Jagd, L. M., Kliebenstein, D. J., & Burow, M. (2018). Comparison of the relative potential for epigenetic and genetic variation to contribute to trait stability. *G3 (Bethesda)*, *8*(5), 1733–1746. https://doi.org/10.1534/g3.118.200127.

Backenkohler, A., Eisenschmidt, D., Schneegans, N., Strieker, M., Brandt, W., & Wittstock, U. (2018). Iron is a centrally bound cofactor of specifier proteins involved in glucosinolate breakdown. *PLoS One*, *13*(11), e0205755. https://doi.org/10.1371/journal.pone.0205755.

Bell, L., Chadwick, M., Puranik, M., Tudor, R., Methven, L., Kennedy, S., et al. (2020). The Eruca sativa genome and transcriptome: A targeted analysis of Sulfur metabolism and glucosinolate biosynthesis pre and postharvest. *Frontiers in Plant Science*, *11*, 525102. https://doi.org/10.3389/fpls.2020.525102.

Benderoth, M., Textor, S., Windsor, A. J., Mitchell-Olds, T., Gershenzon, J., & Kroymann, J. (2006). Positive selection driving diversification in plant secondary metabolism. *Proceedings of the National Academy of Sciences of the United States of America*, *103*(24), 9118–9123. https://doi.org/10.1073/pnas.0601738103.

Bressan, M., Roncato, M. A., Bellvert, F., Comte, G., Haichar, F. Z., Achouak, W., et al. (2009). Exogenous glucosinolate produced by Arabidopsis thaliana has an impact on microbes in the rhizosphere and plant roots. *The ISME Journal*, *3*(11), 1243–1257. https://doi.org/10.1038/ismej.2009.68.

Burow, M. (2016). Chapter two—Complex environments interact with plant development to shape glucosinolate profiles. In S. Kopriva (Ed.), *Vol. 80. Advances in botanical research* (pp. 15–30). Academic Press.

Buxdorf, K., Yaffe, H., Barda, O., & Levy, M. (2013). The effects of glucosinolates and their breakdown products on necrotrophic fungi. *PLoS One*, *8*(8), e70771. https://doi.org/10.1371/journal.pone.0070771.

Castrillo, G., Teixeira, P. J., Paredes, S. H., Law, T. F., de Lorenzo, L., Feltcher, M. E., et al. (2017). Root microbiota drive direct integration of phosphate stress and immunity. *Nature*, *543*(7646), 513–518. https://doi.org/10.1038/nature21417.

Chan, E. K., Rowe, H. C., Corwin, J. A., Joseph, B., & Kliebenstein, D. J. (2011). Combining genome-wide association mapping and transcriptional networks to identify novel genes controlling glucosinolates in Arabidopsis thaliana. *PLoS Biology*, *9*(8), e1001125. https://doi.org/10.1371/journal.pbio.1001125.

Chan, K. X., Wirtz, M., Phua, S. Y., Estavillo, G. M., & Pogson, B. J. (2013). Balancing metabolites in drought: The sulfur assimilation conundrum. *Trends in Plant Science*, *18*(1), 18–29. https://doi.org/10.1016/j.tplants.2012.07.005.

Chen, J., Ullah, C., Reichelt, M., Beran, F., Yang, Z. L., Gershenzon, J., et al. (2020). The phytopathogenic fungus Sclerotinia sclerotiorum detoxifies plant glucosinolate hydrolysis products via an isothiocyanate hydrolase. *Nature Communications*, *11*(1), 3090. https://doi.org/10.1038/s41467-020-16921-2.

Coleto, I., Bejarano, I., Marin-Pena, A. J., Medina, J., Rioja, C., Burow, M., et al. (2021). Arabidopsis thaliana transcription factors MYB28 and MYB29 shape ammonium stress responses by regulating Fe homeostasis. *The New Phytologist*, *229*(2), 1021–1035. https://doi.org/10.1111/nph.16918.

Czerniawski, P., Piasecka, A., & Bednarek, P. (2021). Evolutionary changes in the glucosinolate biosynthetic capacity in species representing Capsella, Camelina and Neslia genera. *Phytochemistry*, *181*, 112571. https://doi.org/10.1016/j.phytochem.2020.112571.

Eisenschmidt-Bonn, D., Schneegans, N., Backenkohler, A., Wittstock, U., & Brandt, W. (2019). Structural diversification during glucosinolate breakdown: Mechanisms of thiocyanate, epithionitrile and simple nitrile formation. *Plant Journal*, *99*(2), 329–343. https://doi.org/10.1111/tpj.14327.

Elfoul, L., Rabot, S., Khelifa, N., Quinsac, A., Duguay, A., & Rimbault, A. (2001). Formation of allyl isothiocyanate from sinigrin in the digestive tract of rats monoassociated with a human colonic strain of Bacteroides thetaiotaomicron. *FEMS Microbiology Letters*, *197*(1), 99–103. https://doi.org/10.1111/j.1574-6968.2001.tb10589.x.

Fernandez-Calvo, P., Inigo, S., Glauser, G., Vanden Bossche, R., Tang, M., Li, B., et al. (2020). FRS7 and FRS12 recruit NINJA to regulate expression of glucosinolate biosynthesis genes. *New Phytologist*, *227*(4), 1124–1137. https://doi.org/10.1111/nph.16586.

Francisco, M., Joseph, B., Caligagan, H., Li, B., Corwin, J. A., Lin, C., et al. (2016a). The Defense metabolite, allyl glucosinolate, modulates Arabidopsis thaliana biomass dependent upon the endogenous glucosinolate pathway. *Frontiers in Plant Science*, *7*, 774. https://doi.org/10.3389/fpls.2016.00774.

Francisco, M., Joseph, B., Caligagan, H., Li, B., Corwin, J. A., Lin, C., et al. (2016b). Genome wide association mapping in Arabidopsis thaliana identifies novel genes involved in linking allyl glucosinolate to altered biomass and Defense. *Frontiers in Plant Science*, *7*, 1010. https://doi.org/10.3389/fpls.2016.01010.

Frerigmann, H. (2016). Chapter four—Glucosinolate regulation in a complex relationship—MYC and MYB—No one can act without each other. In S. Kopriva (Ed.), *Vol. 80. Advances in botanical research* (pp. 57–97). Academic Press.

Friedrichs, J., Schweiger, R., Geisler, S., Mix, A., Wittstock, U., & Muller, C. (2020). Novel glucosinolate metabolism in larvae of the leaf beetle Phaedon cochleariae. *Insect Biochemistry and Molecular Biology*, *124*, 103431. https://doi.org/10.1016/j.ibmb.2020.103431.

Garrido, A. N., Supijono, E., Boshara, P., Douglas, S. J., Stronghill, P. E., Li, B., et al. (2020). Flasher, a novel mutation in a glucosinolate modifying enzyme, conditions changes in plant architecture and hormone homeostasis. *Plant Journal*, *103*(6), 1989–2006. https://doi.org/10.1111/tpj.14878.

Hacquard, S., Kracher, B., Hiruma, K., Munch, P. C., Garrido-Oter, R., Thon, M. R., et al. (2016). Survival trade-offs in plant roots during colonization by closely related beneficial and pathogenic fungi. *Nature Communications*, 7, 11362. https://doi.org/10.1038/ncomms11362.

Halkier, B. A. (2016). Chapter one—General introduction to glucosinolates. In S. Kopriva (Ed.), *Vol. 80. Advances in botanical research* (pp. 1–14). Academic Press.

Hanschen, F. S., Klopsch, R., Oliviero, T., Schreiner, M., Verkerk, R., & Dekker, M. (2017). Optimizing isothiocyanate formation during enzymatic glucosinolate breakdown by adjusting pH value, temperature and dilution in Brassica vegetables and Arabidopsis thaliana. *Scientific Reports*, 7, 40807. https://doi.org/10.1038/srep40807.

Hansen, J. C., Schillinger, W. F., Sullivan, T. S., & Paulitz, T. C. (2020). Decline in soil microbial abundance when Camelina introduced into a monoculture wheat system. *Frontiers in Microbiology*, *11*, 571178. https://doi.org/10.3389/fmicb.2020.571178.

Harper, A. L., Trick, M., Higgins, J., Fraser, F., Clissold, L., Wells, R., et al. (2012). Associative transcriptomics of traits in the polyploid crop species Brassica napus. *Nature Biotechnology*, *30*(8), 798–802. https://doi.org/10.1038/nbt.2302.

Hirai, M. Y., Klein, M., Fujikawa, Y., Yano, M., Goodenowe, D. B., Yamazaki, Y., et al. (2005). Elucidation of gene-to-gene and metabolite-to-gene networks in arabidopsis by integration of metabolomics and transcriptomics. *Journal of Biological Chemistry*, *280*(27), 25590–25595. https://doi.org/10.1074/jbc.M502332200.

Hiruma, K., Gerlach, N., Sacristan, S., Nakano, R. T., Hacquard, S., Kracher, B., et al. (2016). Root endophyte Colletotrichum tofieldiae confers plant fitness benefits that are phosphate status dependent. *Cell*, *165*(2), 464–474. https://doi.org/10.1016/j.cell.2016.02.028.

Ho, C. L., Tan, H. Q., Chua, K. J., Kang, A., Lim, K. H., Ling, K. L., et al. (2018). Engineered commensal microbes for diet-mediated colorectal-cancer chemoprevention. *Nature Biomedical Engineering*, *2*(1), 27–37. https://doi.org/10.1038/s41551-017-0181-y.

Jeon, J., Bong, S. J., Park, J. S., Park, Y. K., Arasu, M. V., Al-Dhabi, N. A., et al. (2017). De novo transcriptome analysis and glucosinolate profiling in watercress (Nasturtium officinale R. Br.). *BMC Genomics*, *18*(1), 401. https://doi.org/10.1186/s12864-017-3792-5.

Jeschke, V., Gershenzon, J., & Vassao, D. G. (2016). A mode of action of glucosinolate-derived isothiocyanates: Detoxification depletes glutathione and cysteine levels with ramifications on protein metabolism in Spodoptera littoralis. *Insect Biochemistry and Molecular Biology*, *71*, 37–48. https://doi.org/10.1016/j.ibmb.2016.02.002.

Jeschke, V., Gershenzon, J., & Vassão, D. G. (2016). Chapter eight—Insect detoxification of glucosinolates and their hydrolysis products. In S. Kopriva (Ed.), *Vol. 80. Advances in botanical research* (pp. 199–245). Academic Press.

Jeschke, V., Kearney, E. E., Schramm, K., Kunert, G., Shekhov, A., Gershenzon, J., et al. (2017). How glucosinolates affect generalist lepidopteran larvae: Growth, development and glucosinolate metabolism. *Frontiers in Plant Science*, *8*, 1995. https://doi.org/10.3389/fpls.2017.01995.

Kaczmarek, J. L., Liu, X., Charron, C. S., Novotny, J. A., Jeffery, E. H., Seifried, H. E., et al. (2019). Broccoli consumption affects the human gastrointestinal microbiota. *The Journal of Nutritional Biochemistry*, *63*, 27–34. https://doi.org/10.1016/j.jnutbio.2018.09.015.

Kakizaki, T., Kitashiba, H., Zou, Z., Li, F., Fukino, N., Ohara, T., et al. (2017). A 2-oxoglutarate-dependent dioxygenase mediates the biosynthesis of glucoraphasatin in radish. *Plant Physiology*, *173*(3), 1583–1593. https://doi.org/10.1104/pp.16.01814.

Katz, E., Bagchi, R., Jeschke, V., Rasmussen, A. R. M., Hopper, A., Burow, M., et al. (2020). Diverse allyl glucosinolate catabolites independently influence root growth and development. *Plant Physiology*, *183*(3), 1376–1390. https://doi.org/10.1104/pp.20.00170.

Kerwin, R. E., Jimenez-Gomez, J. M., Fulop, D., Harmer, S. L., Maloof, J. N., & Kliebenstein, D. J. (2011). Network quantitative trait loci mapping of circadian clock outputs identifies metabolic pathway-to-clock linkages in Arabidopsis. *Plant Cell*, *23*(2), 471–485. https://doi.org/10.1105/tpc.110.082065.

Kittipol, V., He, Z., Wang, L., Doheny-Adams, T., Langer, S., & Bancroft, I. (2019). Genetic architecture of glucosinolate variation in Brassica napus. *Journal of Plant Physiology*, *240*, 152988. https://doi.org/10.1016/j.jplph.2019.06.001.

Klein, A. P., & Sattely, E. S. (2017). Biosynthesis of cabbage phytoalexins from indole glucosinolate. *Proceedings of the National Academy of Sciences of the United States of America*, *114*(8), 1910–1915. https://doi.org/10.1073/pnas.1615625114.

Kliebenstein, D. J., & Cacho, N. I. (2016). Chapter three—Nonlinear selection and a blend of convergent, divergent and parallel evolution shapes natural variation in glucosinolates. In S. Kopriva (Ed.), *Vol. 80. Advances in botanical research* (pp. 31–55). Academic Press.

Kliebenstein, D. J., Kroymann, J., Brown, P., Figuth, A., Pedersen, D., Gershenzon, J., et al. (2001). Genetic control of natural variation in Arabidopsis glucosinolate accumulation. *Plant Physiology*, *126*(2), 811–825.

Kopriva, S., & Gigolashvili, T. (2016). Chapter five—Glucosinolate synthesis in the context of plant metabolism. In S. Kopriva (Ed.), *Vol. 80. Advances in botanical research* (pp. 99–124). Academic Press.

Kroymann, J., Textor, S., Tokuhisa, J. G., Falk, K. L., Bartram, S., Gershenzon, J., et al. (2001). A gene controlling variation in Arabidopsis glucosinolate composition is part of the methionine chain elongation pathway. *Plant Physiology*, *127*(3), 1077–1088.

Kumar, R., Lee, S. G., Augustine, R., Reichelt, M., Vassao, D. G., Palavalli, M. H., et al. (2019). Molecular basis of the evolution of methylthioalkylmalate synthase and the diversity of methionine-derived glucosinolates. *Plant Cell*, *31*(7), 1633–1647. https://doi.org/10.1105/tpc.19.00046.

Lei, J., Jayaprakasha, G. K., Singh, J., Uckoo, R., Borrego, E. J., Finlayson, S., et al. (2019). CIRCADIAN CLOCK-ASSOCIATED1 controls resistance to aphids by altering indole glucosinolate production. *Plant Physiology*, *181*(3), 1344–1359. https://doi.org/10.1104/pp.19.00676.

Li, B., Tang, M., Caseys, C., Nelson, A., Zhou, M., Zhou, X., et al. (2020). Epistatic transcription factor networks differentially modulate Arabidopsis growth and Defense. *Genetics*, *214*(2), 529–541. https://doi.org/10.1534/genetics.119.302996.

Li, B., Tang, M., Nelson, A., Caligagan, H., Zhou, X., Clark-Wiest, C., et al. (2018). Network-guided discovery of extensive epistasis between transcription factors involved in aliphatic glucosinolate biosynthesis. *Plant Cell*, *30*(1), 178–195. https://doi.org/10.1105/tpc.17.00805.

Liou, C. S., Sirk, S. J., Diaz, C. A. C., Klein, A. P., Fischer, C. R., Higginbottom, S. K., et al. (2020). A metabolic pathway for activation of dietary glucosinolates by a human gut symbiont. *Cell*, *180*(4), 717–728 e719. https://doi.org/10.1016/j.cell.2020.01.023.

Liu, S., Huang, H., Yi, X., Zhang, Y., Yang, Q., Zhang, C., et al. (2020). Dissection of genetic architecture for glucosinolate accumulations in leaves and seeds of Brassica napus by genome-wide association study. *Plant Biotechnology Journal, 18*(6), 1472–1484. https://doi.org/10.1111/pbi.13314.

Liu, X. L., Zhang, J., Yan, Q., Miao, C. L., Han, W. K., Hou, W., et al. (2020). The molecular basis of host selection in a crucifer-specialized moth. *Current Biology, 30*(22), 4476–4482 e4475. https://doi.org/10.1016/j.cub.2020.08.047.

Liu, T., Zhang, X., Yang, H., Agerbirk, N., Qiu, Y., Wang, H., et al. (2016). Aromatic glucosinolate biosynthesis pathway in Barbarea vulgaris and its response to Plutella xylostella infestation. *Frontiers in Plant Science*, 7, 83. https://doi.org/10.3389/fpls.2016.00083.

Liu, Y., Zhou, X., Yan, M., Wang, P., Wang, H., Xin, Q., et al. (2020). Fine mapping and candidate gene analysis of a seed glucosinolate content QTL, qGSL-C2, in rapeseed (Brassica napus L.). *Theoretical and Applied Genetics, 133*(2), 479–490. https://doi.org/10.1007/s00122-019-03479-x.

Malinovsky, F. G., Thomsen, M. F., Nintemann, S. J., Jagd, L. M., Bourgine, B., Burow, M., et al. (2017). An evolutionarily young defense metabolite influences the root growth of plants via the ancient TOR signaling pathway. *eLife, 6*, e29353. https://doi.org/10.7554/eLife.29353.

Malka, O., Easson, M., Paetz, C., Gotz, M., Reichelt, M., Stein, B., et al. (2020). Glucosylation prevents plant defense activation in phloem-feeding insects. *Nature Chemical Biology, 16*(12), 1420–1426. https://doi.org/10.1038/s41589-020-00658-6.

Malka, O., Shekhov, A., Reichelt, M., Gershenzon, J., Vassao, D. G., & Morin, S. (2016). Glucosinolate desulfation by the phloem-feeding insect Bemisia tabaci. *Journal of Chemical Ecology, 42*(3), 230–235. https://doi.org/10.1007/s10886-016-0675-1.

Mao, Y. B., Liu, Y. Q., Chen, D. Y., Chen, F. Y., Fang, X., Hong, G. J., et al. (2017). Jasmonate response decay and defense metabolite accumulation contributes to age-regulated dynamics of plant insect resistance. *Nature Communications, 8*, 13925. https://doi.org/10.1038/ncomms13925.

Marino, D., Ariz, I., Lasa, B., Santamaria, E., Fernandez-Irigoyen, J., Gonzalez-Murua, C., et al. (2016). Quantitative proteomics reveals the importance of nitrogen source to control glucosinolate metabolism in Arabidopsis thaliana and Brassica oleracea. *Journal of Experimental Botany, 67*(11), 3313–3323. https://doi.org/10.1093/jxb/erw147.

McKenzie, M. J., Chen, R. K. Y., Leung, S., Joshi, S., Rippon, P. E., Joyce, N. I., et al. (2017). Selenium treatment differentially affects sulfur metabolism in high and low glucosinolate producing cultivars of broccoli (Brassica oleracea L.). *Plant Physiology and Biochemistry, 121*, 176–186. https://doi.org/10.1016/j.plaphy.2017.10.027.

Mitreiter, S., & Gigolashvili, T. (2021). Regulation of glucosinolate biosynthesis. *Journal of Experimental Botany, 72*(1), 70–91. https://doi.org/10.1093/jxb/eraa479.

Nakano, R. T., Pislewska-Bednarek, M., Yamada, K., Edger, P. P., Miyahara, M., Kondo, M., et al. (2017). PYK10 myrosinase reveals a functional coordination between endoplasmic reticulum bodies and glucosinolates in Arabidopsis thaliana. *Plant Journal, 89*(2), 204–220. https://doi.org/10.1111/tpj.13377.

Nakazaki, A., Yamada, K., Kunieda, T., Sugiyama, R., Hirai, M. Y., Tamura, K., et al. (2019). Leaf endoplasmic reticulum bodies identified in Arabidopsis rosette leaves are involved in Defense against herbivory. *Plant Physiology, 179*(4), 1515–1524. https://doi.org/10.1104/pp.18.00984.

Nour-Eldin, H. H., Andersen, T. G., Burow, M., Madsen, S. R., Jorgensen, M. E., Olsen, C. E., et al. (2012). NRT/PTR transporters are essential for translocation of glucosinolate defence compounds to seeds. *Nature, 488*(7412), 531–534. https://doi.org/10.1038/nature11285.

Nour-Eldin, H. H., Madsen, S. R., Engelen, S., Jorgensen, M. E., Olsen, C. E., Andersen, J. S., et al. (2017). Reduction of antinutritional glucosinolates in Brassica oilseeds by mutation of genes encoding transporters. *Nature Biotechnology*, *35*(4), 377–382. https://doi.org/10.1038/nbt.3823.

Olsen, C. E., Huang, X. C., Hansen, C. I. C., Cipollini, D., Orgaard, M., Matthes, A., et al. (2016). Glucosinolate diversity within a phylogenetic framework of the tribe Cardamineae (Brassicaceae) unraveled with HPLC-MS/MS and NMR-based analytical distinction of 70 desulfoglucosinolates. *Phytochemistry*, *132*, 33–56. https://doi.org/10.1016/j.phytochem.2016.09.013.

Pangesti, N., Reichelt, M., van de Mortel, J. E., Kapsomenou, E., Gershenzon, J., van Loon, J. J., et al. (2016). Jasmonic acid and ethylene signaling pathways regulate glucosinolate levels in plants during rhizobacteria-induced systemic resistance against a leaf-chewing herbivore. *Journal of Chemical Ecology*, *42*(12), 1212–1225. https://doi.org/10.1007/s10886-016-0787-7.

Pastorczyk, M., & Bednarek, P. (2016). Chapter seven—The function of Glucosinolates and related metabolites in plant innate immunity. In S. Kopriva (Ed.), *Vol. 80. Advances in botanical research* (pp. 171–198). Academic Press.

Petersen, A., Hansen, L. G., Mirza, N., Crocoll, C., Mirza, O., & Halkier, B. A. (2019). Changing substrate specificity and iteration of amino acid chain elongation in glucosinolate biosynthesis through targeted mutagenesis of Arabidopsis methylthioalkylmalate synthase 1. *Bioscience Reports*, *39*(7), BSR20190446. https://doi.org/10.1042/BSR20190446.

Pfalz, M., Mukhaimar, M., Perreau, F., Kirk, J., Hansen, C. I., Olsen, C. E., et al. (2016). Methyl transfer in glucosinolate biosynthesis mediated by indole glucosinolate O-methyltransferase 5. *Plant Physiology*, *172*(4), 2190–2203. https://doi.org/10.1104/pp.16.01402.

Pislewska-Bednarek, M., Nakano, R. T., Hiruma, K., Pastorczyk, M., Sanchez-Vallet, A., Singkaravanit-Ogawa, S., et al. (2018). Glutathione transferase U13 functions in pathogen-triggered glucosinolate metabolism. *Plant Physiology*, *176*(1), 538–551. https://doi.org/10.1104/pp.17.01455.

Rahikainen, M., Trotta, A., Alegre, S., Pascual, J., Vuorinen, K., Overmyer, K., et al. (2017). PP2A-B'gamma modulates foliar trans-methylation capacity and the formation of 4-methoxy-indol-3-yl-methyl glucosinolate in Arabidopsis leaves. *Plant Journal*, *89*(1), 112–127. https://doi.org/10.1111/tpj.13326.

Rowe, H. C., & Kliebenstein, D. J. (2008). Complex genetics control natural variation in Arabidopsis thaliana resistance to Botrytis cinerea. *Genetics*, *180*(4), 2237–2250. https://doi.org/10.1534/genetics.108.091439.

Salehin, M., Li, B., Tang, M., Katz, E., Song, L., Ecker, J. R., et al. (2019). Auxin-sensitive aux/IAA proteins mediate drought tolerance in Arabidopsis by regulating glucosinolate levels. *Nature Communications*, *10*(1), 4021. https://doi.org/10.1038/s41467-019-12002-1.

Schillheim, B., Jansen, I., Baum, S., Beesley, A., Bolm, C., & Conrath, U. (2018). Sulforaphane modifies histone H3, unpacks chromatin, and primes defense. *Plant Physiology*, *176*(3), 2395–2405. https://doi.org/10.1104/pp.17.00124.

Schnug, E., & Haneklaus, S. (2016). Chapter ten—Glucosinolates—The agricultural story. In S. Kopriva (Ed.), *Vol. 80. Advances in botanical research* (pp. 281–302). Academic Press.

Sharma, S., Popa, O., Kopriva, S., & Ebenhoeh, O. (2019). An integrative approach to investigate natural variation in the accumulation of aliphatic glucosinolates in *Arabidopsis thaliana*. *bioRxiv*, 838227. https://doi.org/10.1101/838227.

Shukla, S. P., & Beran, F. (2020). Gut microbiota degrades toxic isothiocyanates in a flea beetle pest. *Molecular Ecology*, *29*(23), 4692–4705. https://doi.org/10.1111/mec.15657.

Siebers, M., Rohr, T., Ventura, M., Schutz, V., Thies, S., Kovacic, F., et al. (2018). Disruption of microbial community composition and identification of plant growth promoting microorganisms after exposure of soil to rapeseed-derived glucosinolates. *PLoS One*, *13*(7), e0200160. https://doi.org/10.1371/journal.pone.0200160.

Slaten, M. L., Yobi, A., Bagaza, C., Chan, Y. O., Shrestha, V., Holden, S., et al. (2020). mGWAS uncovers Gln-glucosinolate seed-specific interaction and its role in metabolic homeostasis. *Plant Physiology*, *183*(2), 483–500. https://doi.org/10.1104/pp.20.00039.

Sun, R., Jiang, X., Reichelt, M., Gershenzon, J., Pandit, S. S., & Giddings Vassao, D. (2019). Tritrophic metabolism of plant chemical defenses and its effects on herbivore and predator performance. *eLife*, *8*. https://doi.org/10.7554/eLife.51029.

Textor, S., de Kraker, J. W., Hause, B., Gershenzon, J., & Tokuhisa, J. G. (2007). MAM3 catalyzes the formation of all aliphatic glucosinolate chain lengths in Arabidopsis. *Plant Physiology*, *144*(1), 60–71. https://doi.org/10.1104/pp.106.091579.

Tian, S., Liu, X., Lei, P., Zhang, X., & Shan, Y. (2018). Microbiota: A mediator to transform glucosinolate precursors in cruciferous vegetables to the active isothiocyanates. *Journal of the Science of Food and Agriculture*, *98*(4), 1255–1260. https://doi.org/10.1002/jsfa.8654.

Tian, M., Xu, X., Liu, Y., Xie, L., & Pan, S. (2016). Effect of se treatment on glucosinolate metabolism and health-promoting compounds in the broccoli sprouts of three cultivars. *Food Chemistry*, *190*, 374–380. https://doi.org/10.1016/j.foodchem.2015.05.098.

Traka, M. H. (2016). Chapter nine—Health benefits of glucosinolates. In S. Kopriva (Ed.), *Vol. 80. Advances in botanical research* (pp. 247–279). Academic Press.

van den Bergh, E., Hofberger, J. A., & Schranz, M. E. (2016). Flower power and the mustard bomb: Comparative analysis of gene and genome duplications in glucosinolate biosynthetic pathway evolution in Cleomaceae and Brassicaceae. *American Journal of Botany*, *103*(7), 1212–1222. https://doi.org/10.3732/ajb.1500445.

Vela-Corcia, D., Aditya Srivastava, D., Dafa-Berger, A., Rotem, N., Barda, O., & Levy, M. (2019). MFS transporter from Botrytis cinerea provides tolerance to glucosinolate-breakdown products and is required for pathogenicity. *Nature Communications*, *10*(1), 2886. https://doi.org/10.1038/s41467-019-10860-3.

Wang, B., Wu, Z., Li, Z., Zhang, Q., Hu, J., Xiao, Y., et al. (2018). Dissection of the genetic architecture of three seed-quality traits and consequences for breeding in Brassica napus. *Plant Biotechnology Journal*, *16*(7), 1336–1348. https://doi.org/10.1111/pbi.12873.

Wang, W., Yang, J., Zhang, J., Liu, Y. X., Tian, C., Qu, B., et al. (2020). An Arabidopsis secondary metabolite directly targets expression of the bacterial type III secretion system to inhibit bacterial virulence. *Cell Host & Microbe*, *27*(4), 601–613 e607. https://doi.org/10.1016/j.chom.2020.03.004.

Welte, C. U., de Graaf, R. M., van den Bosch, T. J., Op den Camp, H. J., van Dam, N. M., & Jetten, M. S. (2016). Plasmids from the gut microbiome of cabbage root fly larvae encode SaxA that catalyses the conversion of the plant toxin 2-phenylethyl isothiocyanate. *Environmental Microbiology*, *18*(5), 1379–1390. https://doi.org/10.1111/1462-2920.12997.

Wiesner-Reinhold, M., Schreiner, M., Baldermann, S., Schwarz, D., Hanschen, F. S., Kipp, A. P., et al. (2017). Mechanisms of selenium enrichment and measurement in Brassicaceous vegetables, and their application to human health. *Frontiers in Plant Science*, *8*, 1365. https://doi.org/10.3389/fpls.2017.01365.

Wittstock, U., Kurzbach, E., Herfurth, A. M., & Stauber, E. J. (2016). Chapter six—Glucosinolate breakdown. In S. Kopriva (Ed.), *Vol. 80. Advances in botanical research* (pp. 125–169). Academic Press.

Yatusevich, R., Mugford, S. G., Matthewman, C., Gigolashvili, T., Frerigmann, H., Delaney, S., et al. (2010). Genes of primary sulfate assimilation are part of the glucosinolate biosynthetic network in *Arabidopsis thaliana*. *Plant Journal*, *62*(1), 1–11. https://doi.org/10.1111/j.1365-313X.2009.04118.x.

Zhang, L., Kawaguchi, R., Morikawa-Ichinose, T., Allahham, A., Kim, S. J., & Maruyama-Nakashita, A. (2020). Sulfur deficiency-induced glucosinolate catabolism attributed to two beta-glucosidases, BGLU28 and BGLU30, is required for plant growth maintenance under sulfur deficiency. *Plant and Cell Physiology*, *61*(4), 803–813. https://doi.org/10.1093/pcp/pcaa006.

CHAPTER NINE

Last news on plant mitochondria—A follow-up of ABR volume 63: Mitochondrial genome evolution

Anne-Marie Duchêne and Laurence Drouard*
Institut de biologie moléculaire des plantes-CNRS, Université de Strasbourg, Strasbourg, France
*Corresponding author: e-mail address: Laurence.drouard@ibmp-cnrs.unistra.fr

Contents

Abstract

Volume 63 of *Advances in Botanical Research* "Mitochondrial genome evolution," published in 2012, was edited by Laurence Maréchal-Drouard (series editor Jean-Pierre Jacquot, Pierre Gadal). The volume included 12 chapters written by around 30 internationally recognized researchers. In volume 63, the focus was given to the evolution of the mitochondrial genome in photosynthetic organisms, its organization, and its expression. In this chapter, dedicated to volume 100 of ABR, we first summarize the contributions presented in volume 63, then we describe novel findings obtained in the domain of plant mitochondria during the last 10 years.

1. Major focuses of volume 63

In the first chapter, Lang & Burger, 2012 nicely addressed fundamental questions about the origin of mitochondria but also of the eukaryotic cell. It is commonly accepted that what we now call mitochondria are organelles

Advances in Botanical Research, Volume 100
ISSN 0065-2296
https://doi.org/10.1016/bs.abr.2021.01.004
Copyright © 2021 Elsevier Ltd
All rights reserved.

that emerged only once and are relics of an ancestral α-proteobacteria engulfed into a proto-eukaryotic cell around 1 billion years ago. The authors also discuss the possibility of multiple "mitochondriogenesis" events involving different α-proteobacteria, a hypothesis that cannot be fully excluded and would need broader datasets of sequences and phylogenomic studies. Even more intriguing is the question of the exact nature of the host cell that integrated the symbiotic α-proteobacteria. Its origin is still under debate and the root at the tree of life remains an enigma. In this context, the authors deeply present and confront the two main theories: the serial endosymbiont theory and the symbiogenesis one. The first supports the idea of a gradual evolution of a proto-eukaryotic cell that had the opportunity to acquire a proteobacterium that was subsequently transformed into a mitochondrion. In the second theory, eukaryotes are the result of the symbiosis between an archaebacterium and the alpha-proteobacterium. The merge of these two organisms would have resulted in the emergence of eukaryotic cells with higher complexity. The authors present the weaknesses of such a model that remains one of the most attractive hypotheses to explain the origin of eukaryotes. Yet, there is a lack of evidence for each of these theories, and more reliable information is required to strongly sustain one of them.

In addition to the first endosymbiotic event giving rise to mitochondria, a second occurred more recently in plants, that is the integration of a cyanobacterium at the origin of chloroplasts. Both present-day organelles have strongly evolved and their remnant genomes have very little in common with their ancestral bacterial genomes in terms of gene content, organization, and expression. The two following chapters have dealt with the evolution of the mitochondrial genome with a very strong emphasis on gene transfer events. Many gene transfer events from the organellar to the nuclear genome have been identified. In chapter 2, Ueda & Kadowaki, 2012 discussed the different steps required to achieve complete gene transfer, i.e., to the expression of a nuclear-encoded protein of mitochondrial origin that needs to go back to the organelle for work. Gene translocation and integration, acquisition of a promoter and a poly(A) signal, gaining a mitochondrial targeting signal are all steps described here. The authors also explained how gene transfer can lead to gene substitution when the final location of the protein is changed. This is particularly true for genes of mitochondrial and plastidial origins that correspond to crossed functions. This also led to dual-targeted proteins, an energetic benefit to the cell, and likely also a way to facilitate gene loss from the ancestral bacterial genomes. From an evolutionary point of view, the authors introduce the notion of driving forces to fulfill gene transfer and the dynamic of a still ongoing process in particular in angiosperms. Horizontal gene transfer is also very well described in the following chapter (Mower, Jain, & Hepburn, 2012). Introducing many examples, the authors nicely showed that intracellular transfer of DNA portions exists in all directions between the three nuclear, plastidial, and mitochondrial genomes. Then, they explained how such horizontal transfer has shaped the plant mitochondrial genome. Our knowledge of the mechanisms of the various types of transfers is still limited. Nevertheless, the authors tackled this question and discussed the potential importance of direct contact via parasitism, epiphytic interactions, and grafting.

As emphasized above, plant cells have to deal with three compartments each containing its own genetic information. Mitochondrial and plastidial genomes have lost or transferred in the nuclear genome most of their ancestral prokaryotic DNA, and the biogenesis of the two organelles now strongly rely on very strong interactions with all compartments of the cell. Conversely, plant eukaryotic cell survival strongly depends on the interaction with the organelles, e.g., for the production of energy or the biosynthesis of essential compounds. The two following chapters illustrate very well the coordinated action of nuclear and mitochondrial genomes and the cytonuclear co-adaptation process in plants. First, Touzet (2012) addressed this topic by describing the very appropriate example of the trait cytoplasmic male sterility (CMS). CMS has been found and used in many crop species for hybrid production. The CMS trait is due to the expression of the mitochondrial genome but many mitochondrial genes determining CMS are suppressed or counteracted by proteins expressed from nuclear genes known as "restorer of fertility" genes. In his chapter, P. Touzet described in detail how the process named gynodioecy, i.e., the co-occurrence of female and hermaphrodite plants, is under the nucleo-cytoplasmic control implicating CMS genes and unidentified restorer genes. Interestingly, gynodioecy and dynamic evolution of mitochondrial genome appear to be linked and comparison at the species level showed that CMS genomes can evolve faster (e.g., in maize or beet). Two alternative evolutionary dynamics of gynodioecy on mitochondrial diversity and their potential effect are proposed, and the pattern of diversity observed at the gene and genome levels are described. In the other contribution, Budar & Fujii, 2012 presented a more general view of the crucial interactions between mitochondrial-, plastidial- and nuclear-encoded products, thus implying the necessity of a strong co-adaptation between the various partners. How can this co-adaptation be achieved? They addressed this question by providing several lines of experimental evidence and the molecular actors identified 10 years ago are detailed (e.g., the involvement of some pentatricopeptide repeat (PPR) proteins). Then they focused their description, in the context of the cytonuclear co-adaptation process, on the role played by organellar genomes to the adaptation of plants to their environment. Increasing our knowledge on the adaptative features of cytonuclear co-adaptation remains an important field for breeders as it impacts several traits of interest in many crops.

Then an intriguing chapter on the evolution of mitochondrial genomes in photosynthetic Euglenids and Alveolates is reported by (Flegontov & Lukes, 2012). Besides Archaeplastida (land plants, chlorophytes, rhodophytes, glaucophytes), other organisms possess a plastid or a vestige of it. *Euglena gracillis* is a Euglenida, a group of protists belonging to the Excavata and characterized by the presence of a green plastid acquired by a second endosymbiosis. The members of Excavata possess the most varied mitochondrial genomes in terms of structure and gene content. As an example, the authors reported that *E. gracilis* mitochondrial DNA exists as numerous linear fragments of different sizes, contains a significant portion of non-coding DNA with many diverse repeats but only a very few protein-coding genes. Mitochondrial genomes of photosynthetic Alveolates are also extensively described in chapter 6. Alveolates, with Stramenopiles and Rhizaria, belong to the SAR clade. Within Alveolates are found photosynthetic dinoflagellates and parasitic apicomplexans with a relic of plastids. The smallest mitochondrial genomes were found in this group of organisms. They usually have an extremely reduced coding capacity (e.g., only three protein-coding genes in dinoflagellates). These mitochondrial genomes also encounter very peculiar non-canonical features in architecture and expression and the authors deeply described all the various oddities observed in these genomes. They speculated on the convergent evolution of radically reduced mitochondrial genomes.

In the next chapter, the existence of introns, another interesting illustration of the evolutionary history of mitochondrial genomes in plants and algae, is presented by Bonen (2012). The two categories of introns, introns of class I and II, are first described in detail. Thanks to the broad range of sequenced genomes, peculiar behavior of introns has been collected, such as the presence of trans-splicing mechanisms, the highly invasive nature of some group I introns, the degeneration of ribozyme features in higher plants. The relationship between splicing and other events such as RNA editing (C to U and U to C conversions) and the machineries required for their splicing are also discussed. As a whole, this sums up very well the role played by introns in mitochondrial gene expression.

Among the Archeplastidia, green algae (chlorophytes) represent a model of interest to decipher mitochondrial evolution. Rodriguez-Salinas, Remacle, & Gonzales-Halphen, 2012 have devoted their chapter to green algae mitochondrial genomes. Again, the ever-growing number of fully sequenced mitochondrial genomes revealed an extraordinary diversity in genome size, organization structure, and gene content. Part of the chapter is also focused on the interaction with the other compartments of the cell, in particular, the chloroplast and the nucleus. Information regarding how gene transfer events from the mitochondrion to the nucleus occurred is also detailed. The complex exchanges between mitochondria and chloroplasts at the genomic level and the interactions at the metabolic levels are also largely discussed.

Mitochondria are essential to cell life and numerous dysfunctions often lead to severe diseases in human (e.g., myopathies) or to CMS in plants. Another important aspect of plant mitochondrial genome tackled in the following chapter deals with recombination, maintenance, and repair. Kühn & Gualberto, 2012 first described how highly variable the structure of mitochondrial DNA is among plant species. This is mostly due to highly active recombination processes. Large sequence repeats (but also numerous short ones) allow reciprocal recombination, generate multipartite structures, and lead to mitochondrial heteroplasmy, an important source of genetic diversity. The occurrence of alternative configurations and the dynamics of the process are described. These recombination processes appear to play major roles in DNA replication and repair and part of the chapter is devoted to the enzymes and pathways characterized at that time. The complexity of the various mechanisms seems to represent only the tip of the iceberg and deciphering the interconnected pathways in order to better understand how mitochondrial DNA is transmitted to the progeny is challenging.

In chapter 10, the evolutionary history of plant organellar genomes is exemplified by the fascinating discovery of the PPR proteins. Gutmann, Gobert, & Giegé, 2012 highlighted the work dealing with this huge protein family. PPR proteins are also found in the chloroplasts, but the authors focused their contribution to the mitochondrial world. Contrary to animals, higher plant mitochondrial genomes are very large, subject to splicing and RNA editing (C to U and U to C conversions), and the organelle imports a huge number of PPR proteins (there are 80 times more PPR genes in higher plants than in animal nuclear genomes). Gutmann, Gobert, and Giegé deeply related the evolution of PPR proteins in relationship with the occurrence of specific gene expression processes (transcription, splicing, RNA editing). In the second part, they described in detail the different classes of PPR proteins, their structure, and their specific domains. For instance, the DYW domain found in a subgroup of PPR proteins is proposed to support RNA editing activity, and diversification of this subclass of PPR proteins correlates with the occurrence of this process.

The penultimate and final chapters are dedicated to macromolecules (i.e., proteins and RNAs) trafficking. Mitochondria biogenesis requires at least 1000 proteins. As mitochondrial genomes only encode a very limited set of protein-coding genes (e.g., 35 in *Arabidopsis thaliana* mitochondrial DNA), and often an incomplete set of tRNA genes, macromolecules trafficking allowing the come-back of proteins and tRNAs into mitochondria is essential. A similar situation exists for the second endosymbiotic organelle, the chloroplast, that also needs to import most of its proteins from the cytosol. Exchanges and interactions between the two organelles, the cytosol, and the nucleus lead to a complex situation. Murcha et al. (2014) presented a very detailed analysis of the various protein import pathways that have emerged during evolution. Targeting peptides, signal recognition, and processing steps of mitochondrial proteins are deeply described as well as the components and multiple machineries that have been developed at the level of the outer membrane, intermembrane space and inner membrane of the organelle to translocate mitochondrial proteins to their final location. An overview of the mitochondrial precursor processing and presequence degradation is also presented. Finally, in the last contribution by Michaud & Duchêne, 2012, other important aspects of macromolecules trafficking are discussed. First, the emphasis is given on the phenomenon of dual targeting of proteins to mitochondria and chloroplasts that raises questions about inter-organellar communication. Then, they presented an emerging topic that is mRNA trafficking within the plant cell and the subcellular localization of cytosolic mRNAs to the vicinity of mitochondria. This novel area of research, whose mechanisms and roles are still largely unknown, opens new perspectives concerning the biogenesis of mitochondria. Finally, the last part is devoted to tRNA mitochondrial import, another essential process in plant mitochondria, in particular for the mitochondrial translation process. The extent of tRNA import, the still rather enigmatic mechanism, and evolutionary aspects are discussed.

2. Recent developments in plant mitochondria

2.1 Mitochondrial genome organization, dynamic, and evolution

During the last decade, the still growing number of fully sequenced mitochondrial genomes has confirmed their striking diversity in terms of size, content and organization. Nonexhaustively, we can cite the very compact mitochondrial genome (51 Kb) of *Cyanophora paradoxa*, a freshwater alga belonging to glaucophytes, the most basal group of Archeplastidia (Price et al., 2012). This compact genome still contains around 50 protein-coding genes and an almost complete set of tRNA genes (only tRNAThr genes are missing). In contrast, mitochondrial genomes from two Silene species, with sizes of 6.7 and 11.3 Mb, are by far the largest known ones (Sloan et al., 2012). These two mitochondrial genomes have high mitochondrial mutation rates, present complex multichromosomal structures but only code for 25 protein-coding genes and 2 tRNA genes. Surprisingly, two slow evolving mitochondrial genomes from two other Silene species have a smaller size (253 and 427 Kb) than average for Angiosperms, and contain 25 protein-coding genes and even a higher number of tRNA genes (6 and 9, respectively) as compared to the large Silene mitochondrial genomes. Such differences within the same evolutionary species raise interesting questions: why does this genus present such a rapid mitochondrial genome evolution; what are the forces driving mutation rate and genome structure?

Non-photosynthetic (i.e., parasitic and mycoheterotrophic) plants represent another interesting model to study. They possess reduced plastid genomes and the level of gene loss correlates quite well with the timing of the transition to heterotrophy (Barrett et al., 2014). By contrast, mitochondrial genomes of such plants are similar both in terms of size and gene content to that of autotrophic plants. They nevertheless contain peculiar features, for instance, the mycoheterotrophic plant *Hypopitys monotropa* mitochondria harbor a small circular DNA fragment and a large linear fragment with multiple repeats at the extremities that may function as telomeres (Shtratnikova, Schelkunov, Penin, & Logacheva, 2020).

As illustrated above, plant mitochondrial genomes are highly dynamic. This is particularly true in the case of the large genomes from Angiosperms. Another nice example came out recently with the multichromosomal architecture observed in the *Solanum tuberosum* mitochondrial genome

(470 Kb) where at least three autonomous DNA molecules exist in this species (Varre et al., 2019). Isoforms due to recombination were also detected by PCR. Thus, contrary to what was previously thought, many plant mitochondrial genome structures cannot be resolved into a single circular map. Recombination activities involving repeated sequences represent the main cause explaining the dynamics of these genomes. Over the past decade, nuclear genes involved in pathways of DNA repair by homologous recombination and mismatch repair have been identified. They represent the main pathways responsible for the maintenance of plant mitochondrial genome stability. A very deep overview of our present knowledge on the dynamics and mechanisms of mutation has been written by Gualberto and Newton (2017). They also largely discuss the link between mitochondrial DNA maintenance and the appearance of DNA heteroplasmy and explain how sub-genomic DNA molecules that replicate autonomously can relocate into the main mitochondrial genome.

2.2 Mitochondrial genome expression

Although mitochondria biogenesis mostly relies on the expression of the nuclear genome, accurate and optimal expression of the mitochondrial genome is essential to provide functional respiratory chain complexes and ribosomes and therefore the energy required during cell life. This means that mature and functional mitochondrial RNAs must be generated and their expression tightly regulated. RNA metabolism in plant mitochondria is complex and depends on the coordinated expression between the mitochondrial and the nuclear genomes. In the model plant, *Arabidopsis*, numerous transcription initiation sites located all along the mitochondrial genome were identified almost 2 decades ago and the genome is transcribed thanks to nuclear-encoded phage-type RNA polymerases. There is a quite relaxed control of transcription and almost if not all the entire genome is transcribed. This generates multiple precursor RNAs that need to be further processed at various levels, and it is now clear that post-transcriptional RNA processes play a major role in plant mitochondrial genome expression. This includes 5′- and 3′-processing by exonucleases, RNA splicing and RNA editing, RNA modification, and RNA quality control. The multiple roles of PPR proteins in these processes were ever more extensively described (Hammani et al., 2014). As mentioned above, RNA metabolism in plant mitochondria has been deeply described in Gutmann, Gobert, & Giegé, 2012 and is also presented in other reviews (Gagliardi & Binder, 2007;

Hammani & Giegé, 2014). Analyses of the RNA landscapes (Binder, Stoll, & Stoll, 2016; Varre et al., 2019) and of the proteomes (Fuchs et al., 2020) of plant mitochondria confirmed the importance of RNA metabolism for organelle biogenesis. What did we learn from recent work in the field?

A decade ago, the mechanism governing transcription initiation and termination, and involved transcription factors were unknown. The data are still scarce, nevertheless, two potential mitochondrial nucleases, MNU1 and MNU2, are likely involved in 5′ processing of mitochondrial transcripts (Stoll & Binder, 2016). Furthermore, some recent work led to the identification of mitochondrial transcription termination factors (mTERF). Only a few mTERFs were characterized, and correspond to plastidial or mitochondrial proteins (Quesada, 2016). For example, mTERF15 was shown to be involved in the processing of *nad2* pre-RNA in *Arabidopsis* mitochondria (Hsu, Wang, Hsieh, Hsieh, & Jauh, 2014), and mTERF ZmSmk3 in nad4 intron splicing in maize mitochondria (Pan et al., 2019). However, no mTERF directly involved in transcription termination by specifically binding to the 3′ end of a plant mitochondrial gene has been identified yet.

Mitochondrial RNA editing was discovered 30 years ago. It mainly consists, in seed plants, of C to U conversions, and in some plants such as hornworts, lycophytes, and ferns also in U to C conversions. Since then, extensive work has been dedicated to this topic. While the number of PPR proteins implicated in the editing of specific sites increased regularly, a major breakthrough came out with what was called the "PPR code." How PPR proteins, which contain tandem arrays of degenerate 35 amino acid motifs, can specifically deaminate edited C residues was unsolved until recently. First, it was demonstrated by several groups (see references in Small, Schallenberg-Rudinger, Takenaka, Mireau, & Ostersetzer-Biran, 2020) that the DYW motif of a subclass of PPR proteins likely confers the cytidine deaminase activity and the last evidence was given by inducing a targeted C to U editing activity upon expression of a DYW-PPR protein in *E. coli* (Oldenkott, Yang, Lesch, Knoop, & Schallenberg-Rudinger, 2019). Second, modeling PPR structure followed by experimental validation allowed the elucidation of the RNA recognition for PPR proteins. Roughly, each PPR motif forms two antiparallel α-helices and two amino acids in each motif determine base recognition (Barkan et al., 2012; Yan et al., 2019). Elucidation of this RNA recognition code now represents an attractive tool to manipulate specified RNA targets and for engineering approaches. As a proof of concept, designed PPR proteins were used to knockdown targeted mitochondrial transcripts (Colas Des Francs-Small, Vincis Pereira Sanglard, & Small, 2018). Now, trans-acting factors and editosome machineries are being characterized. An up-to-date review on plant mitochondrial RNA editing has been recently published and summarizes the still numerous open questions facing this fascinating process (Small et al., 2020). Among the many other roles attributed to PPR proteins it is worth mentioning the special case of RNase P activity that cleaves the 5′ leader sequence of tRNA precursors. In *Arabidopsis* organelles and nucleus, it has been demonstrated by Gutmann, Gobert, and Giege (2012) that 5′ end tRNA maturation does not require a ribonucleoprotein complex but rather involves PPR proteins called PRORP (for proteinaceous RNase P).

Another post-transcriptional modification recently identified in *Arabidopsis* and cauliflower mitochondrial transcripts is the presence of m6A (N6-methylation of A residues) (Murik et al., 2020). The authors proposed that m6A may play important roles in regulating plant mitochondrial gene expression. Finally, studies on *Chlamydomonas reinhardtii* mitochondria revealed a novel type of post-transcriptional modification, that is the addition of C-rich tails at the 3′ extremities of mRNAs (Gallaher et al., 2018; Salinas-Giegé et al., 2017). This process, specific of the Chlorophyceae group, is mainly found only in mRNAs and is mitochondria specific. Another intriguing feature of all Chlamydomonas mitochondrial mRNAs is the lack of 5′ untranslated regions (UTR). How leaderless mitochondrial mRNAs are translated, and how and why mRNAs are polycytidylated in this organism are questions to be addressed in the future.

The majority of gene expression processes and their regulation rely on protein-protein interactions and on the existence of protein complexes, which is particularly pertinent in plant mitochondria. While several proteome studies have permitted to obtain a well-defined set of plant mitochondrial proteins (see for instance, Rao et al. 2017 as one of the most recent articles on the topic), it is only recently that complexome data were obtained. Thanks to an extensive *Arabidopsis* mitochondrial complexome profiling analysis (Senkler et al., 2017), 1359 unique proteins belonging to 17 complexome fractions were identified. Interestingly, this study allowed the discovery of new complexes, such as Tat or SAM complexes as well as complexes including several PPR proteins. In this vein, the association of PPR proteins with plant mitochondrial ribosomes has been identified through an extensive complexome profiling in combination with chemical crosslinking. These data comfort the nice work performed by Waltz et al. (2019) and Waltz, Soufari, Bochler, Giegé, and Hashem (2020) determining the structure at high resolution of the *Arabidopsis* and cauliflower mitochondrial ribosomes (mitoribosome). Combining biochemical characterization and single-particle cryo-electron microscopy, the authors revealed the very peculiar structure of the plant mitoribosome. This mitoribosome contains 19 plant-specific proteins of which 10 are PPR proteins, presents a small subunit larger than the large one (a mindboggling characteristic by the way), and extra RNA domains. As a whole, plant mitoribosome appears larger than their bacterial, yeast, or animal counterparts. This again nicely demonstrates how mitochondria biogenesis has diverged during evolution. A follow-up of this work is now crucial to address unsolved questions: How is translation initiated and regulated? What are the links between mRNA processing and translation? What are the structure and composition of mitoribosome from the evolutionary divergent photosynthetic species that evolved since the origin of the "green" plant lineage?

Mitochondria originate from the endosymbiosis of a bacterium. Over the course of evolution, most of the bacterial genes have been lost or transferred to the nucleus. In land plants, most of the remaining mitochondrial genomes contain between 20 and 50 protein genes, mainly involved in OXPHOS and translation (Mower, 2020). Thus, to build the active mitochondrion, most proteins (Rao et al., 2017) need to be imported from the surrounding cytosol, as well as numerous RNAs (such as tRNAs) (Kmiec, Glaser, Duncan, Whelan, & Murcha, 2012; Michaud & Duchêne, 2012).

The determination of a mitochondrial proteome can be performed experimentally or using prediction tools, with both approaches having limits. Proteomics analyses have identified roughly 1000 different proteins in *A. thaliana* mitochondria, but the detection of low abundance proteins is difficult. Prediction tools are efficient to identify a particular set of mitochondrial proteins (those with an N-terminal cleavable sequence, see below) but no other types of imported mitochondrial proteins (Rao et al., 2017). Whatever the signals, all these proteins are translated by cytosolic ribosomes before their import into mitochondria, and four major import pathways have been identified (Fig. 1).

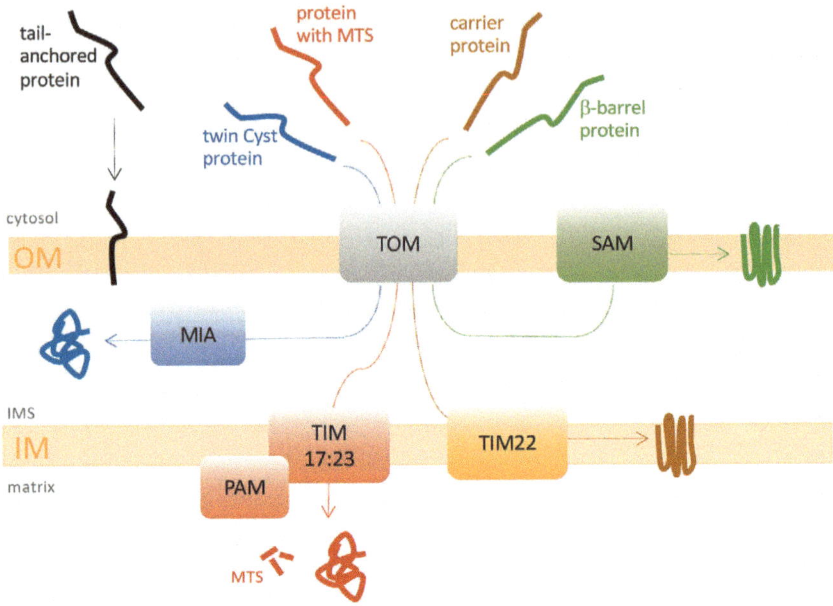

Fig. 1 Proteins are imported into mitochondria through different pathways. Except for some OM proteins, which are directly inserted into the membrane (exemplified with tail-anchored proteins, in black), all imported proteins pass the OM through TOM (translocase of the outer membrane), then are directed to their final destination via different complexes. Twin-cysteine proteins (in blue) are folded in the IMS thanks to MIA (mitochondrial IMS assembly). Proteins with MTS (in orange) use TIM17:23 (translocase of the inner membrane 17:23), and PAM (presequence-assisted motor) to reach the matrix. Carrier proteins (in brown), with their membrane multi-spanning domains, are inserted into IM using TIM22. β-Barrel proteins (in green) are inserted into the outer membrane through SAM (sorting and assembly machinery).

3. Relations within the cell: Targeting and import of proteins and RNAs

3.1 Protein import mechanisms

3.1.1 The protein import signals

Many nuclear-encoded mitochondrial proteins are synthesized as precursors with a cleavable N-terminal extension called presequence or MTS (mitochondrial targeting sequence). Other imported mitochondrial proteins have internal signals. All these signals allow recognition by receptors at the outer membrane and translocation across membranes via specific machineries. In *A. thaliana*, MTSs have an average size of 40–45 amino acids. They are enriched in hydroxylated, hydrophobic, and positively charged amino acid residues, but deficient in acidic amino acids. They also tend to form amphiphilic α-helices (Kmiec, Glaser, Duncan, Whelan, & Murcha, 2012). Due to the presence of chloroplasts in plant cells, which also import most of their proteins, targeting signals to mitochondria need to be specific. However, the differences between mitochondrial and plastidial presequences are not so obvious. Moreover, numerous proteins were found to be dual targeted to both mitochondria and chloroplasts (Ghifari, Gill-Hille, & Murcha, 2018; Michaud & Duchêne, 2012; Pujol, Maréchal-Drouard, & Duchêne, 2007). Moreover, the cleavable targeting sequence can be followed by a second signal for the intra-mitochondrion localization of the protein (bipartite targeting peptide). Last, many mitochondrial proteins have no MTS. The import signals are internal, within the proteins, and in most of the cases poorly identified (Murcha et al., 2014).

According to peculiarities of the proteins, and their final location in the different submitochondrial compartments, specific import machineries are used.

3.1.2 Import machineries at the outer membrane (OM)

3.1.2.1 Translocase of the outer membrane (TOM)

The first contact of most nuclear-encoded mitochondrial proteins with mitochondria is at the level of TOM complex. TOM is found in mitochondria of all eukaryotes. It is composed of a pore, Tom40, receptor proteins, and accessory smalls proteins involved in the complex stability. In yeast, three major receptors have been identified, Tom20, Tom70, and Tom22.

In *A. thaliana*, two essential genes code for Tom40 (At3g20000 and At1g50400), and four accessory proteins have been identified (Tom5, Tom6, Tom7–1, Tom7–2). Concerning receptors, no equivalent of Tom70 could be identified. Tom9 (Tom9–1 and Tom9–2) appear as truncated versions of Tom22. Last, Tom20 has no sequence similarities with yeast Tom20. In *A. thaliana*, four Tom20 have been identified, which seem to have overlapping specificities (Lister et al., 2007).

3.1.2.2 Sorting and assembly machinery (SAM)

SAM is another protein import complex at the OM. SAM is involved in the insertion of outer membrane β-barrel proteins. β-Barrel proteins first pass through TOM, before being directed to SAM (Fig. 1). Sam35, Sam37 (Metaxin), and two Sam50 subunits have been identified in *A. thaliana*

(Murcha et al., 2014). A decrease in import rate of numerous precursor proteins and of the β-barrel proteins TOM40 and VDAC (two β-barrel proteins) was observed in *sam37* mutant, confirming the role of Sam37 in protein import (Lister et al., 2007).

Plant specific receptors at the OM have been also identified. For example, Om64 is a paralogue of Toc34, a plastidial protein import receptor. Om64 seems important for the import of some proteins, but the link with translocation through TOM is not clear (Lister et al., 2007).

Last, several OM proteins are imported through or into the OM without passing through TOM, but the mechanism of their insertion is elusive in plants. This concerns tail-anchored proteins as well as some single and multi-span proteins (Michaud & Duchêne, 2012; Murcha et al., 2014).

3.1.3 Import of proteins into IMS (intermembrane space)

The IMS contains a limited number of proteins. These proteins first pass TOM. Then, those with twin cysteines are assembled by MIA (mitochondrial intermembrane space import and assembly) complex. These proteins typically contain twin CX9C or CX3C motifs. MIA components have been identified in *A. thaliana*, such as Mia40 and Erv1 (Peleh, Zannini, Backes, Rouhier, & Herrmann, 2017). While yeast Mi40 is bound to the IM via its N-terminal sequence, *Arabidopsis* Mia40 lacks its docking domain to IM and is thus soluble.

3.1.4 Import machineries at the inner membrane (IM)

Two major complexes are found at the IM, TIM17:23 and TIM22. Each is specialized in the import of a particular set of proteins (Kolli, Soll, & Carrie, 2018). TIM17:23 is involved in the import of proteins with MTS, which are subsequently targeted into the matrix. Most of Tim proteins exist in multiple isoforms in plants. For example, in *A. thaliana*, three genes for Tim17, Tim21, or Tim23 have been identified (Murcha et al., 2014). The specific role of some of these isoforms has been elucidated: for example, Tim17–1 is expressed in dry seed and regulates germination (Wang et al., 2014). TIM17:23 is associated with the Presequence translocase-Associated Motor (PAM), which drives precursor proteins through the IM and is involved in their folding in the matrix. As for Tim proteins, many paralogues of Pam proteins have been identified in *A. thaliana* (Ghifari et al., 2018; Murcha et al., 2014). The TIM22 complex responsible for the import of carrier-type proteins has been first identified by Haferkamp and Schmitz-Esser (2012). In *A. thaliana*, two identical genes code for Tim22 (Ghifari et al., 2018; Murcha et al., 2014).

3.1.5 Protein processing

Upon import into the matrix, presequences are cleaved and mature proteins are folded. The cleavage of the presequence is performed by the mitochondrial processing peptidase (MPP). In plants, MPP has the particularity to be incorporated into OXPHOS complex III (cytochrome *bc1* complex), showing links between respiration and protein import. Motifs near the cleavage site have been identified, with an Arg in position -2 or -3 respective to the cleavage site. Other peptidases are involved in the protein maturation and presequence degradation, such as octapeptidyl peptidase (Oct1), inner membrane protease (IMP), intermediate cleavage peptidase 55 (ICP55), presequence protease (PreP), organellar oligopeptidase (OOP), and different aminopeptidases (Ghifari, Huang, & Murcha, 2019).

Recently, the protein translocase subunits TatB and C have been identified in *Arabidopsis*, suggesting that plant mitochondria contain a twin-arginine translocation (Tat) pathway (Carrie, Weissenberger, & Soll, 2016). The role of this pathway in mitochondria needs further investigation.

In conclusion, the main import complexes identified in the mitochondria of yeast or mammals are also found in plants. However, a number of these components are in multiple copies in plants, and the exact role of each isoform needs evaluation. Moreover, some complex components appear specific to plants. Besides, plant-specific import systems seem to exist, like TAT, even if their role needs to be explored.

3.2 Post-translational versus co-translational protein import

Before import through the OM and the IM, protein precursors have to reach the mitochondrial surface. Two models, which are not exclusive, have been described for this step. One is a post-translational import. In this case, proteins are translated on free cytosolic polysomes, then targeted and imported into mitochondria. The other model is a co-translational import. In this model, mRNAs are first targeted to the mitochondrial surface, in situ translated, and the newly synthesized proteins are imported into mitochondria (Fig. 2). Consistent with this model, some cytosolic mRNAs and ribosomes were found associated with the mitochondrial surfaces in many organisms such as fungi, animals, or plants (Lesnik, Golani-Armon, & Arava, 2015; Michaud, Maréchal-Drouard, & Duchêne, 2010).

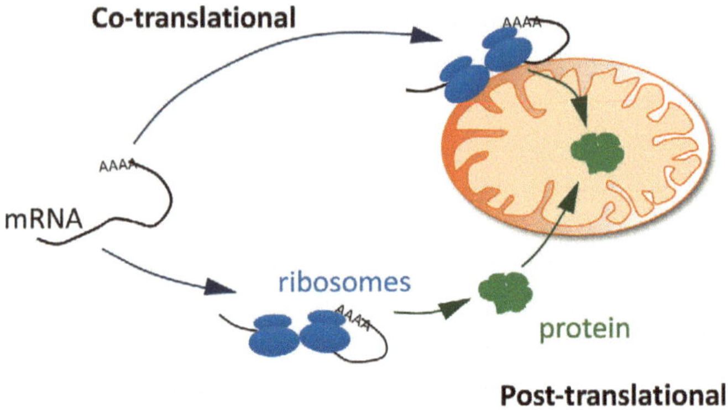

Fig. 2 Two pathways are proposed for protein targeting to mitochondria. The post-translation import proposes that precursor proteins are translated in the cytosol, then targeted and imported into mitochondria. The co-translational process proposes that translation occurs at the mitochondrial surface, and is followed by protein import. In both cases, import through OM and IM is performed via complexes presented in Fig. 1.

Messenger RNA sorting to the plant mitochondrial surface has been first demonstrated in Michaud et al. (2010). Then, RNA-seq analysis of cytosolic mRNAs bound to plant mitochondria was performed, and many cytosolic messengers encoding mitochondrial proteins (MITO mRNAs) were found enriched in the mitochondrial fraction (Vincent et al., 2017). No correlations between mRNA targeting and physico-chemical properties of the encoded proteins were detected (presence of MTS, number of transmembrane domains, hydrophobicity). By contrast, a clear correlation between mRNA targeting and protein function could be established. Messengers corresponding to OXPHOS, TCA, or mitochondrial ribosomal proteins were significantly enriched at the surface of mitochondria. Those corresponding to proteins of mitochondrial fatty acid or DNA metabolisms were lowered in the mitochondrial fraction. Messenger RNA targeting would thus allow a coordinate spatial, and probably temporal, expression of proteins from the same metabolic pathway (Vincent et al., 2017).

Signals allowing or controlling mRNA targeting to the mitochondrial surface were also identified, in 5′ and 3′ UTR of mRNAs. The most studied example of signal in 3′ UTR is that of Voltage-Dependent Anion Channel 3 (VDAC3) in *A. thaliana* (Michaud, Maréchal-Drouard, & Duchêne, 2014; Michaud, Ubrig, et al., 2014). Alternative cleavage and polyadenylation produce two isoforms of VDAC3 transcripts that differ by the length of their

3′ UTR, one being 140 nt longer than the other. Using RT-qPCR and RNA imaging in living cells, it was shown that the longest mRNA was targeted to the mitochondrial surface and that the 140-nt extension (named aUTR for alternative UTR) was necessary for mitochondrial targeting. Moreover, aUTR was sufficient to target an unrelated mRNA to the mitochondrial surface. Finally, VDAC3 mRNA targeting to the mitochondrial surface was shown to impact mitochondria size and number *thaliana* (Michaud, Maréchal-Drouard, & Duchêne, 2014; Michaud, Ubrig, et al., 2014). In 5′ UTR, upstream AUGs appear as signals which modulate mRNA association with mitochondria. Upstream AUGs (uAUGs) and uORFs are known to reduce the translation efficiency of downstream sequences. This was also observed for some mRNAs associated with plant mitochondria. In parallel, the presence of uAUGs was found to reduce mRNAs binding to mitochondria. These points establish a strong connection between the efficiency of translation and mRNAs association with mitochondria (Vincent et al., 2017).

Thus, both the presence of *cis*-element in 3′ UTR and an efficient translation are needed for mRNA localization in the vicinity of plant mitochondria. It seems that the *cis*-element in 3′ UTR is essential for the traffic toward mitochondria, and translation for the strong association with mitochondria (Vincent et al., 2017).

In order to identify mRNAs receptors at the surface of mitochondria, an in vitro system was also developed to study mRNA binding to the mitochondrial surface (Michaud, Maréchal-Drouard, & Duchêne, 2014). This showed that at least some mRNAs were able to bind to isolated mitochondria, and that mRNA docking to the OM is dependent on the TOM complex, and weakly on VDAC. Protein synthesis and import must be tightly regulated together with the synthesis of mitochondrial-encoded proteins, and with the activity of mitochondria and cells. It has been recently shown in yeast that mRNA localization at the surface of mitochondria influence protein synthesis, depending on cell needs (for example during aerobic growth) (Tsuboi et al., 2020). Controlling mRNA localization thus provides a control level of mitochondrial activity.

3.3 tRNA import

Plant mitochondrial genomes contain an incomplete set of tRNA genes. To carry out translation, missing tRNAs are thus nuclear-encoded and imported from the cytosol (Michaud & Duchêne, 2012). The imported tRNAs sets have been initially determined from 2D-acrylamide gels (Maréchal-Drouard et al., 1990). More recently, a systematic analysis could be performed when the nuclear genome is sequenced and contains a limited number of isoacceptor tRNAs families. Indeed, the intracellular localization of members of the 49 nuclear-encoded tRNA families was determined by Northern blot in *Chlamydomonas reinhardtii* (Vinogradova, Salinas, Cognat, Remacle, & Marechal-Drouard, 2009). However, the exact number of mitochondrial imported tRNAs is still unknown in most plants. The new RNA-seq technologies would probably allow deeper analyses of imported tRNAs populations and their evolution among plants (Warren & Sloan, 2020).

3.3.1 Mechanisms of tRNA import

It has been previously shown that two components of the mitochondrial outer membrane, VDAC and TOM, are involved in tRNA recognition and import (Salinas et al., 2006). The interactions between tRNA and VDAC were more precisely determined taking advantage of the particularities of potato VDACs: in vitro, the VDAC34 isoform strongly interacts with tRNAs, but not VDAC36 isoform. This system allowed to show that the whole VDAC34 β-barrel structure and the N-terminal α-helix are essential for tRNA binding (Salinas et al., 2014, 2019). Four VDAC proteins (VDAC1 to 4) have been identified in *A. thaliana*. As in potato, differential interactions were observed in vitro between *At*-VDAC isoforms and tRNA. VDAC4 appeared to strongly interact with a tRNA transcript. Transfer RNA interaction with VDAC1 and VDCA2 was moderate, and that with VDAC3 was very low (Hemono et al., 2020). VDAC1 is the most abundant VDAC in *A. thaliana* (Fuchs et al., 2020), so it's in vivo implication in tRNA import was evaluated. However, no difference in imported tRNA steady-state level was observed in mitochondria from *vdac1* KO plants and wild-type plants (Hemono et al., 2020). Two other proteins of the outer membrane, Tric1 and Tric2, were found to be involved in tRNA import in vitro (Murcha et al., 2016). These two proteins are members of the PRAT family (preprotein and amino acids transporters). However, as for *vdac1*, no change in imported tRNA steady state could be found in mitochondria of the *tric1 tric2* double mutant (Murcha et al., 2016). Last, the phenotype of *vdac4* KO mutant was so strong that mitochondria preparation from this mutant could not be considered (Hemono et al., 2020).

In conclusion, different components of the outer membrane can interact with tRNAs in vitro, but the exact role and importance of each of them are not clear. It is possible that different pathways, more or less redundant, are used in vivo, but it should be underlined that the knockdown of the main tRNA import channel should have a dramatic impact on mitochondria and the whole plant. With these considerations, *At*-VDAC4 deserves further investigations.

In contrast to the outer membrane, no component of the inner membrane has been identified as involved in tRNA import.

3.3.2 Control of tRNA import

Concerning the control of tRNA import, the population of imported tRNAs appears to vary from one plant to another, even between close relatives. Moreover, the identification of mitochondrial genomes with a very limited number of tRNA genes suggests a very large tRNA import in such species (Petersen, Cuenca, Moller, & Seberg, 2015; Sloan, Muller, McCauley, Taylor, & Storchova, 2012; Skippington, Barkman, Rice, & Palmer, 2015). So, import selectivity seems to evolve remarkably fast in the green linage. In parallel, it has been observed that imported tRNAs complement mitochondrial-encoded ones to provide a complete set for mitochondrial translation. For example, the steady-state level of imported tRNAs correlates with codon usage in mitochondria of *C. reinhardtii* (Vinogradova et al., 2009; Vinogradova, Salinas, Cognat, Remacle, & Maréchal-Drouard, 2020). However, the modification of this mitochondrial codon usage does not clearly influence the steady-state level of imported tRNAs, showing that tRNA import cannot immediately adapt to mitochondrial requirements, but rather results from an evolutive adaptation (Salinas et al., 2012, 2020).

3.3.3 Developing tools for RNA import into mitochondria

Given the essential role of mitochondria in the cell, manipulating mitochondrial genomes would be of great interest, to understand mtDNA processes or to complement mt mutations. However, transfection of mitochondria is successful in only a few organisms such as *S. cerevisiae* and *C. reinhardtii* (Niazi et al., 2013). By contrast, all mitochondria can import tRNAs in living cells. This natural way of importing nucleic acids was used by Val et al. (2011). They were able to import into mitochondria a tRNA-like structure (TLS) in fusion with a ribozyme that specifically recognizes the mitochondrial-encoded *atp9* mRNA. A decrease of *atp9* mRNA level was observed in *A. thaliana* plants expressing this construct from a nuclear transgene. Sieber, Placido, El Farouk-Ameqrane, Duchêne, and Maréchal-Drouard (2011) and Sieber, Placido, El Farouk-Ameqrane, Duchêne, and Maréchal-Drouard (2019) have proposed another strategy to import RNA into mitochondria, based on protein import. Taking advantage that DiHydroFolate Reductase (DHFR) can bind RNAs without specificity, they have used it as an RNA carrier. DHFR in fusion with an MTS was able to drive a tRNA precursor and an unedited mRNA into isolated mitochondria. Maturation of these tRNA and mRNA was then observed in mitochondria. These approaches allow to manipulate the mitochondrial RNA content, and thus offer new strategies to manipulate mitochondrial gene expression.

4. Concluding remarks

As a whole, this chapter summarizes part of our recent knowledge in the field of plant mitochondrial research. Broadly, it has covered topics related to genome evolution and dynamics, RNA metabolism and protein synthesis, macromolecule import and trafficking, intercellular connection with the other compartments of the cell. Many unanswered questions remain and are or will be addressed in a near future. Among the most intriguing we can cite: why do we have such a diversity in mitochondrial genome size, organization, and expression? What has shaped such evolutionary traits? How tightly are the links between the different compartments of the cell and to which extent does the mitochondrion play a crucial role in stress response and plant adaptation? Finally, maybe the most challenging question that remains for decades a bottleneck is the following: can we develop a successful higher plant mitochondria transformation system? As in human mitochondria, this possibility would pave the way to characterize the still numerous mitochondrial proteins with unknown functions.

References

Barkan, A., Rojas, M., Fujii, S., Yap, A., Chong, Y. S., Bond, C. S., et al. (2012). A combinatorial amino acid code for RNA recognition by pentatricopeptide repeat proteins. *PLoS Genetics*, *8*, e1002910.

Barrett, C. F., Freudenstein, J. V., Li, J., Mayfield-Jones, D. R., Perez, L., Pires, J. C., et al. (2014). Investigating the path of plastid genome degradation in an early-transitional clade of heterotrophic orchids, and implications for heterotrophic angiosperms. *Molecular Biology and Evolution*, *31*, 3095–3112.

Binder, S., Stoll, K., & Stoll, B. (2016). Maturation of 5′ ends of plant mitochondrial RNAs. *Physiologia Plantarum*, *157*, 280–288.

Bonen, L. (2012). Evolution of mitochondrial introns in plants and photosynthetic microbes. In J. P. Jacquot, & P. Gadal (Series Eds.) & L. Maréchal-Drouard (Vol. Ed.), *Vol. 3. Advances in botanical research* (pp. 155–186). Amsterdam: Elsevier.

Budar, F., & Fujii, S. (2012). Mitochondrial genome evolution. In J. P. Jacquot, & P. Gadal (Series Eds.) & L. Maréchal-Drouard (Vol. Ed.), *Vol. 63. Advances in botanical research* (pp. 99–126). Amsterdam: Elsevier.

Carrie, C., Weissenberger, S., & Soll, J. (2016). Plant mitochondria contain the protein translocase subunits TatB and TatC. *Journal of Cell Science, 129*, 3935–3947.

Colas Des Francs-Small, C., Vincis Pereira Sanglard, L., & Small, I. (2018). Targeted cleavage of nad6 mRNA induced by a modified pentatricopeptide repeat protein in plant mitochondria. *Communications Biology, 1*, 166–173.

Flegontov, P., & Lukes, J. (2012). Mitochondrial genomes of photosynthetic euglenids and alveolates. In J. P. Jacquot, & P. Gadal (Series Eds.) & L. Maréchal-Drouard (Vol. Ed.), *Vol. 63. Advances in botanical research* (pp. 127–154). Amsterdam: Elsevier.

Fuchs, P., Rugen, N., Carrie, C., Elsasser, M., Finkemeier, I., Giese, J., et al. (2020). Single organelle function and organization as estimated from Arabidopsis mitochondrial proteomics. *The Plant Journal, 101*, 420–441.

Gagliardi, D., & Binder, S. (2007). Expression in the plant mitochondrial genome. In J. P. Jacquot, & P. Gadal (Series Eds.) & D. C. Logan (Vol. Ed.), *Plant mitochondria* (pp. 50–84). Amsterdam: Elsevier.

Gallaher, S. D., Fitz-Gibbon, S. T., Strenkert, D., Purvine, S. O., Pellegrini, M., & Merchant, S. S. (2018). High-throughput sequencing of the chloroplast and mitochondrion of Chlamydomonas reinhardtii to generate improved de novo assemblies, analyze expression patterns and transcript speciation, and evaluate diversity among laboratory strains and wild isolates. *The Plant Journal, 93*, 545–565.

Ghifari, A. S., Gill-Hille, M., & Murcha, M. W. (2018). Plant mitochondrial protein import: The ins and outs. *The Biochemical Journal, 475*, 2191–2208.

Ghifari, A. S., Huang, S., & Murcha, M. W. (2019). The peptidases involved in plant mitochondrial protein import. *Journal of Experimental Botany, 70*, 6005–6018.

Gualberto, J. M., & Newton, K. J. (2017). Plant mitochondrial genomes: Dynamics and mechanisms of mutation. *Annual Review of Plant Biology, 68*, 225–252.

Gutmann, B., Gobert, A., & Giege, P. (2012). PRORP proteins support RNase P activity in both organelles and the nucleus in Arabidopsis. *Genes & Development, 26*, 1022–1027.

Gutmann, B., Gobert, A., & Giegé, P. (2012). Mitochondrial genome evolution and the emergence of PPR proteins. In J. P. Jacquot, & P. Gadal (Series Eds.) & L. Maréchal-Drouard (Vol. Ed.), *Vol. 63. Advances in botanical research* (pp. 253–314). Amsterdam: Elsevier.

Haferkamp, I., & Schmitz-Esser, S. (2012). The plant mitochondrial carrier family: Functional and evolutionary aspects. *Frontiers in Plant Science, 3*(2), 1–19.

Hammani, K., Bonnard, G., Bouchoucha, A., Gobert, A., Pinker, F., Salinas, T., et al. (2014). Helical repeats modular proteins are major players for organelle gene expression. *Biochimie, 100*, 141–150.

Hammani, K., & Giegé, P. (2014). RNA metabolism in plant mitochondria. *Trends in Plant Science, 19*, 380–389.

Hemono, M., Ubrig, E., Azeredo, K., Salinas-Giege, T., Drouard, L., & Duchene, A. M. (2020). Arabidopsis voltage-dependent anion channels (VDACs): Overlapping and specific functions in mitochondria. *Cell, 9*, E1023.

Hsu, Y. W., Wang, H. J., Hsieh, M. H., Hsieh, H. L., & Jauh, G. Y. (2014). Arabidopsis mTERF15 is required for mitochondrial nad2 intron 3 splicing and functional complex I activity. *PLoS One, 9*, e112360.

Kmiec, B., Glaser, E., Duncan, O., Whelan, J., & Murcha, M. (2012). Evolution of protein import pathways. In J. P. Jacquot, & P. Gadal (Series Eds.) & L. Maréchal-Drouard (Vol. Ed.), *Vol. 63. Advances in botanical research* (pp. 315–346). Amsterdam: Elsevier.
Kolli, R., Soll, J., & Carrie, C. (2018). Plant mitochondrial inner membrane protein insertion. *International Journal of Molecular Sciences*, *19*, 641–662.
Kühn, K., & Gualberto, J. (2012). Recombination in the stability, repair and evolution of the mitochondrial genomes. In J. P. Jacquot, & P. Gadal (Series Eds.) & L. Maréchal-Drouard (Vol. Ed.), *Vol. 63. Advances in botanical research* (pp. 215–252). Amsterdam: Elsevier.
Lang, B. F., & Burger, G. (2012). Mitochondrial and eukaryotic origins: A critical review. In J. P. Jacquot, & P. Gadal (Series Eds.) & L. Maréchal-Drouard (Vol. Ed.), *Vol. 63. Advances in botanical research* (pp. 1–20). Amsterdam: Elsevier.
Lesnik, C., Golani-Armon, A., & Arava, Y. (2015). Localized translation near the mitochondrial outer membrane: An update. *RNA Biology*, *12*, 801–809.
Lister, R., Carrie, C., Duncan, O., Ho, L. H., Howell, K. A., Murcha, M. W., et al. (2007). Functional definition of outer membrane proteins involved in preprotein import into mitochondria. *The Plant Cell*, *19*, 3739–3759.
Maréchal-Drouard, L., Guillemaut, P., Cosset, A., Arbogast, M., Weber, F., Weil, J. H., et al. (1990). Transfer RNAs of potato (*Solanum tuberosum*) mitochondria have different genetic origins. *Nucleic Acids Research*, *18*, 3689–3696.
Michaud, M., & Duchêne, A. (2012). Macromolecules trafficking to plant mitochondria. In J. P. Jacquot, & P. Gadal (Series Eds.) & L. Maréchal-Drouard (Vol. Ed.), *Vol. 63. Advances in botanical research* (pp. 347–422). Amsterdam: Elsevier.
Michaud, M., Maréchal-Drouard, L., & Duchêne, A. M. (2010). RNA trafficking in plant cells: Targeting of cytosolic mRNAs to the mitochondrial surface. *Plant Molecular Biology*, *73*, 697–704.
Michaud, M., Maréchal-Drouard, L., & Duchêne, A. M. (2014). Targeting of cytosolic mRNA to mitochondria: Naked RNA can bind to the mitochondrial surface. *Biochimie*, *100*, 159–166.
Michaud, M., Ubrig, E., Filleur, S., Erhardt, M., Ephritikhine, G., Marechal-Drouard, L., et al. (2014). Differential targeting of VDAC3 mRNA isoforms influences mitochondria morphology. *Proceedings of the National Academy of Sciences of the United States of America*, *111*, 8991–8996.
Mower, J. P. (2020). Variation in protein gene and intron content among land plant mitogenomes. *Mitochondrion*, *53*, 203–213.
Mower, J., Jain, K., & Hepburn, N. (2012). The role of horizontal transfer in shaping the plant mitochondrial genome. In J. P. Jacquot, & P. Gadal (Series Eds.) & L. Maréchal-Drouard (Vol. Ed.), *Vol. 63. Advances in botanical research* (pp. 41–70). Amsterdam: Elsevier.
Murcha, M. W., Kmiec, B., Kubiszewski-Jakubiak, S., Teixeira, P. F., Glaser, E., & Whelan, J. (2014). Protein import into plant mitochondria: Signals, machinery, processing, and regulation. *Journal of Experimental Botany*, *65*, 6301–6335.
Murcha, M. W., Kubiszewski-Jakubiak, S., Teixeira, P. F., Gugel, I. L., Kmiec, B., Narsai, R., et al. (2016). Plant-specific preprotein and amino acid transporter proteins are required for tRNA import into mitochondria. *Plant Physiology*, *172*, 2471–2490.
Murik, O., Chandran, S. A., Nevo-Dinur, K., Sultan, L. D., Best, C., Stein, Y., et al. (2020). Topologies of N(6)-adenosine methylation (m(6) A) in land plant mitochondria and their putative effects on organellar gene expression. *The Plant Journal*, *101*, 1269–1286.
Niazi, A. K., Mileshina, D., Cosset, A., Val, R., Weber-Lotfi, F., & Dietrich, A. (2013). Targeting nucleic acids into mitochondria: Progress and prospects. *Mitochondrion*, *13*, 548–558.

Oldenkott, B., Yang, Y., Lesch, E., Knoop, V., & Schallenberg-Rudinger, M. (2019). Plant-type pentatricopeptide repeat proteins with a DYW domain drive C-to-U RNA editing in *Escherichia coli*. *Communications Biology*, *2*, 85.

Pan, Z., Ren, X., Zhao, H., Liu, L., Tan, Z., & Qiu, F. (2019). A mitochondrial transcription termination factor, ZmSmk3, is required for nad1 Intron4 and nad4 Intron1 splicing and kernel development in maize. *G3 (Bethesda)*, *9*, 2677–2686.

Peleh, V., Zannini, F., Backes, S., Rouhier, N., & Herrmann, J. M. (2017). Erv1 of Arabidopsis thaliana can directly oxidize mitochondrial intermembrane space proteins in the absence of redox-active Mia40. *BMC Biology*, *15*, 106–120.

Petersen, G., Cuenca, A., Moller, I. M., & Seberg, O. (2015). Massive gene loss in mistletoe (Viscum, Viscaceae) mitochondria. *Scientific Reports*, *5*, 17588–17594.

Price, D. C., Chan, C. X., Yoon, H. S., Yang, E. C., Qiu, H., Weber, A. P., et al. (2012). Cyanophora paradoxa genome elucidates origin of photosynthesis in algae and plants. *Science*, *335*, 843–847.

Pujol, C., Maréchal-Drouard, L., & Duchêne, A. M. (2007). How can organellar protein N-terminal sequences be dual targeting signals? In silico analysis and mutagenesis approach. *Journal of Molecular Biology*, *369*, 356–367.

Quesada, V. (2016). The roles of mitochondrial transcription termination factors (MTERFs) in plants. *Physiologia Plantarum*, *157*, 389–399.

Rao, R. S., Salvato, F., Thal, B., Eubel, H., Thelen, J. J., & Moller, I. M. (2017). The proteome of higher plant mitochondria. *Mitochondrion*, *33*, 22–37.

Rodriguez-Salinas, E., Remacle, C., & Gonzales-Halphen, D. (2012). Green algae genomics: A mitochondrial perspective. In J. P. Jacquot, & P. Gadal (Series Eds.) & L. Maréchal-Drouard (Vol. Ed.), *Vol. 63. Advances in botanical research* (pp. 187–214). Amsterdam: Elsevier.

Salinas, T., Duby, F., Larosa, V., Coosemans, N., Bonnefoy, N., Motte, P., et al. (2012). Co-evolution of mitochondrial tRNA import and codon usage determines translational efficiency in the green alga Chlamydomonas. *PLoS Genetics*, *8*, e1002946.

Salinas, T., Duby, F., Larosa, V., Coosemans, N., Bonnefoy, N., Motte, P., … Remacle, C. (2020). Correction: Co-evolution of mitochondrial tRNA import and codon usage determines translational efficiency in the green alga Chlamydomonas. *PLoS Genet*, *16*, e1008719.

Salinas, T., Duchêne, A. M., Delage, L., Nilsson, S., Glaser, E., Zaepfel, M., et al. (2006). The voltage-dependent anion channel, a major component of the tRNA import machinery in plant mitochondria. *Proceedings of the National Academy of Sciences of the United States of America*, *103*, 18362–18367.

Salinas, T., El Farouk-Ameqrane, S., Ubrig, E., Sauter, C., Duchêne, A. M., & Maréchal-Drouard, L. (2014). Molecular basis for the differential interaction of plant mitochondrial VDAC proteins with tRNAs. *Nucleic Acids Research*, *42*, 9937–9948.

Salinas, T., El Farouk-Ameqrane, S., Ubrig, E., Sauter, C., Duchêne, A. M., & Maréchal-Drouard, L. (2019). Molecular basis for the differential interaction of plant mitochondrial VDAC proteins with tRNAs. *Nucleic Acids Research*, *47*, 1048–1049.

Salinas-Giegé, T., Cavaiuolo, M., Cognat, V., Ubrig, E., Remacle, C., Duchêne, A. M., et al. (2017). Polycytidylation of mitochondrial mRNAs in Chlamydomonas reinhardtii. *Nucleic Acids Research*, *45*, 12963–12973.

Senkler, J., Senkler, M., Eubel, H., Hildebrandt, T., Lengwenus, C., Schertl, P., et al. (2017). The mitochondrial complexome of Arabidopsis thaliana. *The Plant Journal*, *89*, 1079–1092.

Shtratnikova, V. Y., Schelkunov, M. I., Penin, A. A., & Logacheva, M. D. (2020). Mitochondrial genome of the nonphotosynthetic mycoheterotrophic plant *Hypopitys monotropa*, its structure, gene expression and RNA editing. *PeerJ*, *8*, e9309.

Sieber, F., Placido, A., El Farouk-Ameqrane, S., Duchêne, A. M., & Maréchal-Drouard, L. (2011). A protein shuttle system to target RNA into mitochondria. *Nucleic Acids Research*, *39*, e96.

Sieber, F., Placido, A., El Farouk-Ameqrane, S., Duchêne, A. M., & Maréchal-Drouard, L. (2019). A protein shuttle system to target RNA into mitochondria. *Nucleic Acids Research*, *47*, 1045–1047.

Skippington, E., Barkman, T. J., Rice, D. W., & Palmer, J. D. (2015). Miniaturized mitogenome of the parasitic plant Viscum scurruloideum is extremely divergent and dynamic and has lost all nad genes. *Proceedings of the National Academy of Sciences of the United States of America*, *112*, E3515–E3524.

Sloan, D. B., Alverson, A. J., Chuckalovcak, J. P., Wu, M., McCauley, D. E., Palmer, J. D., et al. (2012). Rapid evolution of enormous, multichromosomal genomes in flowering plant mitochondria with exceptionally high mutation rates. *PLoS Biology*, *10*, e1001241.

Sloan, D. B., Muller, K., McCauley, D. E., Taylor, D. R., & Storchova, H. (2012). Intraspecific variation in mitochondrial genome sequence, structure, and gene content in Silene vulgaris, an angiosperm with pervasive cytoplasmic male sterility. *The New Phytologist*, *196*, 1228–1239.

Small, I. D., Schallenberg-Rudinger, M., Takenaka, M., Mireau, H., & Ostersetzer-Biran, O. (2020). Plant organellar RNA editing: What 30 years of research has revealed. *The Plant Journal*, *101*, 1040–1056.

Stoll, B., & Binder, S. (2016). Two NYN domain containing putative nucleases are involved in transcript maturation in Arabidopsis mitochondria. *The Plant Journal*, *85*, 278–288.

Touzet, P. (2012). Mitochondrial genome evolution and gynodioecy. In J. P. Jacquot, & P. Gadal (Series Eds.) & L. Maréchal-Drouard (Vol. Ed.), *Vol. 63. Advances in botanical research* (pp. 71–98). Amsterdam: Elsevier.

Tsuboi, T., Viana, M. P., Xu, F., Yu, J., Chanchani, R., Arceo, X. G., et al. (2020). Mitochondrial volume fraction and translation duration impact mitochondrial mRNA localization and protein synthesis. *eLife*, *9*, e57814.

Ueda, M., & Kadowaki, K.-I. (2012). Gene content and gene transfer from mitochondria to the nucleu during evolution. In J. P. Jacquot, & P. Gadal (Series Eds.) & L. Maréchal-Drouard (Vol. Ed.), *Vol. 63. Advances in botanical research* (pp. 21–40). Amsterdam: Elsevier.

Val, R., Wyszko, E., Valentin, C., Szymanski, M., Cosset, A., Alioua, M., et al. (2011). Organelle trafficking of chimeric ribozymes and genetic manipulation of mitochondria. *Nucleic Acids Research*, *39*, 9262–9274.

Varre, J. S., D'Agostino, N., Touzet, P., Gallina, S., Tamburino, R., Cantarella, C., et al. (2019). Complete sequence, multichromosomal architecture and transcriptome analysis of the *Solanum tuberosum* mitochondrial genome. *International Journal of Molecular Sciences*, *20*, 4788–4810.

Vincent, T., Vingadassalon, A., Ubrig, E., Azeredo, K., Srour, O., Cognat, V., et al. (2017). A genome-scale analysis of mRNAs targeting to plant mitochondria: Upstream AUGs in 5′ untranslated regions reduce mitochondrial association. *The Plant Journal*, *92*, 1132–1142.

Vinogradova, E., Salinas, T., Cognat, V., Remacle, C., & Marechal-Drouard, L. (2009). Steady-state levels of imported tRNAs in *Chlamydomonas* mitochondria are correlated with both cytosolic and mitochondrial codon usages. *Nucleic Acids Research*, *37*, 1521–1528.

Vinogradova, E., Salinas, T., Cognat, V., Remacle, C., & Maréchal-Drouard, L. (2020). Steady-state levels of imported tRNAs in *Chlamydomonas* mitochondria are correlated with both cytosolic and mitochondrial codon usages. *Nucleic Acids Research*, *48*, 8008–8009.

Waltz, F., Nguyen, T. T., Arrivé, M., Bochler, A., Chicher, J., Hammann, P., et al. (2019). Small is big in Arabidopsis mitochondrial ribosome. *Nature Plants*, *5*, 106–117.
Waltz, F., Soufari, H., Bochler, A., Giegé, P., & Hashem, Y. (2020). Cryo-EM structure of the RNA-rich plant mitochondrial ribosome. *Nature Plants*, *6*, 377–383.
Wang, Y., Law, S. R., Ivanova, A., van Aken, O., Kubiszewski-Jakubiak, S., Uggalla, V., et al. (2014). The mitochondrial protein import component, TRANSLOCASE OF THE INNER MEMBRANE17-1, plays a role in defining the timing of germination in Arabidopsis. *Plant Physiology*, *166*, 1420–1435.
Warren, J. M., & Sloan, D. B. (2020). Interchangeable parts: The evolutionarily dynamic tRNA population in plant mitochondria. *Mitochondrion*, *52*, 14156.
Yan, J., Yao, Y., Hong, S., Yang, Y., Shen, C., Zhang, Q., et al. (2019). Delineation of pentatricopeptide repeat codes for target RNA prediction. *Nucleic Acids Research*, *47*, 37283738.

CHAPTER TEN

Traditional use and perspectives for the application of plant latex and its constituents in agriculture, medicine and industry—A follow-up of ABR volume 93 "Latex, laticifers and their molecular components from functions to possible applications"

Robert Nawrot[a,*], Oskar Musidlak[a], Sophia Bałdysz[a], Martyna Węglewska[a], Alicja Warowicka[b], and Anna Goździcka-Józefiak[a]

[a]Molecular Virology Research Unit, Institute of Experimental Biology, Faculty of Biology, Adam Mickiewicz University, Poznań, Poland
[b]Department of Animal Physiology and Developmental Biology, Institute of Experimental Biology, Faculty of Biology, Adam Mickiewicz University, Poznań, Poland
*Corresponding author: e-mail address: rnawrot@amu.edu.pl

Contents

Advances in Botanical Research, Volume 100
ISSN 0065-2296
https://doi.org/10.1016/bs.abr.2021.01.006
Copyright © 2021 Elsevier Ltd
All rights reserved.

Abstract

Latexes of different plants contain a mixture of molecules which exhibit diverse activities against invaders. Among many other functions, they protect plants against herbivores, insects, fungi, parasites, bacteria and viruses. Due to the presence of pharmacologically active substances, mankind for centuries have applied latices and their preparations for their beneficial effect on human health. Most of these substances exert analgesic, wound healing, anti-pathogenic (antiviral and antimicrobial), anti-parasitic, anti-diabetic and other activities, which have been broadly exploited in traditional medicine. Following volume 93, this review presents popular latex applications and discusses perspectives for their use in medicine, agriculture and industry.

1. Latex as a source of useful bioactive compounds

Plant latex ingredients are a treasury of highly concentrated, hence "half purified", active substances (Konno, 2011), like potential medicines, toxins and other bioproducts, produced by plants initially for defense against herbivores and pathogens (Hagel, Yeung, & Facchini, 2008; Ramos, Demarco, da Costa Souza, & de Freitas, 2019). These include alkaloids, terpenoids, cannabinoids, lignans, cardiac glycosides, and tannins (Konno et al., 2011; Lee, Hagel, & Facchini, 2013; Mithöfer & Boland, 2012), which show a range of biological activities, like anticancer, antiviral, antiproliferative, antimicrobial, anti-inflammatory, anti-oxidant, anti-parasitic or insecticidal (Upadhyay, 2011). Latex bioactive compounds have been the subject of interest due to numerous activities such as anti-inflammatory (Kumar, Murugananthan, Nandakumar, & Talwar, 2011), anticancer (Cruz et al., 2018), anti-oxidant (Marino et al., 2015) and antiviral (Nawrot, 2017). From this perspective, numerous latex-bearing plants are important to the global pharmacy and economy. They belong to several families, the most pronounced being Papaveraceae and Euphorbiaceae (Cho et al., 2009). In the Papaveraceae family, opium poppy (*Papaver somniferum* L.) is one of the most important medicinal plants as the source of a few benzylisoquinoline alkaloids (morphine, codeine, papaverine, noscapine and sanguinarine) (Gurkok, Turktas, Parmaksiz, & Unver, 2015; Zulak, Khan, Alcantara, Schriemer, & Facchini, 2009) and serves as a model plant for research on the benzylisoquinoline alkaloid biosynthesis (Beaudoin & Facchini, 2013). The other is California poppy (*Eschscholzia californica*), the cell cultures of which are often used as an alternative method for production of sanguinarine and chelerythrine (Oldham et al., 2010). In Euphorbiaceae, a tropical plant *Hevea brasiliensis* (para rubber tree) is the main source of renewable latex to produce natural rubber (NR), which is used by the industry to produce tires, protective gloves, and different products (Berthelot, Lecomte, Estevez, & Peruch, 2014; Schmidt, Hillebrand, et al., 2010). In other families, for example, glycosidase inhibitors are present in mulberry (Morus spp., *Moraceae*), which are used as insecticidal compounds (Hirayama, Konno, Wasano, & Nakamura, 2007; Kitajima et al., 2016) (Fig. 1).

In this review we are presenting the popular latex applications and discuss as well perspectives for their further use in medicine, agriculture and industry. We are aware of the fact that there are numerous and still growing possibilities for the human use of natural based products, and for sure not all of them are covered in this review. For readers who would like to expand their knowledge regarding this subject please refer to *Advances in Botanical Research* volume 93 "Latex, Laticifers and Their Molecular Components: From Functions to Possible Applications" (Nawrot, 2020).

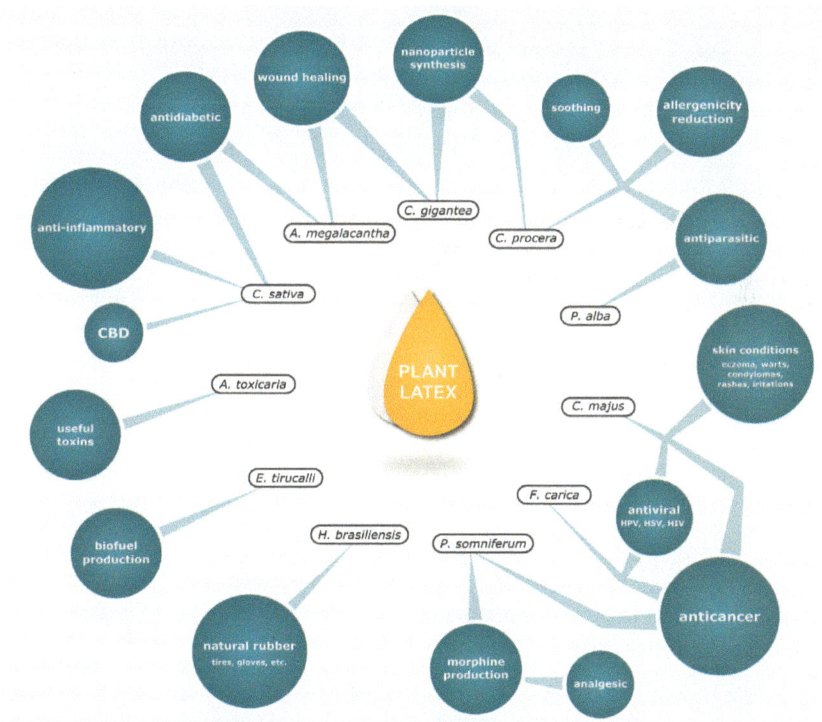

Fig. 1 The map of potential use of latex-bearing plants and their constituents. The chart is based on the information included in the text.

2. Traditional use of latex components and their activities

2.1 Latex-bearing plants and latex-borne defense

Plants which produce latex, also called milky sap, are taxonomically extremely diverse, belonging to over 40 families of angiosperm plants, exceeding 20,000 species (Konno, 2011). Latex is produced by specialized cells called laticifers, distributed across the whole plant, which have evolved several times independently, mostly throughout unrelated plant orders (Hagel et al., 2008). Laticifers have different forms and in the most basic way they are divided into two sub-groups; articulated and nonarticulated laticifers (Agrawal & Konno, 2009; Hagel et al., 2008). Typical families of such latex-bearing plants comprise Asteraceae, Euphorbiaceae, Apocynaceae and Papaveraceae (Cho, Jo, Chu, Park, & Kim, 2014), with higher numbers of tropical than temperate regions' species (Konno, 2011). Such imbalance could be explained by the evolutionary advantage for the latex-bearing plants against herbivores, and particular intense interactions between plant and herbivorous insects in tropical regions (Konno, 2011).

Plant latex comprises the cytoplasm of laticifers, mostly the content of its well-developed vacuoles or the mixture of vacuole contents with the rest of the cytoplasm of degraded laticifer (Konno, 2011; Zhou & Liu, 2010). Finally, a complex emulsion is formed, which contains a range of chemicals like terpenoids, alkaloids, starches, oils, sugars, gums, tannins, and proteins. It exudes from the plant immediately after injury, is highly viscous and mostly coagulates upon air exposure (Agrawal & Konno, 2009; Cho et al., 2014; Konno, 2011). Exuding sap is characteristic "glue-like/milky" usually non-transparent exudate, with different colors, from white for rubber tree (*Hevea brasiliensis*, Euphorbiaceae) and opium poppy (*Papaver somniferum*, Papaveraceae) to orange/yellow for greater celandine (*Chelidonium majus*, Papaveraceae) or other colors. For some species it could also be transparent (Konno, 2011; Nawrot, Barylski, Lippmann, Altschmied, & Mock, 2016).

Latex is exuded immediately by internal pressure after the mechanical injury at its point, for example after insect attack, so it assures fast and precise transport of defense-related molecules to the point of attack. Such substances and mode of action form the latex-borne defense system, which is preformed, containing ready-to-act and highly concentrated chemicals and proteins (Karban & Kuc, 1999; Konno, 2011). Such advantages could explain convergent evolution of laticifers in distant phylogenetic groups, as the system is much more advantageous in terms of production costs than non-laticiferous inducible defense systems (Farrell, Dussourd, & Mitter, 1991; Konno, 2011). Also, the latex coagulation process, namely its clotting and stickiness, is essential for plant defense against possible pathogen attack. The mechanisms involved in this process are diverse in different plants, but it exerts its effect against tiny insects by trapping their whole body or by gluing their mouthparts (Dussourd, 1995; Konno, 2011; Nawrot, 2017; Zalucki et al., 2001).

2.2 Antiviral activities of latex

Latex of the medicinal plant *Chelidonium majus* L. (Papaveraceae) is well known to have a broad spectrum of beneficial properties. In folk medicine it has been used for different skin conditions such as eczema, irritations, allergic rashes. It has been also applied to treat warts and condylomas which are visible manifestations of ongoing human papillomavirus (HPV) infection (Musidlak, Bałdysz, Krakowiak, & Nawrot, 2020; Nawrot, 2020). Warts were usually rubbed with latex exuding from freshly cut *C. majus* stem (Zielińska et al., 2018). Sometimes warts were first scratched off prior to applying the latex (Zielińska et al., 2018). The *C. majus* latex has been used in this or similar manner in many European countries including Poland, Russia, Ukraine, Italy, France, Great Britain, and in the region of Balkans (Barton & Castle, 1845; Behxhet et al., 2012; Koleva, Dragoeva, Stoyanova, Koynova, & Dashev, 2015; Menale, Amato, Prisco, & Muoio, 2006; Tita, Mogosanu, & Tita, 2009; Zevin, Altman, & Zevin, 1997). Latex alkaloids can also affect replication of other viruses. Berberine, an alkaloid found in *C. majus* latex, can partially inhibit the transcription and translation of *Herpes simplex virus* (HSV) genes (Song et al., 2014). Furthermore, berberine blocks the export of influenza type A virus ribonucleoprotein from the nucleus, thus interfering with viral replication (Botwina et al., 2020). *C. majus* contains many different substances such as benzylisoquinoline alkaloids (e.g. chelidonine, berberine, coptisine, protopine, chelerythrine, sanguarine, stylopine, canadine), flavonoids (e.g. rutin, luteolin, quercetin), phenolic acids (gallic acid, chlorogenic acid), unsaturated fatty acids (oleic acid, linoleic acid), carotenoids or saponins (Arora & Sharma, 2013; Zielińska et al., 2018). Other research addressed the defense-related protein content of the *C. majus* latex. Predominant proteins were major latex protein (MLP), polyphenol oxidase (PPO), peroxidase (POX), lipoxygenase (LOX) and enzymes responsible for phenylpropanoid and alkaloid biosynthesis (Nawrot, 2017; Nawrot et al., 2016, 2017). The antiviral activity of plant latex is not only restricted to *C. majus*. Latices from other plants have also been described to have antiviral properties. Latex from *Ficus carica* (Moraceae) acts against HPV infection and cancer (Ghanbari et al., 2019). Latex from *Aloe harlana* (Asphodelaceae) and *Euphorbia lactea* (Euphorbiacae) exert anti-HIV activity (Mahmoud, Gad-Rab, Hussein, & Shoreit, 2010). Based on the recent studies on the predominant and constantly expressed pathogenesis-related proteins of *C. majus* latex, it was hypothesized that antiviral mechanism of plant latex composed of defense-related proteins together with small-molecular compounds, constitutes a novel type of immediate defense response against viral pathogens (Nawrot, 2017; Nawrot et al., 2016, 2017).

3. Applications of latex in medicine, agriculture and industry

3.1 Natural rubber and its applications

Among 20,000 different plant species which produce latex, more than 2500 species are abundant in valuable compounds for use in industry. The most used substance of that type is rubber. Unfortunately, only a few plants that produce large amounts of high molecular weight rubber, efficient for manufacturing purposes, are known (Bauer et al., 2014; Men, Wang, Chen, Zhang, & Xian, 2019). The Amazon rubber tree (*Hevea brasiliensis*) is the most efficient commercial source of natural rubber. Raw material is obtained directly from latex, the colloidal suspension of rubber molecules. The bark of trunk of rubber tree is incised through the vessels without damage. From the wound milky and sticky latex "taps" to the cups. Collected material is used directly or coagulated and further transformed depending on its application.

Natural rubber (NR) is a biopolymer of cis-1,4-poly (isopropene) with small amounts (6%) of other components like proteins, lipids or minerals. The basics of natural rubber biosynthesis are well known but in some details this process is still mysterious. Rubber synthesis is catalyzed by rubber transferases and is conducted in three steps: reaction of initiation, polymerization and termination. The monomer of rubber—isopentenyl pyrophosphate is polymerized and modified by different transferases resulting in species-specific molecular weight end molecules (Cherian, Ryu, & Cornish, 2019; Men et al., 2019). Rubber is biosynthesized and stored in rubber particles, main component of the latex (30–50% volume) in different species-specific tissues. In rubber trees and dandelions rubber molecules are enclosed inside the laticifer cells, in the case of guayule the synthesis is localized in bark parenchymal cells and pith (Men et al., 2019; Mooibroek & Cornish, 2000; Schmidt, Hillebrand, et al., 2010; Schmidt, Lenders, et al., 2010). The physiological role of rubber in plant tissues is still unresolved. Rubber after synthesis remains locked inside the plant cells until their death and there is no known enzyme that could degrade NR. However it is supposed that rubber plays a role in the healing processes, plant defense mechanisms and stress response (Konno, 2011; Men et al., 2019).

The properties of *Hevea* rubber are valued. Rubber trees grow well under cultivation, however the whole process is challenging, labor-intensive and time-consuming, requires very specific temperature and humidity conditions, is prone to fungal infections, and in addition genetic variability of plants is still consistently decreasing (Cornish, 2017; Mazlan et al., 2019; Verheye, 2010). Nowadays, about 90% of natural rubber production is obtained from *H. brasiliensis* (Verheye, 2010). Synthetic rubber substitutes are made from nonrenewable petroleum, so more environment sustainable and eco-friendly alternatives seem to be essential. The main disadvantage of *Hevea* rubber is the fact that approximately 1–6% of global population is allergic to nonremovable protein impurities (van Beilen & Poirier, 2007). Therefore hypoallergic, less-demanding in agriculture plants were introduced for rubber industrial production. Particular attention was paid to Russian dandelion (*Taraxacum kok-saghyz*), guayule (*Parathenium argentatun* Gray) and hardy rubber tree (*Eucommia ulmoides*) (Cherian et al., 2019; Cornish, 2017; Mooibroek & Cornish, 2000; Wahler et al., 2012).

Plants other than *Hevea* appear to be a more sustainable source of rubber, because optimized cultivation gives annually yields, boost farm productivity of biomass and the harvest could be fully mechanized (van Beilen & Poirier, 2007). The industry precious material from Russian dandelions is obtained mainly from root laticifers both in a solid or milky form (Molinu, Piluzza, Campesi, Sulas, & Re, 2019). In addition, guayule rubber, despite its impurity is safer and do not cause allergic reaction in contact, what makes this material suitable especially for biomedical purposes (Cornish, 2017). Natural rubber from all sources is of good quality, has similar characteristics, but because of the harvesting method, biomaterials from alternative sources are more contaminated with other compounds, like lipids or proteins (Cornish, 2017; van Beilen & Poirier, 2007).

The history of the natural rubber application is very long and begins in ancient American cultures like Olmec, Mayans, Incas or Aztec civilizations.

The first writing reports about the use of rubber product by American tribes were made by Christopher Columbus during his second exploration expedition. He reported that they mainly fabricated rubber balls. The first mass application of rubber were shoes, produced in the Amazon and exported to America and Europe. In 1770 an English chemist Charles de la Condamine noticed that cubes from rubber could be used as an eraser of pencil marks. However, the popularization of natural rubber utilization did not start until 1820 when Thomas Hancock invented manufacturing machines to soften and mixing rubber and Charles Mackintosh developed a process for coating fabric with rubber. The breakthrough came in 1893 when Charles Goodyear developed the sulfur vulcanization process to harden the rubber. The vulcanization process is crucial from an economic point of view and allows to obtain unique and industrially important value of rubber (Hurley, 1981; International Rubber Study Group, 2020; Kohjiya & Ikeda, 2014). Sulfur vulcanization in high heat with antioxidants, fillers and accelerators like toxic secondary amine (sulfonamides, thiazoles, guanidines and thiurams) or new developed and eco-friendly like xanthates or starch significantly improve physical and mechanical properties of rubber (Li et al., 2020). During this irreversible reaction cross-linking bonds are formed between polymer chains, resulting in non-sticky, higher elasticity, rigidity, durability and thermostability of end polymeric products. The characteristics of the final products depend on modifiers and filles added during the vulcanization step (Hamilton, 2016; Kohjiya & Ikeda, 2014; Vulcanization and properties of vulcanized rubber, 2021).

Currently on the market about 30% rubber products are made from natural sources. Only less than 1% of total natural rubber consumption are collected from wild trees, the rest coming from growing crops (Kohjiya & Ikeda, 2014). According to reports published by the International Rubber Study Group worldwide rubber consumption is currently quite stable, because of growing economies of developing countries. Natural rubber is mainly used in Asia and Africa, while in Europe and America it is replaced mostly by synthetically produced ones (International Rubber Study Group, 2020). The leading producers are Asian countries Thailand, Malaysia, China, etc., which supply around 90% of the natural rubber. Unfortunately modern market is very competitive and we are observing a reduction in rubber plantations, because of growing demands for palm oil (van Beilen & Poirier, 2007).

Some of NR unique prosperities have never been achieved by synthetic rubber yet. In comparison to synthetic, natural rubber is outstandingly resilient, harder, more flexible, exhibits abrasion, ozone, impact and tear resistance. Moreover, NR shows good tensile strength and high durability, malleability at cold temperatures, easy processability, readily adheres to other materials, is resistant to water and some chemicals. Additionally, NR exhibits the characteristics of an electrical insulator. All of these properties make natural rubber irreplaceable in some industry branches (Cornish, 2017; Kohjiya & Ikeda, 2014; Men et al., 2019).

Natural rubber either alone or with some additives is manufactured to produce over 50,000 different products (Cornish, 2017; Priyadarshan, 2011). The benefits of natural rubber utilization still expand due to the improved processing of raw material. Modern industry mostly relies on rubber composite materials—a composition of two or more materials with different properties to obtain improved final products. Reinforcement of rubber by mixing it with wide range of fillers enhances the desired properties, improves processing and reduces production costs by increasing the weight and volume of the desired material. The best known is carbon black produced from petroleum oil, which enhances mechanical proprieties of rubber. Composites made by mixing these two components are more elastic, gain tensile strength, are resistant to abrasion, oxidation or temperature. Additionally, the finished goods have monotonous black color. This kind of modified rubber compound can be useful especially for shoe soles, gloves and tires (Farida, Bukit, Ginting, & Bukit, 2019; Joy, Chan, Maria, & Thomas, 2013; Omnès, Thuillier, Pilvin, Grohens, & Gillet, 2008). For industrial applications silica, clay, metal particles such as nickel, graphite and many others have been also used frequently (Joy et al., 2013). Growing awareness of environmental protection forces the development and testing of biocomposites that are more eco-friendly and less toxic, cheaper, but without changing the quality of composite. Because of expanding demand, scientists are looking for various alternative possibilities for combining natural rubber with bio-fillers for packaging, automotive, construction industries. In this case rubber is reinforced by mixing with natural fibers, sometimes waste products from cellulose industry, like straw (corn, wheat), grass (bamboo), bast (flax, hemp), seed (cotton), fruit (coconut), wood (pinewood) fibers, nanoscale particles such as nanocellulose and nanocrystals of starch or animal-derived fillers for example chitosan from shrimp and crab shells or mushroom and algae cell-walls (Bokobza, 2018; Ismail, Shaari, & Othman, 2011; Jarnthong, Wang, Wei, Wang, & Li, 2015; Joy et al., 2013; Kohjiya & Ikeda, 2014; Parambath Kanoth, Claudino, Johansson, Berglund, & Zhou, 2015).

70% of natural rubber is used for heavy-duty tire production and other transportation components like belting, engine, mechanical goods. The most valued features of natural rubber are its safety (durability, integrity), reduced environmental impact and comfort of mobility. Currently, natural rubber improved by incorporation of additives and fillers is used to make almost 90–100% of truck and airplane tires and a majority of winter and off-road tires (Kohjiya & Ikeda, 2014).

More than 10% of natural rubber, mainly as latex, is processed into 400 types of medical devices (van Beilen & Poirier, 2007). The consumption of NR has gradually increased since the 1980s, when disposable latex gloves were introduced as a common method of HIV (ang. *Human Immunodeficiency Virus*) and HBV (ang. *Hepatitis B Virus*) infection prevention (Alenius et al., 1995; Joy et al., 2013; Wagner & Breiteneder, 2005). This perfect biomaterial has advantages for medical application such as elasticity, flexibility, resistance against splitting, improved hemocompatibility, stimulation of angiogenesis, cell adhesion, wound healing and tissue replacement and moreover regeneration promotion. Rubber biopolymer is processed into membranes, blood pressure cuffs, tubes, balloons, pacifiers, dental equipment, drug delivery systems, surgical gloves, condoms, catheters, etc. Unfortunately *Hevea* rubber has one weakness, in a small percentage of the population it can cause skin irritation after contact, due to protein impurities (Joy et al., 2013).

Natural rubber is also successfully introduced for building and road constructions. For example, this biopolymer improves asphalt performance. Natural rubber is easily mixed with asphalt and acts as a bio-modifier. The final product is more elastic upon temperature stress, dense, achieve flexibility and stability. What is more, rubber additives counteract deformation of roads, increase resistance to rutting and cracking (Wen, Wang, Zhao, & Sumalee, 2017).

About 6% of the naturally sourced rubber as well as rubber waste are used for production of footwear, boots, shoes soles or heals (Cardoso, Guimarães, Lopes, & Alves, 2019).

Natural rubber properties like abrasion, elastic or electric resistance, damping and shock absorption are mostly valued in other applications such as wire and cable isolation, rubberized fabrics, shock absorbers, washers and gaskets, transmission and conveyor belting, hoses, sports goods, household and hospital supplies, contraceptive appliances, paints, upholstery, mattresses, etc. (van Beilen & Poirier, 2007; Verheye, 2010).

3.2 Latex as a source of healing natural products and drugs

Plant latex has also proven to be beneficial when applied to flesh wounds. Latex has been used in traditional medicine to stop bleeding and to facilitate wound healing (Urs, Manjuprasanna, Rudresha, Yariswamy, & Vishwanath, 2017). Wound healing has been described for *Calotropis gigantea* (Asclepiadaceae) and *Croton lechleri* (Euphorbiaceae) latex (De Marino et al., 2008; Nalwaya, Pokhorna, Deb, & Jain, 2009). Furthermore, the leaf latex of *Aloe megalacantha* Baker (Xanthorrhoeaceae) exhibits wound healing and anti-inflammatory properties (Gebremeskel, Bhoumik, Sibhat, & Tuem, 2018). All these plants have been extensively used in traditional medicine to treat wounds in Africa, South America and Asia. Most of the known plant latices contain proteases, which are responsible in plants for insect- and herbivory-induced defense. These proteolytic enzymes induce clot formation, essential for the first stage of wound healing. They also have clot-hydrolyzing properties, which are crucial for the regenerative phase of wound healing (Urs et al., 2017). Latex proteases assist endogenous proteases in regulating homeostasis, cell proliferation, angiogenesis and in eliminating microbial infection, thus providing proper conditions for the wound healing process. Aforementioned compounds may prove useful in producing safe, plant-based wound healing therapeutics with reduced side effects. However, more studies focusing on functional and physicochemical characterization of isolated latex proteases are required (Urs et al., 2017).

The latex of different plant species has been also used to treat parasitic infections. For example, the chemical constituents of latex of *Plumeria alba* L. (Apocynaceae) have potent activity against scabies, skin ectoparasite (Bhavani & Ravichandran, 2020). Components of other plant latices exert also a toxic effect on endoparasites. Latex from *Calotropis procera* (Apocynaceae) exerts detrimental activity against *Haemonchus contortus*, an endoparasitic nematode living on animals. Latex components had a negative impact on *H. contortus* egg-hatching, larval development and adult motility (Cavalcante et al., 2016). Latex-bearing plants offer a promising source of novel anti-parasitic agents, many of which can be potentially used to develop plant-based therapeutics. However, they first need to undergo thorough testing on animal models and in clinical trials, in order to assess their safety.

Latex also contains useful toxins which have been used by humans e.g. antiarin, which is a toxin in the laticifers of the Upas tree (*Antiaris toxicara*). The aforementioned toxin is used as a poison at the tips of darts for hunting (Carter et al., 1997; Hagel et al., 2008; Roshchina et al., 1993) in eastern

Asia. The latex of the Upas tree can be poisonous, the compounds within the latex being cardio-toxic. Interestingly enough, the fruits of the Upas tree are free from the poison and edible. Latex from some of the species of the Naucleopsis genus are also used to poison darts for hunting, however; unlike the Upas tree latex, it is used in South America (Berg, Corner, & Jarrett, 2006).

People suffering from diabetes mellitus have often adverse side effects related to taking anti-diabetic drugs. Therefore, the search for novel natural anti-diabetic drugs is required. The leaf latex of Mulberry (Moraceae), *Aloe pulcherrima* (Asphodelaceae) and *Aloe megalacantha* (Asphodelaceae), have been reported to contain sugar-mimic alkaloids and other compounds which have anti-diabetic properties. These discoveries are promising for tackling diabetes mellitus, and they may become useful for developing potential anti-diabetic agents. However, deeper and more insightful studies are required in order to identify and isolate the key latex active components and to discover the exact mechanism responsible for the observed anti-diabetic properties (Amare, Meharie, & Belayneh, 2020; Hammeso, Emiru, Getahun, & Kahaliw, 2019; Konno et al., 2016).

Latex constituents can also help relieve pain. Alkaloids found in the latex of *Papaver somniferum* L. (opium poppy) (Papaveraceae), namely morphine and codeine are broadly used in medicine for their analgesic activities (Ramos et al., 2020). Protein fraction from *C. procera* latex also exhibited soothing properties resulting in reduced joint inflammation in rats (Juca et al., 2013; Kumar, Murugananthan, Nandakumar, & Talwar, 2011). Latex components offer a possible source for searching new anti-inflammatory and soothing therapeutics because it is likely that latex constituents of other plants may exhibit similar properties. These novel compounds may be applied in the therapy of arthritis to reduce inflammation and relieve pain (Ramos et al., 2020). Latex anti-inflammatory compounds have already been found in such plants as *Cannabis sativa* and *Macleaya cordata* (Huang, Sun, & Zeng, 2020). Different agents found in *Cannabis sativa* have a wide range of positive properties. *C. sativa* is divided into two groups according to the use: marijuana (used for its psychoactive properties) and hemp (used for its medicinal properties, seed and fiber) (Johnson, 2018). The genus *cannabis* is under strict control of FDA, EMA and other federal agencies of different countries, because of the presence of tetrahydrocannabinol (THC)—a highly psychoactive substance. On the other hand, cannabidiol (CBD) a non-psychoactive compound from *C. sativa* helps treat neurological disorders by controlling signaling pathways and receptors. It is used in therapy of schizophrenia, epilepsy and Alzheimer's disease (Chen, Borgelt, &

Blackmer, 2019; Osborne, Solowij, Babic, Huang, & Weston-Green, 2017; Watt & Karl, 2017). CBD also proved useful in treating other diseases including anxiety (post-traumatic stress disorder, PTSD), obesity, cancer, diabetes and neurodermatitis (Huang et al., 2020). Latex alkaloids such as berberine from *C. majus* latex have also been found useful in treating Alzheimer's disease and other neurodegenerative diseases (Hussain et al., 2018). In 2018 FDA approved the CBD-based drug Epidiolex® for use with patients suffering from Lennox-Gastaut or Dravet syndromes (Chen et al., 2019). Genus *cannabis* comprises hundreds of active compounds and hopefully we will see more of these beneficial substances applied in the fields of pharmacology and medicine to treat various diseases.

As mentioned earlier, latex components have also anticancer effects. Vinblastine, a benzylisoquinoline alkaloid from *Catharanthus roseus* (Apocynaceae) exerts activity against cancer (Hagel & Facchini, 2013). Berberine which can be found in latex of such medicinal plants as *C. majus*, and others from the Papaveraceae family, exhibits anticancer properties by suppressing cell proliferation and inducing apoptosis in different types of cancer cells including cervical, breast and lung cells (Jantova, Cipak, Cernakova, & Kostalova, 2003; Ma et al., 2017; Xiao et al., 2018). Additionally, noscapine, a non-toxic alkaloid from *P. somniferum* has therapeutic activity against prostate cancer when administered orally (Barken, Geller, & Rogosnitzky, 2008).

3.3 Latex in morphine production for its analgesic properties

Latex is a source of secondary metabolites and proteins, which in turn can be used in the industry and agriculture. Laticifers provide an ideal niche to store such components due to the separation and compartmentation of the substances and prevention from harming the plant. Examples of secondary metabolites found in the latex are morphine in that of opium poppy (*Papaver somniferum*) and vindoline (*Catharanthus roseus*) (Hagel et al., 2008; Ziegler & Facchini, 2008). Morphine is the first plant alkaloid ever isolated and first to be used in medicine. Even to this day it has high application value (Sertürner, 1806). Enzymes involved in morphine biosynthesis have been found via immunolocalization in the laticifers in opium poppy (*Papaver somniferum*) (Weid, Ziegler, & Kutchan, 2004). Morphine accumulation itself requires the presence of laticifers, which are present as differentiated cells (Kutchan, Ayabe, Krueger, Coscia, & Coscia, 1983; Onoyovwe et al., 2013; Rush, Kutchan, & Coscia, 1985). Throughout the morphine biosynthesis pathway it is possible to detect spatial distribution during its various stages. Several enzymes which act during the first stages of morphine biosynthesis are localized in the phloem, namely the sieve elements. It has been proposed that early stages of alkaloid biosynthesis occur in cells known as parenchyma phloem cells and are later transported to the laticifers, in which final enzymatic transformation and morphine accumulation occur. An alternative model suggests that all enzymatic steps occurs in the sieve elements, whereas the latex is merely used for storage. Therefore the spatial distribution of enzymes and processes in the morphine biosynthetic pathway remains yet to be discussed. In a study conducted by Onoyovwe et al. (2013) the three final enzymes of the morphine biosynthesis pathway (thebaine 6-O-demethylase (T6ODM), codeinone reductase (COR), codeine-O-demethylase (CODM)) were identified in the laticifers via immunoblot analysis and shotgun proteomics, preceded by immunofluorescence labeling. The presence and relative abundance of the aforementioned enzymes is supported by the detection of the corresponding gene transcripts in the latex (Onoyovwe et al., 2013).

Interestingly enough, the opium poppy and Persian poppy (*Papaver bracteatum*) have been found to contain the neurotransmitter dopamine in their latex (Roberts, McCarthy, Kutchan, & Coscia, 1983). Morphine has been used for centuries as an analgesic drug (Murphy, Bechmann, & Barrett, 2020). The yield of this alkaloid from opium poppy has a direct effect on its availability as a drug on the market, as well as market prices (Eljounaidi & Lichman, 2020). The pharmaceutical industry relies on opium poppy for the production of pain-killers. The World Health Organization reported a shortage of oral morphine in Egypt in 2014, impacting the number of people who could receive analgesia in end-stages of cancer (Alsirafy & Farag, 2016). It has also been reported that the demand for alkaloids from opium poppy remained at a high level in 2013 and the consumption of morphine for severe pain treatment has increased, especially in high-income countries (United Nations Office on Drugs and Crime, 2014). As of 2019 morphine remains on the 21st list of essential medicines issued by the WHO (WHO TEAM Health Product Policy and Standards, 2019).

To increase morphine yield in opium poppy, a study conducted by Frick, Kramell, and Kutchan (2007), in which the aforementioned plant had been genetically engineered via overexpression of the transgene cypb803 cDNA which resulted in the increase of alkaloid yield by almost 450%. Increased production of morphine could be of great significance in this medicinal plant and could be developed for commercial purposes (Frick et al., 2007; Onoyovwe et al., 2013).

Studies on laticifer physiology and structure might assist in reduction of illegal opium acquisition from opium poppy cultivars. Black-market opium, which is later used in heroin production, is collected using traditional lancing techniques. Through limiting latex flow and formation of perforations among laticifer initials, the collection of illegal opium could be decreased (Hagel et al., 2008; Pilatzke-Wunderlich & Nessler, 2001). Opium poppy has been noted to contain several pectin modifying enzymes including pectin methylesterases (Pilatzke-Wunderlich & Nessler, 2001). The use of the aforementioned modifying enzymes could aid the limitation of latex flow and formation of perforations.

3.4 Nanoparticle synthesis

Latex components have been also used for production of nanoparticles which act as drug carriers (e.g. from *E. tirucalli*). Nanotechnology is an important field of science that involves the design, production and modifications of materials having particle size below 100 nm. Nanomaterials have been applied in various areas, ranging from material science and engineering to biotechnology and medicine. In general, nanomaterials can be synthetized using physical, chemical and biological methods. An example of biological plant-based and eco-friendly method applied in the synthesis of nanoparticles is green synthesis also called biosynthesis (Thunugunta, Reddy, & Reddy, 2015).

Latex-bearing plants are a rich source of pharmacologically active natural compounds (e.g. alkaloids, flavonoids, tannins, cardiac glycosides and sterols). Recently, complexing of natural compounds extracted from plant latex with nanomaterials has received great attention. Latex or its active compounds can be utilized in the process of biofunctionalization of nanomaterials, latex compounds can be effectively transported by nanocarriers in different therapies.

In recent times, there has been much attention to nanocarriers of bioactive agents from latex. Despite the pharmaceutical advantages of plant latex compounds, most of them (e.g. alkaloids) possess poor bioavailability and low water solubility. It is essential to improve the efficiency of biologically active latex compounds and reduce these limitations. There are many different types of nanotechnology-based systems, which have been already used in drug delivery (Majidzadeh et al., 2020). Nanomaterials such as nanoparticles, micelles, solid lipid nanoparticles, nonstructural lipid carriers, liposomes, nanodiamonds or nanogels, carbon nanotubes and graphene provide time-controlled and site-specific delivery of biologically active agents. One method to improve beneficial properties and bio-distribution of plant latex compounds is their encapsulation in nanomaterials. It has been shown that plant alkaloids–loaded polymeric nanoparticles have stronger antibacterial activities (Mirhadi, Rezaee, & Malaekeh-Nikouei, 2018). Furthermore, various nanoparticle-based bioactive molecules delivery systems provide a new option in cancer therapy. It is known that nano strategies for plant agent delivery improve the therapeutic efficacy of compounds as an anticancer drug. The encapsulation of alkaloids into different types of nanoparticles improves their bioavailability and cellular uptake. One of the class of polymeric nanoparticles suitable for drug delivery are polydopamine dendrimers (PAMAM). Gupta et al. showed that alkaloid berberine incorporated into PAMAM exhibited a higher anticancer effect on human breast cancer cell lines compared to free, unbounded berberine (Gupta et al., 2017). Moreover, the use of nanomaterials as nanocarriers and incorporating the alkaloids into biocompatible polymer-based nanofibers gives opportunities to improve their stability, bioavailability and release in controlled rate (Khan, Xiangyang, Ahmad, & Mo, 2018). Electro-spun polymer-based nanofibers can be decorated with natural latex compounds and used as a potential material for growth and differentiation of neuronal cells (Requejo-Aguilar et al., 2017). Natural, biodegradable polymer—rubber from the milky sap of the *H. brasiliensis* tree has attracted attention in material science and engineering area. Natural rubber is composed from polymer particles and can be used as non-toxic, natural polymeric material. Natural rubber due to its mechanical properties (elasticity and flexibility) is widely used in various fields of nanotechnology and biomedicine. Rubber latex can be applied in tissue engineering, nano-drug delivery and matrices coating. In some studies latex from plant *Calotropis gigantean* (Apocynaceae) (white mader) has been applied as biocompatible material for assembling biomimetic nanocarriers, encapsulation and release of model anticancer drug, doxorubicine (DOX) (Pradeepkumar, Govindaraj, Jeyaraj, Munusamy, & Rajan, 2017). Moreover, latex from this plant has been used to modify cellulose surfaces and design biomimetic advanced nanomaterials (Hatton et al., 2016). In addition, in material engineering, natural rubber can be used to prepare nanocomposites of natural rubber and polymer nanoparticles (Chaiyasat, 2012). Other studies showed that silver nanoparticles, which are mostly used as antimicrobial agents can be incorporated into natural rubber latex foam to increase antibacterial properties against *Escherichia coli* (Rathnayake, Ismail, Azahari, Bandara, & Rajapakse, 2013). Thus, nanotechnology-based delivery systems can improve bioavailability and the therapeutic effects of natural latex compounds.

Green synthesis of nanoparticles is eco-friendly and cost effective in comparison with physical and chemical methods. Physical and chemical methods are still the most widely used, but the demand for biological methods has risen significantly (Thunugunta et al., 2015). Green biosynthesis of metallic nanoparticles often involves biological molecules from plant latex extract. Then, physical characterization of obtained nanoparticles is determined using spectrophotometric method (UV–vis, Fourier transform infrared spectroscopy—FTIR), transmission electron microscopy (TEM), and X-ray diffraction. Biological properties are usually determined as antimicrobial activities against bacteria and fungi. One example is the green synthesis of silver nanoparticles (AgNPs) using milky latex from plant *Calotropis procera* (Asclepiadaceae). Compounds of the latex may be responsible for the reduction process ($AgNO_3$ to Ag) and can be used as effective, non-toxic, reducing agents (Mohamed, Ismail, Abdel-Mageed, & Mohamed Shoreit, 2019). Another study indicates strong antibacterial and trombolytic activities of *Euphorbia acruensis* (Euphorbiaceae) latex-mediated synthetized AgNPs. This study focused on milky white latex from this succulent plant that is rich in bio-reductants. During the synthesis process bioactive compounds of *E. acruensis* latex mediate the reduction of silver ions leading to the green formation of AgNPs. The green produced AgNPs showed antibacterial activity against Gram-negative bacteria *Pseudomonas putida* and Gram-positive bacteria *Staphylococcus aureus* (Roy, Srivastwa, & Ghosh, 2019). Green synthesis of nanoparticles has attracted considerable attention also in terms of antiviral activities. It is worth to note that the green bioengineered AgNPs showed antiviral activities against herpex simplex virus 1 (HSV-1), Coxsackievirus B4 (CoxB4) and hepatitis A virus (HAV) (Haggag et al., 2019). To sum up, green synthesis provides new therapeutic applications for medicinal plants and bio-synthetized nanoparticles can be a source of new antiviral therapeutic agents. Nevertheless, the mechanisms of antiviral action still require deeper clarifications.

3.5 Allergenicity and applications of latex proteins

Other latex proteins can be applied in decreasing allergenicity of food. A study by Oliveira et al. (2019) found that latex peptidases (cysteine peptidases) from *Calotropis procera* (CpLP), *Cryptostegia grandiflora* (CgLP), and *Carica papaya* (CapLP) hydrolyze proteins in cow's milk. Further administration to mice of the milk post enzyme addition showed no immune reaction. The peptidases would be promising candidates in preparation of formulas for children which have an allergic reaction toward milk (Oliveira et al., 2019). Latex peptidases, isolated from *Calotropis procera* and *Cryptostegia grandiflora* showed milk-clotting activities. These activities are sought for in cheese production to form thick curds and are currently performed using rennet, therefore latex peptidases are a promising alternative to currently used rennet (Freitas et al., 2016). Extract from the latex of *Maclura pomifera* applied to soybean-protein isolate caused hydrolysis of soybean proteins. The hydrolysate could be used as a building block for various foods (Reyes Jara, Liggieri, & Bruno, 2018).

Proteins found in the latex are promising candidates for many industrial applications due to their activity in a wide range of pH and temperatures (Domsalla & Melzig, 2008). One of the applications includes dehairing leather. The aforementioned process is crucial in leather preparation for industrial purposes. Dehairing of leather is currently performed using sodium sulfate, which is toxic for the environment. The latex peptidase from *Calotropis procera* (CpLP) could efficiently remove hair from leather and therefore is a promising candidate to be used in tanneries (Lopéz et al., 2017). Along with CpLP, chymopapain also was used in dehairing leather—however it has also been used in meat tenderization and food processing. This protein belongs to the group of cysteine proteases and can be isolated from the latex of *Carica papaya* L. (Domsalla & Melzig, 2008).

3.6 Latex components used for biofuel production and others

Aside from *Hevea brasiliensis*, natural rubber can also be produced from other plants of the Euphorbiaceae family. An example is *E. characias*, which contains also other biologically active compounds, like bicyclic diterpenes, polycyclic diterpenoids, fatty acids, tocopherols, antioxidants and sterols (Pintus, Medda, Rinaldi, Spano, & Floris, 2010; Spano et al., 2012; Sytwala, Günther, & Melzig, 2015). As one of the Euphorbiaceae plants with high level of terpenoids, *Euphorbia tirucalli*, as well as other species from this family, have been considered as potential biofuel sources. In addition to phytosterols and triterpenes, this plant also contains many additional bioactive components and has been used in traditional folk medicine (Gupta, Vishnoi, Wal, & Wal, 2013; Kitajima et al., 2016). Such high-level latex-yielding Euphorbia have huge potential advantages as feedstock for biofuel production. This perspective for their use was compared with the use of traditional fossil-based fuels. From this point, physical, chemical, and combustive properties of Euphorbia-based biofuels are comparable or even could exceed those of fossil fuels. Euphorbiaceae plants which produce high latex yields, can grow on poor lands under xeric conditions with relatively low freshwater needs (Patan, Bugude, Sake, & Randall, 2018). On the other hand, plant latex may be a promising alternative for the catalysis of transesterification in biodiesel production. Enzymatic catalysis by plant lipases as biocatalysts is an alternative that, unlike chemical catalysis processes, seemed to be easy to perform and can be carried out with a low investment cost (Mazou, Djossou, Tchobo, Villeneuve, & Soumanou, 2016).

Potentially, some latex-bearing plants could also serve for phytoremediation as metal hyperaccumulators. One of the spectacular examples is New Caledonian hyperaccumulator tree *Pycnandra acuminata*, formerly known as *Sebertia acuminate*, which after cut exudes blue-green latex, with a high nickel content (Jaffré, Pillon, Thomine, & Merlot, 2013). The plant, and also many other species have the potential for extracting metallic elements from the soil (phytoextraction) as well as for phytomining of nickel. In such case hyperaccumulator plants are grown on nickel-rich soils, which are however sub-economic for conventional mining. Their harvested biomass is nickel-rich and can be used to derive pure chemicals (Jaffré, Reeves, Baker, Schat, & Ent, 2018).

4. Conclusions

As presented above latex from different plants is composed of diverse molecules which exhibit a broad range of positive properties, beneficial for human health. Many of these activities have been exploited in pharmacology and medicine to develop novel plant-based therapeutics for fighting various diseases. Although broad research on different aspects of latex-bearing plants is still lacking, it has already been shown that latex compounds have been used in various medical applications relieving from pain, healing wounds and tackling viruses and such diseases as diabetes mellitus, Alzheimer's disease, anxiety, and cancer. Therefore, there are high expectations for compounds isolated from latex-bearing plants and their use as modern, safe therapeutics. Additionally, many latex plants contain valuable low-molecular compounds or secondary metabolites, which are widely used in industry, agriculture and other areas, like natural rubber, nanoparticles. Future improvements are expected in breeding crops and pharmacologically important plants, which could be stimulated for a higher expression of defense-related proteins enhancing their defense against pathogens. Increased control of plant viral diseases by stimulation of plant defense response, may help overcome significant barriers in agriculture by avoiding the development of further crop transgenization. Next generation genome sequencing techniques and the reduction of cost, are expected to provide new opportunities for the research on the latex-bearing plants especially in the post-COVID-19 era, which hopefully is approaching.

Acknowledgments

This work was financially supported by the National Science Center of the Republic of Poland (project nos. 2016/21/N/NZ6/00997 to Oskar Musidlak and 2019/35/B/NZ9/03851 to Robert Nawrot).

References

Agrawal, A. A., & Konno, K. (2009). Latex: A model for understanding mechanisms, ecology, and evolution of plant defense against herbivory. *Annual Review of Ecology, Evolution, and Systematics*, *40*, 311–331.

Alenius, H., Kalkkinen, N., Lukka, M., Reunala, T., Turjanmaa, K., Mäkinen-Kiljunen, S., et al. (1995). Prohevein from the rubber tree (Hevea brasiliensis) is a major latex allergen. *Clinical & Experimental Allergy*, *25*(7), 659–665.

Alsirafy, S. A., & Farag, D. E. (2016). A shortage of oral morphine in Egypt. *Bulletin of the World Health Organization*, *94*(1), 3.

Amare, G. G., Meharie, B. G., & Belayneh, Y. M. (2020). Evaluation of antidiabetic activity of the leaf latex of Aloe pulcherrima Gilbert and Sebsebe (Aloaceae). *Evidence-based Complementary and Alternative Medicine*, *2020*, 8899743.

Arora, D., & Sharma, A. (2013). A review on phytochemical and pharmacological potential of genus Chelidonium. *Pharmacognosy Journal*, *5*, 184–190.

Barken, I., Geller, J., & Rogosnitzky, M. (2008). Noscapine inhibits human prostate cancer progression and metastasis in a mouse model. *Anticancer Research*, *28*, 3701–3704.

Barton, B., & Castle, T. (1845). *The British flora medica, or history of the medicinal plants of Great Britain*. London: E. Cox.

Bauer, G., Gorb, S. N., Klein, M.-C., Nellesen, A., von Tapavicza, M., & Speck, T. (2014). Comparative study on plant latex particles and latex coagulation in Ficus benjamina, Campanula glomerata and three Euphorbia species. *PLoS One*, *9*(11), 1–8.

Berg, C. C., Corner, E. J. H., & Jarrett, F. M. (2006). Moraceae genera other than Ficus. *Flora Malesiana—Series 1, Spermatophyta*, *17*(1), 1–146.

Berthelot, K., Lecomte, S., Estevez, Y., & Peruch, F. (2014). Hevea brasiliensis REF (Hev b 1) and SRPP (Hev b 3): An overview on rubber particle proteins. *Biochimie*, *106*, 1–9.

Bhavani, J., & Ravichandran, S. (2020). Investigating the possible mechanism for potency of latex of Plumeria alba L. in treating scabies. *International Journal of Review in Life Sciences*, *9*(1), 1–4.

Bokobza, L. (2018). Natural rubber nanocomposites: A review. *Nanomaterials*, *9*(1), 1–21.

Botwina, P., Owczarek, K., Rajfur, Z., Ochman, M., Urlik, M., Nowakowska, M., et al. (2020). Berberine hampers influenza a replication through inhibition of MAPK/ERK pathway. *Viruses, 12*(3), 344.

Beaudoin, Guillaume, & Facchini, Peter. (2013). Isolation and characterization of a cDNA encoding(S)- cis-N-methylstylopine 14-hydroxylase from opium poppy, a key enzyme insanguinarine biosynthesis. *Biochemical and Biophysical Research Communications, 431*(3), 597–603. https://doi.org/10.1016/j.bbrc.2012.12.129.

Behxhet, M., Hajdari, A., Pajazita, Q., Syla, B., Quave, C. L., & Pieroni, A. (2012). An ethnobotanical survey of the Gollak region, Kosovo. *Genetic Resources and Crop Evolution, 59*(5), 739–754.

Carter, C. A., Forney, R. W., Gray, E. A., Gehring, A. M., Schneider, T. L., Young, D. B., et al. (1997). Toxicarioside A. A new cardenolide isolated from Antiaris Toxicaria latex-derived dart poison. Assignment of the 1H- and 13C-NMR shifts for an antiarigenin aglycone. *Tetrahedron, 53*(40), 13557–13566.

Cavalcante, G. S., de Morais, S. M., Andre, W. P., Ribeiro, W. L., Rodrigues, A. L., De Lira, F. C., et al. (2016). Chemical composition and in vitro activity of Calotropis procera (Ait.) latex on Haemonchus contortus. *Veterinary Parasitology, 226*, 22–25.

Chaiyasat, P. (2012). Preparation and characterization of nanocomposites of natural rubber with polystyrene and styrene-methacrylic acid copolymer nanoparticles. *Express Polymer Letters, 6*(6), 511–518.

Chen, J. W., Borgelt, L. M., & Blackmer, A. B. (2019). Epidiolex (cannabidiol): A new hope for patients with dravet or Lennox-gastaut syndromes. *The Annals of Pharmacotherapy, 53*(6), 603–611, 1060028018822124.

Cherian, S., Ryu, S. B., & Cornish, K. (2019). Natural rubber biosynthesis in plants, the rubber transferase complex, and metabolic engineering progress and prospects. *Plant Biotechnology Journal, 17*(11), 2041–2061.

Cho, W. K., Chen, X. Y., Uddin, N. M., Rim, Y., Moon, J., Jung, J. H., et al. (2009). Comprehensive proteome analysis of lettuce latex using multidimensional protein-identification technology. *Phytochemistry, 70*(5), 570–578.

Cornish, K. (2017). Alternative natural rubber crops: Why should we care? *Technology & Innovation, 18*(4), 244–255.

Cruz, L. S., de Oliveira, T. L., Kanunfre, C. C., Paludo, K. S., Minozzo, B. R., Prestes, A. P., et al. (2018). Pharmacokinetics and cytotoxic study of euphol from Euphorbia umbellata. (Bruyns) Pax latex. *Phytomedicine, 47*, 105–112.

Cho, W. K., Jo, Y., Chu, H., Park, S. H., & Kim, K. H. (2014). Integration of latex protein sequence data provides comprehensive functional overview of latex proteins. *Molecular Biology Reports, 41*(3), 1469–1481. https://doi.org/10.1007/s11033-013-2992-6.

Cardoso, H., Guimarães, M., Lopes, L., & Alves, J. L. (2019). Development of a rubber sole with an integral cushioning system for casual sport shoes. *Proceedings of the Institution of Mechanical Engineers, Part L: Journal of Materials: Design and Applications, 233*(11), 2253–2266. https://doi.org/10.1177/1464420719843154.

De Marino, S., Gala, F., Zollo, F., Vitalini, S., Fico, G., Visioli, F., et al. (2008). Identification of minor secondary metabolites from the latex of Croton lechleri (Muell-Arg) and evaluation of their antioxidant activity. *Molecules, 13*(6), 1219–1229.

Domsalla, A., & Melzig, M. F. (2008). Occurrence and properties of proteases in plant latices. *Planta Medica, 74*(7), 699–711.

Dussourd, D. E. (1995). Entrapment of aphids and whiteflies in lettuce latex. *Annals of the Entomological Society of America, 88*, 163–172.

Eljounaidi, K., & Lichman, B. R. (2020). Nature's chemists: The discovery and engineering of phytochemical biosynthesis. *Frontiers in Chemistry, 8*, 1041.

Farida, E., Bukit, N., Ginting, E. M., & Bukit, B. F. (2019). The effect of carbon black composition in natural rubber compound. *Case Studies in Thermal Engineering, 16*, 100566.

Farrell, B. D., Dussourd, D. E., & Mitter, C. (1991). Escalation of plant defense: Do latex and resin canals spur plant diversification? *The American Naturalist*, *138*, 881–900.
Freitas, C. D. T., Leite, H. B., Oliveira, J. P. B., Amaral, J. L., Egito, A. S., & Vairo-Cavalli, S. (2016). Insights into milk-clotting activity of latex peptidases from *Calotropis procera* and *Cryptostegia grandiflora*. *Food Research International*, *87*, 50–59.
Frick, S., Kramell, R., & Kutchan, T. M. (2007). Metabolic engineering with a morphine biosynthetic P450 in opium poppy surpasses breeding. *Metabolic Engineering*, *9*(2), 169–176.
Gebremeskel, L., Bhoumik, D., Sibhat, G. G., & Tuem, K. B. (2018). In vivo wound healing and anti-inflammatory activities of leaf latex of Aloe megalacantha Baker (Xanthorrhoeaceae). *Evidence-based Complementary and Alternative Medicine*, *2018*, 5037912.
Ghanbari, A., Le Gresley, A., Naughton, D., Kuhnert, N., Sirbu, D., & Ashrafi, G. H. (2019). Biological activities of Ficus carica latex for potential therapeutics in human papillomavirus (HPV) related cervical cancers. *Scientific Reports*, *9*(1), 1–11.
Gupta, L., Sharma, A. K., Gothwal, A., Khan, M. S., Khinchi, M. P., Qayum, A., et al. (2017). Dendrimer encapsulated and conjugated delivery of berberine: A novel approach mitigating toxicity and improving in vivo pharmacokinetics. *International Journal of Pharmaceutics*, *528*(1–2), 88–99.
Gupta, N., Vishnoi, G., Wal, A., & Wal, P. (2013). Medicinal value of Euphorbia tirucalli. *Systematic Reviews in Pharmacy*, *4*(1), 40–46.
Gurkok, T., Turktas, M., Parmaksiz, I., & Unver, T. (2015). Transcriptome profiling of alkaloid biosynthesis in elicitor induced opium poppy. *Plant Molecular Biology Reporter*, *33*, 673–688.
Hagel, J. M., & Facchini, P. J. (2013). Benzylisoquinoline alkaloid metabolism: A century of discovery and a brave new world. *Plant and Cell Physiology*, *54*, 647–672.
Hagel, J. M., Yeung, E. C., & Facchini, P. J. (2008). Got milk? The secret life of laticifers. *Trends in Plant Science*, *13*(12), 631–639.
Haggag, E. G., Elshamy, A. M., Rabeh, M. A., Gabr, N. M., Salem, M., Youssif, K. A., et al. (2019). Antiviral potential of green synthesized silver nanoparticles of Lampranthus coccineus and Malephora lutea. *International Journal of Nanomedicine*, *14*, 6217–6229.
Hamilton, J. L. (2016). *Natural rubber: Properties, behavior and applications*. New York, USA: Nova Science Publishers.
Hammeso, W. W., Emiru, Y. K., Getahun, K. A., & Kahaliw, W. (2019). Antidiabetic and antihyperlipidemic activities of the leaf latex extract of Aloe megalacantha Baker (Aloaceae) in streptozotocin-induced diabetic model. *Evidence-based Complementary and Alternative Medicine*, *2019*, 8263786.
Hatton, F. L., Ruda, M., Lansalot, M., D'Agosto, F., Malmström, E., & Carlmark, A. (2016). Xyloglucan-functional latex particles via RAFT-mediated emulsion polymerization for the biomimetic modification of cellulose. *Biomacromolecules*, *17*(4), 1414–1424.
Hirayama, C., Konno, K., Wasano, N., & Nakamura, M. (2007). Differential effects of sugar-mimic alkaloids in mulberry latex on sugar metabolism and disaccharidases of Eri and domesticated silkworms: Enzymatic adaptation of Bombyx mori to mulberry defense. *Insect Biochemistry and Molecular Biology*, *37*(12), 1348–1358.
Huang, P., Sun, M., & Zeng, J. (2020). Chapter four—Low-molecular compounds of latex-bearing plants and their biosynthesis. In R. Nawrot (Ed.), *Advances in botanical research: Vol. 93. Latex, laticifers and their molecular components: From functions to possible applications* (pp. 99–118). Academic Press.
Hurley, P. E. (1981). History of natural rubber. *Journal of Macromolecular Science: Part A—Chemistry*, *15*(7), 1279–1287.
Hussain, G., Rasul, A., Anwar, H., Aziz, N., Razzaq, A., & Wei, W. (2018). Role of plant derived alkaloids and their mechanism in neurodegenerative disorders. *International Journal of Biological Sciences*, *14*(3), 341–357.

International Rubber Study Group. (2020). http://www.rubberstudy.com/news. last checked: 06.01.2021.

Ismail, H., Shaari, S. M., & Othman, N. (2011). The effect of chitosan loading on the curing characteristics, mechanical and morphological properties of chitosan-filled natural rubber (NR), epoxidised natural rubber (ENR) and styrene-butadiene rubber (SBR) compounds. *Polymer Testing*, *30*(7), 784–790.

Jaffré, T., Pillon, Y., Thomine, S., & Merlot, S. (2013). The metal hyperaccumulators from New Caledonia can broaden our understanding of nickel accumulation in plants. *Frontiers in Plant Science*, *4*, 1–7.

Jaffré, T., Reeves, R. D., Baker, A. J. M., Schat, H., & Ent, A. (2018). The discovery of nickel hyperaccumulation in the new Caledonian tree Pycnandra acuminata 40 years on: An introduction to a virtual issue. *New Phytologist*, *218*(2), 397–400.

Jantova, S., Cipak, L., Cernakova, M., & Kostalova, D. (2003). Effect of berberine on proliferation, cell cycle and apoptosis in HeLa and L1210 cells. *The Journal of Pharmacy and Pharmacology*, *55*(8), 1143–1149.

Jarnthong, M., Wang, F., Wei, X., Wang, R., & Li, J. (2015). Preparation and properties of biocomposite based on natural rubber and bagasse nanocellulose. *MATEC Web of Conferences*, *26*, 01005.

Johnson, R. (2018). *Hemp as an agricultural commodity*. Library of Congress Washington DC Congressional Research Service.

Joy, J., Chan, C. H., Maria, H., & Thomas, S. (2013). Natural rubber-based composites and nanocomposites: State of the art, new challenges and opportunities. In S. Thomas, C. H. Chan, L. Pothen, J. Joy, & H. Maria (Eds.), *Natural rubber materials: Volume 2: Composites and nanocomposites* (pp. 1–33). Royal Society of Chemistry.

Juca, T. L., Ramos, M. V., Moreno, F. B. M. B., de Matos, M. P. V., Marinho-Filho, J. D. B., Moreira, R. A., et al. (2013). Insights on the phytochemical profile (cyclopeptides) and biological activities of Calotropis procera latex organic fractions. *The Scientific World Journal*, 1–9. https://doi.org/10.1155/2013/615454.

Karban, R., & Kuc, J. (1999). Induced resistance against pathogens and herbivores: An overview. In A. A. Agrawal, S. Tuzun, & E. Bent (Eds.), *Inducible plant defenses against pathogens and herbivores: Biochemistry, ecology, and agriculture* (pp. 1–16). St. Paul, MN, USA: American Phytopathological Society Press.

Khan, A., Xiangyang, S., Ahmad, A., & Mo, X. (2018). Electrospinning of crude plant extracts for antibacterial and wound healing applications: A review. *SM Journal of Biomedical Engineering*, *4*(1), 1–8.

Kitajima, S., Miura, K., Aoki, W., Yamato, K. T., Taira, T., Murakami, R., et al. (2016). Transcriptome and proteome analyses provide insight into laticifer's defense of Euphorbia tirucalli against pests. *Plant Physiology and Biochemistry*, *108*, 434–446.

Kohjiya, S., & Ikeda, Y. (2014). *Chemistry, manufacture and applications of natural rubber* (p. 502). USA: Woodhead Publishing.

Koleva, V., Dragoeva, A., Stoyanova, Z., Koynova, T., & Dashev, G. (2015). An ethnobotanical study on current status of some medicinal plants used in Bulgaria. *International Journal of Current Microbiology and Applied Sciences*, *4*, 297–305.

Konno, K. (2011). Plant latex and other exudates as plant defense systems: Roles of various defense chemicals and proteins contained therein. *Phytochemistry*, *72*(13), 1510–1530.

Konno, K., Ono, H., Nakamura, M., Tateishi, K., Hirayama, C., Tamuraet, Y., et al. (2016). Mulberry latex rich in antidiabetic sugar-mimic alkaloids forces dieting on caterpillars. *Proceedings of the National Academy of Sciences of the United States of America*, *103*(5), 1337–1341.

Kumar, R., Murugananthan, G., Nandakumar, G., & Talwar, S. (2011). Isolation of anxiolytic principle from ethanolic root extract of Cardiospermum halicacabum. *Phytomedicine*, *18*, 219–223.

Kutchan, T. M., Ayabe, S., Krueger, R. J., Coscia, E. M., & Coscia, C. J. (1983). Cytodifferentiation and alkaloid accumulation in cultured cells of Papaver Bracteatum. *Plant Cell Reports*, *2*(6), 281–284.

Lee, E. J., Hagel, J. M., & Facchini, P. J. (2013). Role of the phloem in the biochemistry and ecophysiology of benzylisoquinoline alkaloid metabolism. *Frontiers in Plant Science*, *4*, 182.

Li, K., You, J., Liu, Y., Zhu, K., Xue, C., Guo, X., et al. (2020). Functionalized starch as a novel eco-friendly vulcanization accelerator enhancing mechanical properties of natural rubber. *Carbohydrate Polymers*, *231*, 115705.

Lopéz, L. M. I., Viana, C. A., Errasti, M. E., Garro, M. L., Martegani, J. E., Mazzilli, G. A., et al. (2017). Latex peptidases of Calotropis Procera for Dehairing of leather as an alternative to environmentally toxic sodium sulfide treatment. *Bioprocess and Biosystems Engineering*, *40*(9), 1391–1398.

Ma, W., Zhu, M., Zhang, D., Yang, L., Yang, T., Li, X., et al. (2017). Berberine inhibits the proliferation and migration of breast cancer ZR-75-30 cells by targeting Ephrin-B2. *Phytomedicine*, *25*, 45–51.

Mahmoud, S. Y. M., Gad-Rab, S. M. F., Hussein, N., & Shoreit, A. A. M. (2010). Antiviral activity of latex from Ficus nitida against plant viruses. *Global Journal of Biotechnology & Biochemistry*, *5*(3), 198–205.

Majidzadeh, H., Araj-Khodaei, M., Ghaffari, M., Torbati, M., Ezzati Nazhad Dolatabadi, J., & Hamblin, M. R. (2020). Nano-based delivery systems for berberine: A modern anti-cancer herbal medicine. *Colloids and Surfaces. B, Biointerfaces*, *194*, 111188.

Marino, S. D., Gala, F., Zollo, F., Vitalini, S., Fico, G., Visioli, F., et al. (2015). Identification of minor secondary metabolites from the latex of Croton lechleri (Muell-Arg) and evaluation of their antioxidant activity. *Molecules*, *13*, 1219–1229.

Mazlan, S., Md, N., Wahab, A., Sulaiman, Z., Rajandas, H., & Zulperi, D. (2019). Major diseases of rubber (Hevea brasiliensis) in Malaysia. *Pertanika Journal of Scholarly Research Reviews*, *5*(2), 10–21.

Mazou, M., Djossou, A. J., Tchobo, F. P., Villeneuve, P., & Soumanou, M. M. (2016). Plant latex lipase as biocatalysts for biodiesel production. *African Journal of Biotechnology*, *15*(28), 1487–1502.

Men, X., Wang, F., Chen, G.-Q., Zhang, H.-B., & Xian, M. (2019). Biosynthesis of natural rubber: Current state and perspectives. *International Journal of Molecular Sciences*, *20*(1), 50.

Menale, B., Amato, G., Prisco, C., & Muoio, R. (2006). Traditional uses of plants in North-Western Molise (Central Italy). *Delpinoa*, *48*, 29–36.

Mirhadi, E., Rezaee, M., & Malaekeh-Nikouei, B. (2018). Nano strategies for berberine delivery, a natural alkaloid of Berberis. *Biomedicine & Pharmacotherapy*, *104*, 465–473.

Mithöfer, A., & Boland, W. (2012). Plant defense against herbivores: Chemical aspects. *Annual Review of Plant Biology*, *63*, 431–450.

Mohamed, N. H., Ismail, M. A., Abdel-Mageed, W. M., & Mohamed Shoreit, A. A. (2019). Antimicrobial activity of green silver nanoparticles from endophytic fungi isolated from Calotropis procera (Ait) latex. *Microbiology (Reading)*, *165*(9), 967–975.

Molinu, M. G., Piluzza, G., Campesi, G., Sulas, L., & Re, G. A. (2019). Antioxidant sources from leaves of Russian dandelion. *Chemistry & Biodiversity*, *16*(8), e1900250.

Mooibroek, H., & Cornish, K. (2000). Alternative sources of natural rubber. *Applied Microbiology and Biotechnology*, *53*(4), 355–365.

Murphy, P. B., Bechmann, S., & Barrett, M. J. (2020). Morphine. In *StatPearls*. Treasure Island, FL: StatPearls Publishing. https://www.ncbi.nlm.nih.gov/books/NBK526115.

Musidlak, O., Bałdysz, S., Krakowiak, M., & Nawrot, R. (2020). Chapter three—Plant latex proteins and their functions. In R. Nawrot (Ed.), *Advances in botanical research: Vol. 93. Latex, laticifers and their molecular components: From functions to possible applications* (pp. 55–97). Academic Press.

Nalwaya, N., Pokhorna, G., Deb, L., & Jain, N. K. (2009). Wound healing activity of latex of Calotropis gigantea. *International Journal of Pharmacy and Pharmaceutical Sciences*, *1*(1), 176–181.

Nawrot, R. (2017). Defense-related proteins from Chelidonium majus L. as important components of its latex. *Current Protein and Peptide Science*, *18*(8), 864–880.

Nawrot, R. (Ed.). (2020). *Vol. 93. Latex, laticifers and their molecular components: From functions to possible applications*. *Advances in botanical research* Academic Press.

Nawrot, R., Barylski, J., Lippmann, R., Altschmied, L., & Mock, H. P. (2016). Combination of transcriptomic and proteomic approaches helps to unravel the protein composition of Chelidonium majus L. milky sap. *Planta*, *244*, 1055–1064.

Nawrot, R., Lippmann, R., Matros, A., Musidlak, O., Nowicki, G., & Mock, H.-P. (2017). Proteomic comparison of Chelidonium majus L. latex in different phases of plant development. *Plant Physiology and Biochemistry*, *112*, 312–325.

Oldham, J. T., Hincapie, M., Rejtar, T., Wall, P. K., Carlson, J. E., & Lee-Parsons, C. W. (2010). Shotgun proteomic analysis of yeast-elicited California poppy (Eschscholzia californica) suspension cultures producing enhanced levels of benzophenanthridine alkaloids. *Journal of Proteome Research*, *9*(9), 4337–4345.

Oliveira, J. P. B., Candreva, A. M., Rizzo, G., Ramos, M. V., Oliveira, J. S., Oliveira, H. D., et al. (2019). Allergenicity reduction of cow's milk proteins using latex peptidases. *Food Chemistry*, *284*, 245–253.

Omnès, B., Thuillier, S., Pilvin, P., Grohens, Y., & Gillet, S. (2008). Effective properties of carbon black filled natural rubber: Experiments and modeling. *Composites Part A: Applied Science and Manufacturing*, *39*(7), 1141–1149.

Onoyovwe, A., Hagel, J. M., Chen, X., Khan, M. F., Schriemer, D. C., & Facchini, P. J. (2013). Morphine biosynthesis in opium poppy involves two cell types: Sieve elements and laticifers. *The Plant Cell*, *25*(10), 4110–4122.

Osborne, A. L., Solowij, N., Babic, I., Huang, X.-F., & Weston-Green, K. (2017). Improved social interaction, recognition and working memory with cannabidiol treatment in a prenatal infection (poly I:C) rat model. *Neuropsychopharmacology*, *42*, 1447.

Parambath Kanoth, B., Claudino, M., Johansson, M., Berglund, L. A., & Zhou, Q. (2015). Biocomposites from natural rubber: Synergistic effects of functionalized cellulose nanocrystals as both reinforcing and cross-linking agents via free-radical thiol–ene chemistry. *ACS Applied Materials & Interfaces*, 7(30), 16303–16310.

Pilatzke-Wunderlich, I., & Nessler, C. L. (2001). Expression and activity of cell-wall-degrading enzymes in the latex of opium poppy, Papaver Somniferum L. *Plant Molecular Biology*, *45*(5), 567–576.

Pintus, F., Medda, R., Rinaldi, A. C., Spano, D., & Floris, G. (2010). Euphorbia latex biochemistry: Complex interactions in a complex environment. *Plant Biosystems*, *144*, 381–391.

Pradeepkumar, P., Govindaraj, D., Jeyaraj, M., Munusamy, M. A., & Rajan, M. (2017). Assembling of multifunctional latex-based hybrid nanocarriers from Calotropis gigantea for sustained (doxorubicin) DOX releases. *Biomedicine & Pharmacotherapy*, *87*, 461–470.

Priyadarshan, P. M. (Ed.). (2011). *Biology of Hevea rubber* CABI.

Patan, S. S. V. K., Bugude, R., Sake, P. K., & Randall, T. G. (2018). Use of *Euphorbia* sp. (Euphorbiaceae) as biofuel feedstock for semi-arid and arid lands. *Biofuels*. https://doi.org/10.1080/17597269.2018.1501637.

Ramos, M. V., Demarco, D., da Costa Souza, I. C., & de Freitas, C. D. T. (2019). Laticifers, latex, and their role in plant defense. *Trends in Plant Science*, *24*(6), 553–567.

Ramos, M. V., Teixeira Freitas, C. D., Morais, F. S., Prado, E., Medina, M. C., & Demarco, D. (2020). Plant latex and latex-borne defense. In R. Nawrot (Ed.), *Vol. 93. Advances in botanical research. Latex, laticifers and their molecular components: From functions to possible applications* (pp. 1–25). Ireland: Academic Press—Elsevier.

Rathnayake, I., Ismail, H., Azahari, B., Bandara, C., & Rajapakse, S. (2013). Novel method of incorporating silver nanoparticles into natural rubber latex foam. *Polymer-Plastics Technology and Engineering*, *52*(9), 885–891.

Requejo-Aguilar, R., Alastrue-Agudo, A., Cases-Villar, M., Lopez-Mocholi, E., England, R., Vicent, M. J., et al. (2017). Combined polymer-curcumin conjugate and ependymal progenitor/stem cell treatment enhances spinal cord injury functional recovery. *Biomaterials*, *113*, 18–30.

Reyes, J., Milagros, A., Liggieri, C. S., & Bruno, M. A. (2018). Preparation of soy protein hydrolysates with antioxidant activity by using peptidases from latex of Maclura Pomifera fruits. *Food Chemistry*, *264*, 326–333.

Roberts, M. F., McCarthy, D., Kutchan, T. M., & Coscia, C. J. (1983). Localization of enzymes and alkaloidal metabolites in Papaver latex. *Archives of Biochemistry and Biophysics*, *222*(2), 599–609.

Roshchina, V. V., et al. (1993). Intratissular secretion. In V. V. Roshchina, & V. D. Roshchina (Eds.), *The excretory function of higher plants* (pp. 25–66). Berlin, Heidelberg: Springer.

Roy, K., Srivastwa, A. K., & Ghosh, C. K. (2019). Anticoagulant, thrombolytic and antibacterial activities of Euphorbia acruensis latex-mediated bioengineered silver nanoparticles. *Green Processing and Synthesis*, *8*(1), 590–599.

Rush, M. D., Kutchan, T. M., & Coscia, C. J. (1985). Correlation of the appearance of morphinan alkaloids and laticifer cells in germinating Papaver Bracteatum seedlings. *Plant Cell Reports*, *4*(5), 237–240.

Schmidt, T., Hillebrand, A., Wurbs, D., Wahler, D., Lenders, M., & Schulze, C. (2010). Molecular cloning and characterization of rubber biosynthetic genes from Taraxacum koksaghyz. *Plant Molecular Biology Reporter*, *28*, 277.

Schmidt, T., Lenders, M., Hillebrand, A., van Deenen, N., Munt, O., Reichelt, R., et al. (2010). Characterization of rubber particles and rubber chain elongation in Taraxacum koksaghyz. *BMC Biochemistry*, *11*(1), 11.

Sertürner, F. W. A. F. (1806). Darstellung der reinen Mohnsäure (Opiumsäure) nebst einer Untersuchung des Opiums mit vorzüglicher Hinsicht auf einen darin neu entdeckten Stoff und die dahin gehörigen Bemerkungen. Vom Herrn Sertürner in Paderborn. *Journal Der Pharmacie Für Aerzte, Apotheker Und Chemisten*, *14*, 47–93.

Song, S., Qiu, M., Chu, Y., Chen, D., Wang, X., Su, A., et al. (2014). Downregulation of cellular c-Jun N-terminal protein kinase and NF-κB activation by berberine may result in inhibition of herpes simplex virus replication. *Antimicrobial Agents and Chemotherapy*, *58*(9), 5068–5078.

Spano, D., Pintus, F., Mascia, C., Scorciapino, M. A., Casu, M., Floris, G., et al. (2012). Extraction and characterization of a natural rubber from Euphorbia characias latex. *Biopolymers*, *97*, 589–594.

Sytwala, S., Günther, F., & Melzig, M. F. (2015). Lysozyme- and chitinase activity in latex bearing plants of genus Euphorbia—A contribution to plant defense mechanism. *Plant Physiology and Biochemistry*, *95*, 35–40.

Thunugunta, T., Reddy, A. C., & Reddy, L. D. C. (2015). Green synthesis of nanoparticles: Current prospectus. *Nanotechnology Reviews*, *4*(4), 303–323.

Tita, I., Mogosanu, G., & Tita, M. (2009). Ethnobotanical inventory of medicinal plants from the South-West of Romania. *Farmácia*, *57*, 141–156.

United Nations Office on Drugs and Crime. (2014). *Narcotic drugs: Estimated world requirements for 2015; Statistics for 2013*. United Nations.

Upadhyay, R. K. (2011). Plant latex: A natural source of pharmaceuticals and pesticides. *International Journal of Green Pharmacy*, *5*, 169–180.

Urs, A. P., Manjuprasanna, V. N., Rudresha, G. V., Yariswamy, M., & Vishwanath, B. S. (2017). Plant latex proteases: Natural wound healers. In S. Chakraborti, & N. Dhalla (Eds.), *Proteases in physiology and pathology* (pp. 297–323). Singapore: Springer.

van Beilen, J. B., & Poirier, Y. (2007). Establishment of new crops for the production of natural rubber. *Trends in Biotechnology*, *25*(11), 522–529.

Verheye, W. (2010). Growth and production of rubber. In W. Verheye (Ed.), *Land use, land cover and soil sciences. In Encyclopedia of Life Support Systems (EOLSS)* (pp. 1–20). Oxford, UK: UNESCO-EOLSS Publishers.

Vulcanization and properties of vulcanized rubber. (2021). Web Solutions LLC. and its licencors. https://science.jrank.org/pages/7266/Vulcanization.html (last accessed 03.23.2021).

Wagner, S., & Breiteneder, H. (2005). Hevea brasiliensis latex allergens: Current panel and clinical relevance. *International Archives of Allergy and Immunology*, *136*, 90–97.

Wahler, D., Colby, T., Kowalski, N. A., Harzen, A., Wotzka, S. Y., Hillebrand, A., et al. (2012). Proteomic analysis of latex from the rubber-producing plant Taraxacum brevicorniculatum. *Proteomics*, *12*(6), 901–905.

Watt, G., & Karl, T. (2017). In vivo evidence for therapeutic properties of cannabidiol (CBD) for Alzheimer's disease. *Frontiers in Pharmacology*, *8*, 20.

Weid, M., Ziegler, J., & Kutchan, T. M. (2004). The roles of latex and the vascular bundle in morphine biosynthesis in the opium poppy, Papaver Somniferum. *Proceedings of the National Academy of Sciences of the United States of America*, *101*(38), 13957–13962.

WHO TEAM Health Product Policy and Standards. (2019). *WHO model list of essential medicines—21st list, 2019*. WHO.

Wen, Y., Wang, Y., Zhao, K., & Sumalee, A. (2017). The use of natural rubber latex as a renewable and sustainable modifier of asphalt binder. *International Journal of Pavement Engineering*, *18*(6), 547–559. https://doi.org/10.1080/10298436.2015.1095913.

Xiao, Y., Tian, C., Huang, T., Han, B., Wang, M., Ma, H., et al. (2018). 8-Cetylberberine inhibits growth of lung cancer in vitro and in vivo. *Life Sciences*, *192*(1), 259–269.

Zalucki, M. P., Malcolm, S. B., Paine, T. D., Hanlon, C. C., Brower, L. P., & Clarke, A. R. (2001). It's the first bites that count: Aurvival of first-instar monarchs on milkweeds. *Austral Ecology*, *26*, 547–555.

Zevin, I. V., Altman, N., & Zevin, L. V. (1997). *A Russian herbal: Traditional remedies for health and healing*. Healing Arts Press.

Zhou, Y.-F., & Liu, W.-Z. (2010). Laticiferous canal formation in fruit of Decaisnea fargesii: A programmed cell death process? *Protoplasma*, *248*(4), 683–694.

Ziegler, J., & Facchini, P. J. (2008). Alkaloid biosynthesis: Metabolism and trafficking. *Annual Review of Plant Biology*, *59*, 735–769.

Zielińska, S., Jezierska-Domaradzka, A., Wójciak-Kosior, M., Sowa, I., Junka, A., & Matkowski, A. M. (2018). Greater Celandine's ups and downs – 21 centuries of medicinal uses of Chelidonium majus from the viewpoint of today's pharmacology. *Frontiers in Pharmacology*, *9*, 1–29.

Zulak, K. G., Khan, M. F., Alcantara, J., Schriemer, D. C., & Facchini, P. J. (2009). Plant defense responses in opium poppy cell cultures revealed by liquid chromatography-tandem mass spectrometry proteomics. *Molecular & Cellular Proteomics*, *8*(1), 86–98.

CHAPTER ELEVEN

When dietary supplements meet metabolomics: A fast-evolving field—A follow-up of ABR volume 67: Metabolomics coming of age with its technological diversity

Dominique Rolin[a,d], Catherine Bennetau-Pelissero[b], Sarada D. Tetali[c], and Pierre Pétriacq[a,d,*]

[a]UMR1332 BFP, University of Bordeaux, INRAE, Villenave d'Ornon, France
[b]UMR1212 ARNA, University of Bordeaux, Inserm, CNRS, Equipe TAMS, Bordeaux, France
[c]Department of Plant Sciences, School of Life Sciences, University of Hyderabad, Hyderabad, Telangana, India
[d]Bordeaux Metabolome, MetaboHUB, PHENOME-EMPHASIS, Villenave d'Ornon, France
[*]Corresponding author: e-mail address: pierre.petriacq@inrae.fr

Contents

Abstract

Volume 67 of Advances in Botanical Research "Metabolomics coming of age with its technological diversity" edited by Dominique Rolin (series Editors Jean-Pierre Jacquot and Pierre Gadal) was published in 2013 and featured 13 chapters including more than 40 authors with a range of expertise in biology, biochemistry, analytical chemistry, chemometrics and bioinformatics. This volume highlighted the technological diversity of metabolomics approaches that have revolutionized life sciences, more particularly plant biology, providing a deeper understanding of the multifaceted chemical and physiological variety of plant systems. Since its initiation, metabolomics has never failed

Advances in Botanical Research, Volume 100
ISSN 0065-2296
https://doi.org/10.1016/bs.abr.2021.01.009
Copyright © 2021 Elsevier Ltd
All rights reserved.

to deliver further progress in research fields. In this anniversary volume 100 of ABR, a special focus is placed on the contribution of metabolomics to the study of dietary supplements, where Chemistry, Botanics and Physiology cross path. We also highlight the current challenges and future perspectives in this inspiring, fast-evolving area, which holds great promise for unraveling novel mechanisms by which food supplements benefit Human nutrition and health.

1. Introduction

1.1 Salient features of volume 67 on metabolomics technical diversity

Beyond the genomic era, metabolomics has become a functional approach rapidly enhancing our knowledge of complex biological systems by gathering insightful inferences at the different omics levels (Fig. 1) including the

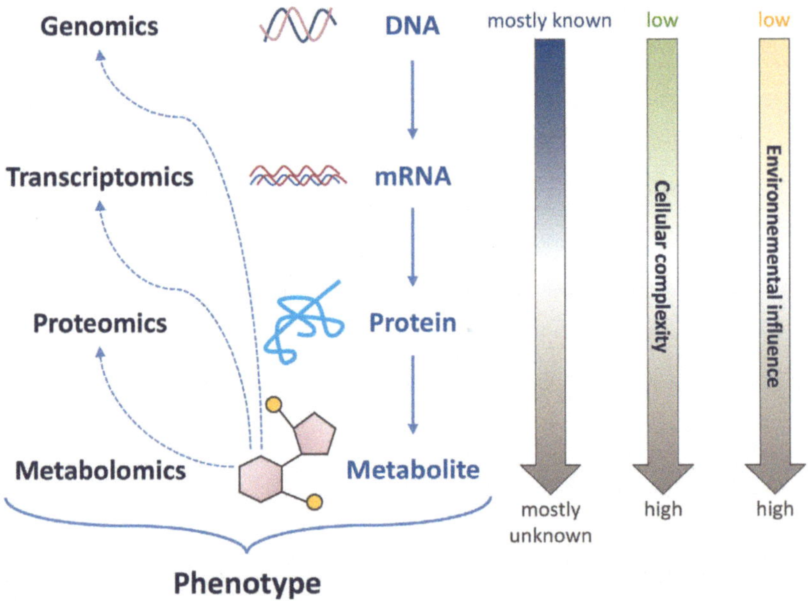

Fig. 1 Metabolomics as an integrative phenotyping strategy. Metabolomics ultimately underlies phenotypes by gathering inferences from the genome, transcriptome and proteome. While DNA and mRNA have direct relationships with proteins, metabolites can feedback on genomic, transcriptomic and proteomic regulations through biochemical interactions. The cellular complexity increases from genomes (low), transcriptomes to proteomes and finally metabolomes (high). Besides, the metabolome, which mostly comprises unknown metabolites, not only integrates genetic determinants but is also greatly influenced by environmental factors.

genome, transcriptome, proteome or enzyme activome (Johnson, Ivanisevic, & Siuzdak, 2016). It thus ultimately underlies phenotypes by integrating both genetic (*i.e.*, internal) and environmental (*i.e.*, external) determinants. This technology has proven particularly useful to study various aspects of plant sciences as it encompasses multiple techniques that can be tailored to the plant biologist's needs (Hall et al., 2002). Besides, plant metabolomes, *i.e.*, the complete set of low molecular weight compounds (<1500 Da), are chemically very complex as they comprise a large number of metabolites harboring diversified structures and abundance that are pivotal to plant developmental processes and environmental interactions.

The term "metabolomics" nowadays refers to both traditional targeted analytical methods that describe metabolomes through the determination of precise concentrations of metabolites characterizing organismal chemistries, and untargeted techniques that provide a global picture of the metabolic profiles (Fig. 2). Briefly, the two main integrated technologies, historically

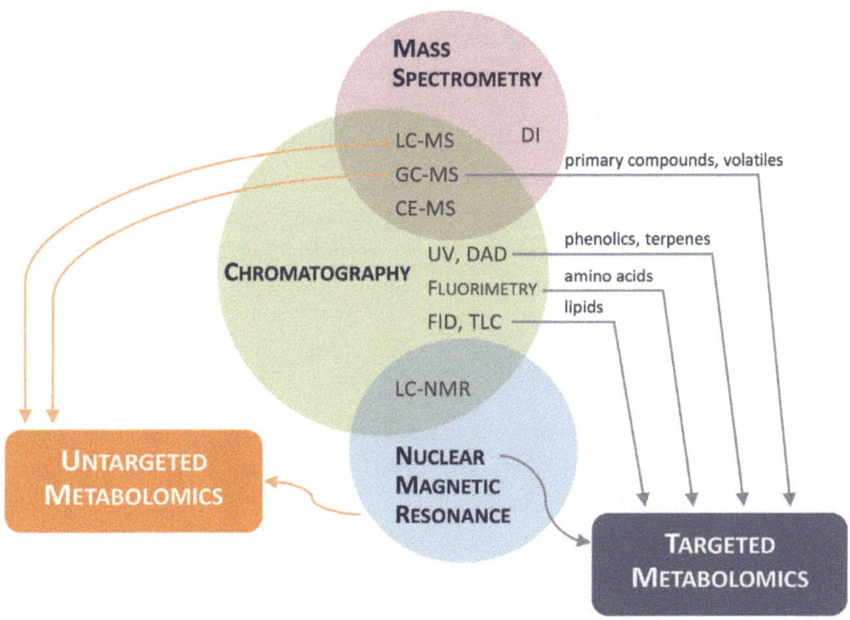

Fig. 2 Main analytical approaches to metabolomics. Metabolomics refers to both targeted, traditional analyses and untargeted methods. Examples are indicated for some specific compounds that are detected by targeted techniques. CE, capillary electrophoresis; DAD, diode array detector; DI, direct injection; FID, flame ionization detector; GC, gas chromatography; LC, liquid chromatography; MS, mass spectrometry; NMR, nuclear magnetic resonance; TLC, thin layer chromatography; UV, ultra violet.

brought by Physics and now used by (bio)chemists then biologists, are (1) non-destructive nuclear magnetic resonance spectroscopy (NMR) and (2) mass spectrometry (MS)-based techniques including GC/LC/CE-MS (gas/liquid chromatography/capillary electrophoresis-MS) and FT-ICR-MS (Fourier transform ion cyclotron resonance-MS) (Deborde et al., 2017; Perez de Souza, Alseekh, Naake, & Fernie, 2019). It is presently impossible to attain a perfect and comprehensive depiction of the whole metabolome in a biological sample using single analytical strategies as they fail to encompass the great chemical diversity of metabolites. Therefore, combined approaches are increasingly used in metabolomics analysis and assisted by other method developments to identify, quantify and localize plant metabolites. Further, due to the multivariate nature of metabolomics datasets, biostatistics and bioinformatics have evolved concomitantly to provide tools and methods to depict the metabolomics complexity, reveal discriminant biomarkers and even predict metabolic fluxes based on mathematical modeling. These technological developments were examined in volume 67 of Advances in Botanical Research (ABR) "Metabolomics coming of age with its technological diversity" through 13 chapters (Rolin, 2013). Three chapters of volume 67 were devoted to NMR-based protocols, four chapters to MS-based techniques, one to lipidomics, one to isotopomics and one to high-throughput biochemical phenotyping, while the last four chapters reviewed data handling, analysis and modeling approaches. A typical metabolomics workflow that applies to the study of Food supplements is also presented in Fig. 3, emphasizing both targeted and untargeted strategies.

1.2 Food supplements

Food supplements are produced nowadays to help keep healthy, relieve from physiological discomforts and ensure well-being (Egan, Hodgkins, Shepherd, Timotijevic, & Raats, 2011). Such effects could be attributed to bioactive compounds of plant or non-plant origin that improve many of the human functions, by influencing, *e.g.*, brain, heart, liver and gut or joints (Fig. 3). Based on the nature of their ingredients, dietary supplements can be classified into two main categories. The first one concerns dietary substances, such as vitamins and minerals, that are normally found in our diet and may be eventually under-consumed. These micronutrients can be either synthesized or extracted through processes that provide pure compounds. Metabolomics analysis is not required in that case although traditional

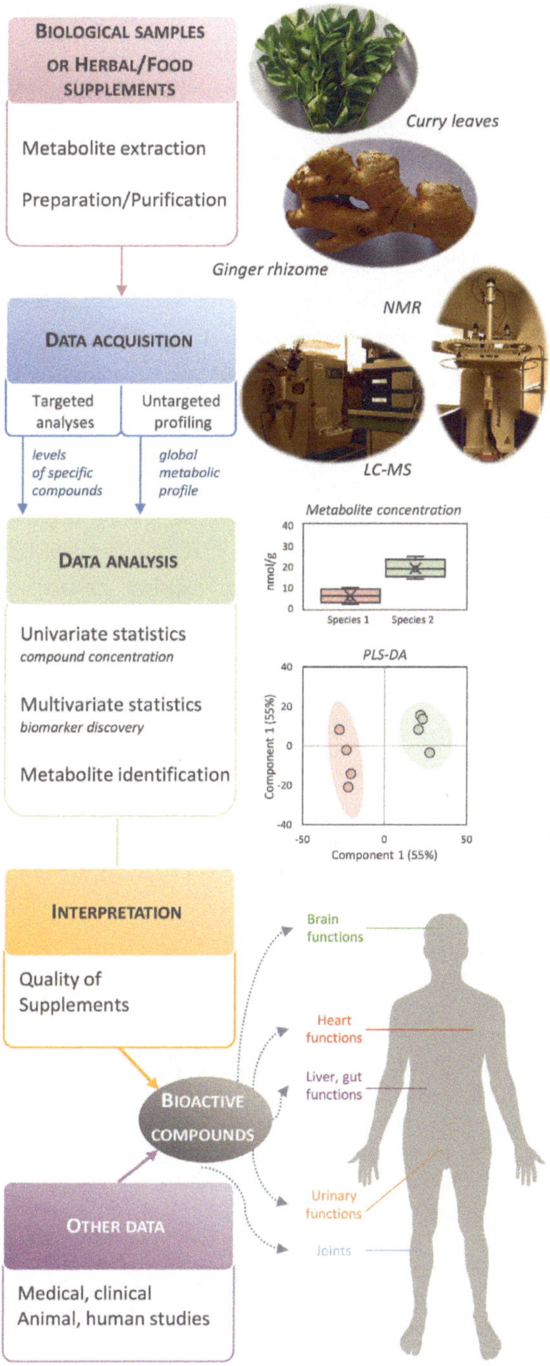

Fig. 3 See legend on opposite page.

chemical analysis helps to determine purity or molecular conformation, *e.g.*, when enantiomeric forms exist.

The second class of food supplements includes ingredients that are not always classical in normal diets. This applies to milk thistle, Californian poppy or non-plant shark cartilage, for instance (Harty et al., 2018). These ingredients are based on natural substances extracted or prepared from natural sources, which may be a source of concern for the European Food Safety Agency (EFSA) ("Compendium of Botanicals Reported to Contain Naturally Occurring Substances of Possible Concern for Human Health When Used in Food and Food Supplements", 2012). Such ingredients are chemically complex and classical analytical methods generally fail approaching such complexity. These substances include plants and plant preparations produced in nature, fields, greenhouses or *in vitro*. Nowadays in Europe, plant preparations also include essential oils and other substances chemically defined such as alginate, collagen, cartilage or chondroitin, which are not of plant origin. These latter ones are polymers of different biochemical units and the bonds between these units as well as the number of units can vary from one source to another. Other complex natural ingredients are algae or cyanobacteria, and fungi. Only some of them are authorized for dietary supplement preparations. Food supplements can contain traditional ingredients like bee products *i.e.*, propolis, royal jelly or pollens. They can also contain probiotics. The latter are increasingly used although their efficiency is not yet fully demonstrated.

In more than 60% of the cases, dietary supplements contain complex ingredients (Fibigr, Šatínský, & Solich, 2018). Therefore, metabolomic tools

Fig. 3 Typical metabolomics workflow applied to the study of herbal/food supplements. Metabolites are extracted and purified from biological samples or herbal/food supplements (here plants presenting health benefits) prior to metabolomics. Targeted analyses involve quantitative measurements of specific compounds and yield data that are processed by univariate statistical methods for comparative or absolute quantification using NMR methods, for instance. Untargeted metabolomics typically relies on LC-MS techniques providing a comprehensive metabolic overview of thousands of metabolic features. Multivariate statistics is used to visualize the response of metabolic profiles and discover discriminant metabolic markers such as potential bioactive compounds. These relevant markers are putatively annotated using metabolite databases, then confirmed by targeted techniques (comparing spectra and retention time data with metabolite standards). Metabolomics can be used to confirm the quality of food supplements or combined to other medical/clinical data based on animal/human models in order to address potential claims on health benefits from the bioactives studied.

are relevant to assess both their quality and standardization (Fig. 3). When demonstrating and claiming a health effect for a supplement, it is crucial to ensure the standardized composition of formula in active substances. Because the bioavailability of the chemical substances can differ from one chemical family to another (Manach, Williamson, Morand, Scalbert, & Rémésy, 2005), the fine determination of a mixture composition should be known far beyond its main components. Indeed, the most concentrated substances in a plant extract may not be the most concentrated in the consumer body fluids (Pujos-Guillot et al., 2013). Additionally, many products on the market combine several plant sources. Their fine composition and metabolic fate in the consumers can be evaluated *via* metabolomic analyses.

Metabolomic strategies now deliver both quantitative and qualitative data that are being implemented in the field of food and dietary supplements. This chapter will present different metabolomics approaches available for the food supplement area and discuss the metabolomics data obtained for well-known ingredients. Metabolomics inputs are first examined for the standardization of the sourcing according to the culture conditions and for quality control. Second, the contributions of metabolomics in deciphering the action of the active substances on the consumers is exemplified in both animals and humans through recent examples. The last section emphasizes the current challenges and future perspectives in this inspiring field.

2. Metabolomics inputs to the dietary supplement field

As mentioned previously, food supplements can be based on complex matrices containing many different compounds. The fine balance between each constituent may be responsible for the beneficial or adverse effect of the ingredients. Below are presented complex ingredients used in dietary supplements and the input of metabolomics on standardization and quality management.

2.1 Standardization of raw material

2.1.1 Influence of the plant environment

The growing conditions of plants deeply influence the composition of dietary supplement ingredients, more particularly for aquatic resources. The use of marine resources in dietary supplements could be challenging since constant quality products are difficult to obtain when the aquatic medium is unstable. This instability is due to seasonal changes, current variations, tide movements, the presence of estuaries and rivers, and human coast activities inducing permanent or episodical pollutions. Such impacts of environmental factors on algae composition have been studied on four *Lobophora* species from the New Caledonian lagoon in addition to the contamination issues (Gaubert, Payri, Vieira, Solanki, & Thomas, 2019). Using an LC-HRMS (high-resolution MS) metabolomic fingerprinting approach, they pointed out to the great variability of the algae metabolome. Based on several MS biomarkers from the polyolefin (alkaline polymer) family, they managed to show the influence of several parameters on the algae metabolome: growing site, month and season of the year, light, temperature, salinity, rainfall and substratum nature. The authors suggest that such variations reflect the plant adaptation to the medium variation. Nowadays, seaweeds are also currently used in food and dietary supplements for their alginic acid or alginate content. Alginic acid is a high molecular weight parietal polysaccharide with thickening properties mainly produced by brown macroalgae. In the latter, its main usage aims at reducing food intake by stomach filling for weight loss purpose. The metabolome of *Sargassum fusiform*, a phaeophycean, was studied by Ito and co-workers using different analytical techniques (*e.g.*, NMR, solid-state FT-IR, thermogravimetric analysis, inductively coupled plasma-optical emission spectrometry) (Ito, Sakata, Date, & Kikuchi, 2014). The spectral data obtained on complex substances were pre-processed using the multivariate statistical analyses showing changes in the chemical compositions of alginates during the mature stage and identified trends in seasonal variations. Correlation network analysis further revealed relationships between organic and mineral components, while structural equation modeling suggested that the structure of alginic acid varied during *Sargassum fusiform* growth. This modified the capacity of alginate to bind mineral ions like iron, aluminum and titanium. Spirulina is another freshwater ingredient, the composition of which can vary with culture conditions. *Arthrospira platensis* is increasingly used as a dietary supplement for its high protein content and iron bioavailability (Lupatini, Colla, Canan, & Colla, 2017; Puyfoulhoux et al., 2001). The freshwater cyanobacteria can be collected either in the wild as in Center Africa or South America, or grown in tanks and bio-fermenters. The standardization of the product quality increases with the control of the growing conditions. Therefore, analyzing spirulina metabolome in specific growing conditions is particularly relevant, as exemplified by the comparison with the cyanobacterial model *Synechocystis* sp. PCC6830 (Hasunuma et al., 2013). This study addressed the metabolic adaptation of spirulina to nitrogen depletion and exploited ^{13}C labeled bicarbonate as a carbon source to demonstrate that in such conditions, spirulina re-orchestrated its metabolome toward glycogen production. The kinetic adaptation of the cyanobacteria to the natrium deficiency was further deciphered. Overall, this work suggests that metabolomics is key to help understand the metabolic adaptation of an organism to several growing situations and thereby optimize these conditions. Besides, other complex ingredients like propolis may have a largely variable composition depending on their production environment. Propolis, an animal-derived ingredient, is increasingly used in dietary supplements for its traditional usage and potential benefits. In France, several cases of severe allergies to bee products were recorded by the nutrivigilance system. Propolis is a complex mixture of wax and resins containing a great variety of polyphenols that are characteristic of the flowers foraged by the insects and eventually responsible for health effects. Saftić and co-workers analyzed 56 samples of propolis collected in different locations of Croatia in order to define their polyphenol compositions (Saftić, Peršurić, Fornal, Pavlešić, & Kraljević Pavelić, 2019). Targeted and untargeted LC-MS, as well as direct injection MS methods, were developed and exploited to demonstrate the variability in propolis samples composition. Targeted metabolomics could identify three main categories of samples based on their geographic location: the Mediterranean samples that corresponded to the island samples, the continental samples that were collected in the North and center of Croatia and the intermediate samples that originated from the Adriatic Croatian coast. These locations most probably host specific flora, which would explain specific polyphenol compositions. In addition, untargeted metabolomics allowed detecting 68 biochemical markers, 22 of which were identified, which greatly helps the construction of a database for propolis analysis.

2.1.2 Search for beneficial molecules

Besides this valuable input of metabolomics in deciphering the environmental factors influencing the global composition of natural ingredients, this analytical approach can also be used to search for potential health benefits of plants obtained under various conditions. As such, Salehi and co-workers analyzed several samples of turmeric looking for the best health benefits. Turmeric and its main chemical compound, curcumin, are central to today's food supplements. As a spice, turmeric consumption is traditional and many health effects were attributed to this ingredient in recent preclinical and clinical studies (Salehi et al., 2019). This spice is known to produce large numbers of metabolites from diverse metabolic pathways. Xie and co-workers tested whether co-regulated metabolite modules do exist in this plant by performing metabolic profiling analysis (Xie, Ma, & Gang, 2009). They analyzed 16 turmeric rhizome samples from different growth and development treatments and for each condition, 249 volatile and non-volatile metabolites. The authors also identified metabolite modules and reported their usefulness for compound identification and biosynthetic investigations. Quite interestingly, such modules could derive from the co-ordinated expressions of enzyme-coding genes, and the expressions were influenced by growing conditions. Thanks to this study, it should be possible to get high-quality turmeric preparations based on plants raised in the most favorable culture conditions. Similarly, green tea leaf extracts are currently used in dietary supplements in Europe. However, the French nutrivigilance system pointed out adverse effects in the past with some preparations. Miyauchi and co-workers used *Camellia sinensis* as an experimental model to study the effect of light quality during the growth phase on plant metabolite content, based on GC-MS profiling (Miyauchi et al., 2017). They built a ranking predictive model based on tea metabolite profile that included several biomarkers: glutamine, glutamic acid, oxalic acid, epigallocatechin, phosphoric acid, and inositol. Once built, the predictive model was applied to tea leaves obtained from plants grown under artificial conditions. The study reported that the sample supplemented with ultraviolet-A (315–399 nm) bore the highest-ranking with high amino acid and low epigallocatechin content. Hence, metabolic profiling can enable us to deepen our understanding and thus help develop high-quality plants cultured under specific conditions. Another example is the one of *Vitis vinifera* which produces edible grapes used to make wine. It is rich in complex polyphenols like anthocyanins and tannins as well as stilbenes like resveratrol, viniferin, piceid, piceatannol, hopeaphenol, isohopeaphenol (van Dorsten et al., 2009). These substances have different pharmacokinetic characteristics and different health effects. Therefore, metabolomics allowing the fine analysis of complex mixtures is useful to characterize dietary supplements formulated from vine extracts. Such food supplements are quite popular nowadays to treat for venous insufficiency and other blood flow-related physiological disorders (Krikorian, Nash, Shidler, Shukitt-Hale, & Joseph, 2010). Besides, LC-MS-based targeted metabolomics analyses of grape stems from 8 cultivars allowed the determination of 42 polyphenols related to phenolic acids, flavonoids, procyanidins and stilbenoids (Billet et al., 2018). The grape stem biochemical profiles discriminated according to their genotype, thus indicating that polyphenol profiles could provide varietal signatures. Using hierarchical clustering, some similarities of polyphenol profiles between grape varieties were also reported, in good agreement with the genetic distance determined using clustering analyses of 22 microsatellite DNA markers. Several polyphenol subclasses were identified suggesting the existence of metabolite correlation networks that are differently controlled.

2.2 Quality control

2.2.1 Search for toxic compounds

Plant-derived food supplements are now widespread in the general population. Reports of adverse effects are not unusual due to the complex and variable nature of botanical ingredients. This led to concern for negative human health impacts following the consumption of these products. Beside traditional bioassay-guided fractionations (BGF) that help determine the toxic compounds of a mixture, metabolomics provides additional data (Roberts et al., 2019). Following an unsupervised approach, the raw material could be applied to a bioassay and the primary and secondary metabolites then be chemically identified. Such a method avoids overlooking substances and their biological response, even when they are low in abundance or unstable. As for BGF, an iterative process of bioassay exposure is required to confirm that identified metabolites retain the biological activity of the whole mixture. In addition to natural toxic compounds, natural ingredients collected in the wild can also contain pollutants. This can be particularly true for marine resources. As mentioned previously, the marine environment is unstable and one factor of instability is anthropomorphic pollution. Pollutants can contaminate animals (especially filter feeder shellfish and seaweeds). In this context, the metabolomic signature of *Ulva lactuca* (*Chlorophytae*) was examined when exposed for several durations to various concentrations of gasoline and diesel oils (Pilatti et al., 2017). Fourier Transform Infrared (FT-IR) spectroscopy and multivariate analyses on the entire mid-infrared spectral window allowed discriminating diesel oil-exposed thalli from the gasoline-exposed ones. Furthermore, clustering analysis enabled to discriminate both time and concentration of exposure on metabolic alterations induced by the xenobiotics, and also revealed that protein, lipid and carbohydrate profiles seemed sensitive to the contaminations investigated.

2.2.2 Origin identification of complex resources

As seen previously propolis composition is complex and highly variable according to environmental factors and harvest season. Therefore, it can be difficult to standardize the dietary supplement formulation. Furthermore, as quality control appears necessary, it can be useful to ensure the origin of the ingredients used. Metabolomics may be of help in such issues. Maraschin et al. used NMR-based metabolomics data analysis workflow and chemometrics including machine learning algorithms, *i.e.,* partial least square discriminant analysis (PLS-DA) and random forest (RF) on Brazilian propolis samples and estimated the variable importance of different parameters for propolis sample classification (Maraschin et al., 2016). This allowed the identification of 21 biomarkers for sample discrimination, further indicating a net effect of the season of propolis harvesting using targeted biomarkers (ferulic acid, pinostrobin). Using 14 biomarkers, the variations with season and origin of the samples could be emphasized, constituting a chemical fingerprinting of the harvest seasons and production regions. Similarly, the identification of the origin of turmeric samples was shown to be possible using metabolomic approaches. As mentioned previously, turmeric or *Curcuma* is increasingly used nowadays in dietary supplements. Four species have been used for ages in Asian traditional medication *i.e., Curcuma phaeocaulis, kwangsiensis, wenyujinin and longa*. Nowadays, the market of this ingredient can be stretched and, in such circumstances, one species can be sold for another. This is lawfully objectionable and fraudulent since the price of each species differs from one another. As such, quality control was assessed by metabolomics through the GC-MS analysis combined with multivariate statistical methods of the 3 turmeric species and their essential oils, based on 62 samples of authentic, introduced and commercial botanicals (Xiang, Wang, Cai, & Zeng, 2011). Principal component analysis (PCA) allowed distinguishing the different species and their ecotypes, and four compounds exhibited sufficient variability to be used as discrimination biomarkers, namely curzerenone, germacrone, curdione and epicurzerenone. These substances can be used further for quality control among different groups of turmeric samples. Equally, metabolomics was used to identify various grape hybrids. As previously mentioned dietary supplement based on vine extract are quite popular nowadays essentially for their antioxidant effect and their ability to improve endothelial function (Breuss, Atanasov, & Uhrin, 2019). It was hypothesized that non-anthocyanic flavonoids are the most powerful compounds in this respect. Therefore, De Rosso et al. investigated the fine composition of five hybrid red grape varieties for their composition in non-anthocyanic flavonoids using LC-MS coupled to targeted and untargeted molecular identification methods (De Rosso, Panighel, Vedova, Gardiman, & Flamini, 2015). This allowed identifying 31 non-anthocyanic flavonoids including 19 flavonols, 2 flavanonols, 3 flavan-3-ols, 2 flavanones, 1 flavone and 4 proanthocyanidins, with an identification score over 90% except in three cases. The compounds were essentially present under their glycosidic and glucuronic forms and there could be several conjugates for each parent compound. This metabolomics approach unraveled the large diversity of non-anthocyanic flavonoids in the five hybrids studied. The exact grape origin can thus be determined based on the places where such hybrids are currently used.

Overall, metabolomic approaches through targeted analyses of specific bioactives or untargeted profiling for global metabolic overview (Fig. 3) may help evaluate the quality of food supplements, including ingredient sourcing, complex ingredient production but also potential breeding programs that aim at selecting better plants to be used in the field of food supplements. Metabolomics is also a valuable tool for the quality analysis of complex plant matrices, targeting the identification of potentially deleterious resources or the limitation of frauds by the determination of typical biomarkers.

3. Metabolomics contributions to consumer analysis

In complement to the analysis of dietary supplement ingredients, plant metabolomics is also exploited to decipher the health impacts on consumers. Such effects are usually studied in preclinical and clinical approaches. We present here data obtained with animal models and others gathered in clinical studies where metabolomics is used to follow appropriate biomarkers.

3.1 Data collected in animal models

As already mentioned, turmeric, which contains curcumin, is nowadays widely used both in traditional Asian medicine and in dietary supplements sold world-wide. One specific application concerns blood stasis. Curcumin is usually believed to be the main active compound despite its low bioavailability, which raises the question of its efficiency both in animal and human. A recent study based on UPLC-Q/TOF-MS analysis of rat plasma samples was conducted to investigate both the blood metabolome including *Curcuma* bioactive chemicals after turmeric treatment and the effect of oral intake on blood stasis biomarkers (Hao et al., 2019). The animals were subjected to four conditions: a control situation with eight Sprag Dawley rats (NC), eight rats on Acute Blood Stasis treatment (ABS), eight rats on ABS + *Curcuma wenyujin* extract treatment (EZhu) and eight rats under ABS + Compound Slavia Tablet (CST) treatment as a positive control. Twenty-one metabolites were associated with *Curcuma* intake by ABS animals. Comparing ABS to NC allowed discriminating 26 biomarkers related to blood stasis. Both CST and EZhu showed a modest reduction of the ABS biomarkers. These biomarkers were then related to metabolic pathways essentially linked to lipid (arachidonic acid and linoleic acid, sphingolipids, glycerophospholipids, ether lipids) and amino acid (tryptophan, serine and choline) metabolisms. According to the authors, the LC-MS metabolomic method and network pharmacology could help explore pathological mechanisms and clarify the mechanisms of action of *Curcuma*. This can be extrapolated to other situations. Besides plant extracts, animal resources also used in dietary supplements can be studied for their potential health benefits, such as glucosamine. Glucosamine belongs to the glucose family and is one constituent of cartilage (Selacians) and chitin (Crustaceans). Glucosamine is used in medications and dietary supplements to relieve from arthrosis joint pains, although its efficiency alone or in combination with chondroitin is still under debate. Chen et al. used the rat model of osteoarthritis to investigate (1) pathology biomarkers and (2) their evolution with time while treating with different traditional plants and drugs (Chen et al., 2017). In addition to the three control groups, four treatments were investigated: glucosamine/Celecoxib, *Epimedii folium*, *Chuanxiong Rhizoma* and Bushen-Huoxue. In total 234 plasma samples were analyzed at different treatment times of the different groups. The authors used chemical isotope labeling and LC-MS targeting the amine/phenol sub-metabolome with 2923 metabolites that could be consistently analyzed in most samples. Histological images of rat joints were analyzed at 6 and 14 weeks of treatment. The metabolome was shown to be influenced by surgery and animal aging. Thanks to appropriate controls including sham-operated animals, it was possible to determine the influence of such parameters on the rat metabolome. Based on 11 main metabolites and especially 2-aminoadipic acid, saccharopine and GABA, differences between groups were observed. Drugs and botanicals also enabled a return to the basal situation after a 14 week-treatment. These results were consistent with the histological analysis. Hence, this approach allowed determining the effect of treatment on acute osteoarthritis based on validated biomarkers. This strategy will further be adopted for human samples.

In some circumstances, *in vitro* approaches can be coupled to metabolomics to simplify the correlation between drugs and efficiency biomarkers. This is exemplified with propolis which is often used for its antibacterial and antioxidant properties. Oral intake of complex mixtures like propolis leads to even more complex metabolite mixtures. However, an *in vitro* approach is the basis for further investigations in fractionation analyses. In this sense, a metabolomic approach of Brazilian propolis samples was coupled to *in vitro* tests to determine the substances in propolis mixtures that can exhibit the best antibacterial and antioxidant effects (Bittencourt et al., 2015). Antibacterial properties were tested with three strains: *Staphylococcus aureus*, *Bacillus subtilis* and *Micrococcus luteus*, and the antioxidant properties were also assessed using a radical scavenging assay. The analysis of the propolis samples was performed using GC-MS with propolis extracts obtained using either ethanol, dichloromethane or hexane, which yielded 29 biomarkers identified in the different extracts. Additionally, a PCA of the Brazilian propolis samples was performed as well as a hierarchal cluster analysis based on the distribution of 64 detected metabolites. After testing the extracts for both their antibacterial and antioxidant effect *in vitro*, a statistical correlation was possible and associated the main 25 metabolites of each extract with the activity recorded *in vitro* in the different systems. This study highlighted that dichloromethane extraction led to higher antioxidant and antibacterial properties with higher levels of phenolic compounds. Hence, metabolomics allowed identifying candidate compounds for the observed biological activities of the extracts, which underlined that the antimicrobial activity might involve synergistic effects between different metabolites of the extracts. Altogether, this suggests that the transposition of an effect recorded using a crude compound to a dietary intake should be done with caution.

3.2 Data collected in human studies

3.2.1 Identification of active metabolites

Metabolomics is key in the identification of active compounds (Fig. 3). This is the case for green tea (*Camelia sinensis*) obtained from unfermented tea leaves and proposed as tablets in dietary supplement mainly for its diuretic properties and potential help in weight loss. Despite a poor bioavailability, epigallocatechin 3-gallate (EGCG) is considered to be the main active compound exhibiting antioxidant, anti-carcinogenic, ant-microbial and neuroprotective properties in animal and *in vitro* studies (Chakrawarti, Agrawal, Dang, Gupta, & Gabrani, 2016). Other minor flavonoids are further present in green tea, *i.e.*, epigallocatechin, epicatechin-3-gallate and epicatechin. In a recent study, 63 post-menopausal women *vs* 61 placeboes were supplemented with green tea polyphenols (GTP) for 12months (Zhou, Zhang, Arikawa, & Chen, 2019). In order to address the impact of GTP on gut flora and its metabolism, LC-MS metabolomics of urine and fecal samples was conducted for GTP-derived and GTP-responsive metabolites using structural elucidation and quantitative analysis comparing the treated and control groups in the multivariate models. Interestingly, no unmetabolized polyphenol was recovered neither in feces nor the urine, which indicated that EGCG might not be the active compound *in vivo*. Instead, the GTP-derived, microbial metabolites 5-(dihydroxyphenyl)-valerolactone sulfate (I), 5-(dihydroxyphenyl)-valerolactone sulfate (II), 5-(dihydroxyphenyl)-valerolactone glucuronide (III), methyl epicatechin sulfate (IV) and methyl epigallocatechin glucuronide (V) were all increased in the urine of the supplemented women after 12months of treatment. Furthermore, some compounds belonging to the aromatic amino acid metabolism and considered as GTP-responsive metabolites (phenylacetylglutamine, hippuric acid and indoxyl sulfate) were depleted, but nevertheless, the gut bacteria microbiome was not deeply modified by the 12-month GTP supplementation. Altogether, these results demonstrate that the microbial metabolites (I, II, III, IV, V) that are prominent in human urine after chronic ingestion of GTP are more likely to be the active compounds *in vivo* rather than EGCG, despite what was previously stated. A similar conclusion can be drawn for turmeric and its main substance curcumin that have been associated with many beneficial health effects, as mentioned previously. Nevertheless, while curcumin is poorly bioavailable, an LC-MS study investigated the metabolisation of curcumin by human gut bacteria to unravel 23 potentially active metabolites in gut bacteria cultures (Lou, Zheng, Hu, Lee, & Zeng, 2015). This work uncovered new metabolizing pathways, thus suggesting that the active compounds might not be curcumin itself but rather some metabolized compounds identified by metabolomics.

Metabolomics studies have further highlighted the role of the gut flora in the health effect potentiation of natural substances, as confirmed by two randomized cross-over trials against placebo involving two grape extracts (red grape juice dry extract: GJX and dry mix of red wine and red grape juice extracts: MIX) for 4-week-duration intakes (van Dorsten et al., 2009). In addition to the effect of the treatment on mild hypertension biomarkers, this study conducted metabolomic analyses of urine samples by NMR and GC-MS which pointed out to only slight metabolic differences between the two extracts. However, the multilevel partial least squares discriminant analysis based essentially on 19 compounds was able to identify variations on the metabolome mainly produced by the gut flora. Considering the dramatic effect of gut flora on the metabolism of natural substances, bacteria flora modulation appears as the next step to improve the health effects of complex resources. While in Europe, EFSA considers that the proof level of efficiency has not yet been reached, the FDA recognizes some effects. In this context, and because gut dysbiosis is nowadays associated with an increasing number of pathologies (Weiss & Hennet, 2017), it can be of interest to confirm the effect of probiotics on any pathology or discomfort. Therefore, Vitellio and co-workers applied a metabolomic approach to a randomized cross-over clinical trial that studied the effect of dietary supplementation on functional gastrointestinal diseases (Vitellio et al., 2019). The patients who had confirmed lactose intolerance were randomly distributed in the two arms of a cross-over design. The treatment included *Bacillus longum* BB536, *Lactobacillus rhamnosus* HN001 and B6 vitamin and was compared to placebo. The treatment lasted for 1 month with a washout of 15 days in between. The gut flora composition was investigated by metagenomics and correlated to gastrointestinal symptoms and metabolome using GC-MS technique. This study revealed several genera that were more abundant in the treated group *i.e., Bifidobacterium, Slackia, Enterococcus* and *Thricocccus*, while *Lactobacillus* genus did not differ between treatments. In parallel, the abundance of some genera decreased with treatment *i.e., Klebsiella, Serratia,* and *Enterobacter*. Eleven volatile compounds were found to be associated with these variations in a PCA analysis. Furthermore, the constipation score was improved with treatment and the evolution of the gut flora composition associated with metabolomic modifications. Although the significance of such metabolomic modification is not fully explained, the effect of the probiotic supplementation is undisputable.

Overall, these recent examples support the idea that metabolomics can reveal the great complexity of metabolite mixtures issued from natural resources ingestion and allows the identification of relevant biomarkers of efficiency both *in vivo, ex vivo* and *in vitro*. It is also clear that metabolomics can be highly valuable in deciphering the identity of an active substance and in settling the question of gut flora and probiotic activity in many circumstances. However, it should be pointed out that when applied to urine samples the metabolome represents the eliminated fraction of the body. In chronic ingestion conditions, it may not fully reflect the plasma profiles due to differences in pharmacokinetic properties of each metabolite (Mathey et al., 2006). Therefore, plasma metabolomic analysis would be more accurate to determine the potential metabolites that could be active at the cellular level.

4. Challenges and perspectives

In recent years, a steep increase in the global market size of dietary/food supplements is observed due to rising concerns about health issues. Supplements are consumed to fill the gaps subsisting in a complete balanced diet. However, it raises the concern over consumers' safety since the regulations for the supplements are not standardized and essentially not as stringent as for drugs. These supplements are sold under various categories, such as dietary supplements in the USA, natural health products in Canada, herbal medicine in the European Union and traditional or complementary medicine in other countries including India and China (Dwyer, Coates, & Smith, 2018). In parallel, the quality control, especially of herbal supplements, is highly challenging due to their high chemical complexity. Several of the herbal supplements, containing multiple biologically active molecules, are not scientifically evaluated but are often consumed not only for maintaining general health but also for therapeutic purposes. Therefore, considerable risk is associated with the intake of dietary supplements. It is further augmented due to over-claiming by the traders and mixing with adulterants or contaminants.

The challenges associated with the dietary herbal supplements are at multilevel like (1) the lack of unanimity in regulatory policies across the nations, (2) botanical misidentifications, (3) chemical variations in the raw material, (4) lack of guidelines for material collection, processing and product development, (5) insignificant efforts in the development of chemical profile signatures of the processed herbal products to be used for quality control checks, and (6) lack of scientific reports on health benefits/toxicity or cross-reactivity with prescription medications. Such challenges and lacunae limit the enforcement of stringent quality control measures and the formulation of definite restrictions on product usage. This can further lead to promoting unscientific claims and unethical business practices.

Challenges in regulating herbal products are being addressed by employing various analytical techniques but are suffering from certain limitations. For instance, botanical identifications involve DNA-based techniques, which help in authenticating the species of interest (Raime, Krjutškov, & Remm, 2020). Various types of specific markers like restriction fragment length polymorphism (RFLP), sequence characterized amplified regions (SCAR) and amplification refractory mutation system (ARMS) have been developed to identify commonly used adulterants and techniques based on DNA barcoding to detect unknown adulterants (Grazina, Amaral, & Mafra, 2020). However, these methods require very high-quality DNA and are often compromised especially for medicinal plants as several protocols are not standardized in the laboratories. Furthermore, these techniques often fail to authenticate processed herbal supplements due to DNA degradation and/or difficulty in good quality DNA preparation due to interference caused by various bioactives present in the samples (Grazina et al., 2020). Another major limitation with DNA-based technologies lies in the identification of diverse types of plant tissue used for product preparation. The phytochemical composition, including bioactives of plant material used for herbal product preparation, is highly variable and cannot be determined using DNA-based techniques. As detailed in this chapter, the essential components of the herbal product could be influenced by various biotic and abiotic factors like, *e.g.*, the geographical location, seasonal variation, developmental stage of the plant tissue, cultivar varieties within a species (Gomes et al., 2019; Kulyal et al., 2016; Ochoa-Velasco, Avila-Sosa, Navarro-Cruz, López-Malo, & Palou, 2017; Tetali, 2019; Voirin, Brun, & Bayet, 1990; Wu, Wang, & Pei-Lin, 1996). It also depends on the type of solvent used for extraction and the protocol employed in product preparation (Silva Junior, Ribeiro Costa, Martins, & Ramos Barbos, 2011). Further, beneficial bioactives can accumulate in different parts of the plants including leaves, flowers, fruits, seeds, bark or belowground parts like roots or rhizomes, and the plant parts used for product preparation cannot be distinguished using DNA-based techniques (Choudhury et al., 2014; Kokkiripati et al., 2011, 2013; Kulyal et al., 2016; Reddi & Tetali, 2019). Therefore, DNA-based authentication can only be employed when material collection for the species identification provides high-quality DNA isolation and standardization with this specific species and its specific plant part (Grazina et al., 2020).

The quality control of food or herbal supplements is highly dependent on the detection and quantification of desired as well as undesired chemical constituents. The latter constituents like adulterants, contaminants and expected/known plant toxins can be detected using standardized LC, GC or TLC techniques (Fig. 2). However, such procedures cannot rule out the possibility of the presence of unknown constituents. Therefore, untargeted chemical profiling using highly sensitive analytical hyphenated techniques like LC-MS, GC-MS or NMR would help to a great extent in addressing several of the aforementioned issues or apprehensions (Sanzini, Badea, Santos, Restani, & Sievers, 2011). Thus, metabolomics can be considered as a suitable alternative to capture the chemical composition of herbal products. This is being achieved using NMR or hyphenated MS technologies (*i.e.*, LC-MS, GC-MS) in order to study the phytochemistry of a given species, the modulatory effect of biotic and abiotic factors including growth conditions and also to establish variations due to cultivars, plant development or even storage conditions (Khin, Cech, Kellogg, & Caesar, 2020). Hence, these tools play a crucial role in the identification of bioactives. However, as of today, there is no single metabolomic technology that yields total chemical information about the biological samples and multiple platforms are thus necessary owing to the great diversity of plant metabolites (Deborde et al., 2017; Ernst, Silva, Silva, Vêncio, & Lopes, 2014). NMR technology is non-destructive and facilitates the quantification of metabolites whereas the sensitivity of MS techniques coupled to liquid or gas chromatography (LC-MS/GC-MS) is much more adapted for the screening of metabolites in complexed mixtures. LC-MS is suitable for the detection of both polar and non-polar compounds, while GC-MS rather addresses volatiles and also non-volatile compounds after derivatization (Emwas, 2015; Johnson & Gonzalez, 2012; Mirsaeidi, Banoei, Winston, & Schraufnagel, 2015; Piasecka, Kachlicki, & Stobiecki, 2019). All these technologies are evolving swiftly and the spectral libraries for many metabolites become available and ever more complete. However, these techniques are far from routine usage for commercial product quality control for several reasons. In brief, metabolomics involves expensive instruments and requires extensive training for data acquisition, analysis and replication. Further, metabolomics data are often not reproducible for several reasons even though the scientific communities intend to address this concern by implementing FAIR (Findable, Accessible, Interoperable, Reusable) strategies (Chaleckis, Meister, Zhang, & Wheelock, 2019; Mendez, Pritchard, Reinke, & Broadhurst, 2019). This is particularly important in the field of food/herbal supplements due to the variation in their phytochemical profile. Moreover, the protocols followed in the preparation of herbal products supplied to the consumer market are quite varied. These protocols are often regionally developed and localized or as described in the respective traditional systems of medicine. Therefore, it is essential to develop metabolomic databases of herbal supplements, which have been shown to have good health benefits. Their chemical profile signatures need to be considered for quality control of the products in order to control unethical business practices.

5. Concluding remarks

Metabolomics is a relatively new approach in the "dietary supplement" world. The last 20 years have seen a rapid development of analytical techniques and savoir-faire in metabolomics with the development of cutting-edge technologies that have offered cost-efficient and high-throughput ways for molecular characterization. All areas of life sciences have benefited from these developments. In plant sciences, metabolomic studies have contributed significantly not only to the understanding of plant physiology and biology through metabolome status that reflects the endpoint of biological activities but also to the depiction of the mechanisms, regulation and evolution of the biosynthesis of specialized plant products that can be interesting for functional food.

With the growing interest in the use of metabolomic technologies in the food supplement industry, there is a strong relevance in stimulating collaboration between analytical platforms developed in the academic world and the R&D departments embedded in the food supplement industry. Indeed, in close future plant metabolomics alone or combined with other « omics » approaches such as genomics or proteomics should boost the search for beneficial molecules and the characterization of health-promoting compounds that can be used for the development of functional food. Even though it has currently some limitations, metabolomics is no doubt an important tool that is revolutionizing the way to find new molecules in the huge diversity of plants, to help the producers maintain quality of complex ingredient or to ensure the origin of the ingredients used in dietary supplements. For this aim, different analytic platforms from the academia world will have to share the same analytical strategy mixing targeted and non-targeted metabolite profiling along with experimental considerations on non-model plants used to prepare food supplements and build dedicated metabolome databases.

Furthermore, only a few popular traditional herbal medicine (*i.e.*, Ginger, Ginseng, Curcuma) coming mainly from Asia have been the subject of a metabolomic analysis. Many plants of the traditional pharmacopeia from Africa, Asia or South America have not yet been exploited and remain to be analyzed before being used as functional food in Western countries. Metabolomics associated with complementary disciplines such as natural chemistry, pharmacology, toxicology, physiology and plant biology could be the most promising way to discover active molecules and to ensure a greater chance of enhancing the functional food area.

In conclusion, while metabolomics holds great promises in forwarding the dietary supplement industry, a pressing specific need exists for this domain (1) to boost academia and industry collaborations in order to fulfill the food supplement market needs and consolidate the scientific basis of this industrial sector, (2) to speed-up the systemic construction of metabolome databases with non-model plants used to prepare dietary supplements, (3) to explore the huge world of the complex and variable nature of botanical ingredients that can be found in plant diversity around the world, and (4) to decipher the plant metabolite activities at the consumer's level.

References

Billet, K., Houillé, B., Dugé de Bernonville, T., Besseau, S., Oudin, A., Courdavault, V., et al. (2018). Field-based metabolomics of *Vitis vinifera* L. stems provides new insights for genotype discrimination and polyphenol metabolism structuring. *Frontiers in Plant Science*, *9*, 798. https://doi.org/10.3389/fpls.2018.00798.

Bittencourt, M. L. F., Ribeiro, P. R., Franco, R. L. P., Hilhorst, H. W. M., de Castro, R. D., & Fernandez, L. G. (2015). Metabolite profiling, antioxidant and antibacterial activities of Brazilian propolis: Use of correlation and multivariate analyses to identify potential bioactive compounds. *Food Research International*, *76*, 449–457. https://doi.org/10.1016/j.foodres.2015.07.008.

Breuss, J. M., Atanasov, A. G., & Uhrin, P. (2019). Resveratrol and its effects on the vascular system. *International Journal of Molecular Sciences*, *20*(7), 1523. https://doi.org/10.3390/ijms20071523.

Chakrawarti, L., Agrawal, R., Dang, S., Gupta, S., & Gabrani, R. (2016). Therapeutic effects of EGCG: A patent review. *Expert Opinion on Therapeutic Patents*, *26*(8), 907–916. https://doi.org/10.1080/13543776.2016.1203419.

Chaleckis, R., Meister, I., Zhang, P., & Wheelock, C. E. (2019). Challenges, progress and promises of metabolite annotation for LC–MS-based metabolomics. *Current Opinion in Biotechnology*, *55*, 44–50. https://doi.org/10.1016/j.copbio.2018.07.010.

Chen, D., Su, X., Wang, N., Li, Y., Yin, H., Li, L., et al. (2017). Chemical isotope labeling LC-MS for monitoring disease progression and treatment in animal models: Plasma metabolomics study of osteoarthritis rat model. *Scientific Reports*, 7(1), 40543. https://doi.org/10.1038/srep40543.

Choudhury, S. S., Bashyam, L., Manthapuram, N., Bitla, P., Kollipara, P., & Tetali, S. D. (2014). Ocimum sanctum leaf extracts attenuate human monocytic (THP-1) cell activation. *Journal of Ethnopharmacology*, *154*(1), 148–155. https://doi.org/10.1016/j.jep.2014.03.049.

De Rosso, M., Panighel, A., Vedova, A. D., Gardiman, M., & Flamini, R. (2015). Characterization of non-anthocyanic flavonoids in some hybrid red grape extracts potentially interesting for industrial uses. *Molecules*, *20*(10), 18095–18106. https://doi.org/10.3390/molecules201018095.

Deborde, C., Moing, A., Roch, L., Jacob, D., Rolin, D., & Giraudeau, P. (2017). Plant metabolism as studied by NMR spectroscopy. *Progress in Nuclear Magnetic Resonance Spectroscopy*, *102–103*, 61–97. https://doi.org/10.1016/j.pnmrs.2017.05.001.

Dwyer, J. T., Coates, P. M., & Smith, M. J. (2018). Dietary supplements: Regulatory challenges and research resources. *Nutrients*, *10*(41), 1–24. https://doi.org/10.3390/nu10010041.

Egan, B., Hodgkins, C., Shepherd, R., Timotijevic, L., & Raats, M. (2011). An overview of consumer attitudes and beliefs about plant food supplements. *Food & Function*, *2*(12), 747–752. https://doi.org/10.1039/C1FO10109A.

Emwas, A.-H. M. (2015). The strengths and weaknesses of NMR spectroscopy and mass spectrometry with particular focus on metabolomics research BT—Metabonomics: Methods and protocols. In J. T. Bjerrum (Ed.), *Metabolomics* (pp. 161–193). https://doi.org/10.1007/978-1-4939-2377-9_13.

Ernst, M., Silva, D. B., Silva, R. R., Vêncio, R. Z. N., & Lopes, N. P. (2014). Mass spectrometry in plant metabolomics strategies: From analytical platforms to data acquisition and processing. *Natural Product Reports*, *31*(6), 784–806. https://doi.org/10.1039/c3np70086k.

European Food Safety Authority. (2012). Compendium of botanicals reported to contain naturally occuring substances of possible concern for human health when used in food and food supplements. *EFSA Journal*, *10*(5), 2663. https://doi.org/10.2903/j.efsa.2012.2663.

Fibigr, J., Šatínský, D., & Solich, P. (2018). Current trends in the analysis and quality control of food supplements based on plant extracts. *Analytica Chimica Acta*, *1036*, 1–15. https://doi.org/10.1016/j.aca.2018.08.017.

Gaubert, J., Payri, C. E., Vieira, C., Solanki, H., & Thomas, O. P. (2019). High metabolic variation for seaweeds in response to environmental changes: A case study of the brown algae Lobophora in coral reefs. *Scientific Reports*, *9*(1), 993. https://doi.org/10.1038/s41598-018-38177-z.

Gomes, A. F., Almeida, M. P., Leite, M. F., Schwaiger, S., Stuppner, H., Halabalaki, M., et al. (2019). Seasonal variation in the chemical composition of two chemotypes of Lippia alba. *Food Chemistry*, *273*, 186–193. https://doi.org/10.1016/j.foodchem.2017.11.089.

Grazina, L., Amaral, J. S., & Mafra, I. (2020). Botanical origin authentication of dietary supplements by DNA-based approaches. *Comprehensive Reviews in Food Science and Food Safety*, *19*, 1080–1099. https://doi.org/10.1111/1541-4337.12551.

Hall, R., Beale, M., Fiehn, O., Hardy, N., Sumner, L., & Bino, R. (2002). Plant metabolomics: The missing link in functional genomics strategies. *The Plant Cell*, *14*(7), 1437–1440. https://doi.org/10.1105/tpc.140720.

Hao, M., Ji, D., Li, L., Su, L., Gu, W., Gu, L., et al. (2019). Mechanism of Curcuma wenyujin rhizoma on acute blood stasis in rats based on a UPLC-Q/TOF-MS metabolomics and network approach. *Molecules*, *24*(1), 82. https://doi.org/10.3390/molecules24010082.

Harty, P. S., Zabriskie, H. A., Erickson, J. L., Molling, P. E., Kerksick, C. M., & Jagim, A. R. (2018). Multi-ingredient pre-workout supplements, safety implications, and performance outcomes: A brief review. *Journal of the International Society of Sports Nutrition*, *15*(1), 41. https://doi.org/10.1186/s12970-018-0247-6.

Hasunuma, T., Kikuyama, F., Matsuda, M., Aikawa, S., Izumi, Y., & Kondo, A. (2013). Dynamic metabolic profiling of cyanobacterial glycogen biosynthesis under conditions of nitrate depletion. *Journal of Experimental Botany*, *64*(10), 2943–2954. https://doi.org/10.1093/jxb/ert134.

Ito, K., Sakata, K., Date, Y., & Kikuchi, J. (2014). Integrated analysis of seaweed components during seasonal fluctuation by data mining across heterogeneous chemical measurements with network visualization. *Analytical Chemistry*, *86*(2), 1098–1105. https://doi.org/10.1021/ac402869b.

Johnson, C., & Gonzalez, F. (2012). Challenges and opportunities of metabolomics. *Journal of Cellular Physiology*, *227*(8), 2975–2981. https://doi.org/10.1002/jcp.24002.

Johnson, C. H., Ivanisevic, J., & Siuzdak, G. (2016). Metabolomics: Beyond biomarkers and towards mechanisms. *Nature Reviews. Molecular Cell Biology*, *17*(7), 451–459. https://doi.org/10.1038/nrm.2016.25.

Khin, M., Cech, N. B., Kellogg, J. J., & Caesar, L. K. (2020). Chemical evaluation of the effects of storage conditions on the botanical goldenseal using marker-based and metabolomics approaches. *The Yale Journal of Biology and Medicine*, *93*(2), 265–275.

Kokkiripati, P. K., Bhakshu, L. M., Marri, S., Padmasree, K., Row, A. T., Raghavendra, A. S., et al. (2011). Gum resin ofBoswellia serrata inhibited human monocytic (THP-1) cell activation and platelet aggregation. *Journal of Ethnopharmacology*, *137*(1), 893–901. https://doi.org/10.1016/j.jep.2011.07.004.

Kokkiripati, P. K., Kamsala, R. V., Bashyam, L., Manthapuram, N., Bitla, P., Peddada, V., et al. (2013). Stem-bark of Terminalia arjuna attenuates human monocytic (THP-1) and aortic endothelial cell activation. *Journal of Ethnopharmacology*, *146*(2), 456–464. https://doi.org/10.1016/j.jep.2012.12.050.

Krikorian, R., Nash, T. A., Shidler, M. D., Shukitt-Hale, B., & Joseph, J. A. (2010). Concord grape juice supplementation improves memory function in older adults with mild cognitive impairment. *British Journal of Nutrition*, *103*(5), 730–734. https://doi.org/10.1017/S0007114509992364.

Kulyal, P., Kuchibhatla, L. N., Maheshwari, K. U., Babu, K. N., Tetali, S. D., & Raghavendra, A. S. (2016). Highly sensitive HPLC method for estimation of total or individual curcuminoids in Curcuma cultivars and commercial turmeric powders. *Current Science*, *111*(11), 1816–1824. Consulté à l'adresse JSTOR.

Lou, Y., Zheng, J., Hu, H., Lee, J., & Zeng, S. (2015). Application of ultra-performance liquid chromatography coupled with quadrupole time-of-flight mass spectrometry to identify curcumin metabolites produced by human intestinal bacteria. *Journal of Chromatography B*, *985*, 38–47. https://doi.org/10.1016/j.jchromb.2015.01.014.

Lupatini, A. L., Colla, L. M., Canan, C., & Colla, E. (2017). Potential application of microalga *Spirulina platensis* as a protein source: Potential application of microalga *Spirulina platensis* as a protein source. *Journal of the Science of Food and Agriculture*, *97*(3), 724–732. https://doi.org/10.1002/jsfa.7987.

Manach, C., Williamson, G., Morand, C., Scalbert, A., & Rémésy, C. (2005). Bioavailability and bioefficacy of polyphenols in humans. I. Review of 97 bioavailability studies. *The American Journal of Clinical Nutrition*, *81*(1), 230S–242S. https://doi.org/10.1093/ajcn/81.1.230S.

Maraschin, M., Somensi-Zeggio, A., Oliveira, S. K., Kuhnen, S., Tomazzoli, M. M., Raguzzoni, J. C., et al. (2016). Metabolic profiling and classification of propolis samples from southern Brazil: An NMR-based platform coupled with machine learning. *Journal of Natural Products*, *79*(1), 13–23. https://doi.org/10.1021/acs.jnatprod.5b00315.

Mathey, J., Lamothe, V., Coxam, V., Potier, M., Sauvant, P., & Bennetau-Pelissero, C. B. (2006). Concentrations of isoflavones in plasma and urine of post-menopausal women chronically ingesting high quantities of soy isoflavones. *Journal of Pharmaceutical and Biomedical Analysis*, *41*(3), 957–965. https://doi.org/10.1016/j.jpba.2006.01.051.

Mendez, K. M., Pritchard, L., Reinke, S. N., & Broadhurst, D. I. (2019). Toward collaborative open data science in metabolomics using Jupyter Notebooks and cloud computing. *Metabolomics: Official Journal of the Metabolomic Society*, *15*(10), 125. https://doi.org/10.1007/s11306-019-1588-0.

Mirsaeidi, M., Banoei, M. M., Winston, B. W., & Schraufnagel, D. E. (2015). Metabolomics: Applications and promise in mycobacterial disease. *Annals of the American Thoracic Society*, *12*(9), 1278–1287. https://doi.org/10.1513/AnnalsATS.201505-279PS.

Miyauchi, S., Yonetani, T., Yuki, T., Tomio, A., Bamba, T., & Fukusaki, E. (2017). Quality evaluation of green tea leaf cultured under artificial light condition using gas chromatography/mass spectrometry. *Journal of Bioscience and Bioengineering*, *123*(2), 197–202. https://doi.org/10.1016/j.jbiosc.2016.07.017.

Ochoa-Velasco, C. E., Avila-Sosa, R., Navarro-Cruz, A. R., López-Malo, A., & Palou, E. (2017). Biotic and abiotic factors to increase bioactive compounds in fruits and vegetables. In *Food bioconversion* (pp. 317–349). https://doi.org/10.1016/B978-0-12-811413-1.00009-7.

Perez de Souza, L., Alseekh, S., Naake, T., & Fernie, A. (2019). Mass spectrometry-based untargeted plant metabolomics. *Current Protocols in Plant Biology*, *4*(4), e20100. https://doi.org/10.1002/cppb.20100.

Piasecka, A., Kachlicki, P., & Stobiecki, M. (2019). Analytical methods for detection of plant metabolomes changes in response to biotic and abiotic stresses. *International Journal of Molecular Sciences*, *20*(2), 379. https://doi.org/10.3390/ijms20020379.

Pilatti, F. K., Ramlov, F., Schmidt, E. C., Costa, C., de Oliveira, E. R., Bauer, C. M., et al. (2017). Metabolomics of Ulva lactuca Linnaeus (Chlorophyta) exposed to oil fuels: Fourier transform infrared spectroscopy and multivariate analysis as tools for metabolic fingerprint. *Marine Pollution Bulletin*, *114*(2), 831–836. https://doi.org/10.1016/j.marpolbul.2016.11.006.

Pujos-Guillot, E., Hubert, J., Martin, J.-F., Lyan, B., Quintana, M., Claude, S., et al. (2013). Mass spectrometry-based metabolomics for the discovery of biomarkers of fruit and vegetable intake: Citrus fruit as a case study. *Journal of Proteome Research*, *12*(4), 1645–1659. https://doi.org/10.1021/pr300997c.

Puyfoulhoux, G., Rouanet, J.-M., Besançon, P., Baroux, B., Baccou, J.-C., & Caporiccio, B. (2001). Iron availability from iron-fortified spirulina by an in vitro digestion/Caco-2 cell culture model. *Journal of Agricultural and Food Chemistry*, *49*(3), 1625–1629. https://doi.org/10.1021/jf001193c.

Raime, K., Krjutškov, K., & Remm, M. (2020). Method for the identification of plant DNA in food using alignment-free analysis of sequencing reads: A case study on Lupin. *Frontiers in Plant Science*, *11*, 646. https://doi.org/10.3389/fpls.2020.00646.

Reddi, K. K., & Tetali, S. D. (2019). Dry leaf extracts of Tinospora cordifolia (Willd.) Miers attenuate oxidative stress and inflammatory condition in human monocytic (THP-1) cells. *Phytomedicine*, *61*, 152831. https://doi.org/10.1016/j.phymed.2019.152831.

Roberts, G. K., Gardner, D., Foster, P. M., Howard, P. C., Lui, E., Walker, L., et al. (2019). Finding the bad actor: Challenges in identifying toxic constituents in botanical dietary supplements. *Food and Chemical Toxicology*, *124*, 431–438. https://doi.org/10.1016/j.fct.2018.12.026.

Rolin, D. (2013). *Metabolomics coming of age with its technological diversity* (1st ed.). Amsterdam: Elsevier.

Saftić, L., Peršurić, Ž., Fornal, E., Pavlešić, T., & Kraljević Pavelić, S. (2019). Targeted and untargeted LC-MS polyphenolic profiling and chemometric analysis of propolis from different regions of Croatia. *Journal of Pharmaceutical and Biomedical Analysis*, *165*, 162–172. https://doi.org/10.1016/j.jpba.2018.11.061.

Salehi, B., Stojanović-Radić, Z., Matejić, J., Sharifi-Rad, M., Anil Kumar, N. V., Martins, N., et al. (2019). The therapeutic potential of curcumin : A review of clinical trials. *European Journal of Medicinal Chemistry*, *163*, 527–545. https://doi.org/10.1016/j.ejmech.2018.12.016.

Sanzini, E., Badea, M., Santos, A. D., Restani, P., & Sievers, H. (2011). Quality control of plant food supplements. *Food & Function*, *2*(12), 740. https://doi.org/10.1039/c1fo10112a.

Silva Junior, J. O. C., Ribeiro Costa, R. M., Martins, F., & Ramos Barbos, W. L. (2011). Processing and quality control of herbal drugs and their derivatives. In Y. Shoyama (Ed.), *Quality control of herbal medicines and related areas*. https://doi.org/10.5772/20346.

Tetali, S. D. (2019). Terpenes and isoprenoids: A wealth of compounds for global use. *Planta*, *249*(1), 1–8. https://doi.org/10.1007/s00425-018-3056-x.

van Dorsten, F. A., Grün, C. H., van Velzen, E. J. J., Jacobs, D. M., Draijer, R., & van Duynhoven, J. P. M. (2009). The metabolic fate of red wine and grape juice polyphenols in humans assessed by metabolomics. *Molecular Nutrition & Food Research*, *54*(7), 897–908. https://doi.org/10.1002/mnfr.200900212.

Vitellio, P., Celano, G., Bonfrate, L., Gobbetti, M., Portincasa, P., & De Angelis, M. (2019). Effects of Bifidobacterium longum and Lactobacillus rhamnosus on gut microbiota in patients with lactose intolerance and persisting functional gastrointestinal symptoms: A randomised, double-blind, cross-over study. *Nutrients*, *11*(4), 886. https://doi.org/10.3390/nu11040886.

Voirin, B., Brun, N., & Bayet, C. (1990). Effects of daylength on the monoterpene composition of leaves of Mentha x piperita. *Phytochemistry*, *29*(3), 749–755. https://doi.org/10.1016/0031-9422(90)80012-6.

Weiss, G. A., & Hennet, T. (2017). Mechanisms and consequences of intestinal dysbiosis. *Cellular and Molecular Life Sciences*, *74*(16), 2959–2977. https://doi.org/10.1007/s00018-017-2509-x.

Wu, T.-S., Wang, M.-L., & Pei-Lin, W. (1996). Seasonal variations of carbazole alkaloids in Murraya euchrestifolia. *Phytochemistry*, *43*(4), 785–789. https://doi.org/10.1016/0031-9422(96)00361-5.

Xiang, Z., Wang, X., Cai, X., & Zeng, S. (2011). Metabolomics study on quality control and discrimination of three Curcuma species based on gas chromatograph–mass spectrometry. *Phytochemical Analysis*, *22*(5), 411–418. https://doi.org/10.1002/pca.1296.

Xie, Z., Ma, X., & Gang, D. R. (2009). Modules of co-regulated metabolites in turmeric (Curcuma longa) rhizome suggest the existence of biosynthetic modules in plant specialized metabolism. *Journal of Experimental Botany*, *60*(1), 87–97. https://doi.org/10.1093/jxb/ern263.

Zhou, Y., Zhang, N., Arikawa, A. Y., & Chen, C. (2019). Inhibitory effects of green tea polyphenols on microbial metabolism of aromatic amino acids in humans revealed by metabolomic analysis. *Metabolites*, *9*(5), 96. https://doi.org/10.3390/metabo9050096.

CHAPTER TWELVE

News on the redox front—A follow-up of ABR volume 52: Oxidative stress and redox regulation in plants

Nicolas Rouhier* and Jean-Pierre Jacquot
Université de Lorraine, INRAE, IAM (Interaction Arbres Microorganismes), Nancy, France
*Corresponding author: e-mail address: nicolas.rouhier@univ-lorraine.fr

Contents

Abstract

Volume 52 of *Advances in Botanical Research* "Oxidative stress and redox regulation in plants" edited by Jean-Pierre Jacquot (series Editors Michel Delseny, Jean-Claude Kader) was published in 2009, so now more than 12 years ago. It featured 15 different chapters and more than 60 authors contributed. Volume 52 has covered many aspects of plant physiology including the generation of oxidative stress, the molecular responses to oxidative stress and redox regulation. In general such a volume is initiated about 2 years before final publication and so it has been now 15 years overall. We believe thus that it is now appropriate for the anniversary volume 100 of ABR to summarize what was described in that volume but also to provide a survey of more recent literature showing how the plant redox field has evolved and expanded since.

1. Salient features of volume 52

Volume 52 of *Advances in Botanical Research* entitled Oxidative stress and redox regulation in plants has dealt in the first two chapters with reactive oxygen and nitrogen species, their generation and roles in protein oxidation but also in plant development. In the contribution by Sweetlove and Møller

Advances in Botanical Research, Volume 100
ISSN 0065-2296
https://doi.org/10.1016/bs.abr.2021.01.010
Copyright © 2021 Elsevier Ltd
All rights reserved.

(2009) the generation of ROS and RNS essentially from the functioning of electron transport chains and their enzymatic degradation was addressed (Sweetlove & Møller, 2009). A large part of their chapter then described the amino acid modifications that can result mostly on cysteine and methionine but also on six additional amino acids. A number of labeling techniques for the detection of modified amino acids was then presented and they have discussed more extensively carbonylation, a subject with which they are very familiar. They have also argumented about whether protein oxidation is necessarily linked to protein dysfunction, how the cell can get rid of oxidized proteins and the energetic cost of protein oxidation. The second chapter by Kim et al. centered less on the damaging effects of ROS but rather on their roles on cell growth and development (Kim, Kim, Hahn, Joung, & Jeon, 2009). ROS (hydrogen peroxide, hydroxyl radicals, superoxide radicals) generation, perception and scavenging is presented and they discussed the importance of antioxidants including glutathione and its derivatives. The interaction of ROS with plant hormones and signaling cascades was then introduced together with the resulting effects on cell growth and development.

Agrawal et al. have paid attention to the damages induced by UV light in plants (Agrawal, Singh, & Agrawal, 2009). Indeed it is believed that the depletion of the ozone layer can be detrimental to biological organisms including plants although recovery seems to be on the way. The more energetic UVB can indeed damage DNA (it is mutagenic) but also proteins, lipid bilayers and the photosynthetic apparatus in plants. In this chapter a number of UV photoreceptors are evaluated including flavins, cryptochromes and phytochromes. UVB induces signal transduction chains featuring phosphorylation cascades and calcium. It also up- and downregulates a number of genes many of them being involved in ROS detoxification. As UVB generates ROS these are then eliminated by enzymes such as SODs, catalases, and peroxidases. Ascorbate and glutathione are also important molecules contributing to those responses.

Then in a very interesting chapter, Ohkama-Ohtsu et al. detailed how glutathione is degraded in plants (Ohkama-Ohtsu, Fukuyama, & Oliver, 2009). In mammals this occurs through the functioning of gamma-glutamyl transpeptidases (GGT) and gamma-glutamyl cyclotransferases (GGC) and these enzymes have been the subject of intense investigation. In plants much less was known and they presented the data available for GGT. The enzymes are present in the apoplast and the vacuole. In the apoplast, they degrade the oxidized form of glutathione (GSSG). In the vacuole they are necessary for the degradation of glutathione conjugates. The sequence evolution of GGT is addressed and models of its 3D structure presented. Based on homology modeling, plant GGTs are predicted to be very similar to the mammalian one and in particular the geometry of the active site is likely conserved. GGC activity is present in plants but at the time the genes had not been identified. More recently Paulose et al. (2013) have shown that plant GGC is involved in the response to heavy metals and Joshi et al. have reported the existence of three GGC genes in Arabidopsis (Joshi, Meyer, Bangash, Zheng, & Leustek, 2019; Paulose et al., 2013).

In the subsequent chapter, Marino et al. reported on the role of the redox state in nitrogen-fixing symbiosis (NFS) (Marino, Pucciariello, Puppo, & Frendo, 2009). ROS are indeed produced at every step of the symbiotic association between Rhizobium and legumes. In turn, protective mechanisms are turned on leading to the synthesis of glutathione and homoglutathione (Fig. 1).

In the nodule there is thus a very delicate interplay between oxidized and reduced species with hydrogen peroxide and glutathione/homoglutathione at the forefront. A recent review focuses on the involvement of the thioredoxin and glutathione/glutaredoxin systems in the two symbiotic partners during symbiosis (Alloing, Mandon, Boncompagni, Montrichard, & Frendo, 2018) including the description of the role of the rhizobial CGFS/monothiol Fe–S cluster-ligating glutaredoxin for nodule development and of the Medicago Trx s1 for bacteroid differentiation (Benyamina et al., 2013; Ribeiro et al., 2017).

Another interesting example of the necessary balance between the generation of oxidants and reductants is found in Morel, Ngadin, Jacquot, and Gelhaye (2009). They have described how lignin is degraded by specialized fungi called ligninolytic fungi, the prototype being *Phanerochaete chrysosporium* (Morel et al., 2009). Lignin is a very inert polymer and its degradation requires radical chemistry. Lignin peroxidases, manganese peroxidases and laccases are involved in this process. So there is an intense production of ROS in these organisms at the level of the mitochondrial respiratory chain but also via NADPH oxidases. The buildup of ROS can be detrimental in the cells of the fungus and Morel et al. have described the many antioxidant systems of this fungus which include glutathione and ascorbate, superoxide dismutases, catalases, thiol peroxidases of various nature and repair enzymes as methionine sulfoxide reductases. The thioredoxin- and glutaredoxin-dependent regeneration systems are also present in this organism. The degradation of lignins liberates a number of smaller molecules that

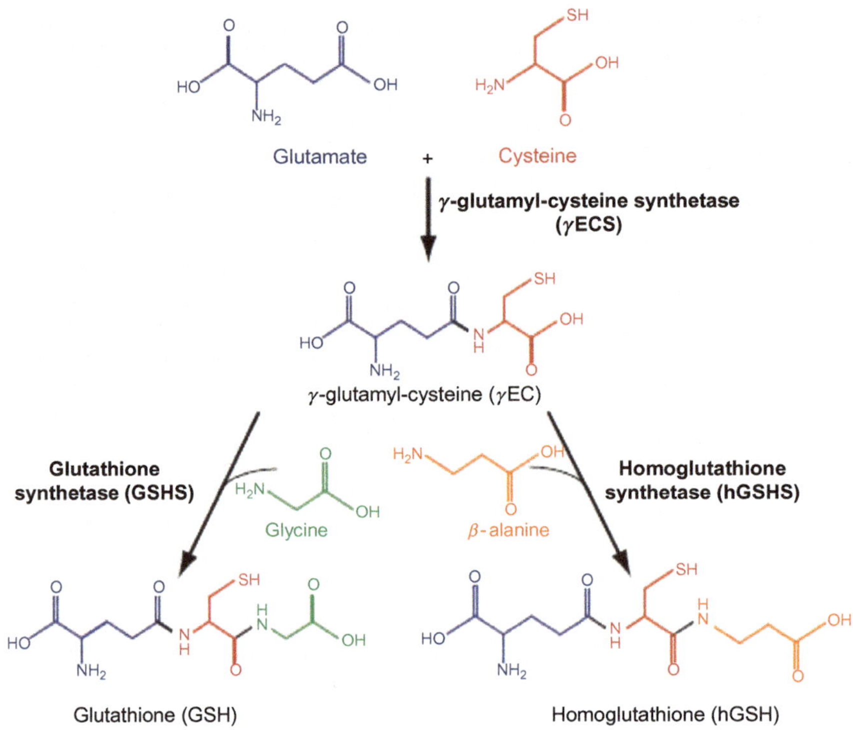

Fig. 1 Synthesis of glutathione and homoglutathione in nitrogen-fixing symbiosis. *From Marino, D., Pucciariello, C., Puppo, A., & Frendo, P. (2009). Chapter 5 The redox state, a referee of the legume–rhizobia symbiotic game. In: Advances in botanical research. Academic Press, Vol. 52, pp. 115–151. https://doi.org/10.1016/S0065-2296(10)52005-6.*

need to be channeled away *via* glutathione transferases. After volume 52 was published Morel-Rouhier and collaborators have done extensive progress on the structure/function of plant and fungal GSTs.

Nishiyama and Hisabori in their chapter have concentrated on the thioredoxin and glutaredoxin systems of cyanobacteria (Nishiyama & Hisabori, 2009). In a first part they have described the procedures to isolate targets of these reduction systems (based on the use of proteomics after trapping the candidate proteins using engineered monocysteinic variants of the redoxins). In the next part of their article, they have discussed extensively how the thioredoxin system controls protein synthesis especially at the level of protein D1 of PSII. They have provided evidence that in cyanobacteria the translation of D1 is regulated by the elongation factor G (EFG) itself controlled by a thioredoxin-dependent disulfide/dithiol switch. Overall they provided evidence that cyanobacteria are an adequate model to study the synchronization between redox and photosynthesis. Additional data concerning cyanobacteria are presented in Chapter 6 of the present volume entitled "Genomics of cyanobacteria: new insights and lessons for shaping our future-Followup of Volume 65: Genomics of Cyanobacteria" by Chauvat and Cassier-Chauvat.

In the next chapter, Hanke et al. have also addressed the question of the link between redox homeostasis and photosynthesis (Hanke et al., 2009). They started by describing the photosynthetic electron (PET) transfer chains, cyclic and linear, and pointed to the importance of ferredoxin and the malate valve for these processes. They have then turned their attention to the network of antioxidant enzymes in the chloroplast. Another section of their chapter concerns the relationship of respiration, photorespiration, and the classical C3 photosynthesis. How all this translates in regulatory signal cascades (containing both oxidants and reductants) is then evaluated. Throughout their chapter they discussed how the use of transgenic plants is required for advancing toward the elucidation of the physiological functions of these components. Of course in recent studies, this type of tool has been since very much used.

Reichheld et al. have concentrated on the interplay between the thioredoxin and glutaredoxin systems in plants (Reichheld et al., 2009). Following a strategy explained in the Hanke et al. chapter, they rely mostly on the generation of mutants of either system. They discussed first the organization of the thioredoxin and glutaredoxin systems in bacteria and yeast and the properties of mutants that had been generated. Then they aimed at perturbing the synthesis or reduction of glutathione, but also of key components of the thioredoxin system including thioredoxin reductase.

In their chapter Balsera et al. have investigated protein import in chloroplasts, detailing in particular its redox control (Balsera, Soll, & Buchanan, 2009). Interestingly a huge majority of the chloroplastic proteins are nuclear encoded and need to be imported to the stroma, the thylakoid membrane, the lumen, the intermembrane space and the envelope (see a scheme detailing the protein import complexes in Fig. 2).

The proteins that need to be imported possess signals at the N-terminus that are recognized by membrane-embedded transport complexes. They presented evidence that TOC 75 possesses a redox regulatory center and TOC 64 apparently possesses also critical cysteines that might be involved in its regulation. Likewise TIC 10, TIC 20 and TIC 40 contain generally conserved cysteines that might also be involved in redox control. Additional regulatory elements are proposed on TIC 110 and SecA. Overall these data

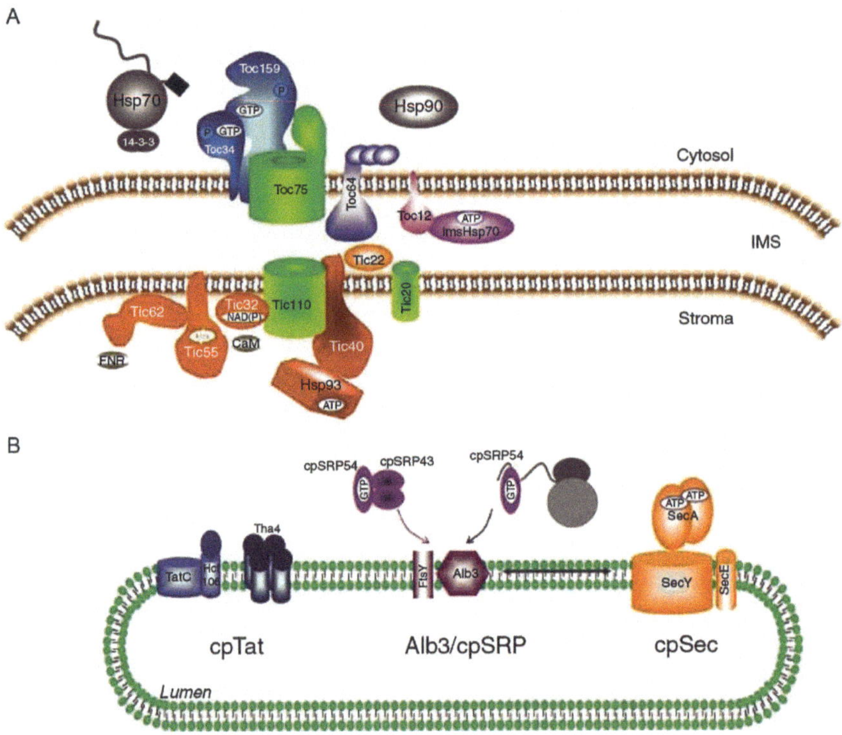

Fig. 2 Chloroplast translocation machineries. *From Balsera, M., Soll, J., & Buchanan, B. B. (2009). Chapter 10 Protein import in chloroplasts: An emerging regulatory role for redox. In: Advances in botanical research. Academic Press, Vol. 52, pp. 277–332. https://doi.org/10.1016/S0065-2296(10)52010-X.*

strongly suggest that chloroplast protein import is heavily redox-controlled. Since TIC55 also binds an iron-sulfur (Fe–S) center and glutaredoxin is involved in the assembly of these centers, a redox control may also be exerted at this level.

The following chapter by Li and Zachgo discusses the roles of glutaredoxins in stress and development (Li & Zachgo, 2009). It starts with a description of the various types of glutaredoxin belonging essentially to the CPYC, CGFS and CC categories. It next zooms on ROXY, a CC type glutaredoxin specific of land plants. ROXY interacts with TGA transcription factors, among them PAN and thus regulates flowering. As several other glutaredoxins, the CC type Grxs are also involved in disease resistance. This chapter also discusses the involvement of CGFS Grxs in Fe–S cluster assembly and of CPYC Grxs in arsenic resistance. They concluded their chapter by discussing GSH-associated developmental processes as well as the relationship between Grx and detoxifying systems and finally the crosstalk with the Trx-dependent systems. Since then biochemistry has shown that CC type Grxs can also bind Fe–S centers and the components of the Fe–S cluster assembly system have been further dissected. All this will be detailed in the second section of this review. Fig. 3 reprinted from Li and Zachgo summarizes the interactions between ROXYs and the transcription factors.

Gao et al. have discussed glutathionylation in photosynthetic organisms in their contribution (Gao, Bedhomme, Michelet, Zaffagnini, & Lemaire, 2009). Glutathionylation consists in the covalent binding of a glutathione molecule on given protein cysteine. The mechanisms leading to glutathionylation and its reverse reaction deglutathionylation are presented and once again glutaredoxins play an important role there. Methods used to identify glutathionylated proteins are exposed and examples of glutathionylated proteins in photosynthetic organisms are provided, GST and glyceraldehyde-3-phosphate dehydrogenase (GAPDH) being among the best described examples. Finally a scheme detailing the interconnections between the thioredoxin/glutaredoxin systems and the Calvin-Benson cycle enzymes is provided including the nitrosylation and glutathionylation sites. It is reproduced in Fig. 4.

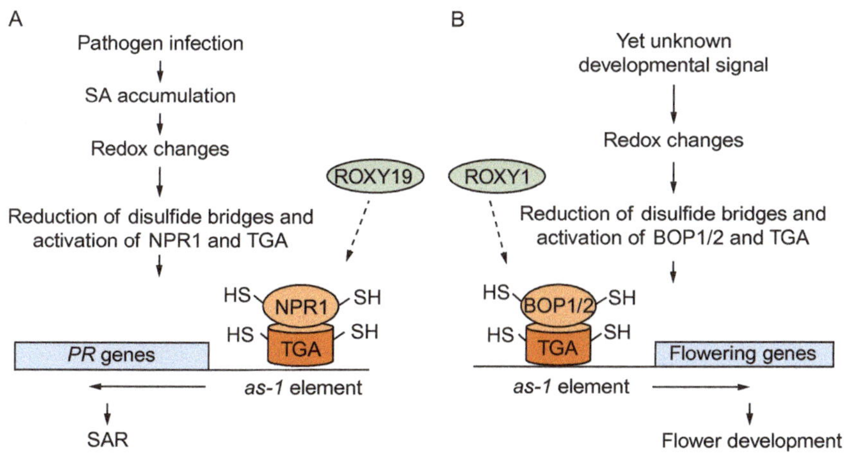

Fig. 3 Interactions between ROXY and TGA transcription factors in flowering. *From Li, S., & Zachgo, S. (2009). Chapter 11 Glutaredoxins in development and stress responses of plants. In: Advances in botanical research. Academic Press, Vol. 52, pp. 333–361. https://doi.org/10.1016/S0065-2296(10)52011-1.*

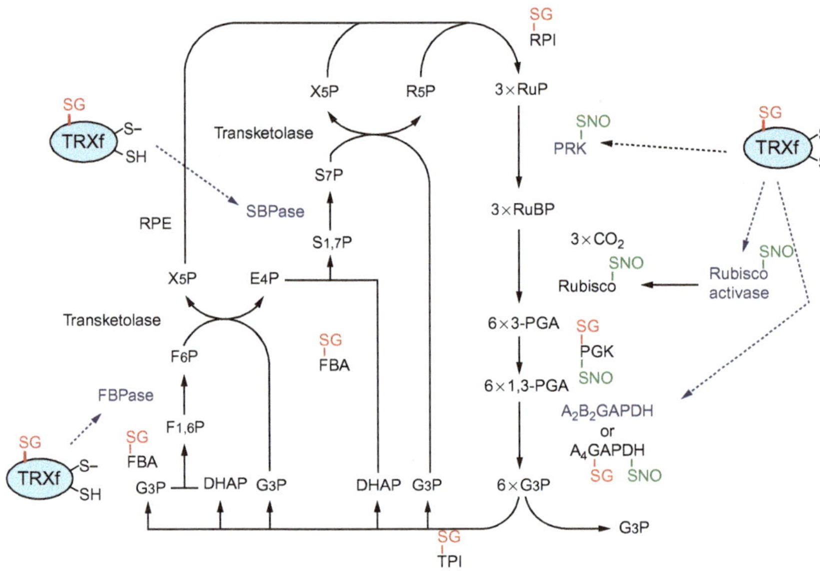

Fig. 4 Redox regulation of Calvin-Benson cycle enzymes. *From Gao, X.-H., Bedhomme, M., Michelet, L., Zaffagnini, M., & Lemaire, S. D. (2009). Chapter 12 Glutathionylation in photosynthetic organisms. In: Advances in botanical research. Academic Press, Vol. 52, pp. 363–403. https://doi.org/10.1016/S0065-2296(10)52012-3.*

Selles et al. discussed the relationships between oxidoreductases of the Trx or Grx type and Fe–S cluster-containing enzymes (Selles et al., 2009). They pointed out that Fe–S clusters are often (but not always) introduced at CXXC positions in the primary sequence of recipient proteins. Interestingly thioredoxin, glutaredoxin and PDI active sites are also mostly of that format (CGPC, CPPC, CPYC, CGYC, CGHC). This coincidence led to question whether this type of active site on a protein with a thioredoxin fold could accomodate Fe–S clusters. A further argument for that possibility was published on a mutant of *E. coli* with a modified active site from CGPC to CACA permissive for Fe–S cluster incorporation (Collet, Peisach, Bardwell, & Xu, 2005). The Selles et al. paper discussed the structural organization of proteins with the thioredoxin fold and provided some evidence for the existence of glutaredoxins containing Fe–S clusters and their involvement in Fe–S cluster assembly. Since then extensive progress has been done in that domain with the description of Fe–S clusters in CC type Grxs, Trxs and PDIs notably. This will be discussed more extensively in the next section of our review.

The penultimate chapter by Pulido et al. discussed seed redox systems and their importance during seed development and germination (Pulido, Domínguez, & Cejudo, 2009). Particular attention was paid to the peroxiredoxins which use thioredoxins as donors to degrade peroxides of various natures. The NTR-Trx system of cereals was also largely discussed. Finally the last chapter by Hägglund et al. (2009) is complementary of the precedent as it centers on the structural properties of NTR and Trx h in cereals using essentially barley and wheat as model species. A detailed comparison of the cereal enzymes vs the Arabidopsis, spinach and Chlamydomonas counterparts was carefully presented.

Following this summary of volume 52 we will now in the second part of this article discuss some important developments that have taken place in the field since.

2. Recent developments in the redox field in plants

The redox field has been moving forward so rapidly in the last years in plants as in non-plant organisms that we will concentrate in this section only on the recent developments about the biochemical and structural properties and physiological roles of plant glutaredoxins and thioredoxins, both families already comprising a large diversity of representatives as highlighted by genomic analyses (Chibani, Wingsle, Jacquot, Gelhaye, & Rouhier, 2009; Couturier, Jacquot, & Rouhier, 2009). Moreover, as it has become impossible to cover such a large field in half of a book chapter, we will only give some flavors of what we believe are the major advances and most promising topics for future research in the plant redox field which has always been at the forefront of this research topic, all organisms considered.

2.1 Latest news about glutaredoxin functions

2.1.1 Iron-sulfur (Fe–S) cluster-containing glutaredoxins

In the thioredoxin protein superfamily, the observation that numerous types of glutaredoxins, but also some other thioredoxin-type proteins, bind Fe–S clusters (Selles, Zannini, Couturier, Jacquot, & Rouhier, 2017; Zannini et al., 2018), has been one of the major developments of the 2000s. In photosynthetic organisms, glutaredoxins are organized into six classes (Couturier et al., 2009). Classes I and II, conserved from cyanobacteria to higher plants, include Grxs with Cxx[C/S] and CGFS active sites, also often referred to as dithiol and monothiol Grxs as a matter of simplification. The first described Fe–S GRXs were class I dithiol GRXs, human GRX2 (Lillig et al., 2005) and poplar GRXC1 (Feng et al., 2006) which differ by their active site signatures, SCSYC and YCGYC, respectively. Another class I GRX in plants, Arabidopsis GRXC5, was also shown to bind a [2Fe-2S] cluster in homodimers using the most N-terminal cysteine of the CxxC signature and cysteines of two non-covalently bound GSH molecules as cluster ligands (Couturier et al., 2011). The structural characterization of these proteins together with the in vitro mutational and spectroscopic analyses of variant proteins allowed determining the importance of the residues immediately preceding and following the protein cysteine ligand (Couturier et al., 2011; Rouhier et al., 2007). However, the exact molecular and physiological functions of class I Fe–S containing Grxs remain elusive so far and understanding if, how and why the Fe–S cluster is required for the in vivo function should become an important aspect of the future research in this specific field.

Soon after, in 2007, the first class II Fe–S GRXs were identified in microbes (Picciocchi, Saguez, Boussac, Cassier-Chauvat, & Chauvat, 2007) and in poplar with the chloroplastic GRXS14 and GRXS16 serving as the first plant representatives (Bandyopadhyay et al., 2008). This property was later confirmed for the mitochondrial GRXS15 and the nucleo-cytosolic GRXS17 (Knuesting et al., 2015; Moseler et al., 2015). It was known since 2002 that the yeast mitochondrial Grx5 ortholog was involved in the maturation of several Fe–S proteins in this organelle (Rodríguez-Manzaneque, Tamarit, Bellí, Ros, & Herrero, 2002) by receiving an Fe–S cluster from the so-called Isu scaffold proteins and transferring it to apoproteins either final targets such as ferredoxin or other Fe–S cluster transfer proteins present in mitochondria (Mühlenhoff, Gerber, Richhardt, & Lill, 2003). However, while good evidence for the function of class II GRXs in the maturation of Fe–S proteins in yeast cells and vertebrates have been obtained over the last two decades, similar evidence for such a function *in planta* have been lacking for a long time. For plastidial GRXS14 and GRXS16, analysis has been hampered by the fact that mutant and over-expressing Arabidopsis lines are viable and do not show strong phenotypes (perhaps explained by partial redundancy) (Rey, Becuwe, Tourrette, & Rouhier, 2017). Nevertheless, the ability of both proteins to complement a yeast mutant that lacks mitochondrial Grx5 strongly hints toward their involvement in the maturation of Fe–S proteins (Bandyopadhyay et al., 2008). Similarly, it has been shown that the cytosolic and nuclear GRXS17 can also complement the *grx5* mutant (Bandyopadhyay et al., 2008; Knuesting et al., 2015). However, Arabidopsis *grxs17* null mutants had only a minor decrease in cytosolic Fe–S enzyme activities even though GRXS17 was found in complex with cytosolic Fe–S cluster maturation factors (Iñigo et al., 2016; Knuesting et al., 2015). A severe developmental phenotype was visible under elevated temperature and extended day light (Cheng et al., 2011; Knuesting et al., 2015). These data indicate that GRXS17 is likely not required for de novo Fe–S cluster assembly in the cytosol. Rather, it may be that GRXS17 functions as an oxidoreductase, regulating the function of its partners as BOLA2, NF-YC11 or other unidentified targets (Couturier et al., 2014; Knuesting et al., 2015). Concerning the mitochondrial GRXS15, in vitro Fe–S cluster reconstitution and transfer assays indicated that GRXS15 coordinates a [2Fe–2S] cluster that can be transferred in vitro to mitochondrial ferredoxin 1 (Moseler et al., 2015). Whereas null mutants are early embryo-lethal showing the essential nature of this protein, a pronounced decrease in aconitase and in lipoic acid-dependent enzymes was measured in viable *grxs15* mutants (RNAi and partially complemented lines) (Moseler et al., 2015; Ströher et al., 2016). This raises the question of whether GRXS15 binds directly a [4Fe–4S] cluster as shown for yeast Grx5 (Zhang et al., 2013) or interacts with ISCA proteins for the reductive conversion of two [2Fe–2S] clusters into a [4Fe–4S] cluster as shown for human proteins (Brancaccio et al., 2014). It is puzzling that GRXS15 is the only class II GRX unable to complement the yeast *grx5* mutant (Bandyopadhyay et al., 2008; Moseler et al., 2015). Also, unlike in yeast, the current data indicate that abrogating *GRXS15* expression has no general effect on mitochondrial Fe–S enzymes but rather specific effects on lipoate-dependent enzymes suggesting a quite specific connection between GRXS15 and the maturation of [4Fe–4S] clusters in LIP1. However, very little is still known about its exact molecular roles and partners. It is established that GRXS15 and most class II GRXs also form hetero-dimeric complexes with BolA proteins, both ligating a [2Fe–2S] cluster but with different lability and spectroscopic signatures than homodimeric complexes (Couturier et al., 2014; Roret et al., 2014). The in vitro studies of these interactions pointed to the capacity of the proteins to interact with or without Fe–S cluster binding (Roret et al., 2014). Moreover, two types of GRX-BOLA holo-heterodimers were distinguished based on the involved BOLA ligands. On the Grx side, the [2Fe–2S] cluster is ligated using the GRX conserved cysteine, a cysteine from glutathione (as in GRX holo-homodimers), and, on the BOLA side, it is ligated using a conserved C-terminal histidine and either an histidine or a cysteine in the $\beta1$–$\beta2$ loop, which was referred to the [H/C] loop and led to distinguish BOLA_C and BOLA_H members (Roret et al., 2014). As for glutaredoxins, an in vivo validation of the involvement of BOLA in Fe–S cluster assembly or in other functions is urgently needed and this should be the focus of future researches.

2.1.2 Involvement of class III GRX in the signaling of the nitrate starvation response

Class III GRXs, also sometimes referred to as CC type GRXs or ROXYs, are specific to land plants as they are present in embryophytes but absent in chlorophytes (Couturier et al., 2009). Since the seminal studies showing the implication of two members, namely ROXY1 and ROXY2, in flower development (Li et al., 2009; Xing & Zachgo, 2008), further studies reported a role in other developmental processes but also in plant stress response (see Gutsche, Thurow, Zachgo, & Gatz, 2015; Rouhier, Cerveau, Couturier, Reichheld, & Rey, 2015 for recent reviews). These effects are mediated by the interaction of these class III GRXs with transcription factors and notably those belonging to the TGA family. A long lasting question is whether this interaction is under redox control.

While this is still far from being clear in most molecular models, a study performed using a TGA factor from the liverwort *Marchantia polymorpha* demonstrated a redox-sensitive binding to a regulatory promoter element alone or together with ROXY1/2 (Gutsche et al., 2017).

Additional recent developments pointed to a role of CC type GRXs in plant nutrition. Several of these GRXs mediate the plant response and long-distance signaling under nitrate starvation in Arabidopsis (Ehrary et al., 2020; Jung, Ahn, & Schachtman, 2018; Ohkubo, Tanaka, Tabata, Ogawa-Ohnishi, & Matsubayashi, 2017; Ota, Ohkubo, Yamashita, Ogawa-Ohnishi, & Matsubayashi, 2020; Patterson et al., 2016). Many genes encoding class III GRXs (AtGRXS3/4/5/8/9/11 and AtGRXC13/C14) are induced by nitrate at the transcriptional level (Ehrary et al., 2020; Ohkubo et al., 2017; Ota et al., 2020; Patterson et al., 2016). The systemic nitrogen acquisition response is triggered by a root-to-shoot mobile peptide, C-TERMINALLY-ENCODED PEPTIDE (CEP), produced in N-starved roots and that acts as a xylem-mobile ascending signal to the shoots. CEPs are recognized by a LRR-receptor kinase CEPR1 (CEP receptor 1) at the level of the phloem cells, leading to the expression of some GRXs referred to as CEP DOWNSTREAM 1 (CEPD1) and CEPD2 or CEPD-like 2 (CEPDL2/GRXC14), acting as phloem-mobile shoot-to-root descending signals (Ohkubo et al., 2017; Ota et al., 2020). These GRXs upregulate the expression of the nitrate transporter gene NRT2.1 in roots when nitrate is present. Accordingly, Arabidopsis lines overexpressing these GRXs exhibit a specific induction of NRT2.1 expression. Using RNAi or knockout lines for single or multiple *GRX* genes allowed to determine that these GRXs act downstream of the CEP-CEPR system and of cytokinins in a signal transduction pathway that negatively regulates plant primary root growth in response to nitrate (Ota et al., 2020; Patterson et al., 2016). Arabidopsis knockout or knockdown lines have longer primary and/or lateral roots. A study focusing on AtGRXS8 showed that it interacts in binary yeast two hybrid with TGA1 and TGA4 transcription factors, two activators involved in the nitrate transcriptional response and in lateral root growth (Ehrary et al., 2020). Overall, this suggests that the regulatory role(s) of class III GRXs in the response to nitrate starvation also implicate(s) an interaction with TGA transcription factors as already shown for their roles in developmental and plant stress response processes.

2.2 Latest news about thioredoxin functions

Historically, thioredoxins were discovered in plants in the 70s as light-activating factors for carbon metabolism enzymes including NADP-malade dehydrogenase (MDH) and some Calvin-Benson cycle enzymes. As all activating systems, switching it off is also a crucial step and it was only very recently that TRX-mediated protein oxidation was proposed to be associated with this step. Another important question that has started to be addressed in 2004 by Balmer and collaborators in a large proteomic study was to assess the extent to which mitochondrial proteins are also subject to redox control. Both aspects will be the focus of the two following sections.

2.2.1 Redox control of the mitochondrial metabolism by thioredoxins

In their study, Balmer and colleagues isolated proteins forming covalent interactions with immobilized cytosolic or chloroplastic TRX variants mutated on the resolving cysteine from mitochondrial extracts of various plants (Balmer et al., 2004). A similar approach was later used by Yoshida and colleagues using Arabidopsis extracts and a TRX o1 cysteinic variant (Yoshida, Noguchi, Motohashi, & Hisabori, 2013). More than 100 proteins involved in more than 15 different processes were identified and almost all enzymes involved in photorespiration and in citric acid cycle and associated reactions were retrieved, strongly arguing for a large redox control of these pathways. In fact, many of these enzymes are also retrieved in some other proteomic approaches aiming at detecting diverse post-translational redox modifications (glutathionylated, nitrosylated or persulfidated proteins), as exemplified for photorespiratory enzymes

(Keech, Gardeström, Kleczkowski, & Rouhier, 2017). The effects of oxidation/reduction on the protein redox state and/or activity were progressively assayed for some proteins present in these pull-down experiments either in vitro or ex vivo using isolated mitochondria. For example, the alternative oxidase forms an intermolecular disulfide-bridged dimer using a conserved cysteine. Its reduction by Arabidopsis TRX o1 and poplar TRX h2 has been one of the first assessed because the monomeric and dimeric forms are easily separated by gel electrophoresis (Gelhaye et al., 2004; Yoshida et al., 2013). However, the redox state of AOX is unchanged in an Arabidopsis *trx o1* mutant possibly suggesting the existence of alternative reducing pathways including a second TRX o2 isoform or TRX h2 (Florez-Sarasa et al., 2019). Noteworthy, evidence for AOX redox changes in vivo are scarce. The best example is the conversion of the inactive, oxidized dimer form of AOX into monomers in leaves of the shade species *Alocasia odora* when plants are shifted from low to high light (Noguchi, Taylor, Millar, Lambers, & Day, 2005). Thus, it was proposed by Nietzel and colleagues that AOX reduction may just occur in specific conditions and that the function of the reducing systems is to avoid AOX oxidation and maintain it constantly reduced but not to regulate its activity (Nietzel, Mostertz, Hochgräfe, & Schwarzländer, 2017). Still, there might be physiological situations where AOX reduction is needed. For instance, a recent redox proteomics study indicated that TRX-mediated reduction of mitochondrial proteins is important during early steps of seed germination and more generally during transitions from metabolically-quiescent to respiring mitochondria (Nietzel et al., 2020).

Although several Krebs/TCA cycle enzymes were identified early as TRX targets, it took about 10 years to obtain the first evidence of a redox control but there is no apparent consistency yet in the way it operates. The formation of intra- and intermolecular disulfide bridges inhibits the activity of the Arabidopsis citrate synthase, and reduction by TRX was shown to enhance its activity (Schmidtmann et al., 2014). Reduction also increased the NAD-malic enzyme (NAD-ME) and aconitase activities (Nietzel et al., 2020). In another study, it was reported that TRX o1 had an opposite effect on mitochondrial succinate dehydrogenase and fumarase activities which were deactivated by reduction in vitro (Daloso et al., 2015). In accordance, the fumarase and a few other enzyme activities were enhanced in the *ntra/b* and *trx o1* mutant backgrounds (Daloso et al., 2015) and a higher TCA cycle flux in relation to other respiratory pathways was observed in *trxo1* mutants (Florez-Sarasa et al., 2019).

Concerning photorespiration, evidence is also accumulating that mitochondrial TRX o1 and the cytosolic/mitochondrial TRX h2 play regulatory roles. Although the *trx h2* knockdown and *trx o1* knockout mutants do not exhibit strong phenotypes when grown under standard conditions, their deficiency impacts mitochondrial photorespiration, both mutants exhibiting alterations in photorespiratory metabolites under some conditions (da Fonseca-Pereira et al., 2020; Reinholdt et al., 2019). However, the defects and metabolic variations seem clearly different in both mutants rather pinpointing the existence of specific roles and targets. The *trx o1* mutant strongly over-accumulated glycine and had decreased serine levels in CO_2 transition experiments, i.e., when plants were transferred from a high CO_2 to a normal air condition (Reinholdt et al., 2019). The steady-state metabolite profile when *trx o1* plants were grown continuously in normal air followed a similar trend but showed also variations in other metabolites. The metabolite profile of the *trx h2* mutant was both quantitatively and qualitatively quite distinct (da Fonseca-Pereira et al., 2020). As mitochondrial serine hydroxymethyltransferase and glycine decarboxylase (GDC) catalyze the conversion of glycine to serine, a decrease in their abundance or activity is expected. A significant decrease in the abundance of SHMT and GDC H and L subunits was indeed observed in the *trx h2* but not *trx o1* mutants. One unifying result between these studies is that both TRX h2 and TRX o1 recombinant proteins reductively deactivated GDC-L in vitro (da Fonseca-Pereira et al., 2020; Reinholdt et al., 2019). However, this may appear somehow contradictory with the observation that activity of a cyanobacterial GDC-P protein, which is the catalytic subunit of the complex, is increased upon reduction (Hasse et al., 2013).

In summary, it seems clear that the mitochondrial TRX and glutathione systems are to a large extent in a reduced state at least in respiration conditions (such as when citrate is produced) favoring the reduction of $NADP^+$ (Nietzel et al., 2020). Even though it may be that all mitochondrial TRXs are not reduced to the same extent or in the same conditions or that they have different targets, the hypothesis is that their catalytic activity is to reduce their target proteins. Here, as already pointed out, the effects differed. Thus, all these evidence about a redox control of mitochondrial enzymes, which remain rather preliminary, will have to be completed by a detailed analysis of the oxidation/reduction effects on each enzyme/enzymatic complexes, notably the GDC complex which is formed by four subunits, each being possibly subject to redox regulation (Balmer et al., 2004; Keech et al., 2017).

2.2.2 Regulatory thiol oxidation in chloroplasts

The light- and dark-dependent redox regulation of the Calvin/Benson cycle, as well as of starch and chlorophyll biosynthesis enzymes ensures metabolic coordination and efficient energy utilization in the chloroplasts in response to dark and light fluctuations. There are at least 20 TRXs in Arabidopsis chloroplasts, i.e., 10 classical TRXs (Trx-f1, -f2, -m1, -m2, -m3, -m4, -x, -y1, -y2, -z), additional TRX-like proteins (the chloroplast drought-induced stress protein of 32 kDa (CDSP32), high chlorophyll fluorescence 164 (HCF164), 5 TRX-Lilium/Atypical Cysteine Histidine-rich TRX (ACHT) proteins and 2 TRX-like2 proteins) and NTRc (Chibani et al., 2009). Current knowledge indicates that classical TRXs may have diversified roles, the TRXs m and f are rather supporting the activity of carbon fixation enzymes, the TRXs x, y, and CDSP32 the activity of antioxidant enzymes such as the thiol peroxidases/peroxiredoxins (PRXs) and methionine sulfoxide reductases, and TRX z is part of the plastid-encoded RNA polymerase complex (Arsova et al., 2010; Rey et al., 2005; Vieira Dos Santos et al., 2007; Yoshida, Hara, & Hisabori, 2015). The activity of all these proteins relies on the ferredoxin-thioredoxin reductase (FTR) reducing system although they are reduced with distinct efficiency (Yoshida & Hisabori, 2017). An alternative reducing system is the NTRC protein, a NADPH thioredoxin reductase fused to a TRX domain which is somehow redundant/complementary with some of the above mentioned TRXs for 2-cys PRXA/B regeneration, as well as starch and chlorophyll biosynthesis (Michalska, Zauber, Buchanan, Cejudo, & Geigenberger, 2009; Pérez-Ruiz, Naranjo, Ojeda, Guinea, & Cejudo, 2017; Richter et al., 2013). Recently it was shown that in *ntrc* plants, TRX f was partially oxidized and as a consequence the fructose 1,6 bisphosphatase (FBPase) was insufficiently reduced and activated in the light (Pérez-Ruiz et al., 2017). This effect was mostly reverted in a *ntrc/2cysprxAB* triple mutants as was the severe growth phenotype of *ntrc* plants. The authors proposed that the lack of NTRC led to an oxidation cascade with 2-Cys Prx A/B oxidation contributing to TRX f oxidation and then to FBPase oxidation. The results indicated that 2-Cys PRXs function as TRX oxidases in vivo which then mediate the inactivation of TRX-dependent target proteins. Further proofs have been obtained by showing that the oxidation/deactivation of NADP-malade dehydrogenase (NADP-MDH), phosphoribulokinase (PRK), FBPase, sedoheptulose 1,7-bisphosphatase (SBPase), glyceraldehyde-3 phosphate dehydrogenase (GAPDH), γ-subunit of ATPase and Rubisco activase that is observed upon light/dark transition in WT plants, was delayed in mutant plants lacking 2-Cys Prxs A and B, and in plants overexpressing NTRC (Ojeda, Pérez-Ruiz, & Cejudo, 2018; Vaseghi et al., 2018; Yoshida, Hara, Sugiura, Fukaya, & Hisabori, 2018). This contribution of 2-Cys-PRXs appears particularly important in darkness but also in low light, and in fluctuating light (Dangoor, Peled-Zehavi, Wittenberg, & Danon, 2012; Vaseghi et al., 2018). No similar delay was observed in *prxIIE* and *prxQ* mutant plants indicating that only 2-Cys PRXs are involved (Ojeda et al., 2018). Additional proofs that this reaction is mediated by TRXs have been obtained in vitro using FBPase and MDH in the presence of their TRX f and m partners (Ojeda et al., 2018; Vaseghi et al., 2018) and using RCA, FBPase, or SBPase using both TRX-like2 members (Yoshida et al., 2018). In fact, it seems that oxidation is primarily mediated by atypical TRXs such as TRXs-like2 and ACHTs with TRXs-like2 showing high 2-Cys PRX reducing activity overpassing that of typical TRXs and with ACHT1 and ACHT2a catalyzing faster oxidation of FBPase than TRX f1 (Yokochi et al., 2019, 2018). In fact, these atypical TRXs are better candidates because the midpoint redox potentials of their catalytic dithiol/disulfide couple is higher, they are poor or inefficient reductants of redox-regulated carbon assimilation proteins and they are not reduced by FTR (Chibani et al., 2012; Dangoor, Peled-Zehavi, Levitan, Pasand, & Danon, 2009; Yoshida et al., 2018). Accordingly, it was shown that TRX-like2 reduction is prolonged in *2cysprxAB* mutant plants during light to dark transitions (Yoshida et al., 2018) and that ACHT1 which is oxidized by 2-Cys PRX is kept in a partially oxidized state in the light (Dangoor et al., 2012). As another evidence, ACHT4/TRX-lilium1 is likely oxidized by a similar disulfide exchange with 2-Cys PRXs (Eliyahu, Rog, Inbal, & Danon, 2015). Moreover, it seems to specifically inactivate the small subunit (APS1) of the ADP-glucose pyrophosphorylase (AGPase), an enzyme catalyzing the first step of the starch synthesis pathway, during the diurnal transition from day to night. Overall, chloroplastic TRX-like2 and ACHT proteins, together with 2-Cys PRXs, form a general regulatory oxidative pathway needed to effectively inactivate or downregulate the assimilation pathways linked to the photosynthetic electron transport chain. In summary, mounting evidence indicate that the in vivo redox state of chloroplastic redox-regulated proteins is controlled by a balance between FTR- and NTRC-dependent reducing signals, and 2-Cys PRXs and atypical TRXs-dependent oxidative signals. This duality allows a fine and dynamic redox control mechanism which confers the capacity for rapid alteration of the protein redox state in darkness or in response to small fluctuations in light intensity during the day.

3. Concluding remarks

Volume 52 of ABR was already indicative of the importance of plants as models for the redox field. Major advances have indeed been made in plants especially concerning the diversification of the thioredoxin and glutaredoxin genes which is not so surprising in retrospect as non-plant organisms contain far less genetic diversity. It is yet not clear why plants have elaborated such a high diversity of antioxidant genes, but the fact that it occurred in photosynthetic O_2 generating organisms is probably not a coincidence. Outside their specific function for regulating photosynthesis in plants, the redox systems have similar targets and functions in plants and non-plant organisms and in particular concerning their relationship with iron metabolism and Fe–S protein maturation. As detailed in the second part of our article Fe–S cluster assembly is indeed a lot more complex in plants as they contain a very large number of those proteins in the chloroplasts. We believe that the developments recorded here in the plant field after volume 52 was published are truly remarkable and that there is still quite a lot of work ahead to understand the connections of those systems. Serendipity might also bring new unexpected avenues of research as those reported earlier for flowering. In short the plant redox field is still vibrant and active more than 50 years after its discovery.

References

Agrawal, S. B., Singh, S., & Agrawal, M. (2009). Chapter 3 Ultraviolet-B induced changes in gene expression and antioxidants in plants. In *Vol. 52. Advances in botanical research* (pp. 47–86). Academic Press. https://doi.org/10.1016/S0065-2296(10)52003-2.

Alloing, G., Mandon, K., Boncompagni, E., Montrichard, F., & Frendo, P. (2018). Involvement of glutaredoxin and thioredoxin systems in the nitrogen-fixing symbiosis between legumes and rhizobia. *Antioxidants (Basel, Switzerland)*, 7(12), 182. https://doi.org/10.3390/antiox7120182.

Arsova, B., Hoja, U., Wimmelbacher, M., Greiner, E., Ustün, S., Melzer, M., et al. (2010). Plastidial thioredoxin z interacts with two fructokinase-like proteins in a thiol-dependent manner: Evidence for an essential role in chloroplast development in Arabidopsis and Nicotiana benthamiana. *The Plant Cell*, *22*(5), 1498–1515. https://doi.org/10.1105/tpc.109.071001.

Balmer, Y., Vensel, W. H., Tanaka, C. K., Hurkman, W. J., Gelhaye, E., Rouhier, N., et al. (2004). Thioredoxin links redox to the regulation of fundamental processes of plant mitochondria. *Proceedings of the National Academy of Sciences of the United States of America*, *101*(8), 2642–2647. https://doi.org/10.1073/pnas.0308583101.

Balsera, M., Soll, J., & Buchanan, B. B. (2009). Chapter 10 Protein import in chloroplasts: An emerging regulatory role for redox. In *Vol. 52. Advances in botanical research* (pp. 277–332). Academic Press. https://doi.org/10.1016/S0065-2296(10)52010-X.

Bandyopadhyay, S., Gama, F., Molina-Navarro, M. M., Gualberto, J. M., Claxton, R., Naik, S. G., et al. (2008). Chloroplast monothiol glutaredoxins as scaffold proteins for the assembly and delivery of [2Fe-2S] clusters. *The EMBO Journal, 27*(7), 1122–1133. https://doi.org/10.1038/emboj.2008.50.

Benyamina, S. M., Baldacci-Cresp, F., Couturier, J., Chibani, K., Hopkins, J., Bekki, A., et al. (2013). Two Sinorhizobium meliloti glutaredoxins regulate iron metabolism and symbiotic bacteroid differentiation. *Environmental Microbiology, 15*(3), 795–810. https://doi.org/10.1111/j.1462-2920.2012.02835.x.

Brancaccio, D., Gallo, A., Mikolajczyk, M., Zovo, K., Palumaa, P., Novellino, E., et al. (2014). Formation of [4Fe-4S] clusters in the mitochondrial iron-sulfur cluster assembly machinery. *Journal of the American Chemical Society, 136*(46), 16240–16250. https://doi.org/10.1021/ja507822j.

Cheng, N.-H., Liu, J.-Z., Liu, X., Wu, Q., Thompson, S. M., Lin, J., et al. (2011). Arabidopsis monothiol glutaredoxin, AtGRXS17, is critical for temperature-dependent postembryonic growth and development via modulating auxin response. *The Journal of Biological Chemistry, 286*(23), 20398–20406. https://doi.org/10.1074/jbc.M110.201707.

Chibani, K., Tarrago, L., Gualberto, J. M., Wingsle, G., Rey, P., Jacquot, J.-P., et al. (2012). Atypical thioredoxins in poplar: The glutathione-dependent thioredoxin-like 2.1 supports the activity of target enzymes possessing a single redox active cysteine. *Plant Physiology, 159*(2), 592–605. https://doi.org/10.1104/pp.112.197723.

Chibani, K., Wingsle, G., Jacquot, J.-P., Gelhaye, E., & Rouhier, N. (2009). Comparative genomic study of the thioredoxin family in photosynthetic organisms with emphasis on Populus trichocarpa. *Molecular Plant, 2*(2), 308–322. https://doi.org/10.1093/mp/ssn076.

Collet, J.-F., Peisach, D., Bardwell, J. C. A., & Xu, Z. (2005). The crystal structure of TrxA(CACA): Insights into the formation of a [2Fe-2S] iron-sulfur cluster in an *Escherichia coli* thioredoxin mutant. *Protein Science: A Publication of the Protein Society, 14*(7), 1863–1869. https://doi.org/10.1110/ps.051464705.

Couturier, J., Jacquot, J.-P., & Rouhier, N. (2009). Evolution and diversity of glutaredoxins in photosynthetic organisms. *Cellular and Molecular Life Sciences, 66*(15), 2539–2557. https://doi.org/10.1007/s00018-009-0054-y.

Couturier, J., Ströher, E., Albetel, A.-N., Roret, T., Muthuramalingam, M., Tarrago, L., et al. (2011). Arabidopsis chloroplastic glutaredoxin C5 as a model to explore molecular determinants for iron-sulfur cluster binding into glutaredoxins. *The Journal of Biological Chemistry, 286*(31), 27515–27527. https://doi.org/10.1074/jbc.M111.228726.

Couturier, J., Wu, H.-C., Dhalleine, T., Pégeot, H., Sudre, D., Gualberto, J. M., et al. (2014). Monothiol glutaredoxin-BolA interactions: Redox control of Arabidopsis thaliana BolA2 and SufE1. *Molecular Plant, 7*(1), 187–205. https://doi.org/10.1093/mp/sst156.

da Fonseca-Pereira, P., Souza, P. V. L., Hou, L.-Y., Schwab, S., Geigenberger, P., Nunes-Nesi, A., et al. (2020). Thioredoxin h2 contributes to the redox regulation of mitochondrial photorespiratory metabolism. *Plant, Cell & Environment, 43*(1), 188–208. https://doi.org/10.1111/pce.13640.

Daloso, D. M., Müller, K., Obata, T., Florian, A., Tohge, T., Bottcher, A., et al. (2015). Thioredoxin, a master regulator of the tricarboxylic acid cycle in plant mitochondria. *Proceedings of the National Academy of Sciences of the United States of America, 112*(11), E1392–E1400. https://doi.org/10.1073/pnas.1424840112.

Dangoor, I., Peled-Zehavi, H., Levitan, A., Pasand, O., & Danon, A. (2009). A small family of chloroplast atypical thioredoxins. *Plant Physiology, 149*(3), 1240–1250. https://doi.org/10.1104/pp.108.128314.

Dangoor, I., Peled-Zehavi, H., Wittenberg, G., & Danon, A. (2012). A chloroplast light-regulated oxidative sensor for moderate light intensity in Arabidopsis. *The Plant Cell, 24*(5), 1894–1906. https://doi.org/10.1105/tpc.112.097139.

Ehrary, A., Rosas, M., Carpinelli, S., Davalos, O., Cowling, C., Fernandez, F., et al. (2020). Glutaredoxin AtGRXS8 represses transcriptional and developmental responses to nitrate in Arabidopsis thaliana roots. *Plant Direct*, *4*(6), e00227. https://doi.org/10.1002/pld3.227.

Eliyahu, E., Rog, I., Inbal, D., & Danon, A. (2015). ACHT4-driven oxidation of APS1 attenuates starch synthesis under low light intensity in Arabidopsis plants. *Proceedings of the National Academy of Sciences of the United States of America*, *112*(41), 12876–12881. https://doi.org/10.1073/pnas.1515513112.

Feng, Y., Zhong, N., Rouhier, N., Hase, T., Kusunoki, M., Jacquot, J.-P., et al. (2006). Structural insight into poplar glutaredoxin C1 with a bridging iron-sulfur cluster at the active site. *Biochemistry*, *45*(26), 7998–8008. https://doi.org/10.1021/bi060444t.

Florez-Sarasa, I., Obata, T., Del-Saz, N. S. F. N., Reichheld, J.-P., Meyer, E. H., Rodriguez-Concepcion, M., et al. (2019). The lack of mitochondrial thioredoxin TRXo1 affects in vivo alternative oxidase activity and carbon metabolism under different light conditions. *Plant & Cell Physiology*, *60*(11), 2369–2381. https://doi.org/10.1093/pcp/pcz123.

Gao, X.-H., Bedhomme, M., Michelet, L., Zaffagnini, M., & Lemaire, S. D. (2009). Chapter 12 Glutathionylation in photosynthetic organisms. In *Vol. 52. Advances in botanical research* (pp. 363–403). Academic Press. https://doi.org/10.1016/S0065-2296(10)52012-3.

Gelhaye, E., Rouhier, N., Gérard, J., Jolivet, Y., Gualberto, J., Navrot, N., et al. (2004). A specific form of thioredoxin h occurs in plant mitochondria and regulates the alternative oxidase. *Proceedings of the National Academy of Sciences of the United States of America*, *101*(40), 14545–14550. https://doi.org/10.1073/pnas.0405282101.

Gutsche, N., Holtmannspötter, M., Maß, L., O'Donoghue, M., Busch, A., Lauri, A., et al. (2017). Conserved redox-dependent DNA binding of ROXY glutaredoxins with TGA transcription factors. *Plant Direct*, *1*(6), e00030. https://doi.org/10.1002/pld3.30.

Gutsche, N., Thurow, C., Zachgo, S., & Gatz, C. (2015). Plant-specific CC-type glutaredoxins: Functions in developmental processes and stress responses. *Biological Chemistry*, *396*(5), 495–509. https://doi.org/10.1515/hsz-2014-0300.

Hägglund, P., Kirkensgaard, K. G., Maeda, K., Finnie, C., Henriksen, A., & Svensson, B. (2009). Chapter 15 Molecular recognition in NADPH-dependent plant thioredoxin systems—Catalytic mechanisms, structural snapshots and target identifications. In *Vol. 52. Advances in botanical research* (pp. 461–495). Academic Press. https://doi.org/10.1016/S0065-2296(10)52015-9.

Hanke, G. T., Holtgrefe, S., König, N., Strodtkötter, I., Voss, I., & Scheibe, R. (2009). Chapter 8 Use of transgenic plants to uncover strategies for maintenance of redox homeostasis during photosynthesis. In *Vol. 52. Advances in botanical research* (pp. 207–251). Academic Press. https://doi.org/10.1016/S0065-2296(10)52008-1.

Hasse, D., Andersson, E., Carlsson, G., Masloboy, A., Hagemann, M., Bauwe, H., et al. (2013). Structure of the homodimeric glycine decarboxylase P-protein from Synechocystis sp. PCC 6803 suggests a mechanism for redox regulation. *The Journal of Biological Chemistry*, *288*(49), 35333–35345. https://doi.org/10.1074/jbc.M113.509976.

Iñigo, S., Durand, A. N., Ritter, A., Le Gall, S., Termathe, M., Klassen, R., et al. (2016). Glutaredoxin GRXS17 associates with the cytosolic iron-sulfur cluster assembly pathway. *Plant Physiology*, *172*(2), 858–873. https://doi.org/10.1104/pp.16.00261.

Joshi, N. C., Meyer, A. J., Bangash, S. A. K., Zheng, Z.-L., & Leustek, T. (2019). Arabidopsis γ-glutamylcyclotransferase affects glutathione content and root system architecture during sulfur starvation. *The New Phytologist*, *221*(3), 1387–1397. https://doi.org/10.1111/nph.15466.

Jung, J.-Y., Ahn, J. H., & Schachtman, D. P. (2018). CC-type glutaredoxins mediate plant response and signaling under nitrate starvation in Arabidopsis. *BMC Plant Biology*, *18*(1), 281. https://doi.org/10.1186/s12870-018-1512-1.

Keech, O., Gardeström, P., Kleczkowski, L. A., & Rouhier, N. (2017). The redox control of photorespiration: From biochemical and physiological aspects to biotechnological considerations. *Plant, Cell & Environment*, *40*(4), 553–569. https://doi.org/10.1111/pce.12713.

Kim, H.-S., Kim, Y.-S., Hahn, K.-W., Joung, H., & Jeon, J.-H. (2009). Chapter 2 Reactive oxygen species: Regulation of plant growth and development. In *Vol. 52. Advances in botanical research* (pp. 25–46). Academic Press. https://doi.org/10.1016/S0065-2296(10)52002-0.

Knuesting, J., Riondet, C., Maria, C., Kruse, I., Bécuwe, N., König, N., et al. (2015). Arabidopsis glutaredoxin S17 and its partner, the nuclear factor Y subunit C11/negative cofactor 2α, contribute to maintenance of the shoot apical meristem under long-day photoperiod. *Plant Physiology*, *167*(4), 1643–1658. https://doi.org/10.1104/pp.15.00049.

Li, S., Lauri, A., Ziemann, M., Busch, A., Bhave, M., & Zachgo, S. (2009). Nuclear activity of ROXY1, a glutaredoxin interacting with TGA factors, is required for petal development in Arabidopsis thaliana. *The Plant Cell*, *21*(2), 429–441. https://doi.org/10.1105/tpc.108.064477.

Li, S., & Zachgo, S. (2009). Chapter 11 Glutaredoxins in development and stress responses of plants. In *Vol. 52. Advances in botanical research* (pp. 333–361). Academic Press. https://doi.org/10.1016/S0065-2296(10)52011-1.

Lillig, C. H., Berndt, C., Vergnolle, O., Lönn, M. E., Hudemann, C., Bill, E., et al. (2005). Characterization of human glutaredoxin 2 as iron-sulfur protein: A possible role as redox sensor. *Proceedings of the National Academy of Sciences of the United States of America*, *102*(23), 8168–8173. https://doi.org/10.1073/pnas.0500735102.

Marino, D., Pucciariello, C., Puppo, A., & Frendo, P. (2009). Chapter 5 The redox state, a referee of the legume–rhizobia symbiotic game. In *Vol. 52. Advances in botanical research* (pp. 115–151). Academic Press. https://doi.org/10.1016/S0065-2296(10)52005-6.

Michalska, J., Zauber, H., Buchanan, B. B., Cejudo, F. J., & Geigenberger, P. (2009). NTRC links built-in thioredoxin to light and sucrose in regulating starch synthesis in chloroplasts and amyloplasts. *Proceedings of the National Academy of Sciences of the United States of America*, *106*(24), 9908–9913. https://doi.org/10.1073/pnas.0903559106.

Morel, M., Ngadin, A. A., Jacquot, J.-P., & Gelhaye, E. (2009). Chapter 6 Reactive oxygen species in Phanerochaete chrysosporium relationship between extracellular oxidative and intracellular antioxidant systems. In *Vol. 52. Advances in botanical research* (pp. 153–186). Academic Press. https://doi.org/10.1016/S0065-2296(10)52006-8.

Moseler, A., Aller, I., Wagner, S., Nietzel, T., Przybyla-Toscano, J., Mühlenhoff, U., et al. (2015). The mitochondrial monothiol glutaredoxin S15 is essential for iron-sulfur protein maturation in Arabidopsis thaliana. *Proceedings of the National Academy of Sciences of the United States of America*, *112*(44), 13735–13740. https://doi.org/10.1073/pnas.1510835112.

Mühlenhoff, U., Gerber, J., Richhardt, N., & Lill, R. (2003). Components involved in assembly and dislocation of iron-sulfur clusters on the scaffold protein Isu1p. *The EMBO Journal*, *22*(18), 4815–4825. https://doi.org/10.1093/emboj/cdg446.

Nietzel, T., Mostertz, J., Hochgräfe, F., & Schwarzländer, M. (2017). Redox regulation of mitochondrial proteins and proteomes by cysteine thiol switches. *Mitochondrion*, *33*, 72–83. https://doi.org/10.1016/j.mito.2016.07.010.

Nietzel, T., Mostertz, J., Ruberti, C., Née, G., Fuchs, P., Wagner, S., et al. (2020). Redox-mediated kick-start of mitochondrial energy metabolism drives resource-efficient seed germination. *Proceedings of the National Academy of Sciences of the United States of America*, *117*(1), 741–751. https://doi.org/10.1073/pnas.1910501117.

Nishiyama, Y., & Hisabori, T. (2009). Chapter 7 Physiological impact of thioredoxin- and glutaredoxin-mediated redox regulation in cyanobacteria. In *Vol. 52. Advances in botanical research* (pp. 187–205). Academic Press. https://doi.org/10.1016/S0065-2296(10)52007-X.

Noguchi, K., Taylor, N. L., Millar, A. H., Lambers, H., & Day, D. A. (2005). Response of mitochondria to light intensity in the leaves of sun and shade species. *Plant, Cell & Environment, 28*(6), 760–771. https://doi.org/10.1111/j.1365-3040.2005.01322.x.

Ohkama-Ohtsu, N., Fukuyama, K., & Oliver, D. J. (2009). Chapter 4 Roles of γ-glutamyl transpeptidase and γ-glutamyl cyclotransferase in glutathione and glutathione-conjugate metabolism in plants. In *Vol. 52. Advances in botanical research* (pp. 87–113). Academic Press. https://doi.org/10.1016/S0065-2296(10)52004-4.

Ohkubo, Y., Tanaka, M., Tabata, R., Ogawa-Ohnishi, M., & Matsubayashi, Y. (2017). Shoot-to-root mobile polypeptides involved in systemic regulation of nitrogen acquisition. *Nature Plants, 3*, 17029. https://doi.org/10.1038/nplants.2017.29.

Ojeda, V., Pérez-Ruiz, J. M., & Cejudo, F. J. (2018). 2-Cys peroxiredoxins participate in the oxidation of chloroplast enzymes in the dark. *Molecular Plant, 11*(11), 1377–1388. https://doi.org/10.1016/j.molp.2018.09.005.

Ota, R., Ohkubo, Y., Yamashita, Y., Ogawa-Ohnishi, M., & Matsubayashi, Y. (2020). Shoot-to-root mobile CEPD-like 2 integrates shoot nitrogen status to systemically regulate nitrate uptake in Arabidopsis. *Nature Communications, 11*(1), 641. https://doi.org/10.1038/s41467-020-14440-8.

Patterson, K., Walters, L. A., Cooper, A. M., Olvera, J. G., Rosas, M. A., Rasmusson, A. G., et al. (2016). Nitrate-regulated glutaredoxins control Arabidopsis primary root growth. *Plant Physiology, 170*(2), 989–999. https://doi.org/10.1104/pp.15.01776.

Paulose, B., Chhikara, S., Coomey, J., Jung, H.-I., Vatamaniuk, O., & Dhankher, O. P. (2013). A γ-glutamyl cyclotransferase protects Arabidopsis plants from heavy metal toxicity by recycling glutamate to maintain glutathione homeostasis. *The Plant Cell, 25*(11), 4580–4595. https://doi.org/10.1105/tpc.113.111815.

Pérez-Ruiz, J. M., Naranjo, B., Ojeda, V., Guinea, M., & Cejudo, F. J. (2017). NTRC-dependent redox balance of 2-Cys peroxiredoxins is needed for optimal function of the photosynthetic apparatus. *Proceedings of the National Academy of Sciences of the United States of America, 114*(45), 12069–12074. https://doi.org/10.1073/pnas.1706003114.

Picciocchi, A., Saguez, C., Boussac, A., Cassier-Chauvat, C., & Chauvat, F. (2007). CGFS-type monothiol glutaredoxins from the cyanobacterium synechocystis PCC6803 and other evolutionary distant model organisms possess a glutathione-ligated [2Fe-2S] cluster. *Biochemistry, 46*(51), 15018–15026. https://doi.org/10.1021/bi7013272.

Pulido, P., Domínguez, F., & Cejudo, F. J. (2009). Chapter 14 Oxidative stress and thiol-based antioxidants in cereal seeds. In *Vol. 52. Advances in botanical research* (pp. 437–460). Academic Press. https://doi.org/10.1016/S0065-2296(10)52014-7.

Reichheld, J.-P., Bashandy, T., Siala, W., Riondet, C., Delorme, V., Meyer, A., et al. (2009). Chapter 9 Redundancy and crosstalk within the thioredoxin and glutathione pathways: A new development in plants. In *Vol. 52. Advances in botanical research* (pp. 253–276). Academic Press. https://doi.org/10.1016/S0065-2296(10)52009-3.

Reinholdt, O., Schwab, S., Zhang, Y., Reichheld, J.-P., Fernie, A. R., Hagemann, M., et al. (2019). Redox-regulation of photorespiration through mitochondrial thioredoxin o1. *Plant Physiology, 181*(2), 442–457. https://doi.org/10.1104/pp.19.00559.

Rey, P., Becuwe, N., Tourrette, S., & Rouhier, N. (2017). Involvement of Arabidopsis glutaredoxin S14 in the maintenance of chlorophyll content. *Plant, Cell & Environment, 40*(10), 2319–2332. https://doi.org/10.1111/pce.13036.

Rey, P., Cuiné, S., Eymery, F., Garin, J., Court, M., Jacquot, J.-P., et al. (2005). Analysis of the proteins targeted by CDSP32, a plastidic thioredoxin participating in oxidative stress responses. *The Plant Journal: For Cell and Molecular Biology, 41*(1), 31–42. https://doi.org/10.1111/j.1365-313X.2004.02271.x.

Ribeiro, C. W., Baldacci-Cresp, F., Pierre, O., Larousse, M., Benyamina, S., Lambert, A., et al. (2017). Regulation of differentiation of nitrogen-fixing bacteria by microsymbiont targeting of plant thioredoxin s1. *Current Biology, 27*(2), 250–256. https://doi.org/10.1016/j.cub.2016.11.013.

Richter, A. S., Peter, E., Rothbart, M., Schlicke, H., Toivola, J., Rintamäki, E., et al. (2013). Posttranslational influence of NADPH-dependent thioredoxin reductase C on enzymes in tetrapyrrole synthesis. *Plant Physiology*, *162*(1), 63–73. https://doi.org/10.1104/pp.113.217141.

Rodríguez-Manzaneque, M. T., Tamarit, J., Bellí, G., Ros, J., & Herrero, E. (2002). Grx5 is a mitochondrial glutaredoxin required for the activity of iron/sulfur enzymes. *Molecular Biology of the Cell*, *13*(4), 1109–1121. https://doi.org/10.1091/mbc.01-10-0517.

Roret, T., Tsan, P., Couturier, J., Zhang, B., Johnson, M. K., Rouhier, N., et al. (2014). Structural and spectroscopic insights into BolA-glutaredoxin complexes. *The Journal of Biological Chemistry*, *289*(35), 24588–24598. https://doi.org/10.1074/jbc.M114.572701.

Rouhier, N., Cerveau, D., Couturier, J., Reichheld, J.-P., & Rey, P. (2015). Involvement of thiol-based mechanisms in plant development. *Biochimica et Biophysica Acta*, *1850*(8), 1479–1496. https://doi.org/10.1016/j.bbagen.2015.01.023.

Rouhier, N., Unno, H., Bandyopadhyay, S., Masip, L., Kim, S.-K., Hirasawa, M., et al. (2007). Functional, structural, and spectroscopic characterization of a glutathione-ligated [2Fe-2S] cluster in poplar glutaredoxin C1. *Proceedings of the National Academy of Sciences of the United States of America*, *104*(18), 7379–7384. https://doi.org/10.1073/pnas.0702268104.

Schmidtmann, E., König, A.-C., Orwat, A., Leister, D., Hartl, M., & Finkemeier, I. (2014). Redox regulation of Arabidopsis mitochondrial citrate synthase. *Molecular Plant*, 7(1), 156–169. https://doi.org/10.1093/mp/sst144.

Selles, B., Rouhier, N., Chibani, K., Couturier, J., Gama, F., & Jacquot, J.-P. (2009). Chapter 13 Glutaredoxin: The missing link between thiol-disulfide oxidoreductases and iron sulfur enzymes. In *Vol. 52. Advances in botanical research* (pp. 405–436). Academic Press. https://doi.org/10.1016/S0065-2296(10)52013-5.

Selles, B., Zannini, F., Couturier, J., Jacquot, J.-P., & Rouhier, N. (2017). Atypical protein disulfide isomerases (PDI): Comparison of the molecular and catalytic properties of poplar PDI-A and PDI-M with PDI-L1A. *PLoS One*, *12*(3), e0174753. https://doi.org/10.1371/journal.pone.0174753.

Ströher, E., Grassl, J., Carrie, C., Fenske, R., Whelan, J., & Millar, A. H. (2016). Glutaredoxin S15 is involved in Fe-S cluster transfer in mitochondria influencing lipoic acid-dependent enzymes, plant growth, and arsenic tolerance in Arabidopsis. *Plant Physiology*, *170*(3), 1284–1299. https://doi.org/10.1104/pp.15.01308.

Sweetlove, L. J., & Møller, I. M. (2009). Chapter 1 Oxidation of proteins in plants—Mechanisms and consequences. In *Vol. 52. Advances in botanical research* (pp. 1–23). Academic Press. https://doi.org/10.1016/S0065-2296(10)52001-9.

Vaseghi, M.-J., Chibani, K., Telman, W., Liebthal, M. F., Gerken, M., Schnitzer, H., et al. (2018). The chloroplast 2-cysteine peroxiredoxin functions as thioredoxin oxidase in redox regulation of chloroplast metabolism. *eLife*, 7, e38194. https://doi.org/10.7554/eLife.38194.

Vieira Dos Santos, C., Laugier, E., Tarrago, L., Massot, V., Issakidis-Bourguet, E., Rouhier, N., et al. (2007). Specificity of thioredoxins and glutaredoxins as electron donors to two distinct classes of Arabidopsis plastidial methionine sulfoxide reductases B. *FEBS Letters*, *581*(23), 4371–4376. https://doi.org/10.1016/j.febslet.2007.07.081.

Xing, S., & Zachgo, S. (2008). ROXY1 and ROXY2, two Arabidopsis glutaredoxin genes, are required for anther development. *The Plant Journal: For Cell and Molecular Biology*, *53*(5), 790–801. https://doi.org/10.1111/j.1365-313X.2007.03375.x.

Yokochi, Y., Sugiura, K., Takemura, K., Yoshida, K., Hara, S., Wakabayashi, K.-I., et al. (2019). Impact of key residues within chloroplast thioredoxin-f on recognition for reduction and oxidation of target proteins. *The Journal of Biological Chemistry*, *294*(46), 17437–17450. https://doi.org/10.1074/jbc.RA119.010401.

Yoshida, K., Hara, S., & Hisabori, T. (2015). Thioredoxin selectivity for thiol-based redox regulation of target proteins in chloroplasts. *The Journal of Biological Chemistry*, *290*(23), 14278–14288. https://doi.org/10.1074/jbc.M115.647545.

Yoshida, K., Hara, A., Sugiura, K., Fukaya, Y., & Hisabori, T. (2018). Thioredoxin-like2/2-Cys peroxiredoxin redox cascade supports oxidative thiol modulation in chloroplasts. *Proceedings of the National Academy of Sciences of the United States of America*, *115*(35), E8296–E8304. https://doi.org/10.1073/pnas.1808284115.

Yoshida, K., & Hisabori, T. (2017). Distinct electron transfer from ferredoxin-thioredoxin reductase to multiple thioredoxin isoforms in chloroplasts. *The Biochemical Journal*, *474*(8), 1347–1360. https://doi.org/10.1042/BCJ20161089.

Yoshida, K., Noguchi, K., Motohashi, K., & Hisabori, T. (2013). Systematic exploration of thioredoxin target proteins in plant mitochondria. *Plant & Cell Physiology*, *54*(6), 875–892. https://doi.org/10.1093/pcp/pct037.

Zannini, F., Roret, T., Przybyla-Toscano, J., Dhalleine, T., Rouhier, N., & Couturier, J. (2018). Mitochondrial Arabidopsis thaliana TRXo isoforms bind an iron–sulfur cluster and reduce NFU proteins in vitro. *Antioxidants*, 7(10), 142. https://doi.org/10.3390/antiox7100142.

Zhang, B., Bandyopadhyay, S., Shakamuri, P., Naik, S. G., Huynh, B. H., Couturier, J., et al. (2013). Monothiol glutaredoxins can bind linear [Fe3S4]+ and [Fe4S4]2+ clusters in addition to [Fe2S2]2+ clusters: Spectroscopic characterization and functional implications. *Journal of the American Chemical Society*, *135*(40), 15153–15164. https://doi.org/10.1021/ja407059n.

CHAPTER THIRTEEN

Nitric oxide synthase in plants—A follow-up of ABR volume 77: Nitric oxide and signaling in plants

David Wendehenne*
Agroécologie, AgroSup Dijon, CNRS, INRAE, University Bourgogne Franche-Comté, Dijon, France
*Corresponding author: e-mail address: david.wendehenne@u-bourgogne.fr

Contents

Abstract

The reactive diatomic gas nitric oxide (NO) has been the subject of extensive investigations in plants. These studies revealed that NO is a signaling compound involved in numerous biological processes. Its main physiological functions, as well as its mode of action at the cellular and molecular levels, have been addressed in the 77th Volume of Advances in Botanical Research entitled "Nitric Oxide and Signaling in Plants". More recently, the discovery and first functional analyses of nitric oxide synthases (NOSs) and NO-forming nitrite reductase in algae allowed a better understanding of NO biosynthesis in photosynthetic organisms. This review updates the NO production mechanisms in algae with a particular focus of NOSs.

1. Introduction

The free radical nitric oxide (NO) has essential physiological functions in mammalian physiology, ranging from neurotransmission to immunity (Lowenstein, Dinerman, & Snyder, 1994). Dysregulation of its production

Advances in Botanical Research, Volume 100
ISSN 0065-2296
https://doi.org/10.1016/bs.abr.2021.01.011
Copyright © 2021 Elsevier Ltd
All rights reserved.

has been implicated in pathological symptoms, making NO signaling an important focus of pharmaceutical industries. Undoubtedly, major progresses in NO research have been achieved following the identification of its enzymatic sources in human in the early 1990s, namely nitric oxide synthases (NOSs) (Nathan & Xie, 1994). NO has also emerged as an integral signaling molecule in plant biology. There too, it exerts a wide range of biological activities and the molecular basis of its signaling functions have been partially discovered (Kolbert et al., 2019).

Important functions for NO in plant biology have been covered in Advances in Botanical Research (ABR) Volume 77: Nitric Oxide and Signaling in Plants. Major research topics including the interaction of NO with reactive oxygen species (ROS) and hydrogen sulfide, the role of NO in metabolism, in the control of gene expression, in root growth and auxin signaling, in the control of nitrogen assimilation, in programmed cell death, in plant immunity and plant pathogen development, in legume-rhizobium symbiosis, in the plant response to abiotic stresses and in the regulation of proteins through post-translational modifications (PTMs) have been well described and discussed. Nevertheless, an essential aspect of NO signaling was not addressed, namely the production of NO in plants. This is not surprising given that this topic of research is still at the center of a number of controversies.

Since the publication of ABR Volume 77, new findings shed light on mechanisms underlying NO production in plants. In particular, the corresponding studies ended the debates on whether plants possess NOSs and also opened new roads of research. I review here these original findings.

2. NO production in plants: A still nebulous situation

It is widely agreed that the production of NO in plants involves two main pathways: the reductive and the oxidative pathways.

2.1 The reductive pathways

The reductive pathways rely on nitrite (NO_2^-) reduction *via* enzymatic and non-enzymatic processes (Gupta, Fernie, Kaiser, & van Dongen, 2011). Numerous studies highlighted a key role for nitrate reductase (NR) in the enzymatic route for NO synthesis. Indeed, *in vitro* assays showed that this cytosolic enzyme reduces nitrate (NO_3^-) into NO_2^- but also NO_2^-

into NO using NADH as an electron donor (Yamasaki, Sakihama, & Takahashi, 1999). The use of plant mutants impaired in NR expression confirmed the contribution of NR as a component of NO biosynthesis *in vivo* (Desikan, Griffiths, Hancock, & Neill, 2002). Nevertheless, NO production appeared to be a small fraction of the NO_3^--reducing capacity of NR (about 1% according to Rockel, Strube, Rockel, Wildt, & Kaiser, 2002) and was strongly inhibited by NO_3^-, thus questioning the mechanisms underlying NR-dependent NO production. The identification of NO-forming nitrite reductase (NOFNiR) in the green algal *Chlamydomonas reinhardtii* partly solved this intriguing question (Chamizo-Ampudia, Sanz-Luque, Llamas, Galvan, & Fernandez, 2017). NOFNiR is a member of the mitochondrial amidoxime reducing component (mARC) family. In mammals, mARC is a molybdenum enzyme acting as the terminal oxidoreductase of a three protein complex in the outer mitochondrial membrane involving also cytochrome *b*5 (CYB5) and cytochrome *b*5 reductase (CYB5R) (Havemeyer et al., 2006). Sparacino-Watkins et al. (2014) reported that the two isoforms of human mARC catalyze the reduction of NO_2^- to NO, CYB5R and CYB5 ensuring the electron flow from NADH to mARC. This process might take place under anaerobic conditions. In *C. reinhardtii*, thanks to ARC and NR mutant strains as well as enzymatic assays, Chamizo-Ampudia (Chamizo-Ampudia et al., 2016) found that NR and ARC (renamed NOFNiR) form a cytosolic complex and are both required for NO synthesis through a two-step process. More precisely, NO_3^- is first reduced intro NO_2^- by NR. Next, NR acts as a substrate and electron supplier for NOFNiR which reduces NO_2^- into NO, even in the presence of NO_3^- concentrations sufficient to block NR-dependent NO_2^- reduction. Plant ARC proteins are well conserved and it has been proposed that the NR-NOFNiR complex might constitute a key system for NO synthesis in plants.

In addition to NR, in hypoxic plant cells cytochrome *c* oxidase emerged as an important mitochondrial site of NO production where it reduces NO_2^- to form NO (Gupta et al., 2011; Meilhoc, Boscari, Bruand, Puppo, & Brouquisse, 2011). The mitochondrial electron transport chain complex III might also display a similar function (Igamberdiev, Ratcliffe, & Gupta, 2014). In addition, the activity of a plasma membrane-bound nitrite-NO reductase (NI-NOR) has been reported in roots (Stohr & Stremlau, 2006). NI-NOR might co-operate with a plasma membrane-bound NR. Nevertheless, the molecular identity of NI-NOR remains unknown. Other enzyme candidates able to catalyze the reduction of NO_2^- into NO have been suggested, such as the peroxisomal xanthine oxidoreductase (Gupta et al., 2011). So far, their involvement as a NO source *in planta* has not been demonstrated.

Besides enzymatic sources, the possibility that NO might be produced from NO_2^- *via* non-enzymatic processes has been underlined. Here too, the corresponding mechanisms as well as their physiological consequences remain obscure. In this regards, it is worth mentioning the investigation of Bethke, Badger, & Jones (2004). These authors found that barley aleurone layers can synthesize NO from NO_2^- non-enzymatically. This production occurred in the apoplast at acidic pH and was accelerated by reducing agents such as phenolics. The biological significance of this pathway for NO synthesis is still unclear. Nevertheless, the observation that it was induced by abscisic acid and gibberellin suggests that it might contribute to hormonal signaling.

2.2 The oxidative pathways

Several studies support a view that NO is also produced through oxidative routes involving mainly L-arginine (L-Arg) and polyamines as substrates (Gupta et al., 2011). As described below, the L-Arg-dependent NO production is thought to be catalyzed by a NOS-like enzyme which identity is still unknown in plants (Corpas, Palma, del Rio, & Barroso, 2009). Concerning polyamines, their synthesis and involvement in NO production are conditioned by the availability of L-Arg as a substrate for the enzymes arginine decarboxylase and L-arginase (Flores et al., 2008). In addition, a recent investigation in *Arabidopsis thaliana* identified the copper amine oxidase AtCuAO8 as a regulator of L-Arg availability, arginase activity and, therefore, NO production (Gross, Rudolf, Thiele, Durner, & Astier, 2017). However, the mechanisms inherent to polyamine-induced NO production remain unknown. Other putative substrates of the oxidative routes, such as hydroxylamine, have been reported although *in vivo* evidence for such a role is lacking (Gupta et al., 2011).

2.2.1 Nitric oxide synthase

In mammals, NO is mainly produced by a specific family of proteins named NOSs (Nathan & Xie, 1994). These enzymes catalyze the O_2- and NADPH-dependent production of NO and L-citrulline using L-Arg as substrate through a two-step catalysis involving N^{ω}-hydroxy-L-Arg as an intermediate (Fig. 1) (Cinelli, Do, Miley, & Silverman, 2020; Santolini, 2019). So far, three major isoforms have been identified, namely neural NOS

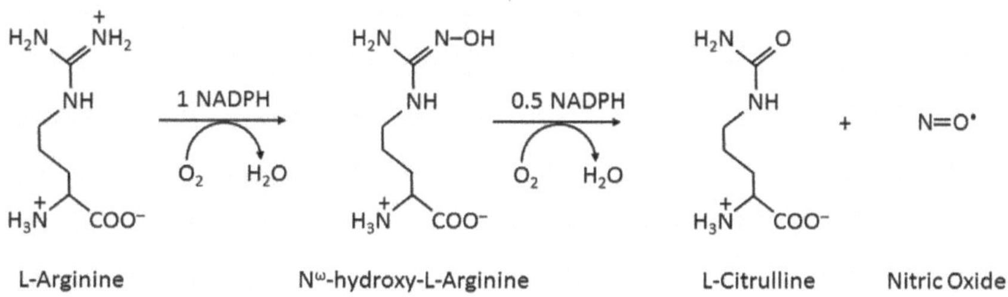

Fig. 1 Catalytic steps of NOS.

(nNOS or NOS1), inducible NOS (iNOS or NOS2) and endothelial NOS (eNOS or NOS3). These isoforms were named according to their predominant tissular distribution and regulation. Indeed, nNOS was initially associated to neurotransmission, eNOS to endothelium-dependent vasorelaxation and NOS2, mainly found in macrophages and neutrophils, to immune responses (Nathan & Xie, 1994). All three enzymes are now recognized to have a wider tissular distribution and to be involved in various physiological processes.

NOSs are active as homodimers. Each monomer comprises an N-terminal oxygenase domain (NOSoxy) and a C-terminal reductase domain (NOSred) (Figs. 2 and 3). The NOSoxy domain binds L-Arg, a Fe-protoporphyrin IX thiolate heme and the redox cofactor tetrahydro-L-biopterin (BH_4) (Stuehr & Haque, 2019). The NOSoxy domain is preceded by an N-terminal extension containing a β-hairpin hook, a zinc ligation motif (Cys-X4-Cys) and BH_4-binding residues (Crane et al., 1999). This extension contributes to the homodimer association and the generation of catalytic activity. The NOSoxy domain contains NADPH, FAD and FMN binding sites and shares structural and functional similarities with cytochrome P450 reductases (Stuehr & Haque, 2019). The NOSoxy and NOSred domains are connected by a helical calmodulin (CaM) binding site (Aoyagi, Arvai, Tainer, & Getzoff, 2003). So far, only the crystal structure of isolated NOSoxy and NOSred domains are available, not of the full length NOSs. This might be explained by the ability of the homodimer to adopt many conformational changes (Campbell, Smith, Potter, Carragher, & Marletta, 2014). Importantly, nNOS and eNOS are constitutively expressed and Ca^{2+}-/CaM-dependent while iNOS is not expressed in resting cells but is induced in response to immunological stimuli. Furthermore, the activity of iNOS is CaM-dependent but Ca^{2+}-independent (Zhang & Vogel, 1994).

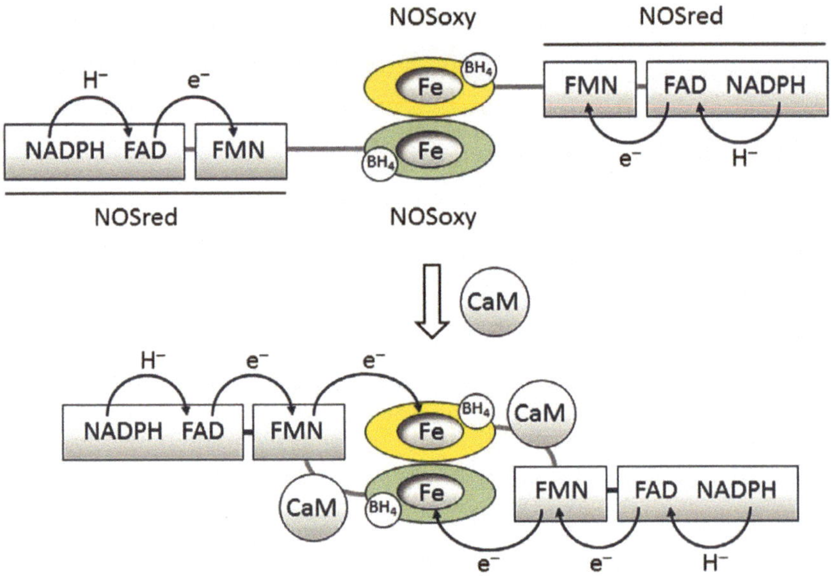

Fig. 2 NOS electron transfer pathway and quaternary structure. The upper image represents the homodimer. The homodimerization involves the NOSoxy domains. Both the FAD/NADPH and FMN subdomains interact, thus allowing the electron transfer. Binding of CaM to each monomer triggers a conformation change bringing nearer the FMN subdomain of one monomer to the heme center of the second monomer (lower image). This allows the electron transfer between FMN and heme.

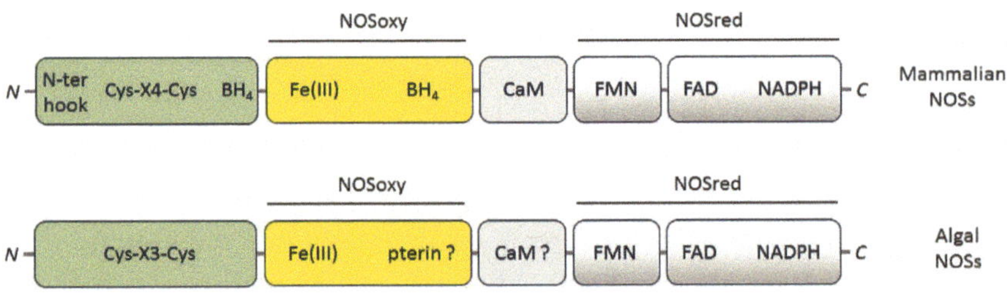

Fig. 3 Comparison of the NOS domain organization between mammalian and algal NOSs. The nNOS isoform of mammalian NOSs also contains an extra N-terminal sequence (PDZ domain) involved in protein-protein interactions (not shown). Compared to mammalian NOSs, in their N-terminal part algal NOSs do not possess the N-terminal hook involved in the dimerization, nor key residues contributing to BH_4 binding. Furthermore, the putative zinc ligation motif is shorter. The requirement of BH_4 in algal NOSs is unlikely and, more generally, of pterin for catalysis is questioned. The ability of algal NOSs to bind CaM is also questioned.

In terms of catalysis, the NOSred domain of one monomer passes electrons from NADPH to the heme domain of the second monomer of the homodimer (Fig. 2) (Santolini, 2011). More precisely, NADPH transfers two electrons to FAD. In turn, FMN accepts one electron from FAD. Then FMN donates one electron to the heme center. This latter step requires a conformational change related to CaM binding. According to Campbell et al. (2014), only a single reductase domain of the homodimer participates in the one electron transfer.

NOS are not specific to mammals and, thanks to increased availability of genome information, they have been found in numerous metazoans, in fungi and also in prokaryotes with some sequence and structural specificities (Santolini, 2019). What is the situation in plants?

2.2.2 Plant nitric oxide synthase

The hypothesis that embryophytes might possess NOS-like proteins was initially raised in the mid-90s (Cueto et al., 1996; Leshem, Y. a. Y., 1996). Since then, many articles have been published, reporting NOS activities in various land plant species, NOSs immuno-localizations in distinct subcellular compartments as well as efficiency of mammalian NOS inhibitors to suppress plant NOS activities and/or NO productions (Astier, Gross, & Durner, 2018; Corpas et al., 2009). Several of these articles provided a rich source of information about plant NOS-like catalytic parameters, pharmacological sensitivity and indicated that these proteins use the same substrate and cofactors than their mammalian counterparts (Corpas et al., 2009) Nonetheless, doubts that land plants possess mammalian-type NOSs were voiced. Indeed:

- Attempts to identify these proteins failed or led to the identification of false positives and even paper retractions (Klessig, Ytterberg, & van Wijk, 2004; Zemojtel et al., 2006),
- The specificity of assays to measure NOS activities was called into question as these assays are skewed by the activity of argininosuccinate lyase, which reversibly converts L-Arg and fumarate to argininosuccinate (Tischner et al., 2007),
- The plant target proteins of mammalian NOS inhibitors have not been identified so far. Furthermore, N-Nitroarginine methyl ester (L-NAME), a well know competitive inhibitor of mammalian NOS, has been shown to suppress NR activity in *A. thaliana* (Rasul et al., 2012),
- It turned out that proteins immuno-reacting with mammalian NOS antibodies were not related to NOSs (Butt, Lum, & Lo, 2003),

- The search for NOS sequences in available embryophyte genomes and transcriptomes failed. In particular, Jeandroz et al. (2016) searched the data set generated by the 1000 Plants (1KP) international consortium for the presence of transcripts encoding NOS1-like proteins. Not a single NOS-like sequence was found in 1087 transcriptomes of land plants, including those in which NOS-like activities were reported.

Overall, these data strongly suggest that the proteins displaying NOS activities in embryophytes do not belong to the NOS family as defined from mammal studies.

2.2.3 And the surprise came from algae

While a general consensus about the absence of mammalian-type NOS in embryophytes was emerging, Foresi et al. (2010) reported the unexpected discovery that this protein was present in the green alga *Ostreococcus tauri*. Several years later, mammalian-type NOS were found in 16 additional algae among 265 species screened (Jeandroz et al., 2016; Kumar, Castellano, Patti, Palumbo, & Buia, 2015; Santolini, Andre, Jeandroz, & Wendehenne, 2017). These proteins were mainly found in the green algae species *Bathycoccus prasinos*, *Bolbocoleon piliferum*, *Chaetosphaeridium globosum*, *Cosmarium subtumidum*, *Golenkinia longispicula*, *Helicodictyon planctonicum*, *Klebsormidium nitens*, *Leptosira obovate*, *Nephroselmis pyriformis*, *Pandorina morum*, *Phacotus lenticularis*, *Planophila terrestris*, *Pleurastrum insigne* and *Scherffelia dubia*, in the red algal *Rhodochaete parvula* and in the brown algal *Sargassum muticum*. These species include marine, freshwater and terrestrial algae. NOSs from green algae share 45–70% identity among themselves but are significantly distant from red and brown algal NOSs, suggesting that NOSs from each photosynthetic group constitute a distinct phylogenetic clade (Santolini et al., 2017).

3. Structure and activity of algal NOSs

3.1 Structure

In silico sequences analyses have indicated that algal NOSs possess both the NOSoxy and NOSred domains (Foresi et al., 2010; Jeandroz et al., 2016) (Fig. 3). In more detail, key residues of the NOSoxy domain contributing to L-Arg, BH_4 and heme binding as well as the NADPH, FAD and FMN binding sites of the NOSred domain are well conserved. These proteins also display singularities. In particular, the β-hairpin hook and BH_4-binding residues of the N-terminal extension are not conserved compared to their mammalian counterparts. Furthermore, the putative zinc ligation motif slightly differs (Cys-X3-Cys instead of Cys-X4-Cys). In addition, the CaM binding site is poorly conserved (Jeandroz et al., 2016). Overall, these basic structural studies brought into question the mechanisms underlying the dimerization of algal NOSs and their capacity to interact with CaM.

Detailed *in silico* analysis of the NOSoxy domain of the NOSs from the green algae *C. subtumidum* (CsNOS) and *O. tauri* NOS (OtNOS) have recently been reported (Foresi et al., 2010; Santolini et al., 2017; Weisslocker-Schaetzel et al., 2017). The resulting 3D models are superimposed over the NOSoxy domain of mammalian iNOS, highlighting a similar global architecture. Nevertheless, interesting features of CsNOS were observed, particularly the presence of an 11-residues insertion located at the surface of the NOSoxy domain and of short gaps/insertions which do no impact the overall structure. Gaps/insertions of 1–6 residues were also found in the other algal NOSs. A detailed analysis of the heme pocket region of the CsNOS NOSoxy domain highlighted two major substitutions as compared to mammalian iNOS (substitution of valine-346 and tryptophan-457 of iNOS to phenylalanine and serine residues in CsNOS, respectively). According to the authors (Santolini et al., 2017), the valine mutation is likely to block the escape of NO during catalysis while the tryptophan substitution could alter BH_4 binding and electron transfers. This later observation raised the question of the requirement of BH_4, as well as of other pterins, for catalysis. Therefore, the functioning of CsNOS might differ from those of mammalian NOSs. Regarding OtNOS, the 3D structural analysis highlighted a looser and more open BH_4 binding pocket and a more exposed heme edge. Again, these observations questioned the binding of BH_4, as well as that of other pterins (Weisslocker-Schaetzel et al., 2017).

The analysis of the primary structure of other algal NOSs pointed out other singularities questioning the reaction that these proteins might be catalyzing (Santolini, 2019; Santolini et al., 2017). For instance, NOS from the red alga *R. parvula* does not seem to be able to bind L-Arg, BH_4 and heme and, consequently, to produce NO. Similarly, it appears unlikely that NOS from the green alga *B. piliferum* binds heme in its catalytic side. In contrast, NOS from the brown algae *S. muticum* appears to be well fit to bind L-Arg, BH_4 and heme.

3.2 Catalytic activity

In their pioneer work, Foresi et al. (2010) provided first evidence that recombinant OtNOS produces both NO and L-citrulline in *in vitro* assays with the following parameters: 0.49 μM NO min^{-1}, 0.53 μM L-citrulline min^{-1}. The enzyme activity was comparable to that of mouse iNOS with a K_m value for L-Arg in the same order to those of previously characterized NOSs (12 ± 5 μM *versus* 1 ± 22 μM). The activity was BH_4-dependent, only partially CaM-dependent and was suppressed by the mammalian NOS inhibitor L-NAME.

This study was recently completed by Weisslocker-Schaetzel et al. (2017) who focused their analysis on the NOSoxy domain of OtNOS. More precisely, using various spectroscopies, these authors characterized the structural and electronic fingerprints of some NOSoxy reaction intermediates. They concluded that OtNOS displays similar spectroscopic fingerprints as compared to mammalian NOSs. Interestingly, OtNOS also showed an ultrafast NO-producing capacity. In particular, kinetic analyses indicated that that the oxygen activation rate was 5–10 times faster compared to those of characterized mammalian and bacterial NOSs. Furthermore, NO dissociation from the ferric heme appeared to be 10–80 times faster than that of other NOSs. The remarkable fast rate of NO release could be explained by a more exposed heme pocket.

As discussed above, the requirement of BH_4 and, more generally, of pterins, is a matter of some debates.

While the activity of recombinant OtNOS was found to be BH_4-dependent (Foresi et al., 2010), the structural analysis of the NOSoxy domain of OtNOS does not support such a role (Weisslocker-Schaetzel et al., 2017). In fact, data-bases searching revealed that embryophytes and algae do not possess the genes encoding the major enzymes for BH_4 biosynthesis, notably 6-pyruvoyl-tetrahydropterin synthase and sepiapterin reductase (Jeandroz et al., 2016). Therefore, it has been proposed that algal NOSs might use other pterins such as tetrahydrofolate (H_4F), a BH_4-related pterin. Supporting this possibility, it has been reported that unlike mammalian iNOS, the recombinant OtNOS effectively uses H_4F for catalysis (Foresi et al., 2015). Furthermore, the addition of H_4F in the growth media of *A. thaliana* seedling overexpressing OtNOS increased NO production in roots. Nevertheless, whether H_4F acts as an electron donor for *in vivo* catalysis has yet to be demonstrated.

4. Physiological functions

Data discussed above highlight that only few photosynthetic organisms contain NOS orthologs, their presence being restricted to algae living in various ecosystems. Whereas extensive background knowledge exists regarding mammalian NOSs, how and why algae NOS are activated, and their biological functions, remain practically unknown. The only reference study, conducted in *O. tauri*, suggested that NO produced by OtNOS contributes to algal growth and may reduce the oxidative damage due to high intensity light irradiation (Foresi et al., 2010). Furthermore, when overexpressed in *A. thaliana*, OtNOS confers an enhanced salt, drought and oxidative stress tolerance (Foresi et al., 2015).

At the cellular level, how NO produced by algal NOSs exerts its effects is still enigmatic. In animal cells, the best-characterized NO-dependent signaling pathway occurs through the second messenger cyclic guanosine 3′,5′-monophosphate (cGMP). More precisely, NO produced by NOS binds and activates soluble guanylate cyclase (sGC) which produces cGMP from guanosine 5′-triphosphate. Then cGMP activates its target proteins including cGMP-dependent protein kinases (PKGs) and cyclic nucleotide gated channels (CNGCs). This signaling process triggers rapid cellular responses and is at the heart of several physiological processes such as smooth muscle relaxation and neurotransmission. It is negatively regulated by cGMP-dependent phosphodiesterases (PDEs) catalyzing the hydrolysis of cGMP to guanosine monophosphate (Cary, Winger, Derbyshire, & Marletta, 2006; Denninger & Marletta, 1999; Francis, Busch, Corbin, & Sibley, 2010). Many studies have endorsed the prevailing opinion that NO signaling is supported by this pathway in plants. Nevertheless, a recent search for plant orthologues of sGCs, PKGs, CNGCs and PDEs minimized this assumption (Astier, Mounier, Santolini, Jeandroz, & Wendehenne, 2019). Indeed, only few homologues of these proteins were found in some algal species and their presence did not correlate with that of NOSs.

Coming back to animals, NO also controls signaling processes through protein PTMs (Nathan, 2004). In the last two decades, S-nitrosation emerged as a main NO-dependent PTMs (Hess, Matsumoto, Kim, Marshall, & Stamler, 2005). This process refers to the reversible covalent attachment of a NO group to the thiol side chain of a cysteine residue. It impacts protein activity, subcellular localization, stability and partner interactions. The reversibility involves mainly two enzymatic systems, namely S-nitrosoglutathione reductase (GSNOR) and thioredoxins (TRX) (Stamler & Hess, 2010). It is now well established that S-nitrosation is a major signaling mechanism through which NO operates in embryophytes (Kolbert et al., 2019). This is also true in algae in which S-nitrosation events and the presence of GSNOR and TRX were confirmed (Astier et al., 2019; Berger et al., 2016; Chen et al., 2016; Morisse, Zaffagnini, Gao, Lemaire, & Marchand, 2014). Whether NO produced by NOS acts through S-nitrosation remains, however, not demonstrated so far.

5. Conclusion

In the past few years, our knowledge about NO biosynthesis in plants has increased at a rapid pace. This has resulted in answers to two major questions: How does NR produce NO under physiological conditions? Do plant possess mammalian-type NOSs? The functional characterization of NOFNiR and the discovery of new members of the NOS family in algae also shed light on the importance of these organisms as biological models to unravel NO synthesis in photosynthetic organisms but also to investigate the evolution of NO signaling. These findings have also generated new questions and challenges. In particular, although well conserved between organisms, the involvement of NOFNiR in embryophytes has not been reported so far. Furthermore, the enzyme catalyzing NOS-like activities in plants remains to be identified. Concerning algal NOSs, this new field of research is almost untapped. Yet it offers promising research beyond plant biology. Indeed, the different NOSs identified in algae display structural singularities that pose questions about their regulation, catalysis, ability to produce NO (and/or derived products) and signaling functions. Furthermore, algae possessing both NOS and NR represent unexpected model organisms to understand the cross-regulation between the reductive and oxidative pathways for NO synthesis. Finally, why NOSs is absent in embryophytes and many algae is an intriguing question which brings another one: do they need NOS? Studies in Human report that gene losses in the course of evolution have provide opportunities for adaptations (Wang, Grus, & Zhang, 2006). In this regard, the possibility that plants have lost NOS seems plausible. Indeed, plants and algae such as *C. reinhardtii* use nitrate as their main source of nitrogen and possess efficient systems for nitrate uptake, signaling and metabolism (Calatrava et al., 2017; Tegeder & Masclaux-Daubresse, 2018). Therefore, the loss of NOS may be due to adaptation of alternative pathways for NO synthesis using nitrite and, indirectly, nitrate, as a main substrate. Inevitably, the hypothesis that NOS seems unnecessary to plants brings the question of why this enzyme is still present in few algal species. Investigating the function of these NOSs and also the benefit they bring to algae represents a fascinating challenge.

Acknowledgments

The author would like to thank the members of the Immunity and Signaling Research Group of the Agroecology Unit of Research, Jean-Pierre Jacquot, Michael Schenker and all the authors contributing to the ABR Volume 77. Work in my lab is currently supported by the Ministère de l'Enseignement supérieur, de la Recherche et de l'Innovation, Investissements d'Avenir program, project ISITE-BFC NOISELESS (contract ANR-15-IDEX-0003; Grant NOISELESS-RA18041.AEC.IS), the Agence Nationale de la Recherche, project ALGAE-NOS (Grant ANR-18-CE20-0022-02) and the Région de Bourgogne, project RHINO (Grant 2018Y-07113).

Abbreviations

ARC	amidoxime reducing component
BH_4	tetrahydro-L-biopterin
CaM	calmodulin
cGMP	cyclic guanosine 3′,5′-monophosphate
CNGC	cyclic nucleotide gated channel
CYB5	cytochrome *b*5
CYB5R	cytochrome *b*5 reductase
GSNOR	nitrosoglutathione reductase
H_4F	tetrahydrofolate
L-Arg	L-arginine
L-NAME	*N*-nitroarginine methyl ester
NO	nitric oxide
NOFNiR	NO-forming nitrite reductase
NI-NOR	nitrite-NO reductase
NOS	nitric oxide synthase
NOSred	NOS C-terminal reductase domain
NOSoxy	NOS N-terminal oxygenase domain
NR	nitrate reductase
PDE	phosphodiesterase
PKG	cGMP-dependent protein kinase
PTM	post-translational modifications
ROS	reactive oxygen species
sGC	soluble guanylate cyclase
TRX	thioredoxin

References

Aoyagi, M., Arvai, A. S., Tainer, J. A., & Getzoff, E. D. (2003). Structural basis for endothelial nitric oxide synthase binding to calmodulin. *The EMBO Journal, 22*(4), 766–775. https://doi.org/10.1093/emboj/cdg078.

Astier, J., Gross, I., & Durner, J. (2018). Nitric oxide production in plants: An update. *Journal of Experimental Botany, 69*(14), 3401–3411. https://doi.org/10.1093/jxb/erx420.

Astier, J., Mounier, A., Santolini, J., Jeandroz, S., & Wendehenne, D. (2019). The evolution of nitric oxide signalling diverges between the animal and the green lineages. *Journal of Experimental Botany, 70*, 4355–4364. https://doi.org/10.1093/jxb/erz088.

Berger, H., De Mia, M., Morisse, S., Marchand, C. H., Lemaire, S. D., Wobbe, L., et al. (2016). A light switch based on protein S-nitrosylation fine-tunes photosynthetic light harvesting in Chlamydomonas. *Plant Physiology, 171*(2), 821–832. https://doi.org/10.1104/pp.15.01878.

Bethke, P. C., Badger, M. R., & Jones, R. L. (2004). Apoplastic synthesis of nitric oxide by plant tissues. *Plant Cell, 16*(2), 332–341. https://doi.org/10.1105/tpc.017822.

Butt, Y. K., Lum, J. H., & Lo, S. C. (2003). Proteomic identification of plant proteins probed by mammalian nitric oxide synthase antibodies. *Planta, 216*(5), 762–771. https://doi.org/10.1007/s00425-002-0926-y.

Calatrava, V., Chamizo-Ampudia, A., Sanz-Luque, E., Ocaña-Calahorro, F., Llamas, A., Fernandez, E., et al. (2017). How Chlamydomonas handles nitrate and the nitric oxide cycle. *Journal of Experimental Botany, 68*(10), 2593–2602. https://doi.org/10.1093/jxb/erw507.

Campbell, M. G., Smith, B. C., Potter, C. S., Carragher, B., & Marletta, M. A. (2014). Molecular architecture of mammalian nitric oxide synthases. *Proceedings of the National Academy of Sciences of the United States of America, 111*(35), E3614–E3623. https://doi.org/10.1073/pnas.1413763111.

Cary, S. P., Winger, J. A., Derbyshire, E. R., & Marletta, M. A. (2006). Nitric oxide signaling: No longer simply on or off. *Trends in Biochemical Sciences, 31*(4), 231–239. https://doi.org/10.1016/j.tibs.2006.02.003.

Chamizo-Ampudia, A., Sanz-Luque, E., Llamas, A., Galvan, A., & Fernandez, E. (2017). Nitrate reductase regulates plant nitric oxide homeostasis. *Trends in Plant Science, 22*(2), 163–174. https://doi.org/10.1016/j.tplants.2016.12.001.

Chamizo-Ampudia, A., Sanz-Luque, E., Llamas, Á., Ocaña-Calahorro, F., Mariscal, V., Carreras, A., et al. (2016). A dual system formed by the ARC and NR molybdoenzymes mediates nitrite-dependent NO production in Chlamydomonas. *Plant, Cell & Environment, 39*(10), 2097–2107. https://doi.org/10.1111/pce.12739.

Chen, X., Tian, D., Kong, X., Chen, Q., Abd Allah, E. F., Hu, X., et al. (2016). The role of nitric oxide signalling in response to salt stress in Chlamydomonas reinhardtii. *Planta, 244*, 651–669. https://doi.org/10.1007/s00425-016-2528-0.

Cinelli, M. A., Do, H. T., Miley, G. P., & Silverman, R. B. (2020). Inducible nitric oxide synthase: Regulation, structure, and inhibition. *Medicinal Research Reviews, 40*(1), 158–189. https://doi.org/10.1002/med.21599.

Corpas, F. J., Palma, J. M., del Rio, L. A., & Barroso, J. B. (2009). Evidence supporting the existence of L-arginine-dependent nitric oxide synthase activity in plants. *The New Phytologist, 184*(1), 9–14. https://doi.org/10.1111/j.1469-8137.2009.02989.x.

Crane, B. R., Rosenfeld, R. J., Arvai, A. S., Ghosh, D. K., Ghosh, S., Tainer, J. A., et al. (1999). N-terminal domain swapping and metal ion binding in nitric oxide synthase dimerization. *The EMBO Journal, 18*(22), 6271–6281. https://doi.org/10.1093/emboj/18.22.6271.

Cueto, M., Hernández-Perera, O., Martín, R., Bentura, M. L., Rodrigo, J., Lamas, S., et al. (1996). Presence of nitric oxide synthase activity in roots and nodules of Lupinus albus. *FEBS Letters, 398*(2–3), 159–164. https://doi.org/10.1016/S0014-5793(96)01232-X.

Denninger, J. W., & Marletta, M. A. (1999). Guanylate cyclase and the .NO/cGMP signaling pathway. *Biochimica et Biophysica Acta*, *1411*(2–3), 334–350.

Desikan, R., Griffiths, R., Hancock, J., & Neill, S. (2002). A new role for an old enzyme: Nitrate reductase-mediated nitric oxide generation is required for abscisic acid-induced stomatal closure in Arabidopsis thaliana. *Proceedings of the National Academy of Sciences of the United States of America*, *99*(25), 16314–16318. https://doi.org/10.1073/pnas.252461999.

Flores, T., Todd, C. D., Tovar-Mendez, A., Dhanoa, P. K., Correa-Aragunde, N., Hoyos, M. E., et al. (2008). Arginase-negative mutants of Arabidopsis exhibit increased nitric oxide signaling in root development. *Plant Physiology*, *147*(4), 1936–1946. https://doi.org/10.1104/pp.108.121459.

Foresi, N., Correa-Aragunde, N., Parisi, G., Calo, G., Salerno, G., & Lamattina, L. (2010). Characterization of a nitric oxide synthase from the plant kingdom: NO generation from the green alga Ostreococcus tauri is light irradiance and growth phase dependent. *Plant Cell*, *22*(11), 3816–3830. https://doi.org/10.1105/tpc.109.073510.

Foresi, N., Mayta, M. L., Lodeyro, A. F., Scuffi, D., Correa-Aragunde, N., García-Mata, C., et al. (2015). Expression of the tetrahydrofolate-dependent nitric oxide synthase from the green alga Ostreococcus tauri increases tolerance to abiotic stresses and influences stomatal development in Arabidopsis. *The Plant Journal*, *82*(5), 806–821. https://doi.org/10.1111/tpj.12852.

Francis, S. H., Busch, J. L., Corbin, J. D., & Sibley, D. (2010). cGMP-dependent protein kinases and cGMP phosphodiesterases in nitric oxide and cGMP action. *Pharmacological Reviews*, *62*(3), 525–563. https://doi.org/10.1124/pr.110.002907.

Gross, F., Rudolf, E. E., Thiele, B., Durner, J., & Astier, J. (2017). Copper amine oxidase 8 regulates arginine-dependent nitric oxide production in Arabidopsis thaliana. *Journal of Experimental Botany*, *68*(9), 2149–2162. https://doi.org/10.1093/jxb/erx105.

Gupta, K. J., Fernie, A. R., Kaiser, W. M., & van Dongen, J. T. (2011). On the origins of nitric oxide. *Trends in Plant Science*, *16*(3), 160–168. https://doi.org/10.1016/j.tplants.2010.11.007.

Havemeyer, A., Bittner, F., Wollers, S., Mendel, R., Kunze, T., & Clement, B. (2006). Identification of the missing component in the mitochondrial benzamidoxime prodrug-converting system as a novel molybdenum enzyme. *The Journal of Biological Chemistry*, *281*(46), 34796–34802. https://doi.org/10.1074/jbc.M607697200.

Hess, D. T., Matsumoto, A., Kim, S. O., Marshall, H. E., & Stamler, J. S. (2005). Protein S-nitrosylation: Purview and parameters. *Nature Reviews. Molecular Cell Biology*, *6*(2), 150–166. https://doi.org/10.1038/nrm1569.

Igamberdiev, A. U., Ratcliffe, R. G., & Gupta, K. J. (2014). Plant mitochondria: Source and target for nitric oxide. *Mitochondrion*, *19*(Pt. B), 329–333. https://doi.org/10.1016/j.mito.2014.02.003.

Jeandroz, S., Wipf, D., Stuehr, D. J., Lamattina, L., Melkonian, M., Tian, Z., et al. (2016). Occurrence, structure, and evolution of nitric oxide synthase-like proteins in the plant kingdom. *Science Signaling*, *9*(417), re2. https://doi.org/10.1126/scisignal.aad4403.

Klessig, D. F., Ytterberg, A. J., & van Wijk, K. J. (2004). The pathogen-inducible nitric oxide synthase (iNOS) in plants is a variant of the P protein of the glycine decarboxylase complex. *Cell*, *119*(3), 445. https://doi.org/10.1016/j.cell.2004.10.007.

Kolbert, Z., Barroso, J. B., Brouquisse, R., Corpas, F. J., Gupta, K. J., Lindermayr, C., et al. (2019). A forty year journey: The generation and roles of NO in plants. *Nitric Oxide*, *93*, 53–70. https://doi.org/10.1016/j.niox.2019.09.006.

Kumar, A., Castellano, I., Patti, F. P., Palumbo, A., & Buia, M. C. (2015). Nitric oxide in marine photosynthetic organisms. *Nitric Oxide*, *47*, 34–39. https://doi.org/10.1016/j.niox.2015.03.001.

Leshem, Y. a. Y. (1996). Nitric oxide in biological systems. *Plant Growth Regulation*, *18*(3), 155–159. https://doi.org/10.1007/bf00024375.

Lowenstein, C. J., Dinerman, J. L., & Snyder, S. H. (1994). Nitric oxide: A physiologic messenger. *Annals of Internal Medicine, 120*(3), 227–237. https://doi.org/10.7326/0003-4819-120-3-199402010-00009.

Meilhoc, E., Boscari, A., Bruand, C., Puppo, A., & Brouquisse, R. (2011). Nitric oxide in legume-rhizobium symbiosis. *Plant Science, 181*(5), 573–581. https://doi.org/10.1016/j.plantsci.2011.04.007.

Morisse, S., Zaffagnini, M., Gao, X. H., Lemaire, S. D., & Marchand, C. H. (2014). Insight into protein S-nitrosylation in Chlamydomonas reinhardtii. *Antioxidants & Redox Signaling, 21*(9), 1271–1284. https://doi.org/10.1089/ars.2013.5632.

Nathan, C. (2004). The moving frontier in nitric oxide-dependent signaling. *Science's STKE, 2004*(257), pe52. https://doi.org/10.1126/stke.2572004pe52.

Nathan, C., & Xie, Q. W. (1994). Nitric oxide synthases: Roles, tolls, and controls. *Cell, 78*(6), 915–918. https://doi.org/10.1016/0092-8674(94)90266-6.

Rasul, S., Dubreuil-Maurizi, C., Lamotte, O., Koen, E., Poinssot, B., Alcaraz, G., et al. (2012). Nitric oxide production mediates oligogalacturonide-triggered immunity and resistance to Botrytis cinerea in Arabidopsis thaliana. *Plant, Cell & Environment, 35*(8), 1483–1499. https://doi.org/10.1111/j.1365-3040.2012.02505.x.

Rockel, P., Strube, F., Rockel, A., Wildt, J., & Kaiser, W. M. (2002). Regulation of nitric oxide (NO) production by plant nitrate reductase in vivo and in vitro. *Journal of Experimental Botany, 53*(366), 103–110.

Santolini, J. (2011). The molecular mechanism of mammalian NO-synthases: A story of electrons and protons. *Journal of Inorganic Biochemistry, 105*(2), 127–141. https://doi.org/10.1016/j.jinorgbio.2010.10.011.

Santolini, J. (2019). What does "NO-Synthase" stand for ? *Frontiers in Bioscience (Landmark Edition), 24*, 133–171.

Santolini, J., Andre, F., Jeandroz, S., & Wendehenne, D. (2017). Nitric oxide synthase in plants: Where do we stand? *Nitric Oxide, 63*, 30–38. https://doi.org/10.1016/j.niox.2016.09.005.

Sparacino-Watkins, C. E., Tejero, J., Sun, B., Gauthier, M. C., Thomas, J., Ragireddy, V., et al. (2014). Nitrite reductase and nitric-oxide synthase activity of the mitochondrial molybdopterin enzymes mARC1 and mARC2. *Journal of Biological Chemistry, 289*(15), 10345–10358. https://doi.org/10.1074/jbc.M114.555177.

Stamler, J. S., & Hess, D. T. (2010). Nascent nitrosylases. *Nature Cell Biology, 12*(11), 1024–1026. https://doi.org/10.1038/ncb1110-1024.

Stohr, C., & Stremlau, S. (2006). Formation and possible roles of nitric oxide in plant roots. *Journal of Experimental Botany, 57*(3), 463–470. https://doi.org/10.1093/jxb/erj058.

Stuehr, D. J., & Haque, M. M. (2019). Nitric oxide synthase enzymology in the 20 years after the Nobel Prize. *British Journal of Pharmacology, 176*(2), 177–188. https://doi.org/10.1111/bph.14533.

Tegeder, M., & Masclaux-Daubresse, C. (2018). Source and sink mechanisms of nitrogen transport and use. *The New Phytologist, 217*(1), 35–53. https://doi.org/10.1111/nph.14876.

Tischner, R., Galli, M., Heimer, Y. M., Bielefeld, S., Okamoto, M., Mack, A., et al. (2007). Interference with the citrulline-based nitric oxide synthase assay by argininosuccinate lyase activity in Arabidopsis extracts. *The FEBS Journal, 274*(16), 4238–4245. https://doi.org/10.1111/j.1742-4658.2007.05950.x.

Wang, X., Grus, W. E., & Zhang, J. (2006). Gene losses during human origins. *PLoS Biology, 4*(3), e52. https://doi.org/10.1371/journal.pbio.0040052.

Weisslocker-Schaetzel, M., André, F., Touazi, N., Foresi, N., Lembrouk, M., Dorlet, P., et al. (2017). The NOS-like protein from the microalgae Ostreococcus tauri is a genuine and ultrafast NO-producing enzyme. *Plant Science, 265*(Suppl. C), 100–111. https://doi.org/10.1016/j.plantsci.2017.09.019.

Yamasaki, H., Sakihama, Y., & Takahashi, S. (1999). An alternative pathway for nitric oxide production in plants: New features of an old enzyme. *Trends in Plant Science*, *4*(4), 128–129.
Zemojtel, T., Frohlich, A., Palmieri, M. C., Kolanczyk, M., Mikula, I., Wyrwicz, L. S., et al. (2006). Plant nitric oxide synthase: A never-ending story? *Trends in Plant Science*, *11*(11), 524–525. author reply 526-528 https://doi.org/10.1016/j.tplants.2006.09.008.
Zhang, M., & Vogel, H. J. (1994). Characterization of the calmodulin-binding domain of rat cerebellar nitric oxide synthase. *The Journal of Biological Chemistry*, *269*(2), 981–985.

MIX

Papier | Fördert gute Waldnutzung

FSC® C083411

Kolibri 360 GmbH
Ferdinand-Jühlke-Straße 7
99095 Erfurt, Deutschland
produktsicherheit@kolibri360.de